HANDBOOK OF
ENVIRONMENTAL
HEALTH
AND
SAFETY
Principles and Practices
VOLUME I

HERMAN KOREN
MICHAEL BISESI

NATIONAL
ENVIRONMENTAL
HEALTH ASSOCIATION

LEWIS PUBLISHERS

Boca Raton   New York   London   Tokyo

Library of Congress Cataloging-in-Publication Data

Koren, Herman.
  Handbook of environmental health and safety : principles and
practices / Herman Koren, Michael Bisesi. —3rd ed.
    2 v.   cm.
  Includes bibliographical references and index.
  ISBN 1-56670-124-4 (alk. paper)
    1. Environmental health.   2. Environmental engineering.
  I. Bisesi, Michael S.   II. Title.
  RA565.K67   1995
  363.7—dc20   95-17131
                                                                                         CIP

© 1996 by CRC Press, Inc.
Lewis Publishers is an imprint of CRC Press

No claim to original U.S. Government works
International Standard Book Number 1-56670-124-4
Library of Congress Card Number 95-17131
Printed in the United States of America 1 2 3 4 5 6 7 8 9 0
Printed on acid-free paper

HANDBOOK OF
ENVIRONMENTAL
HEALTH
AND
SAFETY

VOLUME I
Principles and Practices

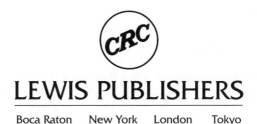

LEWIS PUBLISHERS

Boca Raton   New York   London   Tokyo

Dr. Herman Koren

Dr. Michael Bisesi

Authors

Herman Koren, R.E.H.S., M.P.H., H.S.D., is Professor and former Director of the Environmental Health Science Program, and Director of the Supervision and Management Program I & II at Indiana State University at Terre Haute. He has been an outstanding researcher, teacher, consultant, and practitioner in the environmental health field, and in the occupational health, hospital, medical care, and safety fields, as well as in management areas of the above and in nursing homes, water and wastewater treatment plants, and other environmental and safety industries for the past 41 years. In addition to numerous publications and presentations at national meetings, he is the author of six books, entitled *Environmental Health and Safety*, Pergamon Press, 1974; *Handbook of Environmental Health and Safety*, Volumes 1 and 2, Pergamon Press, 1980 (now published in updated and vastly expanded format by Lewis Publishers, CRC Press, as a 3rd edition); *Basic Supervision and Basic Management*, Parts I & II, Kendall Hunt Publishing, 1987; *Illustrated Dictionary of Environmental Health and Occupational Safety*, Lewis Publishers, CRC Press, 1995. He has served as a district environmental health practitioner and supervisor at the local and state level. He was an administrator at a 2000 bed hospital. Dr. Koren was on the editorial board of the *Journal of Environmental Health* and the former *Journal of Food Protection*. He is a founder diplomate of the Intersociety Academy for Certification of Sanitarians, a Fellow of the American Public Health Association, a 41-year member of the National Environmental Health Association, founder of the "National" Student National Environmental Health Association, and the founder and advisor of the Indiana State University Student National Environmental Health

Association (Alpha chapter). Dr. Koren developed the modern internship concept in environmental health science. He has been a consultant to the U.S. Environmental Protection Agency, the National Institute of Environmental Health Science, and numerous health departments and hospitals, and has served as the keynote speaker and major lecturer for the Canadian Institute of Public Health Inspectors. He is the recipient of the Blue Key Honor Society Award for outstanding teaching and the Alumni and Student plaque and citations for outstanding teaching, research, and service. The National Environmental Health Association has twice honored Dr. Koren with presidential citations for "Distinguished Services, Leadership and Devotion to the Environmental Health Field" and "Excellent Research and Publications."

Michael Bisesi, Ph.D., CIH, was appointed in 1990 by the Medical College of Ohio (MCO) at Toledo as as Associate Professor and is presently the Chairman of the Department of Occupational Health in the School of Allied Health. He also holds faculty appointments in the Departments of Otolaryngology and Pharmacology in the School of Medicine. Dr. Bisesi is responsible for conducting research, teaching, and administration. His areas of specialization are human exposure assessment and environmental toxicology. He has B.S. and M.S. degrees in Environmental Science from Rutgers University and a Ph.D. in Environmental Science from the S.U.N.Y. College of Environmental Science and Forestry in association with Syracuse University. Dr. Bisesi has worked in private industry, consulting, and academia, and has established a solid foundation of practical experience. He is a Certified Industrial Hygienist (CIH), a Registered Sanitarian (RS) and has authored and co-authored several scientific and technical articles and chapters in international publications, as well as the textbook *Industrial Hygiene Evaluation Methods.*

Preface

This handbook, in two volumes, is designed to provide a comprehensive but concise discussion of each of the important environmental health areas, including energy, ecology and people, environmental epidemiology, risk assessment and risk management, environmental law, air quality management, food protection, insect control, rodent control, pesticides, the chemical environment, environmental economics, human disease and injury, occupational health and safety, noise, radiation, recreational environment, indoor environments, medical care institutions, schools and universities, prisons, solid and hazardous waste management, water supply, plumbing, swimming areas, sewage disposal, soils, water pollution control, environmental health emergencies, and nuisance complaints.

Sufficient background material is introduced throughout these books to provide students, practitioners, and other interested readers with an understanding of the areas under discussion. Common problems and potential solutions are described; graphs, inspection sheets, and flow charts are utilized as needed to consolidate or clarify textual material. All facts and data come from the most recent federal government documents, many of which date from the late 1980s and early 1990s. Rules and regulations specified will continue to be in effect into the late 1990s. For rapidly changing areas in which the existing material used is likely to become dated, the reader is referred to the appropriate sources to update a given environmental health area or portion of an area as needed. This enhances the value of the text by providing basic and current materials that will always be needed and secondary sources that will enable the reader to keep up to date.

These books are neither engineering texts nor comprehensive texts in each area of study. The purpose of these books is to provide a solid working knowledge of each environmental health area with sufficient detail for practitioners. The text can be used in basic courses in environmental health, environmental pollution, ecology, and the environment and people that are offered at all universities and colleges in the United States and abroad. These courses are generally taught in departments of Life Science, Geology, Science Education, Environmental Health, and Health and Safety. For general areas of study, the instructor can omit specific details, such as resources, standards, practices and techniques, and modes of surveillance and evaluation. This same approach may be used by schools of medicine, nursing, and allied health sciences for their students. These texts are also suitable for basic introductory courses in schools of public health, environmental health, and sanitary science, as well as junior colleges offering 2-year degree programs in sanitary Science and Environmental Science.

Practitioners in a variety of environmental health and occupational health and safety fields will find these books handy references for resolving current problems and for obtaining a better understanding of unfamiliar areas. Practitioners and administrators in other areas, such as food processing, water-quality control, occupational health and safety, and solid and hazardous waste management, will also find these reference books useful.

High school teachers often must introduce environmental health topics in their classes and yet have no specific background in this area. These books could serve as a text in graduate education courses for high school teachers as well as a reference source.

Public interest groups and users of high school and community libraries will obtain an overall view of environmental problems by reading Chapter 1 and the Background and Status, Problems, Potential for Intervention, Resources, and Control sections in each chapter. This volume will also supply a concise reference for administrators in developing nations, for it explains tested controls and provides a better understanding of environmental problems, various standards, practices and techniques, and a variety of available resources.

The material divides easily into two separate courses. Course I would correspond to the content of Volume I and would include Chapter 1. Environment and Humans; Chapter 2. Food Protection; Chapter 3. Food Technology; Chapter 4. Insect Control; Chapter 5. Rodent Control; Chapter 6. Pesticides; Chapter 7. Indoor Environment; Chapter 8. Institutional Environment; Chapter 9. Recreational Environment; Chapter 10. Occupational Environment; and Chapter 11. Instrumentation.

Course II, corresponding to the content of the Volume II, would include Chapter 1. Air Quality Management; Chapter 2. Solid and Hazardous Waste Management; Chapter 3. Private and Public Water Supplies; Chapter 4. Swimming Areas; Chapter 5. Plumbing; Chapter 6. Private and Public Sewage Disposal and Soils; Chapter 7. Water Pollution and Water Quality Controls; Chapter 8. Environmental Health Emergencies, Nuisance Complaints, and Special Problems; and Chapter 9. Instrumentation.

Since the problems of the environment are so interrelated, certain materials must be presented at given points in order to give clarity and cohesiveness to the subject matter. As a result, the reader may encounter some duplication of materials throughout the text.

With the exception of Volume I, Chapters 1, 10, and 11 and Volume II, Chapters 8 and 9, all of the chapters have a consistent style and organization, facilitating retrieval. The introductory nature of Volume I, Chapter 1 and the unusual nature of Volume II, Chapter 8 do not lend themselves to the standard format. Volume I, Chapter 11 and Volume II, Chapter 9 discuss Instrumentation for the specific areas of each volume and therefore do not follow standard format.

In Volume I, Chapter 1, the reader is introduced to the underlying problems, basic concerns, and basic philosophy of environmental health. The ecologic, economic, energy, toxicologic, and epidemiologic bases provided help the individual to understand his or her relationship to the ecosystem and to the real world of economic and energy concerns; and to understand the relationship between biological, physical, and chemical agents and disease and injury causation. It also provides an understanding of the role of government and the environmental health practitioner in helping to resolve environmental and ecological dilemmas created by humans.

In Volume II, Chapter 8, the many varied facets of environmental emergencies, nuisances, and special problems are discussed. Students may refer to other chapters of the text to obtain a complete idea of each of the problems and the potential solutions.

The general format of Volume I, Chapters 2–10 and Volume II, Chapters 1–7 is as follows:

STANDARD CHAPTER OUTLINE

1. Background and Status (Brief)
2. Scientific, Technological, and General Information
3. Problem
   A. Types
   B. Sources of Exposure
   C. Impact on Other Problems
   D. Disease Potential
   E. Injury Potential
   F. Other Sources of Exposure Contributing to Problems
   G. Economics
4. Potential for Intervention
   A. General
   B. Specific
5. Resources
   A. Scientific and Technical; Industry, Labor, University; Research Groups
   B. Civic
   C. Governmental
6. Standards, Practices, and Techniques
7. Modes of Surveillance and Evaluation
   A. Inspections and Surveys
   B. Sampling and Laboratory Analysis
   C. Plans Review
8. Control
   A. Scientific and Technological
   B. Governmental Programs
   C. Other Programs
   D. Education

9. Summary
10. Research Needs

- The Background and Status section of each chapter presents a brief introduction to, and the current status of, each problem area. An attempt has been made in each case to present the current status of the problem.
- The Problem section is subdivided into several important areas to give the reader a better grasp of the total concerns. To avoid disruption in continuity of the standard outline, the precise subtitles listed may not be found in each chapter. However, the content of the subtitles will be present. The subtitle, Impact on Other Problems, is given as a constant reminder that one impact on the environment may precipitate numerous other problems.
- The Potential for Intervention section is designed to succinctly illustrate whether a given problem can be controlled, the degree of control possible, and some techniques of control. The reader should refer to the Controls section for additional information.
- Resources is a unique section providing a listing of scientific, technical, civic, and governmental resources available at all levels to assist the student and practitioner.
- The section of Standards, Practices, and Techniques is specifically geared to the reader who requires an understanding of some of the specifics related to surveys, environmental studies, operation, and control of a variety of program areas.
- The Modes of Surveillance and Evaluation Section explains many of the techniques available to determine the extent and significance of environmental problems.
- The Control section presents existing scientific, technological, governmental, educational, legal, and civic controls. The reader may refer to the Standards, Practices, and Techniques section in some instances to get a better understanding of controls.
- The Summary presents the highlights of the chapter.
- Research Needs is another unique section that is intended to increase reader awareness to the constantly changing nature of the environment and of the need for continued reading or in-service education on the future concerns of our society.
- The Reference section is extensive and as current as possible. It appears as the last chapter in each volume. It provides the reader with sources for further research and names of individuals and organizations involved in current research.

Acknowledgments

To Boris Osheroff, my friend, teacher, and colleague, for opening the numerous doors needed to obtain the most current information in the environmental health field, and for his many fine suggestions and ideas before and after reading the manuscript; to Ed O'Rourke for his intensive review of the manuscript and his recommendations for revisions and improvements; to Karol Wisniewiski for his review and comments on the manuscript; to Dr. John Hanlon for his encouragement and for helping a young teacher realize his potential; to all the environmental health administrators, supervisors, practitioners, and students who have shared their experiences and problems with me and have given me the opportunity to test many of the practical approaches used in the book; to the National Institute of Occupational Safety and Health, National Institutes of Health, National Institute of Environmental Health Science, U.S. Environmental Protection Agency, U.S. Food and Drug Administration, Cunningham Memorial Library, Indiana State University, Indiana University Library, Purdue University Library, and the many other libraries and resources for providing the material that was used in developing the manuscript; to my wife, Donna Koren, and my student, Evelyn Hutton, for typing substantial portions of the manuscript; to my daughter, Debbie Koren, for helping me organize the materials and for working along with me throughout the night at the time of deadlines to complete the work.

In the second edition, Kim Malone has typed portions of the new manuscript, retyped the entire manuscript, and has been of great value to me. Pat Ensor, Librarian, Indiana State University, has been of considerable value in helping gather large numbers of references in all areas of the book. A very special thanks to my sister-in-law, Betty Gardner, for typing a substantial portion of the new manuscript, despite recurring severe illness. Her cheerfulness during my low periods has helped me complete my work. Finally, thanks to my wife Donna for putting up with my thousands of hours of seclusion in the den, while I was working, and for encouraging me throughout the project and my life with her. She has truly been my best friend.

In the third edition, Alma Mary Anderson, CSC, and her assistants, Carlos Gonzalez and Brian Flynn, redid the existing illustrations and added new ones to enhance the manuscript. Professor Anderson directed the production of all of the new artwork. In addition, thanks to Bill Farms for his assistance with the original computer assisted drawings for the Instrumentation chapters in both volumes. My wife Donna typed much new material, and the previously mentioned libraries,

Centers for Disease Control, and the University of South Florida were most helpful in my research efforts. I would like to recognize Dr. Michael Bisesi, friend and colleague, who became my co-author. He would like to acknowledge his appreciation of his wife, Christine Bisesi, MS, CIH, his parents, Anthony and Maria Bisesi, and his mentors, Dr. Jorge Berkowitz, Dr. Raymond Manganelli, and especially Dr. Ray Harenstein for sharing their knowledge, wisdom, and encouragement at various phases of his academic and professional career.

Contents
Volume I

Contents
Volume II

Chapter 1

ENVIRONMENT AND HUMANS

Health and safety refer to the avoidance of disease and injury through efficient use of the environment, a properly functioning society, and an inner sense of well-being. *Environmental health and safety* is the art and science of protection of human function, the promotion of aesthetic values, and the prevention of disease and injury through the control of positive environmental factors and the reduction of potential hazards — physical, biological, and chemical.

To understand the relationship of the environment to humans and to understand how to protect humans from disease and injury, it will be necessary to discuss the ecosystem, ecosystem dynamics, and energy. Human impact on the environment and the various approaches used to evaluate and resolve environmental problems, including risk assessment, epidemiological, economic, legal, and governmental aspects also will be discussed. In order to understand abnormal physiology, toxicology, and infectious disease, a brief discussion on normal physiology is included. Finally, it will be necessary to understand the role of professional environmental health practitioners, the skills that they need, and how they address the expanding field of environmental problems.

THE ECOSYSTEM

The Earth is divided into the lithosphere, or land masses, and the hydrosphere, or the oceans, lakes, streams, and underground waters. The hydrosphere includes the entire aquatic environment. Our world, both lithosphere and hydrosphere, is shaped by varying life forms. Permanent forms of life create organic matter and, in combination with inorganic materials, help establish soil. Plants cover the land and reduce the potential for soil erosion — the nature and rate of erosion affects the redistribution of materials on the surface of the Earth. Organisms assimilate vast quantities of certain elements and molecules, such as carbon and oxygen. Animals, through respiration, release carbon dioxide into the atmosphere — carbon dioxide affects the heat transmission of the atmosphere. Organisms affect the environment and in turn are affected by it.

Two environments, biotic (living environment or community) and abiotic (nonliving environment), combine to form an ecosystem. An ecosystem can also be subdivided by more specific criteria into the following four categories: abiotic, the nutrient minerals that are synthesized into living protoplasm; autotrophic, the producer organisms (largely the green plants), which assimilate the nutrient minerals using energy and combine them into living organic substances; heterotrophic, the consumers, usually the animals that ingest or eat organic matter and release energy; and the heterotrophic reducers, bacteria or fungi, which return the complex organic compounds to their original abiotic state and release the remaining chemical energy. The biotic group in the ecosystem complex is essentially comprised of the autotrophs, or producer organisms that synthesize organic substances, and the heterotrophs, or consumer or reducer organisms that decompose labile organic substances. The ecosystem is important when considering the food chain, which is in effect a transfer of energy from plants through a series of organisms that eat and, in turn, are eaten. Eventually, decay will start the process all over again.

The ecological niche is the combination of function and habitat of each of the approximately 1.5 million species of animals and a half million species of plants on the Earth. There are many interactions between species in the ecosystem, yet a balance is dictated by nature. The law of limiting factors states that a minimum quantity of essentials, such as nutrients, light, heat, moisture, and space, must be available within the ecosystem for survival of the organisms. In some instances where these limiting factors apply or where pesticides or other environmental elements are introduced into the ecosystem, the organism alters itself in order to exist within the new environment. This change is called mutation. Unfortunately, mutation becomes a serious concern in the area of pest control as well as in disease, because the new organism may be highly resistant to effective control and may therefore cause disease and physical destruction of plants and animals. The ecosystem is always in a dynamic, rather than static, balance — changes in one part of the ecosystem will cause changes in another.

The Biosphere

The biosphere is that part of the Earth in which life exists. However, this definition is not complete, since spores may commonly be found in areas that are too dry, too cold, or too hot to support organisms that metabolize. The biosphere contains the liquid water necessary for life; it receives an ample supply of energy from an external source, which is ultimately the sun, and within it liquid, solid, and gaseous states of matter interface. All of the actively metabolizing organisms operate within the biosphere. The operation of the biosphere depends on photosynthesis, during which carbon dioxide is reduced to form organic compounds and molecular oxygen. Oxygen, the byproduct of photosynthesis, replenishes the atmosphere and most of the free water, which contains dissolved oxygen.

Ecosystem Dynamics

The ecosystem changes frequently. Several of the cycles that are important and that may be affected by humans include the hydrologic cycle, the carbon cycle, the nitrogen cycle, the phosphorous cycle, and energy flow. The hydrologic cycle is the movement of water from the atmosphere to the Earth and back into the atmosphere. This will be more fully discussed in the chapter on water.

The carbon cycle begins with the fixation of atmospheric carbon dioxide by means of photosynthesis performed by plants and certain algae. During this process carbon dioxide and water react to form carbohydrates, and free oxygen is simultaneously released into the atmosphere. Some of the carbohydrates are stored in the plant, and the rest are utilized by the plant as a source of energy. Some of the carbon that has been fixed by the plants is then consumed by animals, who respire and release carbon dioxide. The plants and animals die, decompose by action of microorganisms and other catalysts in the soil, and the carbon in their tissues is then oxidized to carbon dioxide and returned to the atmosphere. The carbon dioxide is recycled through the plants and the process repeats itself. Appropriate diagrams on the carbon cycle may be found in the chapter on sewage.

The nitrogen cycle begins when atmospheric nitrogen is fixed or changed into more complex nitrogen compounds by specialized organisms, such as certain bacteria and blue-green algae. Some fixation may occur as a result of lightning, sunlight, or chemical processes, however, the most efficient nitrogen fixation is carried out by biological mechanisms. Other bacteria, fungi, and algae may also play an important role in nitrogen fixation. Basically the atmospheric nitrogen is changed into a nitrate, which is absorbed by plants, eventually combining with other elements and becomes a plant protein. The plant protein decays when the plant dies, releases nitrogen as ammonia, and through bacterial oxidizing action becomes a nitrite, and through further bacterial action is reduced and released as atmospheric nitrogen. The plant protein may also be eaten by animals, broken down into molecules called amino acids and eventually synthesized into an animal protein. Through decay of the dead animal or breakdown of excreted feces and urine, this protein is changed to ammonia. The ammonia returns to the nitrite stage through bacterial action, and again through bacterial action becomes atmospheric nitrogen. A further description and diagram of the nitrogen cycle will be found in the chapter on sewage.

In the phosphorus cycle, the element moves rapidly through similar stages, becoming locked in sediment or in biological forms such as teeth or bones. The

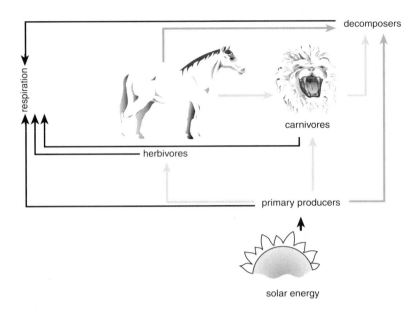

Figure 1.1.   Food Chain

primary sources of phosphorus for agriculture are phosphate rocks and living or dead organisms. Diagrams on the phosphorus cycle may be found in the chapter on sewage disposal.

Food Chain

The cycle of energy flow may also be described as the food web. The food web, or food chain, (Figure1.1) implies that an organism has consumed a smaller organism and is then consumed by a larger organism. Eventually, the microscopic plants and animals become the food supply for the small fish or animals, which become the food supply for humans. The importance of the food chain is illustrated by biomagnification, in which the impurities found in water are concentrated in the lower forms of life and are reconcentrated substantially during the movement of the impurities through the food chain. For example, whereas people might only get 0.001 μg of mercury in drinking water, they might get 30–50 μg of mercury by consuming fish that have been bioconcentrating mercury.

Energy Cycles

Solar energy is absorbed by the Earth and is eventually reradiated into space as heat. The heat is distributed over the surface of the Earth through circulation caused by the atmosphere and the oceans. Diurnal changes or changes occurring in a 24-hr period due to light, wind, temperature, and humidity are important near ground level and at high levels in the atmosphere. In certain localities land masses and sea breezes may affect the overall heat and weather patterns. However, total heat movement is not affected significantly by local conditions.

About 30% of the solar energy entering the atmosphere is deflected or scattered back toward outer space because of the atmosphere, the clouds, or the Earth's surface. This portion of the solar energy is lost. About 50% of the incoming radiation reaches the ground or ocean, where it is absorbed as heat. The properties of the surface that receive the energy will determine the thickness of the layer over which the available heat will be distributed. In the oceans the surface wave motions effectively distribute the heat over a 300 ft layer of air. On land the energy transferred downward into the ground occurs very slowly through the process of molecular heat conduction. The penetration distance is very small. About 20% of the incoming solar radiation is absorbed as it goes through the atmosphere. In the upper atmosphere, oxygen and ozone molecules absorb an estimated 1–3% of the incoming radiation. This absorption occurs in the ultraviolet range and, therefore, limits the penetrating radiation to wavelengths longer than 300 nm. This absorption is very important, since it is the main source of energy for the movement of air in the upper atmosphere. This absorption also shields people on Earth from damaging effects of ultraviolet radiation. Most of the rest of the 20% is absorbed by water vapor, dust, and water droplets in the clouds. The energy that sustains all living systems is fixed in photosynthesis, as previously mentioned. Part of the energy fixed in plants and animals has been compressed over millions of years and is the source for stored energy, namely, coal, oil, and natural gas.

Current Ecosystem Problems

People and advanced technology affect the ecological niche by interfacing with the intricate function and habitat of various species of animals and plants. Discharges to air, water, and soil, including toxic and infectious chemical wastes, biological wastes, and radiological wastes create an environmental pressure that is detrimental to all life forms, including people. From the destruction of the Rain Forest of Brazil, because of the excessive clearing of trees, to the potential destruction of forests in the Northern Hemisphere due to acid precipitation, a huge number of ecosystems may be eliminated. Nature's diversity offers many opportunities for finding new pharmaceuticals, genetic mapping, genetic engineering, and the potential power to improve crops. We are wasting our greatest natural resource on which we depend for food, oxygen, clean water, energy, building materials, clothing, medicine, sociological well-being, and countless other benefits.

Biological diversity includes two related concepts, genetic diversity and ecological diversity. Genetic diversity is the amount of genetic variability among individuals in a single species, whereas ecological diversity is the number of species in a community of organisms. An organism's ability to withstand the challenge of varying physical problems, chemical problems, parasites, and competition for resources is largely determined by its genetic makeup. The more successful organisms that survive pass on the genes to the next generation. An example of this would be in the case of climatic changes where plants have smaller, thicker leaves in order to lose less water in areas that are becoming arid. Genetic diversity is important in developing new crops to meet new conditions. This is how disease-resistant crops can be utilized to sustain life for people. Genetic variation among species is greater than within the species. It is difficult to determine which of these species are of extreme importance and will be essential to people in the future as new supplies of food, energy, industrial chemicals, and medicine. It is anticipated that the human population will increase by nearly 50% in the next 20 years, with much of this increase occurring in very poor nations. The resources needed to feed these individuals will be substantial, and many new forms of food will have to be developed in order to avoid starvation.

There appears to be adequate sources of fossil fuel for the immediate future. However, these fossil fuels, when burnt, can potentially cause severe pollution. Eventually, fossil fuel sources will be depleted. It will then be necessary to utilize plants that can produce energy-rich materials, such as soybean oil. At present this process is very expensive.

Coastal waters contain several major natural systems. Since they directly abut the land, they are affected by activities on the land. Pollutants coming from coastal sewage treatment plants, industrial facilities, the settling of air contaminants, and erosion of land, many hundreds of miles upstream, can cause considerable problems to the coastal waters. Oil slicks are extremely hazardous to ecosystems. Trace metals can be concentrated in the food chain and, thereby, become a hazard to people eating sea life. Toxic metals, such as mercury, lead, and cadmium, have been found in coastal waters.

Molluscs become indicators of environmental contamination and environmental quality. These shellfish, which live in the mud or sand bottoms of the aquatic ecosystems, accurately reflect some of the important characteristics of the water adjacent to the land. These organisms may show a direct adverse effect when they accumulate chemical substances or microorganisms. They are good for systematic monitoring, because they are stationary in nature, except for the early egg and larva stages, and filter large quantities of water in order to feed themselves. Since they are of significant commercial importance, there is reliable quantitative data available on harvesting and other aspects of the molluscs' life cycle.

In the past, the EPA sponsored studies of the accumulations of chlorinated organic pesticides in shellfish. Residues of the pesticide DDT were found for years after the use of the chemical was banned in 1972. The Mussel Watch Program has shown elevated levels of polychlorinated biphenyls (PCBs). High-level concentrations of silver, cadmium, zinc, copper, and nickel have been detected in various bodies of water. Mussels have also shown the presence of radionuclei as a result of weapon testing programs and also from effluent that has come from nuclear reactors. Records of commercial catches of clams and oysters indicate that the number of molluscs in coastal waters has been steadily declining over the years. This may be due to overfishing, loss of habitat, and deterioration due to pollution and natural disasters. It has been shown that the building of a sewage treatment plant has the potential to adversely affect shellfish habitats by altering water salinity.

The continued loss of tropical forests is the single greatest threat to the preservation of biological diversity. Much of Central America's ecosystems may be affected by these problems.

Wetlands have exceptional value because of their location at the land/water interface. They provide ecological, biological, economic, scenic, and recreational resources. Many wetlands are among the most productive ecosystems on the planet, especially estuaries and mangrove swamps. At least one half of the biological production of the oceans occurs in the coastal wetlands. From 60 to 80% of the world's commercially important marine fish either spend time in the estuaries or feed upon the nutrients produced there. Coastal wetlands help protect inland areas from erosion, flood, waves, and hurricanes. The wetlands are an important economic resource, because they provide cash crops, such as timber, marsh hay, wild rice, cranberries, blueberries, fish, and shellfish.

Desertification is the process that occurs in semiarid lands where sterile sandy desert-like ecosystems are created from grasslands, brushlands, and sparse open forests due to a loss of organic carbon in the form of stable humus from the soils. This desertification can be stimulated beyond natural forces by over-grazing of animals and by misuse of the area. This can lead to mass starvation and death.

ENERGY

Energy problems came to the attention of the American public and the world in 1973 as a result of the Arab oil embargo and the actions of the Organization of Petroleum Exporting Countries (OPEC). Unfortunately, in the years preceding these dramatic events, the United States was buying an increasing amount of petroleum abroad and utilizing more natural gas than was readily available. As a result of these shortages, the President authorized the allocation of scarce fuels. By May 1973 the Office of Emergency Preparedness was reporting widespread problems of gasoline stations closing due to lack of fuel. Voluntary guidelines were announced and a variety of emergency actions taken including the lowering of the speed limit to 55 mi/hr. Project Independence was conceived with the goal of self-sufficiency in energy production by 1980. Congress established the Federal Energy Administration in December 1973, and the President created the Federal Energy Office to deal with the immediate crisis. Funding was provided for additional research and development in a variety of energy areas.

Unfortunately, by 1978 the amount of fuel being purchased abroad was escalating. The variety of programs at the federal level were so confusing, complex, and expansive over a variety of agencies that the administration in Washington, DC requested the establishment of a new federal Department of Energy with an administrator of cabinet rank. It was hoped that the new department would bring organization to the chaotic mass of confusing energy problems and move the country toward self-sufficiency.

Conservation and development of new energy sources or the expansion of existing sources were two major phases of energy programs. Energy conservation was of critical importance to the United States; yet despite predictions of severe shortages, conservation attempts had not been entirely successful and great amounts of energy were still wasted. Greater energy efficiency was needed in cars and in electrical and gas equipment. Better insulated homes and a general reduction in energy usage was essential. The United States had to become less dependent on oil and gas as sources of energy.

Oil and gas in 1972 accounted for nearly 78% of U.S. energy consumption. By 1980, oil and gas remained the greatest sources of energy used, despite the fact that the United States had about a 300-year supply of coal available. Although in 1979, Alaska was providing new oil, the United States was more dependent than ever on foreign oil. It had been recommended that natural gas be deregulated at the wellhead, enabling the gas companies to provide adequate funds to further research. However, it was understood that no matter how many sources of oil and gas were developed, eventually the supplies would be depleted; therefore, it was necessary to examine other kinds of energy sources. These sources included oil shale, nuclear energy, geothermal energy, solar energy, fusion energy, and coal. Oil shale was available, but at rather high costs. Additional research was necessary to determine how best to extract the oil from the shale both economically and without creating surface environmental problems.

Nuclear power could certainly provide a continuing source of available energy to the country. However, many permits had been delayed because of consumer groups concerned with the possibility of nuclear accidents and with the disposal of the radioactive wastes. The threat of terrorist groups obtaining wastes from the fusion process and utilizing these materials to terrorize the world was also of great concern. In 1973, the U.S. nuclear electrical generating capacity was 20,000 MW, or over 5% of the nation's total electrical capacity.

Geothermal energy is energy produced by tapping the Earth's heat. It was being used at the geyser sites in Oregon and in California, where the Pacific Gas and Electric Company employed geothermal steam as a power source for a 400-MW electrical generating facility. Geothermal resources include dry steam, hot steam, and hot water. At that time, because of cost factors, only dry geothermal heat was used to drive electric turbines.

Solar energy was both economically and technologically feasible. The benefits of capturing the sun's energy had been long recognized. It was predicted that if research and development programs were successful, solar energy could provide, by the year 2020, 25% of the total U.S. energy requirements. Funding for solar energy research had increased and considerable research was needed. The 1979 federal budget included $137,500,000 for solar research and development. Techniques under investigation included direct thermal applications, solar electric applications, and fuels from biomass. Direct thermal application consisted of solar heating for cooling of buildings and hot water supply, and the use of solar heat for agriculture and industrial processes. Agricultural applications were the drying of food, crop drying, lumber drying, heating and cooling of greenhouses, and heating of animal shelters. The solar electrical applications included research on the generation of electricity from windmills, solar cells (photovoltaic), solar thermal electric systems, and ocean thermal energy conversion. Theoretically, these systems were well understood; however, their applications were not yet economically feasible. The fuels produced from biomass suggested large-scale use of organic materials, such as animal manure, field crops, crop waste, forest crops and waste, and marine plants and animals. The conversion process was technically feasible. During World War II, much of the liquid fuel of France consisted of methanol produced from wood. It is hoped that by the year 2000 biomass will supply the equivalent of 1.5 million barrels of oil per day. Processes requiring study included fermentation to produce methane and alcohol, chemical processes to produce methanol, and pyrolysis to convert organic waste material to low British thermal unit (Btu) gaseous fuels and oils.

Coal was the one major source of energy that could certainly be utilized to reduce our reliance on other nations. The most serious deterrent was the many contaminants, particularly sulfur, found in coal. Sulfur could be removed by utilizing new technologies in coal mining and conversion. Since coal made up 85% of the U.S. fossil energy resource base, it was wise to utilize it as well and as quickly as possible. New coal technology under development promised two very important changes in the potential pollution areas: the reduction of sulfur oxide in particulate matter from direct burning of coal, and the provision of liquid and gaseous fuel substitutes for domestic oil and gas production. The emphasis in research had shifted from conventional combustion to direct fuel to fuel conversion, that is, coal gasification and coal liquification. Sulfur could be extracted even when coal was directly burned using fluidized beds in which crushed coal was injected into the boiler near its base. Coal was fluidized by blowing air uniformly through a grid plate. Sulfur dioxide was then removed by injecting the crushed coal with limestone particles less than $1/_8$ in. in diameter. The limestone particles were noncombustible and would bind the sulfur dioxide. This system also reduced the nitrogen oxides below emission standards for utility boilers of 0.7 lb of nitrogen dioxide per million Btu. The nitrogen dioxide could actually be reduced to 0.2–0.3 lb per million Btu. The gasification process involved the reaction of coal with air, oxygen, steam, or a mixture of these that yielded a combustion product containing carbon monoxide, hydrogen, methane, nitrogen, and little or no sulfur. The energy of the product ranged from 125–175 Btu per standard cubic foot for a low Btu gasification process to about 900–1000 Btu per standard cubic foot for a high Btu gasification process. The high Btu gasification process was comparable with the energy content of natural gas. The liquification transformed the coal into a liquid hydrocarbon fuel and simultaneously removed most of the ash and sulfur. Hydrogen was frequently added in many of the direct catalytic liquification techniques.

The Energy Dilemma

The 1980s

The trauma of the 1970s led to a more active role by government in the 1980s to try to resolve energy problems in the United States. The Energy Security Act and the Crude Oil Windfall Act of 1980 gave the President additional power to encourage the development of oil and gas reserves and a synthetic fuels industry. Since coal was our greatest fuel resource, there was an attempt to substitute coal for oil in various industries. Energy was used more efficiently because of a variety of conservation programs. The federal conservation and solar programs were passed by Congress, and additional funds were allotted for weatherization assistance for low-income people and tax breaks for others who made their homes more efficient. Funds were provided by the Solar Energy and Conservation Bank to finance the use of solar energy systems. Solar wind and geothermal tax credits were made available for residential users. Funds were provided for research and development in the use of energy sources other than fossil fuels. The biomass program and the alcohol tax credits provided 1.6 billion dollars thru 1990 to promote the conversion of grain, farm residues, and other biomasses to alcohol fuels. Sixteen and one half billion dollars were allotted through 1990 to improve transportation efficiency by enhancing public transportation programs. A fuel economy program was put in place and funds were provided for enforcement of the 55 mi/hr speed limit.

Auto efficiency standards were raised from 20 mi/gal for the 1980 model year vehicle to 27.5 mi/gal by 1985. A gas guzzler tax was added to new vehicles with mileage below 15 mi/gal. A building energy conservation code and an appliance efficiency standard code was established to improve fuel conservation and efficiency. As a result of all this effort, the growth of energy usage slowed substantially from the previous years.

By 1984 the national energy picture had continued to brighten substantially. Prices of gasoline had dropped and the price of crude oil had dropped substantially. This progress was due in part to the decontrol of the domestic petroleum markets and the removal of restrictions on energy use. Potential new sources of energy were being developed in the oceans abutting the United States. This area is larger than the country's land mass and includes many of the deep water areas, as well as the Outer Continental Shelf. Further, the OPEC nations were in disarray because of the long war between Iran and Iraq, and the propensity for cheating on the amount of oil that each country was permitted to produce in order to maintain their high income. Continued conservation of energy because of the many permanent changes made in energy use over the years helped keep the price of oil at lower levels.

In 1984, 1608 geothermal leases were in effect, in which 30 were capable of production and 12 actually were in production.

There was an estimated 2 trillion barrels of shale oil in the Green River formation in northwestern Colorado, northwestern Utah, and southwestern Wyoming. Of this amount, it was estimated that 731 billion bbl of oil could be extracted commercially. It was also estimated that in Utah, there were approximately 2 billion barrels of oil recoverable with current technology. Approximately 100–200 million barrels could be recovered by surface mining. The question was not the availability of these energy resources, but the cost of developing them.

Since the Three Mile Island debacle of March 1979, there have been numerous public protests about the building and use of nuclear power as a source of energy. There still is a need for facilities for the permanent disposal of the high radioactive waste that is generated during the production of electricity in the nuclear power plants. Until very recently it was assumed that the nuclear waste was being handled appropriately on site. However, it has been disclosed that there have been numerous leaks of nuclear waste from the nuclear power plants over the decades. A prime example of this is the serious problems occurring in Ohio, which must be corrected to prevent disease. In 1990 the cleanup of these sites were estimated to cost 31.4 billion dollars.

The 1990s

In 1988, the United States consumed more energy than in any previous year in its history. The demand for petroleum was about 17 million barrels/day which was the highest point since 1980. However, the strong underlying energy efficiency trend continued, except for the use of energy in the transportation sector. The projection for energy use by the year 2000 is 90 quads, with the current usage of 81.5 quads in 1991. A quad is equivalent to one quadrillion British thermal units. The U.S. domestic oil production, which was on the rise in the 1980s will start to decline again, and it is anticipated that by the year 1994 over half of the oil used by the United States will be imported. This may once again leave the country vulnerable to the demands and needs of OPEC, wars, and other catastrophes, such as the running aground of the Exxon Valdez in Prince William Sound, Alaska. This ecological debacle temporarily cut the amount of oil coming from Alaska down to 20% of its normal flow. Although oil flow recovered and prices then declined, one wonders about future catastrophes and their effect.

Because of lower oil prices during 1989, U.S. exploration decreased sharply. In 1989, there were only 740 drilling rigs, whereas in 1981 there were 4500 functioning drilling rigs. There will be a need for a long lead time to simply get ready to do further exploration for oil in the United States. The environmental and conservation groups will attempt to limit further oil exploration in Alaska because of the incident that occurred in March of 1989.

On August 2nd, 1990, Iraq invaded Kuwait. By the end of February 1991, Iraq had set on fire or damaged 749 oil facilities in Kuwait. Although there is currently an over supply of oil now that Operation Desert Storm is over, there will be a drop in oil production. The restoration of oil production from Kuwait will be a long, complicated, and expensive process.

The demand for natural gas could expand between 10 and 15% before the year 2000. This would mean gas production in excess of 20 trillion ft^3/year again. Electricity, despite increased efficiency in usage, is projected to rise at a yearly average 3.2%. Since there are not any newly ordered nuclear power plants to be on line by the year 2000, fossil-fuel plants must continue to provide most of the generating capacity.

Despite the hope of the early 1970s, nuclear power has never achieved anywhere near the amount of electrical output that had been anticipated. This is in part due to the concern of citizens, a variety of lawsuits, and in a large part due to the Three Mile Island incident in 1979. The shoddy construction and need to obtain multiple permits, such as in Indiana, have further aggravated the situation. The U.S. Nuclear Regulatory Commission (NRC) has listed a series of unresolved safety issues. These include systems interactions, seismic design criteria, emergency containment, containment performance, station black out, shutdown decay, heat-removal requirements, seismic qualification of equipment in operating plants, safety implications of control systems, hydrogen control measures and effects of hydrogen burns on safety equipment, high-level waste management, low-level waste management, and uranium recovery and mill tailings.

With the increasing problems related to oil and gas, as well as nuclear power, electric utilities and others may turn increasingly to coal as a source of fuel and power. This occurs at a time when acid precipitation, in the form of acidic rain and snow, continues to be an issue. The current administration in the White House will find it hard to maintain the line against mandated actions related to acid rain. The damage caused by acid rain is increasing, as well as the pressure from Canada to do something about it. Although new energy technologies related to coal, as well as other sources, have been put pretty much into a holding pattern, it seems that it will now be necessary to explore these further and attempt to utilize other means of securing energy.

While it is true that the Clean Air Act of 1990 puts numerous additional constraints on the use of coal as a source of energy, at some time in the future it may be economically feasible to clean up the coal and use this abundant energy resource.

Health Implications of New Energy Technologies

Coal, which is our most abundant fossil fuel, presents a potential threat to human health and the environment. Coal miners excessively exposed to coal dust may suffer from lung diseases known as pneumoconioses, including black lung disease, silicosis, and emphysema. The fossil fuel byproducts include two major groupings: coal-tar, which comes from the combustion or distillation of coal, and polycyclic aromatic hydrocarbons. Polycyclic aromatic hydrocarbons (PAHs) are potential carcinogens found in mixtures within fossil fuels and their byproducts. Coal tar products can cause skin and lung cancer through the cutaneous and respiratory routes, respectively.

Human exposure to complex mixtures of PAHs has been extensive, and carcinogenic effects following long-term exposure to them have been documented. Coke-oven emissions and related substances, such as coal-tar, have long been associated with excess disease. The first observation of occupational cancer, that is, scrotal cancer among London chimney sweeps, was made by Percival Pott in 1775. More recently, lung and genitourinary cancer mortality have been associated with coke-oven emissions. Human tumorigenicity also has been associated with the exposure to creosote. *Creosote* is a generic term that refers to wood preservatives derived from coal-tar creosote or coal-tar oil, and includes extremely complex mixtures of liquid and solid aromatic hydrocarbons.

Bituminous coal is the starting point for coal liquification and coal gasification. The mere process of breaking up the coal's structure appears to release a large number of potentially carcinogenic compounds. Some of these compounds include the PAHs benz(a)anthracene, chrysene, and benzopyrene. In the coal gasification process, where coal is exposed to molecular oxygen and steam at temperatures of 900°C or higher, much of the hydrocarbon structure is destroyed, and therefore, the levels of carcinogenic compounds can be drastically reduced. However, at these temperatures the hazards due to carbon monoxide increase, as well as the potential hazards due to other toxic agents such as hydrogen sulfide and other carbon-nitrogen products that may be released to the environment.

Coal liquification occurs in the temperature range of 450–500°C. This is the most economical temperature range to produce pumpable liquid. During the initial conversion reaction, hydrogen is added to the coal, which produces a wide range of compounds containing condensed aromatic configurations. Many of these are carcinogenic. A potential also exists for the production of other toxic compounds, including carbon monoxide.

Oil shale is a generic name for the sedimentary rock that contains substantial organic materials known as kerogen. The kerogen content of oil shale may vary between 5 and 80 gal of oil per equivalent ton. The extraction of the oil from the oil shale occurs when the shale is heated to 350–550°C under an inert atmosphere. The products include oil vapor, hydrocarbon gases, and a carbonaceous residue.

There are major concerns related to the retorting shale process. (1) The production of one ton of spent shale per barrel of oil. (2) The volume of the shale increases by more than 50% during the crushing process and a considerable amount of alkali minerals contained in the oil shale could reach into the groundwater supply. (3) Since a large amount of water is needed for controlling dust and reducing the alkalinity of spent shale, this would create a problem in areas where water is in short supply.

During the processing of the oil shale, organic materials containing nitrogen, oxygen, and sulfur are produced in large quantities. Further, waste gases containing hydrocarbons, hydrogen sulfide, sulfur dioxide, and nitrogen oxides are formed.

Geothermal energy is a general term that refers to the release of the stored heat of the Earth that can be recovered through current or new technologies. The only available technologies are those that extract heat from hydrothermal convection systems, where water or steam transfer heat from the deep parts of the system to areas where this energy can be tapped. The amount of pollutants found in the energy source varies dramatically from area to area. In any case, geothermal fluids may contain arsenic, boron, selenium, lead, cadmium, and fluorides. They also may contain hydrogen sulfide, mercury, ammonia, radon, carbon dioxide, and methane. Wastewater management is an environmental problem if the spent liquids are put into surface water. They may contaminate the surface water and also cause an elliptical dish-shaped depression, such as occurred in New Zealand. If the liquids are injected under high pressure back into the geological formation, they may enhance seismic activities. Another problem that is related to the production of geothermal energy is noise. Some noise levels reach 120 dBA. Blowouts can occur and produce accidents with potential injuries. Airborne emissions and noise can affect human health. The most significant potential air emission problem is hydrogen sulfide gas. Hydrogen sulfide is a very toxic gas that produces immediate collapse with respiratory paralysis at concentrations above 1000 ppm by volume. Serious eye injuries can occur at 50–100 ppm by volume. Further, hydrogen sulfide has a very offensive odor that can be detected by the human nose at very low concentrations.

The photovoltaic solar cell is an efficient, direct energy-conversion device. The sun falls on this device and liberates electrons, which flow as an electrical current. Solar cells are used for powering most satellites. The major health hazards involved relate to the mining and the various steps needed in the production of the silicon cells. Other cells utilized include cadmium sulfide and gallium arsenide. The health hazards involved relate to the production and use of these chemicals in the occupational setting. If satellite solar power stations come on line in the future, there will need to be an extensive research effort to determine if there will be potential biohazards due to the microwave beams being brought back to Earth from the solar space stations.

Biomass is plant material and animal waste used as a source of fuel. These plant and animal materials include forest products and residues, animal manure, sewage, municipal wastes, agriculture crops and residues, as well as aquatic plants. The value of biomass is that there is a continuous renewable source of energy. The problems of using biomass for energy purposes are many.

Health hazards may be associated with the emissions from biomass combustion residuals, biomass gasification residuals, biomass liquification, and the related air and water pollutants. The burning of biomass creates the same primary air pollutants as in the burning of coal. Unburned hydrocarbons, sulfur oxides, and high-fugitive dust levels are created. Organic acids and minerals that affect water quality may be found in boiler-water treatment chemicals, in leachates from ash residues, and in biomass storages piles. Residential wood burners, because of incomplete combustion, release carbon monoxide and unburned hydrocarbons, which may include photochemically reactive chemicals as well as carcinogens. Emissions from biomass gasification may originate from the process stack, waste ponds, storage tanks, equipment leaks, and storage piles. These pollutants include oxides of nitrogen, hydrogen cyanide, hydrocarbons, ammonia, carbon monoxide, and particulates. The process water and condensates also may contain phenols and trace metals. The tars produced by the thermochemical decomposition of organic substances may contain PAHs, some of which are known carcinogens. Anaerobic digestion may lead to the production of odors, hydrogen sulfide, and ammonia gases. The effluent contains large quantities of biochemically oxygen-demanding organic materials, organic acids, and mineral salts.

Nuclear power plants have the potential for providing abundant supplies of electricity without contributing substantial amounts of pollutants to the environment. (Figure 1.2) The industry, however, has failed to deliver on the promise because the costs of making nuclear energy safe have spiraled out of control. The nuclear cycle starts with the mining of uranium and the resulting risk of lung cancer in miners attributable to the alpha radiation from the decay of radon-22 daughter products. Further, there are injuries associated with the mining and

Fig 1.2.   Nuclear Energy Site

quarrying of uranium. During the milling of uranium ore, there are small airborne releases of radon and the residues or "tailings" still contain most of the radioactive species of the original ore and require careful disposal. Next, the uranium oxide is converted to uranium hexafluoride. This enrichment increases the percentage amount of fissionable uranium-235. Finally, fuel rods are fabricated during the processing of the uranium after it has been mined. (The potential amount of worker radiation exposure that comes from gaseous solid wastes is small.) The reactor then produces energy. Finally, after the fuel rods are spent, the irradiated rods are removed from the reactor cooler and stored for long periods of time to permit decay of the short-lived fission products before reprocessing. Radioactive water and waste must be contained.

Engineers can build reactors that are safer than those now in operation. The basic technology has been available for more than 20 years. This technology has been ignored in favor of water-cooled reactors, which have already been proven in nuclear submarines. However, these reactors are particularly susceptible to the rapid loss of coolant, which led to the accidents at Three Mile Island and Chernobyl. All nuclear reactors split large atoms into smaller pieces and thereby release heat. It is necessary to keep the core of nuclear fuel from overheating and melting into an uncontrollable mass that may breach the containment walls and release radioactive material. One way to prevent a meltdown is to make sure that the circulating coolant, which is water, will always be present in adequate quantities. To prevent mechanical failures from interrupting the transfer of heat, most reactors employ multiple backup systems. This technique is known as "defense in depth." The problem with this technique is that it can never be 100% safe against a meltdown.

The U.S. Department of Energy wanted to use a new strategy in Idaho Falls, Idaho. They wanted to build a series of four small-scale modular reactors that use fuel in such small quantities that their cores could not achieve meltdown temperatures under any circumstances. The fuel would be packed inside tiny heat-resistant ceramic spheres and cooled by inert helium gas. The whole apparatus would be buried below ground. The main problem is that the smaller units produce less electrical output and, therefore, are less economical initially. However, over time the units may become more efficient.

During the early morning hours on March 28, 1979, the most serious accident in the history of U.S. nuclear power took place. A series of highly improbable events involving both mechanical failure and human error led to the release of a considerable amount of radioactivity and the evacuation of preschool children and pregnant women within 5 mi of the Three Mile Island nuclear power plant, located near Middletown, Pennsylvania. The accident occurred at unit #2 of the complex. The accident was initially triggered by the failure of a valve and the subsequent shutting down of a pump supplying feed water to a steam generator. This in turn led to a "turbine trip" and a shutdown of the reactor. This in itself was not the problem, because of the backup systems that were used anticipate such failures.

However, a pressure relief valve failed to close, which led to the loss of substantial amounts of reactor coolant, usually in quench tanks. Auxiliary feed water pumps were nonoperational because of closed valves, in violation of the Nuclear Regulatory Commission (NRC) regulations. The operator failed to respond promptly to the stuck relief valve. There were faulty readings on the control room instruments, which led the operators to turn off the pumps for the emergency core cooling system. The operator shut down the cooling pumps more than an hour after the emergency began. Two days after the initial problems, a large bubble of radioactive gas, including potentially explosive hydrogen, was believed to have formed in the top of the reactor. The President of the United States established a 12-member committee to review the accident and make recommendations. The major findings of the commission after a 6-month study were (1) the accident was initiated by mechanical malfunction and exacerbated by a series of human errors; (2) the accident revealed very serious shortcomings in the entire government and private sectors systems used to regulate and manage nuclear power; (3) the NRC was so preoccupied with the licensing of new plants that it had not given adequate consideration to overall safety issues; (4) although the training of power plant operators at Three Mile Island conformed to NRC standards, it was extremely inadequate; (5) the utility owning Three Mile Island and the power plant failed to acquire enough information on safety to make good judgements; (6) there was an extremely poor level of coordination and a lack of urgency on the part of all levels of government after the accident occurred. The NRC had not made mandatory, state emergency or evacuation plans.

Other potential emergency situations have occurred in the past. On December 12, 1952, the accidental removal of four control rods at an experimental nuclear reactor at Chalk River, Canada near Ottawa led to a near meltdown of the reactor's uranium core. A million gallons of radioactive water accumulated inside. Fortunately, there were no accident-related injuries. On October 7, 1957, a fire occurred in the reactor north of Liverpool, England. Like the Chernobyl facility, the Windscale Tile #1 plutonium production plant used graphite to slow down neutrons emitted during nuclear fission. Two hundred square miles of countryside were contaminated. Officials banned the sale of milk from the cows grazing the area for more than a month. It was estimated that at least 33 cancer deaths could be traced to the effects of the accident. On January 3, 1961, a worker's error in removing control rods from the core of an SL-1 military experimental reactor near Idaho Falls, Idaho caused a fatal steam explosion, and three servicemen were killed. On March 22, 1975, a worker using a lighted candle to check for air leaks in the Brown Ferry reactor near Decatur, Alabama touched off a fire that damaged electrical cables connected to the safety system. Although the reactors coolant water dropped to dangerous levels, no radioactive material escaped into the atmosphere. On March 8, 1981, radioactive wastewater leaked for several hours from a problem-ridden nuclear power station in Tsuruga, Japan. Workers who mopped up the wastewater were exposed to radiation. The public did not become aware of the accident for 6 weeks until radioactive materials were detected. On January 4, 1986, one worker at the Kerr-McGee Corporation, a uranium processing plant in Gore, Oklahoma, died from exposure to a caustic chemical that formed when an improperly heated, overfilled container of nuclear materials burst. Some of the radiation flowed out of the plant. More than 100 people went to local hospitals.

April 26, 1986, the worst disaster in history relating to a nuclear plant occurred in Chernobyl, Russia. A loss of water coolant seemed to trigger the accident. When the water circulation failed, the temperature in the reactor core soared over 5000°F, causing the uranium fuel to begin melting. This produced steam that reacted with the zirconium alloy of the fuel rod to produce explosive-type hydrogen gas. Apparently, a second reaction produced free hydrogen and carbon monoxide. When the hydrogen combined with the oxygen, it caused an explosion, which blew off the top of the building and ignited the graphite. Next, a dense cloud of radioactive fission products discharged into the air as a result of the burning graphite. In subsequent days, radioactive materials were found in Finland, Sweden, Norway, Denmark, Poland, Romania, and Austria. In Italy, freight cars loaded with cattle, sheep, and horses from Poland and Austria were turned back because of the concern due to abnormally high levels of radiation found in many of the animals. Britain cancelled a spring tour of the London Festival Ballet, which was to go to the Soviet Union. In West Germany, officials insisted that children be kept out of sand boxes to avoid contamination. Slight amounts of radiation were found in Tokyo, in Canada, and in the United States. In the immediate vicinity of the reactor and up to 60 mi^2 from the reactor, the topsoil may be contaminated for decades. The residents of Kiev, 80 mi from the disaster and the surrounding areas, were told to wash frequently and to keep their windows closed. They were warned against eating lettuce and swimming outdoors. Water trucks were used to wash down streets to wash away radioactive dust. It will be many years before there is an accurate determination as to the number of people who will die as a result of this catastrophe. Approximately 100,000 people will have to be monitored for the rest of their lives for signs of cancer. In 1995, large populated areas surrounding the reactor site in the Ukraine and in nearby Belorussia remain contaminated with high levels of radioactivity. The cost of the cleanup could run as high as $358 billion dollars.

It should be understood that the nation will have to make a myriad of difficult decisions concerning the conservation and use of fuel. Obviously the burning of carbonaceous fuels has significantly contributed to air pollution and accompanying health problems. The energy problem has affected our entire economy by the purchase of energy from abroad. An adequate program of development and use of energy sources within our own country will provide us with the needed energy and yet prevent the many potential hazards that may occur as a result of the burning of fossil fuels.

Environmental Problems Related to Energy

The major environmental problems related to energy are caused by pollutants created by fossil fuels, destruction of the natural environment by removal or spillage of fossil fuels, and the effect of the continued rise in cost of fossil fuels on the economy, and therefore, on our way of life. The pollutants will be detailed in the chapter on air pollution. With an increased need for fossil fuels comes an increase in the destruction of the natural environment. Improper fuel removal, fuel storage, or fuel transportation degrades the environment, destroys the aesthetic value of many of our most beautiful areas, and causes destruction of fish, wildlife, and plant life. The economics of energy is extremely important, since the world continues to shift huge sums of money from industrialized societies to oil-producing nations. If the industrialized nations are unable to pay for the energy being provided, mass unemployment and recessions may result. In the Third World, the nations burdened by huge costs for energy will be faced with the choice of literally shutting down their countries or allowing their people to starve because of the tremendous financial drain. The problems related to health, aesthetics, and economics are critical, since they are totally interrelated, and therefore must be considered as a unit.

TRANSPORT AND ALTERATION OF CHEMICALS IN THE ENVIRONMENT

Humans are part of the flow of energy; in the biosphere we interact with thousands of plants and animals. Because of our power and productivity, we have the ability to alter the ecosystems of the Earth, of which we are a part, in a beneficial or harmful manner. Technological advancement since the early 1800s has been revolutionary rather than evolutionary. In the great movement forward to improve our own lifestyle, we have destroyed many of the natural ecosystems and severely polluted much of the air, water, and soil. To better understand the nature of the environmental alterations, it is important to understand the role of chemicals in the environment and how they are transported and altered.

Much of the environment is made up of chemicals, the largest amount of which are occurring naturally and have few detrimental effects on humans. However, we change the form, distribution, and concentration of these chemicals and synthesize or produce, either as products or as byproducts, chemicals that are naturally found in the environment, as well as chemical compounds that are not. It is necessary to understand the movement of those chemical compounds that are not natural and to understand the movement of all of these chemicals through the environment, their concentrations, the degree of human exposure, and the potential hazards to human health (Figure 1.3).

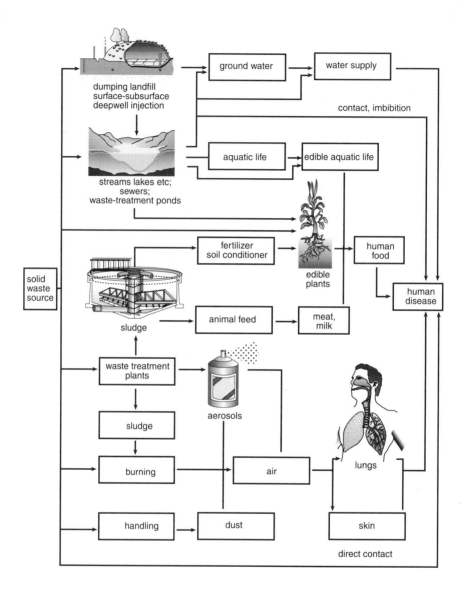

Figure 1.3 Chemical waste/human disease pathways (postulated).

Once a pollutant is released into the environment, the chemicals are transported and transformed in a variety of complex ways. Chemicals are transported widely throughout the environment by active and passive movement through air, water, biota (biological life), and soil. In addition, they serve as vectors of transport. Chemicals are transformed by chemical or biochemical reactions, and diluted, diffused, or concentrated by physical or biological processes. Some substances

Chemicals in the Environment

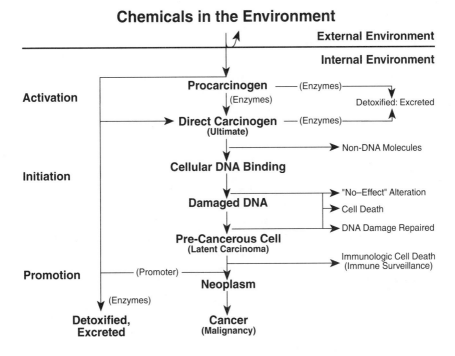

Figure 1.4 Chemicals in the environment. *Source: Environmental Toxicology and Risk Assessment: An Introduction, Student Manual* (Chicago, IL: U.S. Environmental Agency, Region V), Visual 3.8.

diffuse upward, where they are degraded by ultraviolet light, or diffuse downward, and are adsorbed onto the surfaces of suspended particulate matter. Other substances are dissolved in water droplets and returned to earth by rainfall. Weather conditions may set up cycles of elimination of toxic substances from air onto land and then into water, and eventually they are reintroduced into the air. Where sources of pollution are in close proximity to sources of water usage, an extensive and potentially hazardous series of chemicals may be transported directly into the water source and may cause significant health problems following human exposure (Figure 1.4).

Humans are exposed to chemicals or pollutants through inhaled air, ingested water and food, absorption through the skin, and at times through a combination of these modes. We may be continuously, intermittently but repeatedly, or sporadically exposed. We may be exposed to chemicals present in substantial or in minute quantities. In some instances, chemicals present in minute quantities are more hazardous than those in large quantities. There is also direct intentional exposure through chemicals added to foods and cosmetics, and those involved in the preparation of drugs. Good sense dictates limiting the level of exposure of any chemical that may cause adverse responses in an individual.

Human health is endangered not only at the point of use or discharge of a specific chemical, but also at varying points in the ecosystem. Unfortunately, many of the fundamental changes occurring in the environment are poorly understood. Reconcentration, specifically bioconcentration, is a process involving potentially harmful human exposure to chemicals. Plants and animals accumulate certain chemicals at levels higher than those of ambient environment. Chemicals also may be absorbed by air- or waterborne particles that are inhaled and concentrated in the lungs.

Dispersion of Contaminants

Contaminants tend to spread continuously within the environment. The characteristics of the medium affect dispersion of the contaminants. Contaminants are actively dispersed via convective or turbulent manner. Convection is the circulatory motion that occurs in a fluid at a nonuniform temperature owing to the variation of its density and the action of gravity. Turbulence in the atmosphere is a complicated phenomenon that depends upon the wind velocity, the direction of the wind, the altitude, and the friction of the air over the surface of the Earth, the temperature of the air, the pressure in the air and the presence or absence of bodies of water, mountains, and flat areas. Dispersion from ground-level sources are seen almost immediately downwind from the source. Dispersion from tall smoke stacks or any elevated discharge may be found for considerable distances from the source. Contaminants also are passively dispersed through the atmosphere. The passive process is known as diffusion and implied movement of contaminants based on a concentration gradient. The contaminants diffuse through the air from areas of high to low concentration. This is most applicable to gases and vapors.

Dispersion in the hydrosphere is much more varied and complex than dispersion in the atmosphere. The dilution volumes and mixing characteristics, as well as the rate of transport of the contaminants, vary with rivers, lakes, estuaries, coastal waters, and oceans. The contaminants in the water may further contaminate the land, the groundwater supply, and the air.

The environment of the soil is very complex with regard to the dispersion of contaminants. The dispersion depends upon the specific nature of the chemicals, the type, texture, and structure of soil, and other factors, such as moisture, pH, and temperature. Most chemical contaminants do not move readily through the soil once they enter it. Gases may easily diffuse through air pockets and pores in the soil. Soil water is the major means of moving chemicals downward through pores and channels to the lower horizons of the soil or upward to the point of evaporation. Water also may move the contaminants in a lateral manner or help in surface runoff depending upon the slope of the soil. The more rapidly water percolates into and leaches through the soil, the more readily the chemicals may diffuse into the pores between the soil particles.

Alteration of Contaminants

Contaminants are removed from the atmosphere by a variety of means. The contaminants may either be in the original form or may have altered during the dispersion process. Particles are physically removed from the atmosphere by gravity, impacting on objects, and through rain, snow, or sleet. The larger the particle, the more rapidly it will settle out. The particles may attract gases, vapors, and/or viruses and bacteria. Gases and vapors also may be dissolved in moisture, or through chemical reactions in the atmosphere they may be chemically converted into other compounds. Photochemical interaction is the most significant catalyst for altering physical and chemical air contaminants. Ultraviolet energy from the sun can interact with air contaminants resulting in oxidation and transformation of chemical structures.

Chemicals found in water may be bioconcentrated in fish or plant life, deposited on surfaces, or converted into other products by the self-purification process that is related to chemical and biochemical oxidation. Physical self-purification may occur when the chemicals bond to suspended particles in water.

The persistence of chemicals in the soil is a complex function, related to physical, chemical, and biological factors. The major means of removal of contaminants in soil are by degradation due to biotic microbial action, abiotic chemical degradation, evaporation, volatilization from the surface, and uptake by vegetation.

In order to get some better understanding of transport and alteration of chemicals in the environment, seven chemicals have been chosen for discussion. These seven all have been identified on the CERCLA (Superfund) National Priorities List. They are part of a large group that pose a significant potential threat to human health, as determined by the Agency for Toxic Substances and Disease Registry. These chemicals are chrysene, chloroform, cyanide, lead, 2,3,7,8-tetrachloro-dibenzo-*p*-dioxin (2,3,7,8-TCDD), tetrachloroethylene, and trichloroethylene.

Chrysene is found in the environment from natural sources, such as forest fires and volcanos, and from people-made sources. Chrysene, like all PAHs, is formed during high-temperature pyrolytic processes. Combustion is the major source of environmental chrysene. Virtually all direct releases into the environment are to the air, although small amounts may be released to the water and land. Chrysene is removed from the atmosphere by photochemical oxidation and dry or wet deposition onto land or water. It is very persistent in either soil or sediment. Biodegradation is a slow process. The half-life of chrysene in the soil is estimated to be 1000 days.

Incomplete combustion of carbonaceous materials is the major source of chrysene in the environment. Residential heating and open burning are the largest combustion sources. The inefficient combustion process creates uncontrolled emissions, with heating in the home as the greatest contributor to chrysene emissions. Hazardous waste sites can be concentrated sources of PAHs on a local scale and, therefore, may contain larger quantities of chrysene. Chrysene in the atmosphere is expected to be associated primarily with particulate matter, especially soot. Chrysene in aquatic systems is expected to be strongly bound to suspended particles or sediments. Chrysene is expected to be strongly sorbed to surface soils.

Chloroform is a colorless or water-white liquid used to make fluorocarbon-22. It is also used as a pesticide and as a solvent in the manufacture of pesticides and dyes. Chloroform is discharged into the environment from pulp and paper mills, pharmaceutical manufacturing plants, chemical manufacturing treatment plants, and from the chlorination of wastewater. Most of the chloroform released ends up in the air. It may be transported for long distances before being degraded by reacting with photochemically generated hydroxyl radicals. Significant amounts of chloroform from the air may be removed by precipitation. However, the chloroform may reenter the air by volatilization. When released to soil, chloroform will either volatilize rapidly from the surface or leach readily through the soil to the groundwater, where it may persist for relatively long periods of time. It is estimated that its half-life in the atmosphere is 70–79 days. However, when chloroform is found in photochemical smog, its half-life is 260 days.

Volatilization is the primary mechanism for removal of chloroform from water. Its half-life in water may vary from 36 hr to 10 days. Although chloroform has been found to adsorb strongly to peat moss and less strongly to clay, typically it will volatilize rapidly in either dry or wet soil.

Although cyanides are naturally occurring substances found in a number of foods and plants and produced by certain bacteria, fungi, and algae, the greatest amount found in the environment comes from industrial processing. Hydrogen cyanide is used primarily in the production of organic chemicals. Cyanide salts are used primarily in electroplating and metal treatment. The major sources of cyanide released to water are reported to be discharges from metal finishing industries. The major source of the cyanide released to air is vehicle exhausts, and the major sources of cyanide released to soil appear to be disposal of cyanide waste in landfills and the use of cyanide-containing road salt. Cyanide released to air appears to be transported over long distances before reacting with photochemically generated hydroxyl radicals. The half-life of cyanide is 334 days. In water cyanide may volatilize, or if in an alkali metal salt form, it may readily disassociate into anion and cations. The resulting cyanide ion may then form hydrogen cyanide or react with metals to form metal cyanides. Insoluble metal cyanides are expected to adsorb to sediment and possibly bioaccumulate in aquatic organisms.

In soil, cyanides may occur in the form of hydrogen cyanide, alkali metal salts, or immobile metallocyanide complexes. At soil surface with a pH less than 9.2, it is expected that hydrogen cyanide will volatilize. In subsurface soil, where the cyanide present is at a low concentration, it is probably biodegradable. In soil with a pH less than 9.2, hydrogen cyanide is expected to be highly mobile, and where cyanide levels are toxic to microorganisms such as in landfills or in spills, the cyanide may reach the groundwater.

Lead is a naturally occurring element that may be found in the Earth's crust and in all parts of the biosphere. It is released into the atmosphere by a variety of industrial processes and from leaded gasoline. The deposition of lead found in both soils and surface waters generally comes from the atmosphere. Lead is transferred continuously among air, water, and soil by natural chemical and physical processes, such as weathering, runoff, precipitation, dry deposition of dust, and stream/river flow. Long distance transport of up to about 600 mi may occur. Lead is extremely persistent in both water and soil. In the atmosphere, lead exists primarily in the particulate form. The chemistry of lead in the water is highly complex, because it may be found in a multitude of forms. The amount of lead that remains in solution depends upon the pH of the water and the dissolved salt content. In most soils, the accumulation of lead is primarily a function of the rate of deposition from the atmosphere. Most lead is maintained strongly in the soil and very little is transported into surface water or groundwater.

The chemical 2,3,7,8-tetrachlorodibenzodioxin (TCDD), or simply dioxin, which is an inaccurate colloquial name except when it is used as a reference standard, is not intentionally manufactured by any industry. It is inadvertently produced in very small amounts as an impurity during the manufacture of certain herbicides and germicides. It is also produced during the incineration of municipal and industrial wastes. The environmental fate of 2,3,7,8-TCDD is not understood with certainty for air, water, and soil, but nonetheless is considered a potent toxin.

Tetrachloroethylene is a nonflammable liquid solvent widely used for dry cleaning fabrics and textiles, and for metal-degreasing operations. It is used as a starter chemical for the production of other chemicals. It is also called perchloroethylene, perc, PCE, perclene, and perchlor. Most of the tetrachloroethylene used in the United States is released into the atmosphere by evaporative losses. The dry cleaning industry is the major source. When the chemical is released in water, it rapidly volatilizes and returns to the air. However, rainwater will dissolve it, and once again it falls to the land. The usual transformation process for tetrachloroethylene in the atmosphere is the result of the reaction with photochemically produced hydroxyl radicals. The degradation products of this reaction include phosgene and chloroacetyl chlorides (phosgene is a severe respiratory irritant).

Trichloroethylene is used as a solvent for removing grease from metal parts. It is released into the atmosphere by evaporative losses. It may also be released into the environment through gaseous emissions from waste-disposal landfills, and it may leach into groundwater from waste-disposal landfills. It is dissolved in rainwater. The dominant transformation process for trichloroethylene in the atmosphere is its reaction with sunlight-produced hydroxyl radicals. The degradation products of this reaction include phosgene, dichloracetyl chloride, and formyl chloride.

ENVIRONMENTAL PROBLEMS AND HUMAN HEALTH

This volume concentrates on problems related to human health, rather than to the various other systems coexisting on Earth. The decay of the environment and the resultant effect on human health grows daily. The sun is shrouded by the smoke of industry and dwellings. Our eyes and respiratory system are irritated daily by air pollutants. Emphysema and lung cancer are increasing sharply. Our noses are offended by acrid, noxious doses of chemicals, overflowing sewage, and other obnoxious odors. Our auditory system is damaged by the noise of traffic, industry, construction, and jet aircraft.

Pesticides essential to food production and protection against the severe epidemics of the past, such as malaria, yellow fever, and plague, are misused and commonly abused. Too many pesticides are being applied carelessly by uninformed or indifferent individuals. Research on the ultimate effects of pesticides on humans and their environment has not kept pace with the sharp increase in usage in the last 20 years. Over three billion pounds of pesticides are applied annually in the United States alone.

We continue to use dangerous, misbranded, or adulterated food produced under unsanitary conditions despite several governmental enforcement acts, including the Pure Food and Drug Act of 1906, the Food, Drug, and Cosmetic Act of 1938, and its numerous amendments in the 1950s, 1960s, 1970s, 1980s, and 1993. *S. aureus* is present in most supplies of pooled raw milk. *Salmonella* organisms are found in 15–30% of raw dressed poultry and in many commercial egg products. *Clostridia perfringens* is present on more than half of the red meat sold. *C. botulinum*, type E, is frequently found in raw fish. *S. aureus* is present on the mucous membranes or skin of 30–50% of the population. Chemical residues, pesticides, and additives are components of food that may cause subtle problems. Although the overall incidence of food-borne disease is unknown, it is estimated from statistics of actual outbreaks that 5–10 million cases of gastroenteritis occur each year and that 200,000–1,000,000 of these cases are salmonellosis. In smaller communities, food-service programs are either inadequate or totally lacking.

The residential environment for 6 million poor American families is the hazardous blighted slums of 19th century England. The healthful, pleasant, attractive, comfortable housing of the make-believe world of television is replaced in reality by tin and tarpaulin shacks; crowded, ramshackle houses; and tenements overflowing with garbage, flies, roaches, rodents, and sewage. Walls and floors deteriorate, lights fail, windows disintegrate, heating is poor, and ventilation is almost nonexistent. Lead poisoning, tuberculosis, and infectious diseases are endemic in this environment.

In the suburban areas, inadequate planning and poor land utilization have led to the subdivision nightmare, where children play in overflowing sewage and houses crack, settle, and are subject to flooding. Home accidents related to housing have sharply increased during the last 30 years.

Insects and rodents spread disease, cause annoyance, destroy crops, and ruin property. During the outbreaks of bubonic plague in Europe in the 14th and 15th centuries, an estimated 25 million people died of this rat-borne disease. In the 1990s, bubonic plague was still present in southwestern United States. Yellow fever affected 23,000 out of a population of 37,000 people in Philadelphia in 1793. Four thousand people died. In addition, malaria, encephalitis, spotted fever, tick fever, tularemia, and rickettsial pox have occurred regularly during the 19th and 20th centuries. In 1959, an epidemic of eastern encephalitis occurred in New Jersey, and in 1962 an epidemic of St. Louis encephalitis occurred in the St. Petersburg-Tampa Bay area of Florida. In 1975 and 1976, St. Louis encephalitis spread to many states. In 1986, Houston, Texas experienced its largest outbreak since 1980. It has also reoccurred in Long Beach, California. Further, from 1984 to present, there has been a substantial amount of Lyme's disease from ticks. In 1993, hantavirus carried by deer mice caused a deadly disease called hantavirus pulmonary syndrome. Tens of thousands of rat bites occur each year, notably to very young children.

Many individuals spend part of each day or most of their lives in a variety of institutions. Each institution, particularly if it provides sleeping accommodations, is in effect a small community, with all of the environmental health problems discussed in this book, and the added problem of a rapidly shifting mobile population that is highly susceptible to disease and accidents.

Noise, or unwanted sound, causes temporary or permanent hearing loss, physical and mental disturbances, breakdowns in the reception of oral communications, reduced efficiency in performing work-related tasks, irritability, disruption of sleep and rest, and increased potential for accidents. Steady exposure to 90 dB of sound can cause eventual hearing loss. Heavy traffic can reach 90 dB, and the noise of jet planes, rock-and-roll bands, motorcycles, power mowers, auto horns, heavy construction, and farm equipment exceed this level.

Occupational hazards occur in all industries. In 1992, 8500 workers died from occupational hazards and 3,300,000 were the victims of disabling injuries. Billions of dollars are spent each year on worker's compensation cases for injuries and work-related illnesses. These problems are caused by accidents and various health hazards, such as toxic chemicals, including dusts, gases, fumes, mists, vapors; physical agents, including noise, pressure, ionizing radiation, and severe temperature variations; biological hazards, including insects, bacteria, fungi, and viruses; and other hazards, including unusual work-related posture, boredom, fatigue, repetitive motion, and monotony.

Low and high levels of radiation may adversely affect human health by causing genetic damage, burns, destruction of tissue, reduced life span, and cancer. Old X-ray and medical fluoroscopic machines may contain inadequate devices to attenuate unneeded or stray radiation. Improper use, storage, and/or disposal of radioactive material and wastes could lead to serious potential hazards.

Outdoor recreation is a complex and rapidly expanding area of the economy. Americans spend over 30 billion dollars a year on recreation and over 100 million Americans vacation each year. The environmental health problems resulting from these activities are enormous. They include all areas discussed in these books, plus the problems of mass migrations; inadequate housing, water, and sewage facilities; and a high potential for accidents.

Soil is the single most important natural factor in choosing a site for construction of dwellings and for developing an effective, operating, on-site sewage disposal system. Unfortunately, lack of knowledge in this area, poorly designed tests, and inadequate supervision have led to severe settlement cracks in houses, flooded basements, flooded homes, slippage of hillsides and houses, soil erosion, contaminated or inadequate water supplies, and overflowing sewage. The disease and accident potential for humans is substantial.

Solid wastes generate enormous economic, aesthetic, social, and health problems. Our affluent society, improved technology, new packaging methods, and disposable items have increased the amount of solid waste per person per day from 2.65 lbs in 1960 to 3.60 lbs in 1994 and growing. At this rate, our society will become inundated by waste unless changes in waste production and disposal are made immediately. Unsatisfactory storage, collection, and disposal of solid waste lead to insect and rodent problems, air pollution, offensive odors, accidents, fires, explosions, contamination of the water supply, degradation of the landscape, destruction of fish, and conversion of bodies of water into open sewers. Hazardous waste has complicated the problem enormously.

Water, our most precious resource, is used to sustain life, support the growth of food, develop business and industry, and provide recreation. Despite these essential human needs, individuals, municipalities, industries, commercial establishments, and agriculture continue to pollute our water supply. Historically many of our worst epidemics have been caused by contaminated water. Although in 1994 the epidemics are gone, water-borne outbreaks of disease still occur and the potential for disease outbreaks continues as occured in Milwaulkee, Wisconsin where there was a very large outbreak of cryptosporidosis. Approximately 43,000 people became ill and 4400 were hospitalized. In addition, a new hazard has been added: chemical pollution. Mercury has caused food poisoning in Japan and Sweden. Oil spills have caused the deaths of birds and fish, and have befouled once lovely beach areas. Insecticides have caused millions of fish to die and could well cause serious harm to people. Fertilizers are helping to destroy our streams and lakes by transforming them into open sewers.

Potable water used for drinking and swimming may be unsafe because of physical, chemical, and bacteriological hazards. Physical deficiencies exist because of inadequate groundwater sources, inadequate design of treatment plants, inadequate disinfection capacity, inadequate system capacity, improper design of swimming pools, and inadequate training of water plant and swimming pool operators. Poor, outmoded, unsafe plumbing can easily lead to contamination of potable water by nonpotable water through cross-connections and submerged inlets.

HUMAN SYSTEMS AND TOXICOLOGICAL PRINCIPLES

Human Systems

Physiology is the basic biomedical science dealing with the function of living organisms. To understand function it also will be necessary to understand something about structure or anatomy. A major concern of physiology is how environmental factors influence the function of individuals. The intracellular processes in humans proceed properly if the fluid environment surrounding each cell is maintained in a nearly constant state. Temperature, oxygen supply, acidity, and nutrients must be at a nearly constant level. The external environment that the individual is exposed to, however, varies considerably. There are wide ranges of temperature, humidity, ionizing and nonionizing radiation, pressure, and a variety of chemicals, as well as microorganisms, in this environment that can affect people. All of these factors alone or in combination can elicit a biological response that can be harmful to the individual. In order to understand the adverse physiological reactions caused by chemical, physical, and biological agents, it is necessary to have some knowledge of the cell and a select group of organ systems.

The Cell

The cell is the basic unit of all living organisms. The cell consists of the following components: lipid bilayer, endoplasmic reticulum (ER), Golgi apparatus, mitochondria, lysosomes, ribosomes, nucleus, and nucleoli. (Figure1.5) The cell membrane is the boundary of the cell between intracellular and extracellular water and maintains cellular integrity by regulating influx and efflux of various substances. The endoplasmic reticulum serves as the cell's circulatory system and is a site of protein and lipid synthesis. The Golgi apparatus synthesizes carbohydrates and then combines with protein to produce glycoprotein. The mitochondria function in cellular respiration and adenosine triphospate molecules (ATP) synthesis. The lysosomes are the cell's digestive system. The ribosomes synthesize proteins. The nucleus dictates how protein synthesis occurs. Therefore, it directs the other cell activities, such as active transport, metabolism, growth, and heredity. The nucleoli are essential in the formation of ribosomes. The nucleus stores, transcribes, and transmits genetic information. This information is stored in the DNA molecules present in the nucleus. The sequence of the DNA base pairs is transcribed into the sequence of base pairs in mRNA molecules. This mRNA is known as the messenger RNA. The mRNA then acts as a carrier of a coded message that is translated at the ribosome into the specific proteins they synthesize.

Figure 1.5 The Human Cell

Metabolism consists of catabolism, which is a decomposition process, in which relatively large food molecules are broken down to yield smaller molecules and energy, and anabolism, in which energy is used to synthesize relatively small molecules into larger molecules. Enzymes, hormones, and antibodies are all produced during the process of anabolism. Enzymes catalyze both catabolic and anabolic chemical reactions. There are intracellular enzymes that are synthesized within the cell and also function within the cell. There are extracellular enzymes that are synthesized in cells and function outside of the cells. Energy changes accompany metabolic reactions. Catabolism releases energy as heat or chemical energy. Chemical energy released by catabolism is first converted into, and stored as, adenosine triphosphate (ATP) molecules at the cellular level. The ATP molecules, which are easily broken or hydrolyzed, then release the energy, which can be utilized in anabolism.

The Blood

The primary function of blood is the transport of various substances to and from the body cells and the exchange of materials, such as oxygen, nutrients, and waste materials, between the respiratory, digestive, and excretory organs. Its secondary function contributes to the homeostasis of fluid volume, pH, and temperature. It is also necessary for cellular metabolism and defense against microorganisms. (Homeostasis is the maintaining of the internal environment of the body in a relatively uniform manner.) The blood consists of the erythrocytes or red blood cells, leukocytes or white blood cells, and thrombocytes or platelets. The red blood cells transport oxygen and carbon dioxide. The white blood cells are important in the immune response. The platelets initiate blood clotting and homeostasis, which is the arrest of bleeding. The plasma is the liquid part of the blood minus its cells.

The immunological response starts when bacteria or viruses invade the body. Scavengers, including blood cells called neutrophils, are among the first line of defense at the site of the infection. Neutrophils are produced in the bone marrow. They survive for a few days. Next, the complement system, which is a group of at least 20 proteins circulating in the blood, stick to the microorganism, setting off a chain reaction that eventually will destroy it. Macrophages, which are long-lived scavengers, migrate through the body and engulf foreign matter, such as microorganisms, as well as cellular debris. They signal other cells in the immune system to confront and destroy the microorganisms. A macrophage, after ingesting the microbe, shows specific markers on its surface. These markers or antigens signal other immune system cells, called helper T-cells, to come forth. T-cells have their own set of receptors that can recognize specific antigens of a given microbe. The T-cell grows and divides. The helper T-cells start to reproduce when a protein, which is called interleukin-1, is released by a macrophage. The helper T-cells then produce a variety of interleukins that activate other T-cells and B-cells. They also produce gamma interferon, which activates the macrophages. The B-cells that are stimulated by the helper T-cells divide and mature into plasma cells. The plasma cells produce antibodies that are directed against the specific antigens. (Antibodies are proteins that recognize and bind to a specific microbe.) This either stops the microbe from going further or it makes the microbe more vulnerable to macrophages and neutrophils. Antibodies also activate the complement system. It takes about 1 week for the immune system to be at its highest level of functioning. Killer T-cells recognize and destroy virus-infected cells or cancer cells. They use lethal proteins that punch holes in the cell's membrane and cause it to rupture. The suppressor T-cells probably work by sending out chemical signals that either slow or stop the immune reaction after the organisms have been destroyed. This keeps the body from attacking itself. Memory cells are types of B- and T-cells that circulate through the body after an infection has occurred. The next time the antigens are released, the memory cells start the process of destroying the antigens. Cancer cells, whether they are created by chemicals, radiological agents, or possibly viruses, are surrounded by killer T-cells because they are attracted by the surface antigens. These cells can release chemicals that break down the membrane of the cancer cells and cause the cells to die.

The Lymphatic System

The lymphatic system is a specialized part of the circulatory system. It consists of a moving fluid, lymph. The lymph comes from the blood and tissue fluid, and returns to the blood. Lymph is a clear, watery fluid that is found in the lymphatic vessels. The interstitial (intercellular) fluid fills the spaces between the cells. The clear watery fluid, in fact, is complex and organized material. It makes up the internal environment of the body. Both lymph and interstitial fluids closely resemble blood plasma in composition. The main difference is that they contain a lower percentage of proteins than does the plasma. The lymphatic system, like the blood system, can carry contaminants throughout the body. The lymphatics return water and proteins from interstitial fluid to the blood from where they came. Lymph nodes have two major functions. They filter out injurious substances and phagocytose them. Unfortunately, sometimes the number of microorganisms entering the lymph nodes are greater than the phagocytes can destroy, and therefore the node becomes infected. Also, cancer cells that break away from a malignant tumor may enter the lymphatics, travel to the lymph nodes, and set up new sites of cancer growth. The lymphatic tissue of the lymph nodes forms lymphocytes and monocytes, which are the nongranular white blood cells and plasma cells. This is called hemopoiesis.

Tissue Membranes

Tissue membranes are thin sheets of tissues covering or lining the various parts of the body. Four important membranes are mucous, serous, synoval, and cutaneous (skin). The mucous membranes line the cavities of passageways of the body that open to the exterior. This includes the lining of the gastrointestinal tract, respiratory tract, and genitourinary tract. Mucous membranes protect the underlying tissue, secrete mucus, and absorb water, salts, and other solutes. The serous and synoval membranes line the closed cavities of the body. The pleura is the serous membrane that lines the thoracic cavity. The mucous film of this lining consists of a superficial gel layer that traps inhaled particles in the respiratory system. The ciliated mucous membrane contains goblet cells. The cilia propel the mucus upward in the respiratory tract to the pharynx via mucociliary escalation, which causes a coughing reflex. Another serous membrane is the pericardium, the sac in which the heart lies. The synovial membrane has smooth, moist surfaces that protect against friction. It lines the joints, tendon sheaths, and bursae.

The cutaneous, or dermal, membrane (skin) protects the body against various microorganisms, sunlight, and chemicals. The skin helps to maintain the normal body temperature by regulating the amount of blood flowing through it and by sweat secretion. Skin consists of an epidermis, dermis, and accessory organs of the skin, such as the hair, nails, and skin glands. The very top layer of the epidermis is composed of dead cells and is virtually waterproof. Because of the innumerable microscopic nerve endings throughout the skin, the body is kept informed of changes in the environment. Finally, beneath the dermis is the subcutaneous fatty tissue, which provides cushioning.

Nervous System

The nervous system is a major interface between humans and the environment. It is involved in maintaining homeostasis. It controls posture and body movements. It is the center of subjective experience, memory, language, and the thought process that is peculiar to human activity. The fundamental unit of the nervous system is the neuron, which consists of a cell body, dendrites, and the axon, which may be several feet in length. The dendrites conduct impulses to the cell body of the neuron. Receptors that receive the stimuli that initiate conduction are the distal ends of dendrites of sensory neurons. (Distal is the furthest point away from the point of origin.) The neuron axon conducts impulses away from the cell body.

The connections between the nerve cells are called synapses. They play a key role in the transmission of impulses in the nervous system. A single cell may be connected to as many as 15,000 other cells by means of synapses. The transmission of impulses occur from the depolarization of the cell membrane brought about by chemical or mechanical events. The nerve impulse or action potential is a self-propagating impulse of electrical negativity that travels along the surface of the neuron's cytoplasmic membrane. A stimulus to the neuron greatly increases the membrane's permeability to sodium ions. Sodium ions rush into the cell at the point of stimulation. The membrane becomes depolarized at the point of stimulus. This occurs for an instant. As more sodium ions come into the cell, they produce an excess of positive ions inside the cell and leave an excess of negative ions outside the cell. The negatively charged point of the membrane sets up a local current, with the positive point adjacent to it. The local current acts as the stimulus. Within a fraction of a second, the adjacent point on the membrane becomes depolarized and its potential reverses from positive to negative. The cycle keeps repeating itself (polarization then depolarization) until the electrical current, in a wave motion, moves the full length of the neuron.

Many nerve impulses travel a route called the reflex arc consisting of two or more neurons arranged in a series that conduct impulses from the periphery or from the outside portion to the central nervous system. (Figure 1.6) The receptors are stimulated, and the nerve impulse flows into the spinal cord and brain and back out to the effectors, which aid in the carrying out of the intended action. The synapse is the place where the nerve impulses are transmitted from one neuron to another neuron. For instance, if there is an environmental stimulus to the receptor neuron, it then goes along the afferent nervous pathway to carry the action potentials to the central nervous system. Within the spinal cord or brain, there is an integrating center where the action potentials are now sent out along different pathways from the central nervous system, the effector is activated, and the body will now respond.

The synapse is the place where the nerve impulses are transmitted from one neuron to another neuron. A synapse consists of a synaptic knob, a synaptic cleft, and the cytoplasmic membrane of the neuron dendrite or cell body. Since an action potential cannot cross the synaptic clefts, a chemical mechanism operates, where the chemical, which is called a neurotransmitter, is released from the synaptic knobs into the synaptic cleft. The neurotransmitter molecules diffuse rapidly across the microscopic width of the synaptic cleft and bind to specific protein

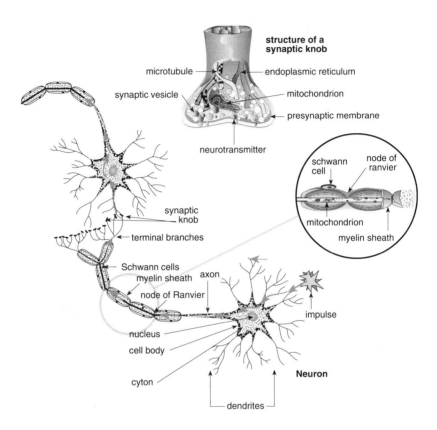

Figure 1.6.   Neuron.

molecules called neurotransmitter receptors. This leads to an opening of channels in the membrane through which sodium ions diffuse into and potassium ions diffuse out of the interior of the postsynaptic neuron. The message now moves forward.

Several different compounds serve as neurotransmitters. Acetylcholine is released at neuromuscular junctions with skeletal muscle cells. Acetylcholine is also released at neuromuscular junctions with smooth muscle cells, cardiac muscle cells, and in neuroglandular junctions. Norepinephrine is released at other neuroglandular junctions and at some neuromuscular junctions with smooth muscle and cardiac muscle cells. The enzyme acetylcholinesterase rapidly halts the stimulation of muscle cells by acetylcholine. The enzyme hydrolyzes acetylcholine to acetate and choline.

The nervous system is composed of the central nervous system and the peripheral nervous system. (see Figure 1.7) The CNS consists of the brain and spinal cord. The cranial nerves, spinal nerves, and ganglia make up the peripheral nervous system. The nervous system is also classified by the effectors innervated,

Figure 1.7 The Nervous System: Comparative Chart

that is, the somatic nervous system and the autonomic nervous system. The somatic nervous system consists of the brain, spinal cord, cranial nerves, and spinal nerves. It innervates the skeletal muscles. The autonomic nervous system is composed of the autonomic or visceral motorneurons, which innervate the cardiac muscle, smooth muscle, and glandular epithelial tissues.

The autonomic system is divided into two parts by function. The sympathetic part is an emergency system that greatly increases sympathetic impulses to most visceral effectors during stress and prepares the body for the use of maximum energy. It increases the secretion of epinephrine by the adrenal medulla, which in turn increases and prolongs the effects of norepinephrine. The parasympathetic nervous system regulates many visceral effects under normal conditions. Its neurotransmitter, acetylcholine, stimulates the digestive juices, insulin secretion, and contraction of the smooth muscle of the digestive tract. The autonomic system does not function autonomously or independently of the CNS.

The CNS is intimately related to the endocrine system. The two systems perform the same general function for the body of communication, integration, and control. Whereas the nervous system sends nerve impulses conducted by neurons from one specific structure to another, the endocrine system sends tiny quantities of chemical messengers, known as hormones. Nerve impulses produce rapid, short-lasting responses, whereas hormones produce slower and longer lasting responses. The cerebral cortex sends impulses into the hypothalamus, which controls the production and secretion of six separate hormones of the interior pituitary. The hypothalamus secretes the antidiurectic hormone (ADH), which regulates the reabsorption of water into the kidney. The hypothalamus also controls the autonomic nervous system, which in turn controls the secretion of epinephrine.

The Respiratory System

The function of the respiratory system is to distribute air and to act as a means for exchange of gases, so that oxygen may be supplied to body cells and carbon dioxide may be removed from them. The circulatory system is jointly responsible for meeting the respiratory needs of the body, since it provides a mechanism for oxygen transport to all of the body cells. The nose, pharynx, larynx, trachea, bronchi, and bronchioles act as a conduit for air into the alveoli, where the actual gas exchange takes place with the blood capillaries. The nose also filters out impurities and warms, moistens, and chemically examines the air for substances that might become irritating to the mucous lining of the respiratory tract. It is also an organ of smell, and acts as an aid in phonation (the production of vocal sounds, especially speech). The pharynx is a corridor for the respiratory and digestive tracts. Both air and food pass through the structure before reaching the appropriate tubes. It is also important in phonation. The larynx or voice box is another part or the corridor to the lungs. It protects the airway from solids or liquids during swallowing. It is also the organ of voice production. The trachea is a continuation

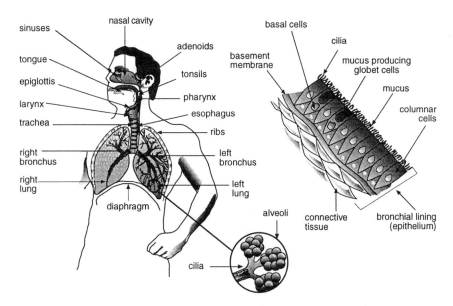

Figure 1.8 Conductive airways of the respiratory systems, with details of the ciliated epithelial lining. Adapted from: *Environmental Toxicology and Risk Assessment: An Introduction*, Student Manual (Chicago, IL: U.S. Environmental Protection Agency, Region V), Visual 2.6.

of the corridor through which air passes to the lungs. The bronchi is a continuation of the trachea and aids in the distribution of air to the interior of the lungs. The alveoli are surrounded by networks of capillaries. This is where the main function of the lungs occurs, that is, gas exchange between the air and the blood. The lungs contain the bronchus, bronchioles, alveoli, pleura (sac around the lungs), and alveolar sacs (Figure 1.8).

The conducting portion of the respiratory system provides little resistance to the movement of gases to the alveolar surface. Further, this area conditions the air and protects the lungs from the largest of infectious or potentially toxic particles in the air. The convoluted, moist, and richly vascular mucosa of the nose protects nose-breathers from inhaling particles larger than 5–10 μ in diameter. Soluble gases may be removed by absorption. The air is warmed and moistened, or cooled when the conditions are hot and dry. The trachae-bronchial system is lined with cilia that constantly force mucus toward the larynx. The mucus carries with it the microorganisms and particles that have been trapped in it, as well as the macrophages that move out of the alveoli with material that have been scavenged from the alveolar surface. When the mucus reaches the pharynx it is usually swallowed or may be spit out. These pulmonary mechanisms help prevent problems in the respiratory system due to the inhalation of dusts, fumes, and other materials that can cause disease or injury. The coughing mechanism helps in the removal of these materials.

Pulmonary ventilation is brought about by changes in the size of the thorax (chest). The contraction of the diaphragm and chest-elevating muscles leads to an expansion of lungs that decreases the alveolar pressure and creates an inspiration. Expiration is a passive process that begins when pressures change. This occurs when the muscles are relaxed, which leads to a decrease in size of the thorax, a decrease in the size of the lungs, and a pressure gradient in the alveoli that is greater than the atmosphere.

The exchange of gases in the lungs take place between the alveolar air and venous blood that flows to the lung's capillaries. The alveolar-capillary membranes allow gases to move back and forth. Oxygen enters the blood because the partial pressure of oxygen in the alveolar air is greater than in the venous blood. At the same time, the partial pressure of carbon dioxide in the venous blood is much higher than in the alveolar blood. This causes carbon dioxide to flow from the venous blood to the alveolar blood. Once the oxygen enters the blood, a small percentage is transported as a solute in the blood, whereas much more is transported as oxyhemoglobin in the red blood cells. The carbon dioxide is transported in the blood as a true solute in small quantities. A large amount of the carbon dioxide is transported as bicarbonate ions in the plasma and a moderate amount of the carbon dioxide is transported in the red blood cells as carbaminohemoglobin.

Carbon dioxide is the major regulator of respiration. An increase in blood carbon dioxide to a certain level may stimulate respiration, and a decrease may cause a decreased level of respiration. If the oxygen level of blood is low, this also may stimulate respiration to a certain point.

The Gastrointestinal System

The digestive tract is largely a tube that is open at both ends. In a real sense, the contents of the gastrointestinal tract is exterior to the body (Figure 1.9). In order to get into the tissues or cells, the ingested materials have to go through an extremely acidic environment in the stomach, as well as being broken apart by enzymes. Some of the large molecules, such as cellulose, are unaltered and therefore are excreted in the feces. Almost all of the digestion and absorption of food and water take place in the small intestine. The three basic food groups—the carbohydrates, proteins, and fats—are broken down in different ways. The carbohydrates, most of which are ingested in the form of starch, are split into disaccharides by the enzyme amylase, found in the saliva and in the pancreas. They are then split into monosaccharides by enzymes in the small intestinal mucosa. The sugar molecules are then transported into the blood. The proteins are broken down further into peptides and into free amino acids, which are transported across the intestinal cells. Most of the fat digestion occurs in the small intestine from the combined actions of pancreatic lipase and bile salts secreted by the liver. The bile salts act chiefly as emulsifying agents. Fatty acids are resynthesized to triglycerides in the intestinal cells. They are then secreted into the lymphatics as small lipid droplets.

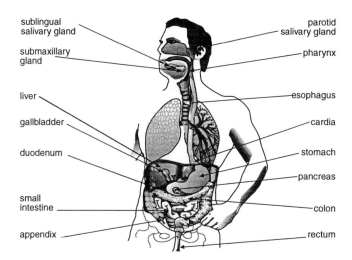

Figure 1.9   The gastrointestinal tract. *Source: Environmental Toxicology and Risk Assessment: An Introduction, Student Manual,* Chicago, IL: U.S. Environmental Protection Agency

The function of the digestive system is to accept raw materials in the form of different foods, minerals, vitamins, and liquids, and to prepare them for absorption into the capillaries or lymphatics for distribution through the body.

The main organs of the digestive system are the mouth, pharynx, esophagus, stomach, and intestine. The accessory organs that are open to the main organs are the salivary glands, liver, gallbladder, and pancreas.

The stomach serves as a food reservoir that expands and contracts depending on the amount of food present. The glands secrete most of the gastric juice, which is fluid make up of mucus, enzymes, and hydrochloric acid. The epithelial cells that form the surface of the gastric mucosa secrete mucus. The chief cells (zymogenic cells) secrete the enzymes of gastric juice. The parietal cells secrete hydrochloric acid and are also thought to produce a protein known as intrinsic factor. When the muscular coat of the stomach contracts, it churns up food and breaks it into small particles, which allows for mixing with the gastric juice. In time it moves the gastric content into the duodenum. The stomach, in a limited quantity, absorbs water, alcohol, and certain drugs. It also produces the hormone gastrin in cells that are in the pyloric region.

The small intestine, which is about 20 ft in length, has three major divisions: the duodenum, the jejunum, and the ileum. The function of the small intestine is to complete the digestion of food, to absorb the end products of digestion into the blood and lymph, and to secrete hormones that help control the secretion of pancreatic juice, bile, and intestinal juice. The intestinal juice contains mucus from the intestine, digestive enzymes from the pancreas, and bile from the liver.

The lower part of the alimentary canal is called the large intestine. It is divided into the cecum, colon, and rectum. The main functions of the large intestine are the absorption of water, secretion of mucus, and elimination of the wastes of digestion.

The peritoneum is a large, continuous sheet of serous membrane that lines the walls of the entire abdominal cavity and forms the serous outer coat of the organs.

The liver, which is the largest gland in the body, is one of the most vital organs. Liver cells detoxify a variety of substances. They secrete about a pint of bile a day. The liver carries on a number of important steps in the metabolism of proteins, fats, and carbohydrates. Liver cells store iron, vitamins A, B_{12}, and D. Poisonous substances that enter the blood from the intestine are circulated to the liver, where a series of chemical reactions occur that turn the substances into nontoxic compounds. The bile secreted by the liver is made up of bile salts, bile pigments, and cholesterol. The bile salts that are formed in the liver from the cholesterol are the most essential part of the bile that aids in the absorption of fats.

The gallbladder is a storage organ for the bile, which is produced by the liver. When digestion is going on in the stomach and intestines, the gallbladder contracts, ejecting the concentrated bile into the duodenum.

The pancreas is composed of two different types of glandular tissue, one exocrine and one endocrine. Embedded between the exocrine units of the pancreas are clusters of exocrine cells called islands of Langerhans or pancreatic islets. The pancreas secretes the digestive enzymes found in pancreatic juice. Its beta cells secrete insulin, and its alpha cells secrete glucagon.

The last step in the digestive process is defecation. The stomach is emptied in about 2–6 hr after a meal, depending upon what is eaten. Chyme, which is gastric juices mixed with food to form a milky white material, is ejected about every 20 sec into the duodenum. Defecation is brought about by a reflex that is caused by stimulation of the receptors in the rectal mucosa. Constipation occurs when the material in the lower colon in the rectum moves at a rate that is slower than normal. This causes extra water to be absorbed from the fecal mass and thereby hardens the stool. Diarrhea occurs when chyme moves too quickly through the small intestine, and therefore, not enough water and electrolytes are absorbed. It is the loss of water, leading to dehydration and loss of electrolytes, that makes diarrhea in infants such a serious problem.

The Urinary System

The urinary system consists of the kidneys, ureters, bladder, and urethra (Figure 1.10). The functions of the kidneys are to excrete urine, which eliminates various toxins and metabolic wastes. They also help maintain fluid, electrolytes, acid-base balance, and the proper level of blood pressure in the body. The ureters

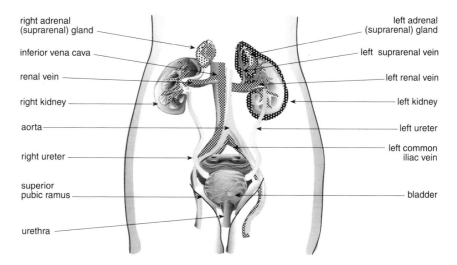

right adrenal (suprarenal) gland

inferior vena cava

renal vein

right kidney

aorta

right ureter

superior pubic ramus

urethra

left adrenal (suprarenal) gland

left suprarenal vein

left renal vein

left kidney

left ureter

left common iliac vein

bladder

Figure 1.10   The urinary system

collect urine and drain it into the bladder. The bladder is a reservoir for storing urine and also acts as the organ that expels the urine through the urethra which is the passageway for expulsion from the body.

TOXICOLOGICAL PRINCIPLES

Toxicology is the study of adverse human biochemical, morphological, and physiological effects resulting from exposure to harmful agents, such as chemicals and types of radiation, that are present in many environmental settings. The fate of these agents is dependent upon the toxicokinetics or movement of toxic elements and molecules into, through, and out of the human body. The impact or toxicodynamics is dependent on factors related to the toxic agent, the exposure, and the individual exposed.

Toxicokinetics

Toxicokinetics refers to the fate or, more specifically, the *absorption, distribution, biotransformation, and elimination* of toxic agents and/or metabolites (Figure 1.11). Toxic agents enter and absorb into the body via three major modes and routes of entry: (1) inhalation into the respiratory system; (2) ingestion into the gastrointestinal system; and, (3) contact with the dermal (skin) system. The toxic agent may deposit and interact at the site of initial contact causing local effects and/or absorb across respiratory, gastrointestinal, and dermal cell membranes and enter the circulatory system eventually causing systemic effects.

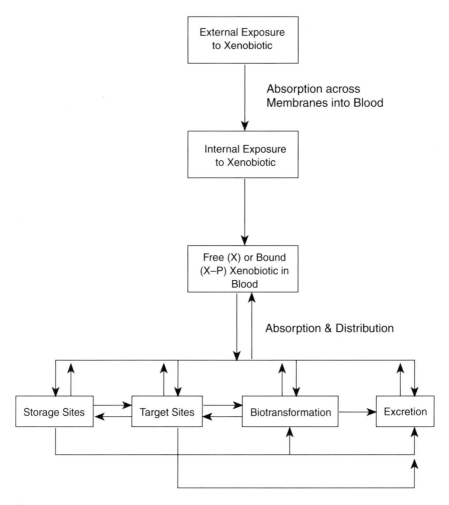

Figure 1.11 Possible fate of a chemical agent following external exposure.

The lipid bilayer of cells that comprise the tissues of organ systems regulates the absorption of toxic agents into the body. Small water soluble or polar molecules, and both small and larger lipid soluble or nonpolar molecules, most readily cross cell membranes. Indeed, based on the premise of solvent chemistry that "like dissolves like," nonpolar "lipophillic" agents can most readily cross the lipid bilayer that constitutes membranes. As a result, some of the most hazardous toxic chemicals are those considered fat soluble or lipophillic.

Inhalation brings chemicals into contact with the lungs. Most of the inhaled chemicals are gases or vapors generated from volatile liquids. Absorption in the lungs can be high, both because of the surface area and because the blood vessels are in close proximity to the exposed surface of the respiratory system. Gases and vapors cross the lung by means of simple diffusion. The rate of absorption depends on the solubility of the toxic agent in the blood. Chemicals may also be inhaled in solid or liquid form as dusts or mists. Mists, if lipid soluble, may cross the cell membranes by passive diffusion. The absorption of solid particulates depends on the size and chemical nature of the particles. The rate of absorption of the particles from the alveoli is determined by the solubility of the chemical in lung fluids. Certain small insoluble particles may remain in the alveoli indefinitely. Particles of 2 to 5 microns are deposited into the tracheobronchiolar regions of the lungs. They are typically cleared by coughing and sneezing, or are swallowed and deposited in the gastrointestinal tract. Particles of 5 microns or larger are usually deposited in the nasopharyngeal region, where they are either expectorated or swallowed.

Ingestion brings the chemicals into contact with the tissues of the gastrointestinal system. The normal function is absorption of foods and fluids that are ingested. The gastrointestinal system is affected by absorbing toxic chemicals that are contained in food or water. The degree of absorption usually depends on whether the chemical is easily soluble in water or easily soluble in organic solvents or fats. The lipophilic compounds, such as organic solvents, are usually well absorbed, since the chemicals can easily diffuse across the membranes of the cells that line the gastrointestinal track. The hydrophilic compounds, such as metal ions, cannot cross the cell lining as easily as the lipophilic agents and have to be transported by systems in the cells. Many chemicals may bond to ingested food and, therefore, are not absorbed as efficiently as when ingested in water. Further, some chemicals may be altered by digestive enzymes or intestinal bacteria, and yield different chemicals with altered toxicological properties.

Absorption of toxic chemicals through the epidermal layer of the skin is hindered by the epidermal cells. These cells are densely packed layers of horny keratinized material. Absorption of the chemicals through the skin will occur more readily if the skin is scratched or broken. In addition, the absorption of the chemicals by the skin is increased if the agent is lipid soluble. A more rapid exposure to the body occurs when the chemicals are accidentally or purposefully injected through the skin into the body.

Once an agent absorbs into the blood, it may interact with blood cells and/or plasma proteins. In addition or alternatively, the agent can be distributed via the circulatory system and absorbed into other organ systems such as the liver, kidneys, and brain. Some agents, such as lead and fluorides, may bond to and be stored temporarily in bone tissue. Others, such as fat-soluble organic solvents, may distribute to and be stored in adipose or fat tissue.

The originally absorbed agent or "parent" compound can be biotransformed into a more toxic or less toxic metabolite prior to, during, or following distribution. A parent compound that must be biotransformed or bioactivated into a toxic or active metabolite can be called a protoxicant (Figure 1.12). In addition, nonpolar molecules that absorb into the body most readily can be biotransformed into more polar molecules that are excreted from the body more readily. The liver is the major site

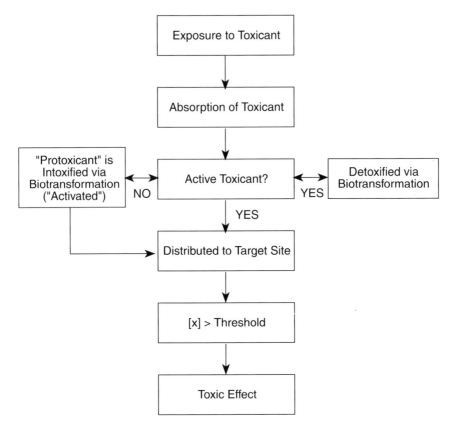

Figure 1.12   Protoxicant vs. Toxicant

of enzymes responsible for catalyzing biotransformation reactions, but there are numerous other organ systems that are important sites for biotransformation reactions as well.

Toxic parent compounds and/or toxic metabolites are eliminated via biotransformation reactions that result in detoxification. In addition, toxic substances are eliminated via excretion, mainly in urine and feces. Toxic agents and metabolites also can be eliminated from the human body via mammary milk, perspiration, exhaled breath, hair, and nails.

Toxicodynamics

Toxicodynamics refers to the impact that a toxic agent may have on the human body. Several factors influence the potential interaction and related adverse effects that may be manifested following exposure. The dose–response relationship of a chemical and effect is the most fundamental concept in toxicology. The dose–response curve describes the relationship that exists between the degree of exposure to a certain dose and the magnitude of the effect, which is the response, by an exposed organism. No response is seen if the chemical is not present. As the amount of the chemical exposure increases, the response becomes more apparent and therefore increases. Many chemicals produce responses that show a threshold value. That is an exposure below which no response can be detected. The No Observed Adverse Effect Level (NOAEL) is the dose at which no adverse effects are observed. The lowest value where a significant adverse effect is first seen is the Lowest Observed Adverse Effect Level (LOAEL). NOAEL and LOAEL values depend upon the effect (endpoint) being measured.

Some chemicals produce adverse effects that show a dose–response curve with no threshold. The reason for this is that the cells that are affected have little or no defense against the chemical and have little or no ability to repair or compensate for the damage that is done. It appears that there is no threshold for the effects of lead on the nervous system in infants and children. Chemicals that are carcinogenic are also considered to be part of the group that do not have thresholds. Therefore, any chemicals in these groups are considered to create a degree of risk at any exposure.

Dose–response curves also help show the toxic properties of the chemical and may be useful in comparing the toxicity of several chemicals. The midpoint of the dose–response curve is called the ED_{50}. This is the dose that produces an effect in 50% of the test population. When a toxic effect is being measured, the term used is TD_{50}. When a lethal effect is being measured the term used is LD_{50}.

The slope of a dose–response curve is an important variable in assessing the toxicity of the chemical. The steeper the dose–response curve, the more cautious the individual must be in being exposed to a given chemical, because a small difference in the dose may produce a serious effect. In order to assess the toxicity of a chemical, it is important to determine the route of exposure, the length of exposure, the species and individual characteristics of the exposed organism, and the nature or endpoint of the toxic effect being measured. In order for a chemical to show a toxic effect, it must first gain access to the cells and tissues of the organism. In humans, the major routes of exposure or entry into the body are through ingestion, inhalation, and dermal adsorption.

In toxicity assessments using animals in comparison to human beings, it is important to understand that there is a difference in the toxic effect of a chemical in different species. The difference may relate to the absorption of the chemical, biotransformation or metabolism of the chemical, or the differences in anatomical function. The absorption of chemical across the skin, lungs, or gastrointestinal tract is determined by the properties of the cells at the surfaces of these tissues. Biotransformation of the chemical in the liver and kidney or other tissues is different in different organisms. The rate of metabolism of chemicals differs, as well as the kind of metabolic byproducts.

The toxicity of many chemicals depends on the time of exposure and frequency of exposure. This time period is significant because some chemicals will accumulate in the body over long periods of time and will not be excreted. Toxic effects may depend on the duration of the exposure as it relates to the ability of these cells to repair themselves. Some adverse effects, such as some cancers and lung diseases, require an extended period of time to develop.

Individual characteristics within a given group, such as human beings, may vary and cause different types of responses to chemicals. Some of these differences may be related to sex, race, or age. Nutritional status and dietary factors also may contribute to the differences in people and their reaction to different types of chemicals.

Adverse or toxic effects can involve a variety of possible target sites and endpoints such as the liver (hepatotoxicity); kidneys (nephrotoxicity); lung (pneumotoxicity); blood (hematotoxicity); nervous system (neurotoxicity); DNA (genotoxicity, mutagenicity); structural or functional abnormality usually to developing embryos or fetuses (teratogenicity), and growth of malignant tumors (carcinogenicity). Related observations include neurotoxicity, where there is damage to reflexes, coordination, intelligence, memory, nerve impulses, and other behavioral types or concerns; carcinogenicity, where cancerous tumors are produced; hematological toxicity where there are changes in hemoglobin levels, erythrocytes, leukocytes, platelets, and plasma components; hepatoxicity, where there are malfunctions in the liver, lipid metabolism, protein metabolism, carbohydrate metabolism, and metabolism of foreign compounds; inhalation toxicity, where there is a gross or microscopic breakdown in the anatomy or function of the lungs; mutagenicity, where there are alterations in chromosomes, or DNA damage; renal toxicity, where there is damage to the kidneys; reproductive toxicity, where fertility levels are lower and more stillborns occur; and teratogenicity, where there are functional or physical defects in neonates.

Carcinogenesis

Cancer is a general term for a group of related diseases that cause uncontrolled growth of certain types of cells and related tissues. Chemical carcinogenesis generally means the induction by chemicals of malignant neoplasms or tumors not usually observed. The response to the chemical carcinogen varies with the species, strain, and sex of the experimental animal. Chemical carcinogens interact with other environmental agents that sometimes enhance and other times decrease their effect. Chemical carcinogens have a persistent and delayed biological effect, may be more effective in divided doses than in an individual large dose, and may have distinct interactions with host genetic elements. Carcinogens may either be direct initiators of the genetic change or act only as a promoter in combination with another initiator. (Figure 1.13) At times, one carcinogen may be required to cause a second carcinogen to promote a response. They would then be called co-carcinogens. One of the major problems in determining the possibilities of carcinogenic activity are the long periods of latency, which may extend for many years. Typically, time periods for the ultimate production of the cancer may take from 15 to 40 years after the onset of the exposure and prior to disease

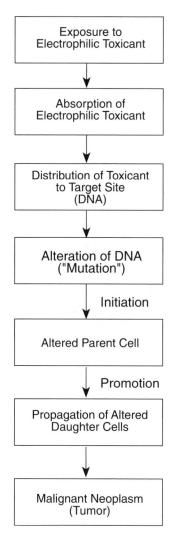

Figure 1.13 Schematic summary of genotoxic carcinogenesis

manifestation. The site of the development of a malignant tumor may vary considerably. It is based either on the portal of entry or on the site where biotransformation takes place. Asbestos causes cancer in the lung. Azo dyes produce tumors or cancer in the liver, which is the site of biotransformation. Radium produces cancer in the bone, where the radium is stored. The aromatic amines produce cancer in the bladder, the site at which excretion takes place.

The testing, assessing, and regulation of carcinogens is carried out by a variety of agencies, based on a large number of health, safety, and environmental laws that have been passed in the last 17 years. These agencies have issued guidelines and policies on how they intend to identify, evaluate, and regulate carcinogens. The Office of Technology Assessment has been given the task of gathering together all the information and techniques used by these agencies in making their decisions. There are many important issues in assessing potentially carcinogenic chemicals. They are based on interpretation of test data and the use of assumptions. These assumptions are derived from theories about how cancer is caused and decisions about what is considered appropriate public policy. The OTA has identified four important kinds of assumptions: (1) assumptions used when the data are not available for a particular case; (2) assumptions that are potentially testable but have not yet been tested; (3) assumptions that probably cannot be tested because there are certain experimental limitations; (4) assumptions that cannot be tested because there are ethical considerations.

There are four kinds of evidence that may be used for qualitatively identifying carcinogens. They are epidemiological studies, long-term animal bioassays, short-term tests, and structure-activity relationships.

Epidemiological studies are used to collect information about human exposures and diseases. Reports of individuals or clusters of cases help to generate hypotheses for later study. Many of the chemicals now known to be human carcinogens were first identified by physicians. Larger epidemiological studies were then devised and conducted. Descriptive epidemiological studies correlate risk factors, exposures, and diseases that are causes of death in specific groups. They are useful in establishing hypotheses for further study in providing clues about the hazards. Analytical epidemiological studies are used in comparing populations between a group exposed to the agent and a group that is not exposed. In case-control studies, the comparison is made between people with a given disease and those who do not have the disease.

Long-term animal bioassays are laboratory studies in which animals are exposed to suspected carcinogens for long periods of time (about 2 years). The animals are then examined for tumors. The tissues of animals of those who survive and those who die are compared. Short-term tests are used to examine genetic change in laboratory cultures of cells, in humans or other animals. These tests may take days or weeks to conduct.

Structure-activity relationships (SARs) are used to determine the chemical structures of substances and carcinogenicity. This is a comparison that helps to predict whether or not chemicals of a class closely related to carcinogens may also be carcinogenic.

The political considerations related to carcinogenesis are considerable. In regulatory proceedings, industry, labor, environmental groups, public interest organizations, and government may voice opinions that are frequently and substantially different. These groups place different values on the harm caused by the unnecessary regulation of a chemical that may later turn out to be safe. Even when accepting the value of animal data, there are many discussions on whether a particular animal study is reliable and if the data apply appropriately to people.

The Food and Drug Administration (FDA) was the first agency to establish guidelines for toxicity. The FDA, under the 1958 Food Additives Amendment to the Food, Drug and Cosmetic Act, which included the "DELANEY Clause", prohibited the intentional use of food and color additives that were determined to be carcinogenic in animals or humans. The "DELANEY Clause" does not apply to all food ingredients, because some were federally sanctioned prior to the 1958 amendment and some were considered to be generally safe. In the 1970s the FDA began using quantitative assessments for certain environmental contaminants found in food. By the 1980s, the FDA began applying these techniques to food and color additives. Much discussion has centered around what is an approved analytical technique to determine if the food and color additives are contaminated with small amounts of carcinogenic impurities, or when the additive itself has been determined to be carcinogenic. In 1985, it was decided that an approved analytical technique could be defined as one that could detect residue contents as low as the level associated with the upper portion of human risk estimate of one cancer for every one million persons exposed. This technique requires a risk assessment to estimate what residue levels correspond to this risk level.

The Environmental Protection Agency (EPA) began to develop carcinogenic assessment guidelines during the regulatory proceedings on the suspension and cancellation of several pesticides. The attorneys for the EPA summarized the expert testimony and developed summaries that were called *cancer principles*. There was considerable criticism of these cancer principles. In 1977, the Environmental Defense Fund petitioned the EPA to establish a policy on classifying and regulating air pollutants that were carcinogenic. In response to a court order to assess the hazards and risks of a large group of substances related to the Clean Water Act, the EPA set a methodology for assessing human risk. In 1984, the EPA published a proposed revision of its Carcinogen Assessment Guidelines, and also published proposed guidelines for assessing exposure to agents for mutagenicity and the developmental toxicants. It also published proposed guidelines for risk assessments of chemical mixtures. The final version of these guidelines were published in 1986.

The Carcinogen Assessment Guidelines published in the Federal Register described the general framework to be used in assessing carcinogenic risks and some of the principles to be used in evaluating the quality of the data and in making judgements concerning the nature and magnitude of the risk of cancer from suspected carcinogens. The various steps of risk assessment include hazard identification, dose-response assessment, exposure assessment, and risk characterization. The policy also presents a *weight-of-the-evidence* classification system. The Environmental Protection Agency lists five groups. (1) Group A includes human carcinogens. This group is only used when there is sufficient evidence from epidemiological studies to support an association between exposure from agents and cancer. (2) Group B is the probable human carcinogens. In this group the weight of the evidence of human carcinogenicity as based on epidemiological studies is limited. However, the weight of evidence based on animal studies is sufficient. (3) Group C is the possible human carcinogens. In this group, the evidence of carcinogenicity in animals is limited and there is also an absence of human data. (4) Group D is not classified as to human carcinogenicity. Here, there is inadequate human and animal evidence of carcinogenicity. (5) In Group E, there is evidence of noncarcinogenicity for humans.

The Consumer Product Safety Commission (CPSC) published carcinogen guidelines in 1978. The guidelines were challenged in court and were thrown out. Subsequently, CPSC decided to use the guidelines adopted by the Inter Agency Regulator Liaison Group (IRLG).

The Occupational Safety and Health Administration (OSHA) and numerous other agencies have written regulations on carcinogens. There have been a lot of arguments by various scientific groups, and there have been numerous legal challenges of a variety of carcinogen guidelines.

In 1993, the agency for Toxic Substances and Disease Registry developed a "Cancer Policy Framework" based on an assessment of the current practice across programs within the agency. Most of the policies established by the various federal agencies declare that well-conducted positive epidemiological studies provide conclusive evidence for carcinogenicity. Some of the factors used in evaluating epidemiological studies include the strength of association, the level of statistical significance, information on dose-response relationship, biological plausibility, temporal relationships, accuracy of exposure and cause-of-death classification, adequacy of follow-up, and a determination of the amount of time needed for latent effects to show up. A "nonpositive" study cannot be used to indicate an absence of a carcinogenic hazard. It can be used to establish upper limits of risk. There were nonpositive studies for arsenic, benzene, coke-oven emissions, petroleum refinery emissions, and vinyl chloride. Subsequently, these chemicals were proven to be carcinogenic. Long-term animal bioassay data have been accepted as a predictor for human beings. Substances shown to be carcinogenic in animals are presumed to present carcinogenic risk to humans. The reason for this conclusion is twofold. One, a number of chemicals that were first identified as animal carcinogens were subsequently confirmed as human carcinogens. Two, all chemicals accepted as human carcinogens, if properly studied in animals, have been shown to be carcinogenic in at least one species of animals. Although this cannot establish that all animal carcinogens also cause cancer in humans, in the absence of data on humans, it is biologically plausible. However, one must consider the amount of substance that is necessary to cause the potential carcinogenic effect in people and then determine if the chemical should be banned from use. Further, the route of administration in animal studies must be taken into account when determining the potential health effects in people. A big question is what do you do when you have conflicting animal data or conflicting animal and human data?

Quantitative risk assessment relates to dose-response determination, exposure estimation, and risk characterization. Quantitative estimation is a serious problem because often a series of untestable assumptions must be made from an extrapolation of animal to human cases. The exposure levels are deliberately set at high quantities to maximize the probability of detecting a carcinogenic effect. Extrapolating from a high concentration to a low concentration, even when using human data, can be difficult. Therefore, it can be seen how difficult it must be to reach appropriate conclusions when using animal data. The relationship between the dose or exposure of an agent and the biological response of the human is one of the most fundamental in the fields of toxicology and epidemiology. The dose-response curve may have several different shapes, ranging from a straight line to different curves. The data from animals usually represent dose levels substantially higher than the range of human exposure. An important question is whether or not threshold limits exist for exposure to carcinogens. These limits would be set at a level where there are no possible effects from the carcinogenic agent.

The development of cancer takes place in stages. These are typically called initiation, promotion, and progression. Initiation involves an alteration or mutation of the cell's genetic material or DNA. This mutation of DNA can remain latent (that is, without apparent manifestation of disease) for years. Promotion involves the expression of mutated DNA and the transformation of latent initiated cells into tumors. Progression consists of the growth of the tumors and the development of metastasis, which is the transfer of cancerous cells to other tissues. Some chemicals are primarily initiators. Other chemicals act only as promoters. Some chemicals are both initiators and promoters and are called complete carcinogens. It is possible that the mechanism of promotion involves an alteration in body chemistry, cellular growth, and repair, and other processes.

Mutagenesis

Mutagenesis is brought about by a change in DNA that was transmitted during cell division. (Figure 1.14) This change can cause deleterious effects to the individuals exposed or their offspring. If the change occurs in either the sperm and/or egg cells, the mutagenesis may occur in future generations and is referred to as genetic mutation. The effect of the mutagen may either be long term or delayed, however, high doses may cause toxicity to occur rapidly. Mutagenesis has a number of characteristics in common with carcinogenicity. Indeed, most carcinogens are mutagens as previously discussed. Apparently, there is an absence of a threshold of a given chemical that may cause the mutagenesis, as well as carcinogenicity. The Environmental Mutagen Society in 1975 revealed that a variety of active compounds in all chemical classes could cause mutagenesis. There is a clear and present hazard for the human population now, as well as for

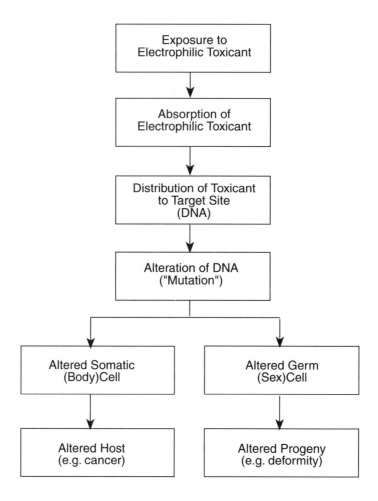

Figure 1.14 Schematic summary of mutagenesis

succeeding generations. A variety of test systems need to be used in order to determine if any given chemical is mutagenistic. Further monitoring studies of populations at great risk contribute significant information to determine what might be mutagenic and what might not. Genetic hazards to humans are typically predicted by bringing together the chemical and biological data generated from various tests and animal models. The degree of hazard or risk is usually a function of the perceived relevance of a test system that provides the positive responses and the total body of evidence that is available.

Reproductive Toxicity and Teratogenesis

A wide range of physical, chemical, and even biological agents can adversely affect reproductive outcomes. At least 50 chemicals, including heavy metals such as lead and cadmium, glycol ethers, organohalyde pesticides, organic solvents, and chemical intermediates, have been shown to produce impairment of reproductive functions in animals. The methyl mercury poisoning in 1970 of the residents of Minamata, Japan demonstrated the teratogenic potential for chemical exposures. Occupational exposures can produce a wide range of adverse affects on reproduction. Paternal exposure before conception may be shown in reduced fertility, unsuccessful fertilization or implantation, or an abnormal fetus. Maternal exposure after conception may result in death of the fetus or structural and functional abnormalities in newborns. It is estimated that there are 560,000 infant deaths, spontaneous abortions, and stillbirths each year in the United States. It is also estimated that 200,000 live infants are born with some birth defect each year. The causes of most of the adverse outcomes are unknown. No recognized anatomical or physiological abnormalities are found in 6–30% of the infertile couples. The causes for 65–70% of the birth defects are not known.

The studies of occupational reproductive hazards have been related to exposures to lead, ethylene oxide, and anesthetic gases. Adverse effects have been found on the quality of semen in workers exposed to lead or ionizing radiation. NIOSH estimates that approximately 200,000 workers are potentially exposed to various glycol ethers, several of which show marked testicular toxicity in animals. An estimated nine million workers are exposed to radiofrequency-microwave radiation, which has been shown to cause embryonic death and impaired fertility in animals. Personnel in hospital operating rooms are potentially exposed to waste anesthetic gases, and other hospital and industrial workers may be exposed to ethylene oxide. Both of these chemicals have been linked to an increased risk of spontaneous abortions in humans.

Teratology is the study of malformations induced during the development of an animal embryo or fetus from conception to birth. The teratogen causes a change to occur during the early embryonic development of the fetus, which results in anatomical defects, physiological defects, biochemical developmental errors, or behavioral effects. Teratogens also may cause the death of the embryo or fetus. Teratogenesis is usually considered to be a chronic effect, although the toxicity that caused it appeared over a short period of time during the developmental stage. The severity of the condition relates to the kind of exposure, the amount of the exposure, and when the exposure occurs during the developmental process.

Exposure and Risk Assessments

Exposure assessments are based on several factors. They include what chemicals are present; if the chemicals are present in air, soils, water, food or a combination of these environmental areas; what concentration of each chemical is present; what living organisms are exposed to these chemicals; through what routes these organisms are exposed; and what degree of exposure is expected in the future. The data for the exposure assessments is collected by doing physical assessments of the site of the chemical exposure and taking appropriate samples for analysis. The major exposure–related factors that influence toxicity are the route of entry, the duration of exposure, and the frequency of exposure. Samples may be taken of air, if the exposure is occupationally related in an industry, or of the air, land, and water, if the exposure is environmental or occupationally related at a hazardous waste site or other chemical release area. It is often necessary to consider the effect of being exposed to multiple chemicals at any given time, along with the other types of environmental conditions related to excessive heat, cold, pressure, inadequate lighting, dust, etc.

Risk assessment depends on toxicological evaluation of the chemical, dose-response evaluation of the chemical, exposure assessment, and risk characterization. The toxicological evaluation and dose–response evaluation determine if the chemical has an adverse effect and in what quantity the chemical would be present to cause this effect. The exposure assessment determines the amount of exposure to a chemical by individuals. The risk characterization, which is the final step in risk assessment, estimates the amount of adverse health effects that will occur when the individual is exposed to certain chemicals for given a period of time. This must be divided into noncarcinogenic risks and carcinogenic risks.

Medical Surveillance

The adverse effects following toxic exposure may either be biochemical, structural, or functional in nature. Although external environmental monitoring devices can be set up close to the individual to evaluate external exposure (see Chapter 11), they do not provide data regarding the internal exposure or absorbed dose and/or potential signs of adverse impact. Fortunately, medical surveillance includes biological monitoring to detect and measure the parent chemical or its metabolites in tissues or excreta, or pathological effects of the toxin to the exposed individual. The individual's blood, urine, breath, hair, neurological response, and physical condition can be evaluated, however, to determine whether exposure to certain agents has occured, and if so, at what levels. Biological monitoring data, in turn, can be compared to various recommended biological exposure indices.

Elevated liver enzymes may be due to exposure to and absorption of chemicals. Kidney function may be determined by abnormal elevations of normal blood constituents, such as microglobulins, urea, and creatinine. A standard blood count that evaluates the number and distribution of blood cell types and morphology helps access the effects of chemicals on the blood and blood–forming tissues. This test is relatively insensitive for most toxic effects. However, a bone marrow examination may be more informative, but it is very traumatic and expensive. A blood cholinesterase inhibition may be found in toxicity due to organophosphate exposure. Elevated carboxyhemoglobin in blood may be indicative of carbon monoxide poisoning.

Changes in skin texture, pigmentation, vascularity (changes in the channels in the body carrying fluids), hair, and nails, as well as the existence of lesions, can be detected with physical examinations. Neurological examinations can be used to detect disorders of intellect, memory, coordination, reflex changes, and motor and sensory problems. The EEG (electroencephalogram) records activity in the brain and the response to visual, auditory, or tactile sensations. This is particularly useful in determining any limiting of the function of the brain.

The respiratory system can be examined and analyzed by the use of pulmonary function tests. These tests can help differentiate between the normal lung volume and the restricted lung volume. The restrictions may be due to a chest–wall dysfunction or a disease such as pneumoconiosis. A reduction in the flow rate would be consistent with asthma or chronic bronchitis. By measuring the air remaining in the lung after a maximum expiration of air, it is possible to make some determination about emphysema being present. Chest X–rays are useful in measuring chronic pulmonary disease and cancer once the condition has occurred. Saliva may contain abnormal cells, which may be a means of determining if "asbestos bodies" are present in the saliva. Although this is not a diagnostic procedure that will confirm lung disease such as asbestosis, it still indicates asbestos exposure.

EPIDEMIOLOGICAL PRINCIPLES AND APPROACHES TO ENVIRONMENTAL HEALTH PROBLEMS

Epidemiology is the study of the factors that determine the occurrence and frequency of disease in a population. Epidemiology draws upon the knowledge and skills of clinical medicine, microbiology, pathology, zoology, sociology, anthropology, toxicology, and environmental health science. Although originally the epidemiological technique was applied only to disease and disease causation, more recently it has been applied to accident causation and injuries and diseases caused by chemicals and environmental factors.

Two major categories encompass epidemiology. Descriptive epidemiology is the study of the amount and distribution of disease within a given population by person, place, and time; analytical epidemiology is the study of causes determining the relatively high or low frequency of disease in specific groups. Age, sex, ethnicity, race, social class, occupation, and marital status are the principal personal variables in epidemiological research.

Age is the most important determinant. Mortality and morbidity rates often show a relationship to age. Acute respiratory infections occur with highest frequency among the very young, and upper respiratory infections, in general, are of greatest concern to the very young and the very old. Sex or gender is another important epidemiological determinant. The mortality rate for emphysema and respiratory cancer is higher in males than in females, whereas the frequency of occurrence of diabetes is higher in females than in males. Ethnic groups and racial groups exhibit definite differences in morbidity and mortality; blacks have a substantially higher age-adjusted death rate from hypertensive heart disease, accidents, and tuberculosis. Cancer of the cervix is higher in blacks than in whites. Whites have a higher death rate from suicide and leukemia. Social class is an important determinant in the epidemiological study of a specific disease. The death rate for unskilled workers is considerably higher than for the professional classes; infant mortality rates follow the same pattern. Schizophrenia occurs more frequently in the lowest social classes of the urban community. Occupation influences susceptibility to given diseases. Pulmonary fibrosis is found among workers exposed to free silica; cancer is found among workers exposed to aniline dyes.

The physical and biological environment influences the occurrence of disease in a given area. Multiple sclerosis is known to occur most frequently in northern climates; malaria and yellow fever are most frequently found in southern climates; lung cancer is more prevalent in urban areas.

Time is another extremely important variable in the study of disease occurrence. If it is possible to determine the cycle of diseases, there may be clues concerning the mutation of various organisms and also the potential effect of various chemicals on arthropods and on human beings. When the time element is determined in an outbreak of disease, it may be possible to quickly recognize the cause of the outbreak. Time is particularly important in outbreaks of food-borne disease. It is also important in determining degrees of exposure. Time, temperature, and concentration are the keys to many environmental health problems and often to the resolution of these problems or to the protection of the individual.

Descriptive epidemiology systematically summarizes the basic data available on health and the major causes of disease and death. These data permit the evaluation of trends in health and the comparison of these trends from subgroup to subgroup and from country to country. Descriptive epidemiology provides a basis for the planning and evaluation of health services and for the identification of health problems.

Analytical epidemiology seeks to discover the causes of problems by formulating hypotheses based on actual observations of existing diseases, conducting studies of outbreaks of disease, and evaluating the data to determine the accuracy of original hypotheses. The classic example of this method was the determination of the epidemiology of cholera by John Snow, the English physician, who realized through observation that cholera was spread through water from contaminated sources.

Germ Theory of Disease

John Snow hypothesized that a specific microorganism caused cholera, yet this concept was not generally accepted during his time. The germ theory of disease was postulated on the basis of Louis Pasteur's research. Pasteur demonstrated that microorganisms caused fermentation and that these microorganisms were not spontaneously generated but could be found in the air. His work was followed by Robert Koch, who first isolated the organisms causing tuberculosis and Asiatic cholera. Koch introduced scientific rigor to the proof of the cause of disease. Koch's postulates were that the parasite had to be shown to be present in every case of the disease by isolation and pure culture; it could not be found in other diseases; once the parasite was isolated, it had to be capable of reproducing the disease in experimental animals; the parasite had to be recovered from the diseased animal. Unfortunately, Koch's postulates did not apply to viruses, since they could not be cultivated in pure culture and were host specific. His postulates also did not apply well to certain pathogenic organisms that infected a host but did not necessarily produce a recognizable disease.

Infectious Disease Causation

In order to understand the spread of infectious or communicable disease, it is necessary to understand the host-parasite relationship and the reservoir of disease. The reservoir of disease is defined as the living organisms or inanimate objects in which infectious agents live and multiply. A parasite is the microorganism causing a specific disease. A host may be either people or animals. The chain of infection of a given disease is usually illustrated as follows:

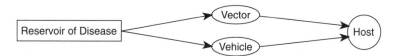

In the environmental health field, the practitioner attempts to halt the transmission of disease by eliminating the vector, which may be an arthropod, or controlling or purifying the vehicle, which may be an inanimate object — water, air, or solid waste. Generally, the reservoir of disease is difficult to control, with the exception of those diseases where hosts can be "vaccinated." Eliminating the culex mosquito, which in turn eliminates St. Louis encephalitis, is an example of breaking the chain of infection at the vector state. Purification and chlorination of water supplies, which destroys the microorganisms causing cholera, is an example of halting the spread of infection at the vehicle stage.

Humans are not only the recipients, but often the carriers, of disease. Carriers discharge microorganisms into the environment through feces and urine, and from the skin, nose, or pharynx. Since carriers may not exhibit symptoms of disease, they are dangerous in the transference of disease. Carriers may have a subclinical case of a disease, or may be in the incubation or in the convalescent period of disease. Some examples of diseases transmitted through carriers are polio, chicken pox, measles, hepatitis, diphtheria, salmonella, and staphylococcus infections. Disease may be transmitted directly from person to person through kissing, sexual intercourse, coughing, sneezing, direct contact of the skin, or through indirect transmission by means of vehicles or vectors. Some examples of vehicles, also called fomites, include bedding, toys, surgical instruments, surgical dressings, contaminated food or water, and needles. Indirect transmission of disease may occur when dust and droplet nuclei enter the airstream through coughing, sneezing, talking, dusting, walking, or through a variety of housekeeping or medical procedures. Q fever, histoplasmosis, tuberculosis, psittacosis, staphylococcus infections, and other diseases are transferred in this manner. Disease may be spread from a common source or from secondary sources. Generally the initial outbreak of an epidemic spreads from a common source; the secondary outbreak, which may occur days or weeks later, spreads from infected secondary sources, or the secondary environment, which became infected during the primary outbreak of disease. The scope of an outbreak of disease depends upon the virulence of the organism, its pathogenicity, the number of unprotected individuals at risk, various environmental factors, and the intervening factors used to control the disease. Virulence refers to the degree to which an infectious agent can produce serious illness. Pathogenicity refers to the ability of an agent to cause disease in an infected host.

The cause of the spread of infectious disease is generally complex, even when the primary cause, or agent, is known. The many contributing factors, or secondary causes, including the environmental factors, sharply influence the rate of disease occurrence. When the disease has run a specific course and new cases are not occurring, and if this is a spontaneous type of leveling off of the disease process, it is recognized that an equilibrium has been established between the parasite and the host, and this equilibrium will last until new hosts and/or a more virulent form of the organism appears, or the environmental factors are changed.

Emerging Infectious Diseases

In 1994, the Centers for Disease Control and Prevention developed a prevention strategy on emerging infectious disease threats for the U.S. This strategy has four goals for dealing with infectious diseases that affect or threaten to affect the health of the public. These goals are:

1. Surveillance: detect, investigate, and monitor emerging pathogens and disease.
2. Applied research: combine laboratory findings and epidemiological studies to enhance disease prevention.
3. Prevention and Control: improve communication of public health information concerning emerging diseases and how to prevent and control them.

4. Infracture: strengthen support and personnel at local, state, and federal agencies and how they interact.

Emerging infectious diseases are diseases of infectious origin whose incidence in humans has increased within the past twenty years, or there is a threat of increase. Examples of these emerging infectious diseases in the U.S. include:

1. *Escherichia Coli* 0157:H7 disease
2. Cryptosporidiosis
3. Coccidiodomycosis
4. Multidrug-resistant pneumococcal disease
5. Vancomycin-resistant enterococcal infections
6. Influenza A/Beijing/32/92
7. Hantavirus infections

Examples of these emerging infections outside of the U.S. include:

1. Cholera in Latin America
2. Yellow fever in Kenya
3. Vibrio Cholerae 0139 in Asia
4. *E. coli* 0157:H7 in South Africa and Swaziland
5. Rifvall fever in Egypt
6. Multidrug-resistant Shigella dysenteriae in Burundi
7. Dengue in Costa Rica
8. Diptheria in Russia

Emerging infections are especially a problem in people with a limited immunity, such as those with HIV infection and those receiving medication for cancer or transplants. Because of widespread use and misuse of antimicrobial drugs, these drugs are beginning to become useless, and there is a development of drug resistant organisms. Lack of appropriate surveillance techniques, laboratory tests, and knowledge by the public and physicians of these serious health problems have complicated the situation and increased the potential for disease.

Environmental Contaminants and Disease

Cancer caused by environmental contaminants is a major problem in our society today. Cancer attributed to occupational or other environmental factors is higher among adult males because of a higher proportion of males employed in industry. More than half of the cancers observed in the third national cancer survey were caused or suspected to have been caused by an external agent. However, in evaluating cancer problems, it is important to recognize that cancer is also associated with socioeconomic status, cigarette smoking, diet, alcohol, and radiation.

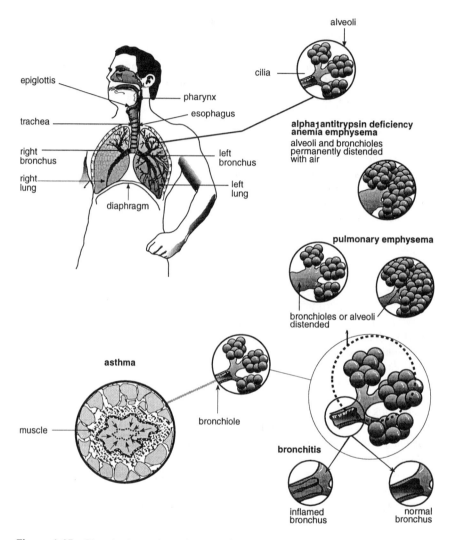

Figure 1.15   Chronic obstructive pulmonary disease

There has been a sharp increase in morbidity and mortality for bronchitis, pulmonary emphysema, and to a lesser extent for asthma. These three conditions, which are collectively termed *chronic obstructive pulmonary disease*, (Figure 1.15) are found more frequently among smokers than nonsmokers. A higher mortality rate exists among the low-income urban population compared with the upper-income or rural population. Studies indicate that environmental conditions contribute to the increase in these diseases.

A sharp increase in cardiovascular conditions, particularly coronary heart disease, has been attributed to diet, cigarette smoking, and problems of the environment. Clinical experimental evidence suggests that exposure to carbon monoxide exacerbates symptoms of angina. There appears to be a strong statistical correlation between the season of the year, the yearly temperature change, and mortality attributed to coronary heart disease. Although infectious agents contribute to this problem, the environment is also involved. Numerous, although controversial, publications concerning the relationship between water hardness and cardiovascular disease have appeared. Some association appears to exist between the level of trace metals found in water and the susceptibility to sudden death from arrhythmia. It is also suspected that carbon disulfide and freon gas contribute to arrhythmia in the occupational environment.

Many investigators are now studying the effect of environmental conditions on the outcome of pregnancy. During pregnancy the internal organism is more vulnerable to problems, since the detoxification mechanisms are altered and changes occur in protein, carbohydrate, and lipid metabolism. Concern exists today regarding teratogens, or substances producing abnormal variants, such as congenital malformations and spontaneous abortions. While further research is needed in this area, certain chemicals are conclusively embryopathic, including methyl mercury, aminopterin, thalidomide, iodine deficiency, steroid hormones with androgenic activity, and carbon monoxide. Certain chemicals are suspected of affecting human prenatal development, including cortisone, vitamin A deficiency, and diethylstilbestrol.

Infants are susceptible to methemoglobinemia caused by the ingestion of nitrates or nitrites with drinking water. Chemicals may also affect the CNS during postnatal development, affecting particularly the cerebrum. Lead poisoning in children produces irreparable brain damage and permanent retardation and eventual death. Long-term exposure to air pollutants by infants and children impairs respiratory health and aggravates chronic bronchitis, asthma, and other lung conditions. From 1981 to 1990, 250 waterborne outbreaks of disease occurred in the United States. The frequency of occurrence of waterborne disease is increasing. In addition to waterborne infections, an increasing number of deaths occurred due to chemical poisoning through the ingestion of water. In the reuse of water, problems are created by the presence of microorganisms and toxic minerals, which ordinarily are not found in the initial pure water supply. Some laboratory data suggest that mutations occur as a result of environmental pollutants. Although this is not conclusively demonstrated, concern exists, and mutation must be considered one of the potential disease problems related to environmental contaminants.

Epidemiological Investigations and Environmental Health

An epidemiological investigation is a systematic study of available current data, past records and data, and on-site investigations, when feasible. Epidemiological investigations vary with the type of study being conducted. Specific techniques will be discussed in the chapters on water and food. A typical epidemiologic study consists of a preliminary review of material, further investigations, analysis of the data, a hypothesis concerning cause, and in some cases, the testing of the hypothesis through the review of new data. In the initial gathering of materials, standardized forms must be used to obtain consistent and complete information. The data are verified through diagnosis of the specific diseases, conditions, and injuries. The investigator determines the time, place, and individuals involved, and plots the cases by time and condition. The investigator asks the questions: how?, why?, what?, when?, where?, and who?. A test hypothesis concerning the cause of the outbreak or the cause of the disease is theorized; additional cases are sought that either fit or do not fit the pattern. If the cases do fit the pattern, the hypothesis is strengthened. Comprehensive, accurate analysis of data is essential in determining the causes of diseases, conditions, or injuries. One of the principal problems facing the investigator is failure on the part of victims to report the disease to the proper authorities, which results in an incomplete picture of disease conditions. Determining the actual cause, or causes, of a given condition can be problematic, however, the trained investigator, in collaboration with the environmentalist and medical nursing team, uses available data in an attempt to control the problem. The work of an epidemiologist is similar to that of a detective: it is long and tedious, and frequently gives poor results. However, where the results are satisfactory, the population is saved from debilitating or death-causing diseases.

Epidemiological analysis of health data is made by age, race, sex, place, and time. Reports include incidence, prevalence, and case fatality, or morbidity of a given disease or condition. Variation by time and place reflects exposure and response to different environmental pollutants. Epidemiological analysis also includes variations in pulmonary function, cholinesterase levels, and potential toxic effects based on blood analysis of various chemicals.

Epidemiological methods are used to compare exposed and controlled groups. For example, the amount of lung disease in a specific group of workers is compared to the general population. The greatest problem in this type of study is controlling the significant variables, such as age, sex, economic status, ethnic group, and smoking habits for the two groups.

Another technique is a longitudinal study of a specific group over a period of time to determine the rate and degree of occurrence of a specific disease within varying time periods. This technique is particularly effective in determining the acute effects of exposure to given environmental pressures. The determination of community exposures, including not only the occupational but other environmental settings, is difficult, since groups are usually poorly defined, vary within the geographic area, and receive different levels of exposure to substances.

In attempting to study the relationship between health and environmental exposure in a community, the subjects must be selected either by geographic comparisons, comparisons over time, or dose-response relationships. In geographic comparison, the exposure is defined by a geographic area, for example, health problems experienced by individuals living in heavily air-polluted areas. A problem with this type of study is that a single measure or combinations of measures of exposure is used to represent the entire group exposure. In interpreting data from this type of study, it is important to account for possible difference in factors other than those under study. These differences include migration, age, and ethnic group.

In time-comparison studies, a specific time period is compared to health effects over that time period. For instance, a community and its health problems are evaluated for a given period and then are reevaluated after a month, a year, or longer. If the time study period is less than a year, seasonal variation must be introduced. An attempt is made to understand differences in health effects in the population between the initial time period and the subsequent one. Occasionally geographic and time comparison methods are combined, producing a more reliable comparison than if the two were made separately.

In a surveillance study, available data are collected and analyzed, and departures from the norm are noted regarding the severity of environmental problems and health effects. This type of study is used to take corrective action as quickly as possible. In all data collection and epidemiological studies of environmental problems, recognition and control or elimination of variables contributing to illness are essential. Without this, data are of little or no value. In examining environmental influences on health, rare occurrences of illnesses, such as the sudden occurrence of liver cancer in the United States, are important indicators. The outbreak of liver cancer aided in the discovery of the problems related to occupational exposure of vinyl chloride.

Death certificates and morbidity data records from state compensation agencies provide information regarding occupational or other exposures to given environmental factors. This information is of value in developing the initial epidemiological study for a given environmental factor and its relationship to poor health. Other record sources include cancer registers, union records, social security records, and interviews with surviving employees or next of kin. The National Center for Health Statistics is another source of data, and is involved in epidemiological analysis by utilizing data available through various other components of the Department of Health and Human Services, the Social Security Administration, and other branches of government and industry. Although death certificates are required in all states, a National Death Index is needed that would provide the epidemiologist with information on a yearly basis. Great Britain and the states of California and Washington analyze death by occupation. This would be invaluable epidemiological data if provided on a national scale by the National Center for Health Statistics. The National Health Interview survey is a source of data on mortality, but until recently inadequate information on morbidity was provided. Better morbidity data are needed. A study that should provide better data is the National Health Examination Survey, developed to determine the possible role of water hardness on cardiovascular disease. The national census could be used to provide mortality and morbidity data for men and for women by occupation and social class. Each yearly census contributes an additional source of information that should be analyzed for environmental problems. The state health departments also gather and provide data on environmental health problems.

At present, data are available from the Environmental Protection Agency, the National Institute of Occupational Safety and Health, and the Food and Drug Administration. As can be seen, adequate data are available in some areas but are lacking in others. However, there is a tremendous need not only to provide data, but also to coordinate data and to make them available to epidemiological, medical, and environmental support services.

Environmental Epidemiology

The establishment of contamination criteria and/or exposure limits can be done through epidemiological studies or toxicological studies. The epidemiological studies are concerned with the exposure of populations at their occupational settings or through the contamination of food, drinking water, or air. These studies provide a statistical association between the levels of contaminants and the reported effects. The toxicological studies are of groups of animals exposed intentionally in a variety of controlled laboratory experiments. In this situation, it is possible to define the doses, frequency of application, and metabolic pathways, areas of storage, and types and amounts of biological damage created by the chemical agent.

The advantage of using epidemiological data for establishing safe limits of human exposure to chemicals is that you do not have to have the compounding factor of interpreting animal data and extrapolating from this data. The epidemiological studies try to determine whether or not correlations exist between the frequency or prevalence of a disease or health condition in humans and some specific factor, such as the concentration of a toxic chemical in the environment. Another advantage to this type of determination is that typically large numbers of humans are involved in the study and the exposure levels are usually subclinical. Therefore, the data are directly relevant to the study and determination of the dosage. The disadvantages are (1) the human exposure to chemicals is fortunately very limited, (2) controlled studies of exposure in humans are neither feasible nor legal, and (3) the application of the available epidemiological data may be limited because of its quality or because there are many risk factors involved in the study that cannot be truly limited or measured. These risk factors might relate to the ages of the individuals, their educational backgrounds, occupational histories, prior smoking habits, drugs or alcohol consumed, dietary practices, general health, and sexual and racial factors. Another major problem is determining what is the health endpoint in a clear and concise manner. Death, of course, is a reliable and a definable statistic, but how do you measure the level of good health or poor health related to a combination of chemicals to which the individual has been exposed? Further, there are a group of chronic nonspecific diseases with nonspecific causations that may occur. Even if measurements are made of the appropriate chemicals, are they accurate and do they show the overall population exposure to the chemicals? The most effective types of epidemiological studies are of working populations where known groups of individuals are exposed to chemicals.

The two classes of experimental design in epidemiology are descriptive and analytical. The two types of descriptive studies include the case study and ecological study. The case study provides information for a single individual about the relationship of an exposure and disease. Many times case studies are the starting point for in-depth investigations. The ecological study is used to determine how a single factor is distributed in the population. The ecological study describes the disease in terms of prevalence, incidence, and mortality rates for the population. The ecological study compares the trends of disease in two or more populations or over time. It is used to generate hypotheses about the impact of an agent on the disease found in human populations. These hypotheses are then used to help design the analytical studies. The two types of analytical studies are the retrospective and prospective. In the retrospective study, the population being studied is determined from death certificates, hospital records, and other sources of data indicating morbidity. A control group that is free of the disease is then chosen for comparison. The exposure history of the study group has been assessed and the causal relationship between exposure to the chemical or other substance and the level of disease is established.

In the prospective design, a cohort (a group) of disease-free patients are followed over time, and the development of disease is monitored. Of the pairs of individuals one half are exposed to the test agent while the other half is not exposed. The development of disease is measured in both groups, and a determination is made of the causal relationship between the agent and disease. Other problems with the cohort study relate to diseases that rarely occur, the sample size of the study, design management problems when the disease has a long latency period, and death of the subjects.

To get around some of the difficulties that have been mentioned, the Environmental Epidemiology Branch of the National Cancer Institute did comprehensive studies of 3056 counties in the United States for a 20-year period between 1950 and 1969. They were trying to do a human risk assessment for cancer. From the studies and maps that were drawn, they were able to identify cancer hot spots. They found Salem County, New Jersey had the highest rate of bladder cancer mortality in the United States. They then did ecological studies and generated maps of industries in that state. The maps included the total population of the county employed in a specific area. Eighteen major industrial categories were examined. They were able to use existing vital records and census data that eliminated extensive data collection. In the comparison rates of mortality and morbidity, it was critical to adjust for confounding factors. (A confounding factor or confounder is a risk factor for the disease under study.) If you do not adjust for the confounders, then it is possible to get a mistaken assessment of the risk for disease as a result of specific agents. In cancer, age always is a confounder. Further, many cancers are sex or race specific. Smoking is a confounder in many cancers, including lung and bladder, when the individual is also exposed to environmental or industrial sources. Another problem is called bias. (Bias is a systematic error in the design, conduct, or analysis of a study that causes a mistaken estimation of the relationship of the study factor to the disease.) Some of the biases include a nonrespondent bias, a diagnostic suspicion bias, an exposure suspicion bias, and a family information bias.

Epidemiological investigations, when adjusted for bias and the confounding factors, can be of enormous importance because they provide information about humans under actual conditions to exposure to a specific agent. The surveillance systems used in epidemiology need to be set up in a very careful manner, since these systems vary widely in methodology, scope, objectives, and characteristics. The strength of the evaluation depends on the evaluator, how each of the items in the framework flow, and how they are established and assessed.

The public health significance of the health event must be established. In the concern about a few persons being exposed to highly toxic chemicals, a cluster of cases occurring, or a general outbreak of a specific disease, the health event may be measured by (1) the total number of cases, incidence, and prevalence; (2) the case-fatality ratio, which is an index of severity; (3) the mortality rate; (4) the index of loss productivity; (5) the index of the years of potential life loss; (6) the medical costs involved; and (7) the degree of preventability. Next, what are the objectives of the system? Describe the health event under surveillance and state a case definition for each health event. Draw a flow chart of the system. Describe the components and operation of the system, including (1) what is the population under surveillance? (2) what is the period of time of data collection? (3) what information is collected? (4) who provides the information? (5) how is it transferred? (6) how is it stored? (7) who analyzes the information? (8) how is this data analyzed and how often? (9) how often are reports disseminated and distributed and to whom are they sent? and (10) what is the level of usefulness of the data from the surveillance system?

Each system has to be simple, flexible, acceptable, sensitive (collects considerable amounts of data), predictive of problems, representative, and timely.

Epidemiological studies that are adjusted for bias and confounding factors and meet the guidelines previously stated may be extremely beneficial in providing information about humans under actual conditions of exposure to specific agents. A well-designed, properly controlled study can provide competent results in health areas and the determination of risk assessment. The weight of evidence needed to determine whether or not something is potentially carcinogenic in humans or whether other kinds of health problems exist will be based on a combination of epidemiological studies, long-term animal studies, pharmacokinetic studies, and relevant toxicological studies.

RISK ASSESSMENT AND RISK MANAGEMENT

Risk is the probability of illness or injury. It is the potential realization of unwanted consequences of an event. Both the probability of the event occurring and the magnitude of the consequences are involved in the term risk. Risk is a function of the hazard involved, the related dose-response data, and the magnitude and duration of human exposure.

Risk Assessment

Risk assessment is the determination of what the problems are. Risk management is the process of deciding what to do about the problems. Risk assessment is made up of four steps: hazard identification, dose-response assessment, exposure assessment, and risk characterization. Hazard identification involves the gathering of data on the substance, including information about the link between the substance and adverse health effects. This step determines that such a link exists. Unfortunately when the hazard identification is based on the extrapolation from animal data to humans, it is difficult at times to reasonably assert that the substance is a significant hazard. Once it is determined that a chemical is likely to cause a specific health effect in humans, it is necessary to establish the relationship between the amount of exposure (dose) and the effect that is produced. It is not entirely clear that safe levels or thresholds exist for toxic chemicals, since short exposures at low doses may cause damage to health in the future. This is a very complex area. It is easy enough to establish a health endpoint when the individual is subject to high doses of a chemical and an immediate result is seen in the body.

Exposure assessment is a determination of how much of the chemical is available to the individuals in a specific area. In addition, the duration of exposure is also considered. Its two components include an analysis of the path of transportation and an assessment of the impact of transformation, as well as the available dose and an analysis of the population characteristics. The best way of measuring human exposure is to do direct measurement or monitoring of the ambient conditions. The degree of exposure may vary from chemical to chemical. Human data typically are quite limited because of the types of monitoring, the expense, and the time required to gather information. Modeling often is used as a substitute. In this technique, data are fed into the computer on pollutant releases, release characteristics, meteorology, hydrology, geography, and other information. The exposed population is estimated through the use of census data, etc. There is considerable uncertainty about these estimates. Risk characterization requires an evaluation of the information from the first three steps. For noncarcinogens, the margin of safety is estimated by dividing the experimental No Observed Adverse Effect Level (NOAEL) by the estimated daily human dose. For carcinogens, the risk is estimated at the human dose by multiplying the actual human dose by the risk per unit of dose projected in dose-response modeling. It may be a range of risk produced by using different models and assumptions about dose-response curves and the relative susceptibility of humans and animals.

Hazard assessment has been based on the underlying principles that animal bioassays are indicative of probable human response and that there is no threshold of response to carcinogens. Technical guidelines have been established by various agencies in risk assessment as they relate to carcinogens.

Toxic substances can lead to adverse effects in the liver, kidneys, lung, and other body systems. These systems have certain thresholds at which they will respond to the chemicals. This further complicates the problem of risk assessment.

Most risk assessments are for individual chemicals. However, in reality, people are subjected many times to chemical mixtures that would alter the risk assessment picture considerably. Are there carcinogenic or additive effects to the mixtures of the chemicals?

The exposure assessment is a combination of field monitoring, mathematical modeling, measurement of actual concentrations of chemicals in tissues, and laboratory modeling data. A complication in this exposure assessment is determining how many people are exposed to the chemicals through the air, soil, water, drinking water, or food. It is then important to calculate the rate of uptake of the chemicals through breathing, eating, drinking, or absorption through the skin. Then it is essential to track the chemicals through the body and through the process of metabolism.

To summarize risk analysis it is necessary to (1) determine the source of the chemical and how long it is released to the environment; (2) the pathways through which the chemical travels through the environment, such as the air, water, food, etc.; (3) the behavior of the substances in the body and the pathways that the chemical takes during metabolism, as well as the toxicity of the metabolite products; (4) estimate the concentration or dose at the specific site in the organs and how long the organ is exposed to the chemical; (5) determine the persistence of the substance in the environment and how long it can be taken into a human being and then passed on to the specific organ; (6) determine the relationship between the dose the individual gets and the effect on the individual; and (7) estimate the risk to the exposed population.

Risk Management

Risk management helps set priorities. Many of the priorities set for the federal agencies, especially the EPA, have been decided by Congress. Because of potential emergencies, it is best that the agencies look at what the potential problems are and decide what is of the most pressing nature. Risk management produces a basis for balanced analysis and decision making. Because we are exposed to a complex, highly dilute mixture of chemicals that come through the air, water, land, and food, it is frequently difficult to determine which area must receive immediate attention. Rational decisions can be made on the basis of good scientific data. Risk management produces more efficient and consistent risk reduction policies. By doing complete evaluations, it is possible to determine how best to approach risk management. Risk management is the complex of judgement and analysis that uses the results of risk assessment to produce the decisions that are necessary to bring about an environmental action.

Inherent in risk management is the idea of comparability. The EPA has a variety of goals, some of which may be in conflict with others. For instance, deep ocean dumping of sewage sludge may reduce the human health risk in comparison with incineration or land spreading, but may have an adverse effect on marine ecosystems. Pollution control, and therefore the reduction of risk, is an incremental process. Consistency must be used here because of the cost involved. At some point when increasing the effectiveness of the process, the cost needed to improve the process by a small increment could be beyond reasonable levels of expenditure. Risk management programs have been set up by the EPA for a variety of problem areas, such as hazardous waste, air pollution, water pollution, etc.

ENVIRONMENTAL HEALTH PROBLEMS AND THE ECONOMY

The environment and the economy are not mutually exclusive; together they form an integral system. As the public demands a higher quality environment, the economy must adjust to meet this need. In the past, Americans have placed a high priority on convenience and consumer goods. There seemed to be no need for concern over the air, water, or the Earth. The economic system was established such that incentives were given to individuals to better their position in our society without taking into account the problem of environmental degradation. The rich endowment of natural resources; the large amounts of investment, technology, and education; the influx of skilled immigrants; and the political and economic systems in the United States have combined to produce an economy with enormous material wealth. American business and industry, through its technological developments, produce huge amounts of goods and services. Consumers continually demand new products, and business and industry have accommodated consumers. Old products are discarded as quickly as new ones become available.

However, a new demand for better environmental quality and cleaner surroundings has emerged in the United States. The American public is demanding cleaner water, cleaner air, relief from noise and congestion, and a more aesthetically pleasing environment. Concurrently, the supposedly unlimited supply of water and air is befouled with huge quantities of pollutants, creating severe health, economic, and aesthetic problems. Exotic chemicals have entered the environment and moved through the food chain to a point where they are becoming a potentially dangerous health hazard.

During the last 15 years the production of synthetic organic chemicals, containers, packaging, and electronic equipment has increased sharply. As production has increased, consumer demand has increased. Unfortunately, although the products are available to the consumer at reasonable prices, these prices do not fully represent the cost to our society, since they do not include the cost of waste of our natural resources or the cost of environmental degradation. Due to the unfair economic incentive for use of raw materials vs reuse of recycled materials, there is further destruction of our environment. The dumping of wastes into the air, water, and land must end. The burden of the cost of the environmental cleanup and environmental improvement must be borne by all. Pollution costs include higher prices due to damage to crops, materials, and properties, and due to disease and injury. The actual costs of control measures are difficult to assess, since one must take into account abatement programs, expenditures by business and industry, expenditures by government, research and development, monitoring devices, and a variety of equipment and service purchases.

Dollars spent improving the environment are an investment in life and human society, although the benefits of these investments may not be readily seen. Imposing environmental controls results in the reallocation of resources, which in the short run may cause adverse economic effects, such as higher prices, temporary unemployment, and plant dislocations. The decrease in medical bills, the increase in recreational activities, the decrease in damage to materials, and the provision of a better society must be weighed against the temporary cost increases. Marginal firms that fail to meet these costs will be forced to close. The ultimate good to the greatest number of people should be the deciding factor in determining what should or should not be done to improve the environment. However, in any given area where changes are undertaken, adequate additional funds must be made available if the area is likely to suffer economic dislocation. In some cases where health is not endangered by pollution, it may be best to carefully and cautiously make temporary exceptions to certain standards in an effort to ease the strain on local economies. These exceptions should only be made with the greatest of care.

RISK-BENEFIT ANALYSES

In all areas, but particularly with regard to chemicals, it is necessary to compare the social benefits of a given environmental substance to the risk it causes to our society. Many of the pesticides are potentially extremely dangerous, and yet without the availability of pesticides the country and the world would be rampant with disease and would probably suffer from famines and starvation. It is therefore necessary to evaluate each of the environmental substances to determine the relative safety of the substance, the relative efficiency in production of the substance compared to the use of natural resources, and the relative impact on the environment at various stages of manufacture, use, or disposal. In all cases, risk-benefit analyses must be completely considered. The benefits include the value of the use of the substance to the consumer, its aesthetic value, the conservation of natural resources and energy resulting from the production of the substance, and the economic impact of employment, regional development, and balance of trade in the production of the substance. The risks include the adverse effect on health; the potential for death occurring as a result of use of the substance; environmental damage to air, water, land, wildlife, vegetation, aesthetics, and property; and the misuse of natural resources and energy. Establishing risk-benefit analyses for each substance utilized within the environment is difficult. However, with analyses of this type, society can determine the necessary trade-offs between potential hazard and employment, health, gross national product, balance of trade, conservation of natural resources, and environmental quality.

Benefit Cost Analysis

The four case studies that will be discussed include the following: lead in gasoline, air toxics issues, incineration at sea, and municipal sewage sludge reuse and disposal.

Lead in Gasoline

In the case of the lead-in-gasoline rule, a careful quantitative analysis was used. Lead in gasoline has been regulated by the EPA for more than 15 years. There was a concern that leaded gasoline was being used in cars that required unleaded gasoline. This would affect the pollution control catalysts. There was a further concern that the lead was a potential risk to children's health. Although previous rules had reduced lead in gasoline, in March 1984, the EPA set a very tight phase-down schedule. The final rule issued in March 1985 was more stringent than the original proposal. It required that lead in gasoline be reduced from 1.1 g per leaded gallon to 0.5 g per leaded gallon by July 1985, and to 0.1 g per leaded gallon by January 1986. Not all of the lead was removed from leaded gasoline, because of the potential effect it would have on older engines. Since the 1920s, refineries have added lead to gasoline as an inexpensive way of boosting octane. With the removal of lead, refineries had to do additional processing or use other additives, which tended to raise the cost of gasoline. The EPA used a computer model of the refining industry that had been developed by the Department of Energy to estimate the cost of the rule. The computer model included various equations that showed how different inputs could be turned into different end products at varying costs and the constraints on the industry's capacity. To estimate the cost of the rule, the EPA first ran the model specifying the then current lead limit of 1.0 g per leaded gallon and computed the cost of meeting the demand for the refined products. It then ran the model specifying the lower lead limits and recomputed the overall cost. The difference between the original and anticipated lead limits was the estimated cost of the tighter standard. Based on that analysis, the EPA estimated that the rule would cost less than 100 million dollars in the second half of 1985 to meet the 0.5 g per leaded gallon standard and approximately 600 million dollars in 1986 to meet the 0.1 g per leaded gallon standard. They then ran extensive tests using more pessimistic assumptions, such as an unexpected high demand for high-octane unleaded gasoline, increased downtime for equipment, and a reduced availability of alcohol additives. The 0.1 g per leaded gallon rule met virtually all conditions. It was extremely unlikely that a combination of these conditions would occur at the same time.

The benefits of the rule were estimated for the maintenance of good health in children and the improvement in educational efforts, compared to the efforts of children who were exposed to lead. Further, benefits were estimated based on the reduction in damage caused by excessive emissions of pollutants from vehicles using the wrong gasoline, the impacts on vehicle maintenance, and the impacts on fuel economy. The EPA also used existing data to study the relationship between levels of lead in the blood and blood pressure to make some estimate of the health effects that adults would have from this rule. In each category, the EPA first estimated the impact of reduced lead in physical terms. In the case of children's health effects, it used statistical studies relating to lead in gasoline and blood lead levels projected in children. It also estimated the cost to people and society and the environment by allowing the lead levels to stay the same vs. the reduced lead levels. It was determined that the benefits of the lead rule exceeded the cost by a 3:1 ratio and that if the potential benefits in reductions in blood pressure, illness, and death were to be counted, the ratio would rise to 10:1.

Air Toxics

Risk assessment technique were used to determine the environmental hazards of toxic air emissions. A study was made to determine how much of the air toxics problems could be controlled by using existing EPA programs. A new toxic strategy was then developed and put into place. It had three main parts: (1) direct federal regulation of significant nationwide problems, (2) state and local control of significant pollutant problems that were national in nature, and (3) an increased study of geographical areas that were subjected to particularly high levels of air pollution. In the initial study, 42 air toxic compounds were evaluated, basically for their potential to cause cancer. A determination was made of how many people were exposed across the country. An estimate of the risk associated with each compound in this national exposure was determined. The estimate was of long-term cancer incidents (a 70-year time frame) associated with these compounds. The analysis suggested that the air toxics problem was complex, caused by many pollutants and sources, varied significantly from city to city, and even varied within a city. In the 1990s, the EPA will attempt to reduce by 50% the 191 toxic pollutants listed in the Clean Air Act Amendments of 1990.

Apparently, preexisting EPA air toxic strategies that focused on regulating each pollutant as part of national emissions standards were too narrowly set up to be effective. A comparison of air quality data for 1970 and 1980, however, showed that significant reductions in national cancer incidents related to air pollution had occurred as a result of the reduction of air pollutants. Air toxic problems were examined in detail in Baltimore, Baton Rouge, Los Angeles, Philadelphia, and Phoenix. Fourteen compounds that were identified in the original study as being of the greatest concern for cancer risk for the general population and how these compounds could be controlled were studied. Again, cancer was the indicator of health effects, since other health effect indicators were more difficult to determine. Emission sources were identified in each of the five cities for these 14 compounds. Human exposure modeling was used to determine the degree of risk experienced by the people of these selected areas and the cancer incidents expected from this risk. The assumption was made that there would be a full implementation of the criteria of the pollutant programs.

The greatest incidents of cancer are associated with vehicles or area sources, such as numerous small pollution sources, like wood stoves. Point-source pollution sources, which are large, relatively identifiable sources of pollution, such as utilities and steel plants, appear to account for little of the total incidents.

All this data suggests that the health problems of air toxics may require a targeted strategy to make the best use of available resources. The data also suggest that additional focus must be made on small and nontraditional sources in order to get the greatest benefit for the health of the individuals. Additional attention needs to be given to wood smoke, vehicles, waste oil burning, and gasoline stations. The risk assessment here has helped to highlight the important areas where the greatest amount of good results will occur from the smallest amount of dollar expenditures.

Incineration at Sea

Industry generates more than 70 billion gal of hazardous waste each year in the United States. This waste must be safely managed through treatment, storage, and disposal. Because the EPA is restricting the use of management practices through permitting, regulation, and enforcement programs, there is a need to determine how best to get rid of this hazardous waste. Incineration is a technique that is being used to help destroy this hazardous waste. Public opinion, however, is strongly against the issuing of permits for new incinerators.

Between 1974 and 1982, the EPA issued permits for four series of burns conducted by the incinerator ship Vulcanus. Three of the burns were in the Gulf of Mexico and one in the Pacific Ocean. Public opposition to incineration at sea has intensified greatly. The EPA undertook a number of studies to determine whether incineration on land or the oceans is more desirable.

The EPA developed a risk assessment study that compared the human and environmental exposure most likely to occur from releases of land-based incineration vs. ocean incineration. The study used existing information and added new analysis of the emissions, transport, fate, and alternate effects. Both systems used land transportation, transfer and storage operations, and final incineration. The ocean system also included an ocean transportation step. For the purposes of the study, the waste was assumed to be a combination of 35% PCBs by weight and 50% ethylene dichloride (EDC) by weight. These two chemicals were used to simplify the waste stream. The analysis was to determine the statistically expected amount of pollutant released from accidental spills and air emissions. The transportation and handling accounted for less than 15% of the expected releases, while the incineration accounted for about 85%.

During land transportation, there were two types of potential losses. They were vehicular accidents and spills from the containers in route. Vehicular accidents were expected to occur on the average of once every 4–5 years and container failure once every 3–4 years. The analysis of transfer and storage of the waste was considered and three types of releases were anticipated: (1) spills when unloading the waste from tank trucks, (2) spills from equipment at the waste transfer and storage facilities, (3) and fugitive emissions in transfer and storage. Spills from transfer and storage components are infrequent events that are estimated to occur at about 0.50% per year and from the tank trucks, loading of waste to ship in the ocean system is approximately 0.002% per year. It was recognized that spills of this low quantity would likely be contained at the facility.

About 320 voyages of incineration ships have been made in the North Sea since 1972, and no casualties or spills have occurred. The spill rates from ships were based on worldwide historical data. It was estimated that the frequency of all spills for the Vulcanus would be about 1 per 12,000 operating years.

The study estimated and compared the possible human health and environmental effects due to incinerator releases and fugitive releases from the transfer and storage equipment.

The analysis of human health risk was based on the most exposed individual, who lives at the location, and the persons at the highest overall risk due to air concentrations from the incinerator stack and transfer storage facility. This risk is based on the cancer potential for an individual who has 70 years of continuous exposure. The risk on land-based incineration was 3 chances in 100,000, whereas the risk on ocean burning was in a range of 1 in 1,000,000 to 6 in 10,000,000. The relative risk on land for incineration of PCB waste is about 40 times more than on the ocean. For EDC waste, the ratio of land/ocean risk is about 30:1.

The conclusions of the incineration study were as follows: (1) incineration, whether on land or at sea, is a valuable and environmentally sound option; (2)

there is no clear preference for ocean or land incineration as it relates to human health risk and the environment; (3) future demands will significantly exceed the capacity for disposal of hazardous waste; (4) continuing research is needed to improve current knowledge of combustion processes and effects; (5) the EPA needs to improve its public communications effort in the area of hazardous waste management.

Municipal Sludge Reuse and Disposal

Sludge management is an intrical part of municipal sewage disposal. A determination was made of the cost effectiveness of the disposal of sludge through various media. Municipalities generate about $6\frac{1}{2}$ million dry tons of wastewater sludge a year. Sludge production is expected to double to approximately 13 million dry tons a year by the year 2000.

The five major sludge use/disposal options currently available are land application, distribution and sale of sludge products, landfill, incineration, and ocean disposal.

Municipal sewage sludge contains over 200 different substances, such as toxic metals, organic chemicals, and pathogenic organisms.

There are many regulations controlling the disposal of sludge. However, the EPA, in order to achieve better results, is consolidating existing sludge management authorities under the Clean Water Act. The Sludge Policy requires a consideration of risks, benefits, and costs of all sludge use and disposal practices. Since there is no methodology available to evaluate and compare the various impacts of sludge disposal, the EPA established the Integrated Environmental Management Division (IEMD) to conduct a major study of human health risks and costs related to the different disposal options used for sludge. The purpose of this approach is to identify high-risk areas and to develop a profile of the disposal options that provide a cost-effective way to reduce these risks. The high-risks disposal options include the exposure route, contaminants, and wastewater treatment plant types.

The IEMD model has four key components. They are (1) model plants, (2) risk profiles, (3) cost, and (4) cost-effectiveness analysis. There are three model plant sizes, each with one or two different types of sludge. These sludge types vary by the amount of industrial liquid waste added to the plants. The risk profiles are based on the four exposure routes of inhalation, groundwater, surface water, and ingestion through food, as they relate to potential carcinogenicity and renal effects. Costs have been estimated for each of the disposal options associated with each of the model plants. Cost-effectiveness analysis are made on a dollar basis for switching from one disposal option to another. The effectiveness of the reduction in the potential for disease is determined. Three major results have come from the IEMD analysis of sludge disposal. (1) There are aspects of sludge management that present very little risk and, therefore, provide very large gains if these aspects could be regulated properly. (2) Some options may be better, with lower cost and lower risk, while others may be worse, with some cost reduction but an increase in risk. (3) With some refinement and the use of site-specific data, the model may be used as an effective decision-making tool at the regional or local level.

The EPA is attempting to use its very best judgement based on scientific findings and projection of data to develop good risk management programs. Considerable additional research will be needed in many areas in order to ultimately reach a proper management program that will reduce the risk to the health of people in a cost-effective manner.

Pollution Prevention

Pollution prevention is a variety of techniques with a single objective — to achieve the most efficient use of resources to reduce or eliminate waste. Pollution prevention reduces or eliminates the generation of pollutants or waste, minimizes or eliminates the use of toxic materials in manufacturing, substitutes less harmful materials for toxic ones, reduces the chance of moving pollutants to a different media (example, air to water), and maximizes the efficient use of resources. Pollution prevention starts in the public sector with changes in a variety of public departments and their operations. It then includes the private sector and people at large. Better management of transportation saves fuel and reduces emissions. This is an example of an effective program. The use of calcium magnesium acetate for sodium chloride as a road salt reduces well contamination. Energy efficiency reduces the need for fuel to produce energy and reduces pollutants which come from burning hydrocarbons.

ENVIRONMENTAL HEALTH PROBLEMS AND THE LAW

The environmental health practitioner works for a unit of the local, state, or federal government, or, if working for industry, is constantly dealing with various levels of government. (For a detailed discussion on the functions of government, the reader is encouraged to refer to one of the many publications on local and state government.) For the purposes of this book, a brief discussion on federal-state relationships, legislative procedure, pressure groups, local government, and state and local finances is included.

The governmental system in a democratic society performs two distinct functions. It provides a foundation for debate on issues and a vehicle for the solution of problems. It also provides a service and regulatory function, since the individual in a complex society cannot provide for all of his or her needs. These needs include adequate police, fire, and health protection. For the society to operate properly, the individual or group must adhere to rules and regulations formulated and enforced by the elected government.

According to the U.S. Constitution and Amendments thereto, federal powers include the control of interstate and foreign commerce, conduct of foreign policy, and national defense. The federal government cannot levy direct taxes, other than income taxes, except in proportion to the population of the states. The federal government cannot abridge civil rights. The state is sovereign. The powers not delegated to the federal government or prohibited by the constitution are reserved for the states or the people. The state powers include the administration of elections, the establishment and operation of local government, education, intrastate commerce, the creation of corporations, police force, and the promotion of health, safety, and welfare. Although the federal government cannot dictate organization and administration of programs, or the establishment of policy, it has gained some control via its use of funds. In order for a state to receive federal funds, it must adhere to certain requirements established by the federal government for the use of these funds. These requirements include preparation and submission of plans, approval of plans by a central federal agency, the establishment of necessary state agencies, provision for matching state funds, and the supervision and auditing of state programs by the appropriate federal agencies.

Public policy is a result of the interaction of groups and individuals. The formulated policy is not necessarily ideal for the public. Major pressure groups — including business, industry, agriculture, labor, medicine, and religion, try to influence the legislature to pass laws beneficial to their own self-interest. These groups also pressure administrative agencies to make or change decisions on policies that are not in the best interest of the public. These groups compete with each other for public support. They use sizable sums of money to influence opinions on legislation. Citizens groups properly educated and stimulated by trained professional public health workers could use the techniques of the pressure groups to establish public policy that would be beneficial to the citizens and would improve the environment.

The state is divided into local government, including counties, townships, villages, cities, and boroughs. Local government functions may include all the functions of the state government, such as the powers of taxation, establishment of budgets, licensing, and the administration of health, safety, and environmental programs. The local government is expected to deliver quality service and to protect the public. It should prevent the duplication of efforts, budgets, and facilities of the state government. The environmentalist must recognize that although concern for the environment is foremost in his or her mind, the county commissioners or township supervisors may be more concerned with other programs, such as road building, recreational facilities, police and fire service, etc. The environmentalist has to compete for budget monies with each of the other operating departments.

State and local governments obtain their funds for operation from taxes levied on sales, automobile licensing, gasoline, corporations, personal income, alcohol and tobacco, gross receipts, property, and establishment licensing. In many cases, tax rates become oppressive and yet income derived from taxes is inadequate to properly finance the necessary programs for the community. Then it becomes necessary for the federal government to supply funds; in this way, the federal government exercises control in some states even at the township level.

Law and Public Health

Public health policies will be discussed briefly, since the environmental health practitioner should have some understanding of the law to function in his or her capacity as a government official. (For more detailed discussion, many books are available describing current laws relating to the environment.)

Law is the rule of civil conduct prescribed by the supreme power in a state commanding what is right and forbidding what is wrong. Law should represent the community's desires or commands, apply to all members of the community, be backed by the full power of the government, and provide the administration of justice under the law for all people. The purpose of law is to protect by the regulation of human conduct of the individual from other individuals, groups, or the state, and vice versa. Statutory laws are basically legislative acts passed by a legislative body; common laws are established customs of the community; equity is a decision by a judge in a given situation where rules and regulations have been violated; and administrative laws are rules and regulations in a given area established by an agency authorized by the legislature. The value of administrative laws is that the rules and regulations established can be changed readily when new scientific data are available. If regulation were part of a statute, the legislature would have to go through the complex process of rewriting the act.

Public health law is that body of statutes, regulations, and precedents that protects and promotes individual and community health. Public health law is founded on the Preamble to the Constitution, which ordains that the government shall "... promote the general welfare," and Section 8, Article 1 of the Constitution, which "... provide(s) for the common defense and general welfare." The interpretation of these clauses and other clauses of the Constitution by the U.S. Supreme Court has established the legal basis for public health. The law of eminent domain empowers the state with the authority to seize, appropriate, or limit the use of property in the best interests of the community. This power is reflected in the establishment of zoning and land use regulations that are so essential to the preservation of a good environment.

Nuisance laws, originating in the Middle Ages, are used repeatedly in environmental health work. Basically they state that the use of private property is unrestricted only as long as it does not injure another person or his or her property. If an injury occurs, a nuisance exists. Large numbers of public health officials use nuisance laws to eliminate problems caused by sewage, solid waste, air pollution, and insects and rodents.

The public health law owes its effectiveness to the police power of the state. In times of great stress, such as severe fires, floods, and outbreaks of diseases, the private property of an individual might be summarily appropriated, used, or even destroyed if the ultimate relief, protection, or safety of the community demands that such action be taken. The legislature can and does delegate police power to an administrative agency for use in the event of an emergency. If a public official fails to use the delegated police power, he or she is guilty of nonfeasance of office. The use of this police power is determined by the chief administrative office, usually the State Secretary of Health, in consultation with the governor of the state.

Administrative law is that series of rules, regulations, and standards needed to implement statutes. The development and use of sound rules, regulations, and standards based on strong scientific data puts the burden of proof on the defendant, who has to show that the rules have not been fairly applied, rather than that they have limited value. Through judicial presumption, the judge presumes that the laws are correct, because they are based on sound scientific criteria or on the judgment of a group of experts. The judge only decides if the law is applied fairly.

Licensing is an important means of control and enforcement of environmental standards. If an individual or business does not meet the standards, a license is denied or revoked. An individual operating without a license is subjected to severe penalties.

Environmental health practitioners must be conscious of those actions that may occur in the enforcement of environmental health laws. These actions are (1) misfeasance, which is the performance of a lawful action in an illegal or improper manner; (2) malfeasance, which is wrongful conduct by a public official; and (3) nonfeasance, which is an omission in doing what should be done in an official action.

Environmental Law

For historical purposes, the Rivers and Harbors Act of 1899 was passed by Congress to prevent the illegal discharge or deposit of refuse or sewage into navigable waters of the United States. This law was used before 1972 to penalize some corporations for polluting the water.

The National Environmental Policy Act of 1969 signed into law on January 1, 1970 (Public Law 91-190), amended by PL 94-52 and PL 94-83, in 1975 set forth the continuing responsibility of the federal government to

1. Fulfill the responsibilities of each generation as trustee of the environment for succeeding generations.
2. Assure safe, healthful, productive, aesthetically, and culturally pleasing surroundings for all Americans.
3. Attain the widest range of beneficial uses of the environment without degradation, risk to health or safety, or other undesirable and unintended consequences.
4. Preserve important historic, cultural, and natural aspects of our national heritage, and maintain, wherever possible, an environment that supports diversity and variety of individual choice.
5. Achieve a balance between population and resource use that will permit high standards of living and a wide sharing of life's amenities.
6. Enhance the quality of renewable resources and approach the maximum attainable recycling of depletable resources.

The National Environmental Policy Act (NEPA) also states that all federal agencies shall

1. Utilize a systematic, interdisciplinary approach ensuring the integrated use of the natural and social sciences and the environmental design arts in planning and in decision making when planning may be an impact on the human environment.
2. Identify and develop methods and procedures, in consultation with the Council on Environmental Quality, ensuring that other environmental amenities and values be given appropriate consideration in decision making, along with economic and technical considerations.
3. Include in every recommendation or report on proposals for legislation and other major federal actions significantly affecting the quality of the human environment, a detailed statement by the responsible official on
   a.   the environmental impact of the proposed action
   b.   any adverse unavoidable environmental effects should the proposal be implemented
   c.   alternatives to the proposed action
   d.   relationship between local short-term uses of the environment and the maintenance and enhancement of long-term productivity
   e.   any irreversible and irretrievable commitments of resources involved in the proposed action should it be implemented
4. Study, develop, and describe appropriate alternatives to recommended courses of action in any proposal involving unresolved conflicts, as well as alternative uses of available resources.
5. Recognize the worldwide and long-range character of environmental problems and, where consistent with the foreign policy of the United States, lend appropriate support to initiatives, resolutions, and programs designed to maximize international cooperation in anticipating and preventing a decline in the quality of the world environment.
6. Make available to states, counties, municipalities, institutions, and individuals advice and information useful in restoring, maintaining, and enhancing the quality of the environment.
7. Initiate and utilize ecological information in the planning and development of resource-oriented projects.
8. Establish the Council on Environmental Quality, whose functions are as follows:
   a.   to assist and advise the President in the preparation of the environmental quality report, which will be submitted to the Congress starting in 1970 (this report is an excellent digest of various environmental health issues)
   b.   to gather timely and authoritative information concerning the conditions and trends in the quality of the environment, currently and for the future
   c.   to review and appraise the various programs and activities of the federal government
   d.   to develop and recommend to the President, national policies to foster and promote the improvement of environmental quality

e. to conduct investigations, studies, surveys, research, and analyses relating to ecological systems and environmental quality

f. to document and define changes in the natural environment

g. to report at least once every year to the President on the state and condition of the environment

h. to make and furnish studies, reports, and recommendations with respect to matters related to policy and legislation

Since 1970 when the NEPA Act was passed and a Council on Environmental Quality was established, Congress has passed a large number of laws and amendments to the laws, related to a variety of environmental issues that potentially affect the health of people and affect the environment and/or ecosystems. These laws include the Clean Air Act (CAA), the Comprehensive Environmental Response, Compensation, and Liability Act (CERCLA), the Consumer Product Safety Act (CPSA), the Emergency Planning and Community Right-to-Know-Act, the Energy Supply and Environmental Coordination Act, the Federal Insecticide, Fungicide, and Rodenticide Act (FIFRA), the Food, Drug, and Cosmetic Act Amendments (FDCA), the Occupational Safety and Health Act (OSHA), the Noise Control Act, the Radon Gas and Indoor Air Quality Research Act, the Resource Conservation Recovery Act (RCRA), the Super Fund Amendments and Reauthorization Act (SARA), and the Toxic Substances Control Act (TSCA).

There have been a number of laws related to water and water pollution that have been enacted since the original Rivers and Harbor Acts of 1899. These laws include the Clean Water Amendments of the Federal Water Pollution Control Act (CWA), the Coastal Zone Management Act, the Deep Water Port Act, the Marine Protection, Research, and Sanctuaries Act, the National Ocean Pollution Planning Act, the Outer Continental Shelf Lands Act, the Port and Tanker Safety Act, the Shore Protection Act, the Safe Drinking Water Act (SDWA), the Water Resources Planning Act, and the Water Resources Research Act.

The major amendments to the Clean Air Act were in 1977, updated in 1983, and again a major amendment was passed in 1990. There had been a substantial debate for the last 8 years concerning how to change the law. The congressional findings included: (1) that the predominant part of the nation's population was located in rapidly expanding metropolitan and other urban areas that generally cross the boundary lines of local jurisdictions and expand into multiple states; (2) that the growth and the amount of complexity of air pollution brought about by urbanization, industrial development, and increased use of motor vehicles has resulted in growing dangers to the public health and welfare, crops, property, and air and ground transportation; and (3) that the prevention and control of air pollution at its source was the primary responsibility of the state and local governments and that federal financial assistance and leadership was necessary for the improvement of air quality.

The law provides an elaborate federal-state scheme for controlling conventional pollutants, such as ozone and carbon monoxide. The 1990 amendments create tighter controls on tailpipe exhaust, reduction of acid rain, reduction of nitrogen oxides, reduction of air toxics, which may be carcinogenic, and protection of the ozone layer by phasing out chlorofluorocarbons, carbon tetrachloride, methylchloroform, and hydrochlorofluorocarbons.

The Comprehensive Environmental Response Compensation and Liability Act of 1980 (CERCLA), also known as SuperFund, was designed to handle the problems of cleaning up the existing hazardous waste sites in the United States. The act, which was originally passed in 1980, was updated numerous times until 1988. The hazardous waste problems range from spills that need immediate attention to hazardous waste dumps that are leaking into the environment and posing long-term health and environmental hazards.

The Emergency Planning and Community Right-to-Know Act of 1986 established state emergency response commissions, emergency planning districts, emergency planning committees, and comprehensive emergency plans. A list of extremely hazardous substances has been prepared and published.

The Energy Supply and Environmental Act of 1974 was updated in 1978. The purposes of this act were to provide for a way to assist in meeting the needs of the United States for fuels in a consistent, practical manner, and to protect and improve the environment. It allowed for coal conversion or coal derivatives to be used in place of oil in power plants. This, of course, can contribute to greater levels of air pollution.

The Consumer Product Safety Commission Act (CPSA) of 1970, updated in 1984, established the Consumer Product Safety Commission as an independent regulatory agency. The CPSA gives the commission the power to regulate consumer products and to oppose unreasonable risks of injury or illness. They also regulate consumer products, except for foods, drugs, pesticides, tobacco and tobacco products, motor vehicles, aircraft and aircraft equipment, and boats and boat accessories. The law authorized the commission to publish consumer product safety standards in order to reduce the level of unreasonable risks. The commission has recalled hair dryers containing asbestos because of potential hazards. They also may regulate carcinogens.

The Federal Insecticide, Fungicide, and Rodenticide Act (FIFRA) was originally passed by Congress in 1947 and was last updated in 1988. It provides for the registration of new pesticides; the review, cancellation, and suspension of registered pesticides; and the reregistration of pesticides. It is concerned with the production, storage, transportation, use, and disposal of pesticides. It also includes areas of research and monitoring.

The Food, Drug and Cosmetic Act, which is an amendment to the Food and Drug Act of 1906, and the Federal Food, Drug, and Cosmetic Act of 1938, has been further amended numerous times. It was the first federal statute to regulate food safety. The amendment of 1938 established the general outlines for the authority of the FDA. Food additives, food contaminants, naturally occurring parts of food or color additives to food, drugs, or cosmetics, as well as potential carcinogens, have been regulated by various parts of the law. In 1958, the Food Additives Amendment, better known as the Delaney clause, stated "... that known additives shall be deemed to be unsafe if found to induce cancer when ingested by man and/or animal, or if it is found, after tests which are appropriate for the evaluation of food additives, to induce cancer in man or animal ... ."

The Noise Control Act of 1972 was amended last in 1978. The findings of Congress were as follows: (1) that inadequately controlled noise presents a growing danger to the health and welfare of the population of the United States; (2) that the major sources of noise include transportation vehicles and equipment, machinery, appliances, and other products used in commerce; (3) that, although the primary responsibility for control of noise belongs to the state and local governments, federal action is needed to deal with major noise sources in commerce.

Since aircraft contribute considerable amounts of noise to the environment, two laws were passed and further amended to reduce these noise sources. The first law was entitled "An Act to Require Aircraft Noise Abatement Regulation." This law was signed in 1968 and amended several times until it became the Quiet Communities Act of 1978. The Congress decided to control and abate aircraft noise and sonic booms. The second law was the Aviation Safety and Noise Abatement Act of 1979, enacted in 1980 and amended in 1982. The purpose of the act was to provide assistance to airport operators to prepare and carry out noise compatibility programs, to provide assistance to assure continued safety in aviation, and for other purposes. It helped to establish a single system of measuring noise, which was reliable, and also a single system for determining the exposure of individuals to noise at airports.

The Occupational Safety and Health Act of 1970 established the Occupational Safety and Health Administration or OSHA, and the National Institute for Occupational Safety and Health (NIOSH). OSHA sets and enforces regulations to control occupational health and safety hazards, including exposure to carcinogens. NIOSH is responsible for research; for various evaluations, publications, training; and for the regulation of carcinogens by supporting epidemiological research and recommending changes in health standards to OSHA.

The Occupational Health and Safety Act provides three mechanisms by law for setting standards to protect employees from hazardous substances. The act initially authorized OSHA to adopt the health and safety standards already established by federal agencies or adopted as national consensus standards. This authority was for the first 2 years in 1972 and 1973. It also authorizes OSHA to issue emergency temporary standards (ETS) that require employers to take immediate steps to reduce workplace hazards. The ETS may be issued by OSHA when they determine employees are exposed to grave danger and that the emergency standard is necessary. Although the public has not had a chance to comment on a standard, because of the nature of the potential hazard, it must be enforced.

However, the final standard must be issued within 6 months of the emergency standard. The third way that OSHA sets standards is to issue new permanent exposure standards and to modify or revoke existing ones. However, the informal rulemaking that goes along with modifying or revoking existing standards is subject to court review. OSHA's approach to rulemaking can result in requirements in monitoring and medical surveillance, workplace procedures and practices, personal protective equipment, engineering controls, training, recordkeeping, and new or modified permissible exposure limits (PELs). The permissible exposure limits are the maximum concentration of toxic substances allowed in the workplace air.

Through 1988, NIOSH has published 35 occupational safety and health guidelines with technical information about chemical hazards for workers. In addition, they publish the criteria documents (CD), alerts, current intelligence bulletins (CIB), health and safety guides (HSG), symposium or conference proceedings, NIOSH administrative and management reports, scientific investigations, data compilations, and other worker-related booklets. The CDs are recommended occupational safety and health standards for the Department of Labor. Usually, as part of the recommended standard is a recommended exposure limit (REL). The CIBs relate important public health information and recommend protective measures to industry, labor, public interest groups, and the academic world. The HSGs provide basic information for employers and employees to help them have a safe and healthful work environment.

The Radon Gas and Indoor Air Quality Research Act of 1986 included the following findings by Congress: (1) high levels of radon gas pose a serious health threat in structures in certain areas of the country; (2) certain scientific studies suggest that exposure to radon, including naturally occurring radon and indoor air pollutants, may cause a public health risk; (3) existing federal radon and indoor air pollution research programs are fragmented and underfunded; (4) there is a need for adequate information concerning exposure to radon and indoor air pollutants; and (5) this need should be met by appropriate federal agencies.

The Resource Conservation and Recovery Act of 1976 (RCRA) was updated through 1988. This law replaced the previous Solid Waste Disposal Act. The Used Oil Recycling Act of 1980 amended the RCRA Act and then was incorporated into the main text of the act.

The RCRA provides for regulating the treatment, transportation, and disposal of hazardous waste. Hazardous waste is defined as solid waste that may cause death or serious disease, or may present a substantial hazard to human health or the environment if it is improperly treated, stored, transported, or disposed. Solid waste includes solid, liquid, semisolid, or contained gaseous materials from a variety of industrial and commercial processes. This definition excludes solid or dissolved materials found in domestic sewage or related to irrigation, industrial discharges subject to the Clean Water Act, or mining wastes. RCRA requires that the EPA develop and issue criteria for identifying the characteristics of hazardous wastes. Defining characteristics of hazardous wastes are (1) it poses a present or potential hazard to human health and environment when it is improperly managed; and (2) it can be measured by a quick, available, standardized test method, or it can be reasonably detected by generators of solid wastes through their knowledge of their wastes: ignitability, corrosivity, reactivity, and extraction procedure toxicity. RCRA also includes the regulation of underground storage tanks that may be used for a variety of storage processes. Further, it includes a program in medical waste tracking.

The Superfund Amendments and Reauthorization Act of 1986 (SARA) of CERCLA established an Alternative or Innovative Treatment Technology Research and Demonstration Program. Its function is to promote the development, demonstration, and use of new or innovative treatment technologies and to demonstrate and evaluate new, innovative measurement and monitoring technologies. The SARA amendments require that an information dissemination program be established along with the research efforts. The three types of technologies include (1) available alternative technology, such as incineration; (2) innovative alternative technology, which is any fully developed technology for which cost or performance information is incomplete and the technology needs full-scale field testing; and (3) emerging alternative technologies that are in their early stage of development and the research has not yet fully passed the laboratory or pilot testing phase.

The Toxic Substances Control Act (TSCA) was enacted in 1976 and amended through 1986. TSCA was enacted by Congress to allow for the regulation of chemicals in commerce, as well as before they even entered commerce. This policy includes (1) that chemical manufacturers and processors are responsible for developing data about the health and environmental effect of their chemicals; (2) that the government regulate chemical substances that pose an unreasonable risk of injury to health or the environment and act promptly on substances that pose imminent hazards; and (3) that regulatory efforts should not unduly hinder industrial innovation. TSCA is directed at hazardous substances, wherever they occur. They do not have to be within a special environment. As long as they may cause a danger to the public — including a significant risk of cancer, genetic mutations, birth defects, or other potential serious hazards — they may be restricted or even banned by the EPA. TSCA permits the EPA to regulate new or existing toxic substances by requiring the testing of new or existing chemicals and by requiring their restriction in production and use, or even outright banning of substances that pose an unreasonable risk to health or the environment. The Asbestos Emergency Response Program is part of TSCA.

There are 11 separate acts of Congress related to water quality and water pollution. These range from the Clean Water Act to proper management of the oceans near the continental United States and the improvement of water quality.

The Federal Water Pollution Control Act was amended by the Clean Water Act of 1977 and was amended numerous times through 1988. The initial act was passed in 1948 and was then referred to as the Federal Water Pollution Control Act. In 1972, Congress set the goals of achieving fishable, swimmable waters by 1983 and prohibiting the discharge of toxic pollutants in toxic amounts by 1985. In the 1977 amendments, Congress endorsed a new means for regulating toxic pollutants, which had been developed to settle a lawsuit between environmental organizations and the EPA. In 1987, Congress continued its emphasis on the control of toxic pollutants. Although the Clean Water Act Provisions are less directly related to human health than the provisions of the Safe Drinking Water Act, it still aims at regulating human exposure to carcinogens and other toxic materials. An important part of the CWA is the National Pollution Discharge Elimination System (NPDES), which creates permits for direct discharges into the waters. It is lawful to discharge a pollutant only if the discharge is in compliance with the NPDES permit, which has to be issued by the EPA or by states whose permit programs are approved by the EPA. To obtain a permit, a facility must have the following information submitted: (1) a list of pollutants that must be regulated according to federal or state law, along with a permissible amount of each pollutant that may be discharged per unit of time; (2) monitoring requirements and schedules for implementing the pollution concentration requirements; and (3) special conditions regarding the pollutants that the agencies feel should be imposed on the polluters. These may include additional testing and procedures for spills of pollutants into the water.

The Coastal Zone Management Act of 1972 was updated through 1986. Congress determined that it was in the national interest to have effective management, beneficial use, protection, and development of the coastal zone. The coastal zone included the fish, shellfish, and other living marine resources and wildlife found in this ecologically fragile area. It considered the increasing and competing demands of economic development, population growth, and harvesting of the fish and shellfish.

The Deep Water Port Act of 1974 was amended through 1984. Congress decided to authorize and regulate the location, ownership, construction, and operation of deep-water ports in waters beyond the territorial limits of the United States. The act provides for the protection of the marine and coastal environment beyond these territorial limits.

The Marine Protection Research and Sanctuaries Act of 1972 was amended through 1988. Congress determined that unregulated dumping of material into the ocean waters endangered human health, welfare, and amenities, and the marine environment, ecological systems, and economic potential. It decided that it was the policy of the United States to regulate the dumping of all types of materials into ocean waters and to prevent or strictly limit the dumping of any material into the ocean waters that would adversely effect health, welfare, and the ecosystem. It regulated the transportation of the material and dumping at sea.

The National Ocean Pollution Planning Act of 1978 was updated through 1988. Congress determined that the activities of people in the marine environment can have a profound short-term and long-term impact on such an environment and can greatly affect ocean and coastal resources. It stated that a comprehensive federal plan should be developed for ocean pollution research and for the development and monitoring of the material that had been dumped, the fate of the material, and the effects of the pollutants on the marine environment.

The outer Continental Shelf Lands Act of 1953 was updated through 1988. The *Outer Continental Shelf* means all the submerged lands lying seaward and outside of the areas of lands beneath navigable waters. This includes the subsoil and seabed that is attached to the United States. The Outer Continental Shelf is of vital importance to the United States, since it is a national resource reserve of a variety of potential minerals, oil, etc. Congress decided that since the exploration, development, and production of the minerals of the Outer Continental Shelf will have significant impact on the coastal and noncoastal areas of the United States and on other affected states, it was in the national interest that this exploration be controlled by the federal government. The Secretary of the Interior, under this act, has the authority to authorize the leasing of the Outer Continental Shelf and to enforce safety, environmental, and conservation laws and regulations, and to formulate and promulgate such regulations, as necessary, to prevent problems from occurring.

The Port and Tanker Safety Act of 1978 was updated through 1986. Congress decided that the navigation and vessel safety and protection of the marine environment were matters of major national importance. It said that increased vessel traffic in the nation's ports and waterways created a substantial hazard to life, property, and the marine environment, and that there was an increased need for supervision of the vessel and port operations. It was particularly interested in the handling of dangerous articles and substances on, or immediately adjacent to, the navigable waters of the United States. It stated that advanced planning is critical in determining proper, adequate, and protective matters for the Nation's ports and waterways, and the marine environment.

The Shore Protection Act was passed in 1988. Congress stated that a vessel may not transport municipal or commercial wastes in coastal waters without a permit from the Secretary of Transportation and without displaying a number or other marking on the vessel, as prescribed by the Secretary. The permit included the name, address, and telephone number of the vessel owner and operator, its transport capacity, and its history of cargo transportation during the previous year, including wastes. The Secretary of Transportation had the right to enforce regulations concerning loading, securing, offloading, and cleanup. The secretary can also refuse the permit.

The Safe Drinking Water Act (SDWA) of 1974 was updated through 1988. It was originally passed to ensure a safe drinking water supply. The Clean Water Act was designed to control water pollution, but it did not provide authority to regulate polluted water discharged into nonnavigable waters, such as groundwater, which often is a source of drinking water. The SDWA is used primarily to regulate water provided by public water systems, and it contains several provisions that may be used to regulate hazardous substances, including carcinogens in drinking water. The SDWA is more directly concerned in protecting human health than the CWA. Under the SDWA, the EPA regulates contaminants that may have an adverse effect on the health of people. It then establishes the steps that the agency must go through, over time, to protect drinking water. The EPA published national interim drinking regulations in 1975. The regulations were to protect the health of the people to the extent feasible using current technology treatment techniques and other means that the administrator determined were generally available. Congress also required that the EPA request a National Academy of Sciences study to determine the potential adverse health effects of contaminants in the water and to help establish recommended maximum contaminant levels (RMCLs). These RMCLs were the recommended maximum level to be set for contaminants to prevent the occurrence of any known or anticipated adverse effect. It had to include an adequate margin of safety, unless there was no safe threshold. In that case, the recommended maximum contaminant level could be set at 0. The RMCLs are not enforceable health goals but are used as guidelines for establishing enforceable drinking water standards. The MCLs or maximum contaminant levels were to be as close to the RMCLs as feasible. The enforcement of the MCLs rested with the states, whereas EPA sets the MCLs (see 1986 SDWA Amendments).

The Water Resources Planning Act of 1965 was updated through 1988. The act stated that in order to meet the rapidly expanding demand for water throughout the nation, it is declared to be the policy of the Congress to encourage the conservation, development, and utilization of water and related land resources of the United States on a comprehensive and coordinated basis by the federal government, states, counties, and private enterprises with the cooperation of all of the various agencies and governments, individuals, and businesses involved.

The Water Resources Act was passed in 1984. The Congress declared that the existence of an adequate supply of water of good quality for the production of materials and energy for the nation's needs and for the efficient use of the nation's energy and water resources is essential to national economic stability and growth and to the well-being of the people. It also stated that the management of water resources is closely related to maintaining environmental quality and social well-being, and that there should be a continuing national investment in water and related research in technology that was commensurate with growing national needs. This research and development of technology should include the development of a technology for the conversion of saline and other impaired waters to a quality usable for municipal, industrial, agricultural, recreational, and other beneficial uses.

The "Pollution Prevention Act" of 1990 identified pollution prevention as a national objective and established the waste management techniques of prevention, recycling, treatment, and as a last resort, disposal. Source reduction is the technique of choice for protecting the quality of the environment and reducing potential disease and injury. Also, it was recognized that the removal of a pollutant from one media, such as sludge from air or water, could cause another problem in land disposal. Non-point source pollution was also emphasized.

How a Bill Becomes a Law

Although a passage of bills in the state legislatures and the federal government varies somewhat, they all follow an approximate route from the introduction of the bill to the final passage by both houses of the state government or the federal government. They are then signed or vetoed by the governor of the particular state or by the President of the United States. In order to get a better understanding of this process of how a bill becomes a law, the procedure used in the State of Indiana will now be discussed. Recognize that this path varies somewhat with each of the kinds of government throughout the country. For a bill to become a law in Indiana, it must follow a prescribed path with carefully planned steps. The path has many detours, lost turns, stumbling points, barriers, and frustrations. The reason for all these problems is because of the unique system of checks and balances that we find in government in the United States. The process in Indiana is as follows: (1) a representative or senator decides to introduce legislation; (2) a legal specialist draws up a bill; (3) the bill is introduced and assigned to a committee; (4) the committee approves the bill or amends the bill and approves it or rejects it; (5) the house or senate votes to accept the committee recommendation; (6) the bill is printed; (7) the bill remains on each representative's desk or senator's desk for 24 hours and then it is placed on the House or Senate calendar for a second reading; (8) the bill is ordered "engrossed" or reprinted to show amendments; (9) the bill passes a second reading. Amendments may be added at this time; (10) the third reading of the bill occurs; (11) speeches are made for or against the bill; (12) the vote is taken (in Indiana, the constitutional majority needed for either approval or rejection is 51 votes in the House or 26 votes in the Senate); (13) the bill is now delivered to the opposite house of the state or federal government, where either a senator or representative has promised to sponsor it; (14) the bill goes through a similar process in the opposite house, and it may also be amended here; (15) if the bill is amended, it must be approved by the House-Senate Conference Committee; (16) it is then submitted to the house and senate for passage; (17) when passed, it is signed by the House Speaker and the President of the Senate; (18) the Attorney General checks its constitutionality; (19) it is signed by the governor or vetoed and sent to the Secretary of State; (20) if it is signed, it becomes an enrolled act and is printed and bound in the "Act of Indiana"; (21) if it is vetoed, it goes back to the house and senate, where a constitutional majority would be a two-thirds vote in order to overrule the veto of the governor; and (22) it is either finally overruled or, if it has been enrolled, it is an enrolled act; it now takes effect when distributed among the circuit court clerks of all the counties of the state (Figure 1.16).

As stated previously, although the procedures in the federal government and the other states vary, the approach to becoming a law is about the same.

ENVIRONMENTAL IMPACT STATEMENTS

The National Environmental Policy Act (NEPA), which established the council on environmental quality (CEQ), requires each federal agency to prepare a statement of environmental impact in advance of each major action, recommendation, or report on legislation that may significantly affect the quality of the human environment. An environmental impact statement is the heart of a federal administrative process designed to ensure achievement of national environmental goals. Each statement must access in detail the potential environmental impact of a proposed action. All federal agencies are required to prepare statements for matters under their jurisdiction. The impact statement is called an EIS. The purpose of the statement is to disclose the environmental consequences of the proposed action and to be continually conscious of environmental considerations.

The actions that the statement cover must be major and environmentally significant. The actions may be a precedent for much larger actions, which may have considerable environmental impact, or they may alter the future course of other environmental actions.

Each environmental impact statement must include (1) a detailed description of the proposed action, including information and technical data adequate to permit a careful assessment of environmental impact; (2) discussion of the probable impact on the environment including any impact on ecological systems and any direct or indirect consequences that may result from the action; (3) any adverse environmental effects that cannot be avoided; (4) alternatives to the proposed action that might avoid some or all of the adverse environmental effects, including an analysis of costs and environmental impacts of these alternatives; (5) an assessment of the cumulative, long-term effects of the proposed action, including its relationship to short-term use of the environment vs. the environment's long-term productivity; (6) any irreversible or irretrievable commitment of the resources that might occur from the action or that would curtail beneficial use of the environment; and (7) a final impact statement must include a discussion of problems and objections raised by other federal, state, and local agencies, private organizations, and individuals during the draft statement's review process.

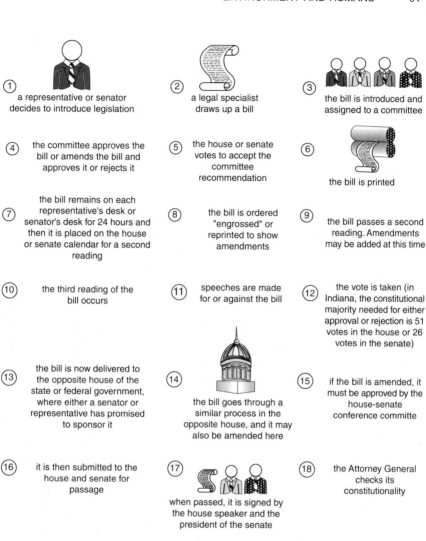

1. a representative or senator decides to introduce legislation

2. a legal specialist draws up a bill

3. the bill is introduced and assigned to a committee

4. the committee approves the bill or amends the bill and approves it or rejects it

5. the house or senate votes to accept the committee recommendation

6. the bill is printed

7. the bill remains on each representative's desk or senator's desk for 24 hours and then it is placed on the house or senate calendar for a second reading

8. the bill is ordered "engrossed" or reprinted to show amendments

9. the bill passes a second reading. Amendments may be added at this time

10. the third reading of the bill occurs

11. speeches are made for or against the bill

12. the vote is taken (in Indiana, the constitutional majority needed for either approval or rejection is 51 votes in the house or 26 votes in the senate)

13. the bill is now delivered to the opposite house of the state or federal government, where either a senator or representative has promised to sponsor it

14. the bill goes through a similar process in the opposite house, and it may also be amended here

15. if the bill is amended, it must be approved by the house-senate conference committee

16. it is then submitted to the house and senate for passage

17. when passed, it is signed by the house speaker and the president of the senate

18. the Attorney General checks its constitutionality

19. it is signed by the governor or vetoed and sent to the Secretary of State

20. if it is signed , it becomes an enrolled act and is printed and bound in the "Act of Indiana"

21. if it is vetoed, it goes back to the house and senate, where a constitutional majority would be a two-thirds vote in order to overrule the veto of the governor;

22. and it is finally overruled or, if it has been enrolled, it is an enrolled act; it now takes effect when distributed among the circuit court clerks of all the counties of the state

Figure 1.16   How a bill becomes a law in Indiana

A draft statement must be prepared and circulated for comment at least 90 days before the proposed action. A final statement must be made public at least 30 days before the proposed action. Any agency unable to meet these requirements must consult with the CEQ.

NEPA requires each federal agency to consult with and obtain the comments of any other agency — state, local, as well as federal — that has jurisdiction by law or special expertise related to the environmental impact.

In its guidelines for preparing statements, CEQ lists the agencies that must be consulted in the following areas: air quality and air pollution control; weather modification; environmental aspects of energy generation and transmission; toxic materials; pesticides; transportation and handling of hazardous materials; coastal areas, including estuaries, waterfowl refuges, and beaches; historic and archaeological sites; floodplains and watersheds; mineral land reclamation; parks, forests, and outdoor recreational areas; noise control and abatement; chemical contamination of food products; food additives and food sanitation; microbiological contamination; radiation and radiological health; sanitation and waste systems; shellfish sanitation; transportation and air quality; transportation and water quality; congestion in urban areas; housing and building displacement; environmental effects with special impacts on low-income neighborhoods; rodent control; urban planning; water quality and water pollution control; marine pollution; river and canal regulation and stream channelization; and wildlife.

The guidelines also require that these statements be made available for public comment. Many individual agencies ask for such comments from interested parties and private organizations.

The impact statement procedure allows the public an opportunity to participate in federal decisions as they affect the human environment. The statements are announced in the *Federal Register*, and many agencies have other procedures to reach interested citizens.

A study of environmental impact statements for a 6-year period made by the CEQ indicated the following findings and recommendations: (1) the impact statement process can and should be more useful in agency planning and decision making; (2) agencies with major EIS responsibilities should support high-level, well-staffed offices charged with implementing NEPA and the EIS process effectively; (3) guidelines from CEQ are necessary to help agencies draw up EISs for broad federal programs or groups of projects; (4) procedures that agencies use to notify other federal, state, and local agencies and the public of important EIS actions need improvement; (5) to improve EIS review, agencies in consultation with CEQ should clearly define their expertise and jurisdiction for purposes of commenting on EISs; (6) the EIS process has contributed much to interagency coordination in environmental matters, but this coordination can be strengthened further; (7) the quality and content of EISs need continuing improvement; (8)

several special issues arising from the EIS process require attention and remedy (among these are procedures for impact statements and federal permit actions); and (9) the adequacy of public participation in the EIS process needs thorough study and evaluation.

The environmental impact statements and their resulting conclusions can and have been challenged in the courts. Each year, there are numerous cases in the courts concerning the statements. It is, therefore, necessary that the statements be completed in a thorough and comprehensive manner.

Cases have been filed in widely varying situations. Calvert Cliff's Coordinating Committee vs the NEC was an early landmark case that set the direction of federal agencies' NEPA responsibilities. The NEC was required to revise its activities to systematically analyze environmental impacts. Since then courts have handed down numerous decisions expanding federal agency responsibilities under NEPA. The court ruled in a case involving the Army Corp of Engineers that the findings by the Council on Environmental Quality indicated that inadequate considerations was given to the safety and water quality effects of a project being carried out by the Army Corp of Engineers. The court required the Corp to revise its impact statement and ordered the lower court to review this revised statement in light of CEQ's findings. Thus, for the first time, CEQ's findings as the principal overseer of the NEPA process were considered by a federal court.

Half of all of the environmental impact statements have been written about road-building actions undertaken by the Federal Highway Administration. The statements have resulted in significant planning changes. The second largest number of statements have been prepared for watershed protection and flood-control projects. Statements have also been prepared in coal development and in strip mining and its effect on existing use of land and water. As a result of the evaluation and the review process of environmental impact statements, several federally sponsored projects have been suspended or modified in recent years. Many states have followed the federal approach and have passed environmental impact statement laws.

THE ENVIRONMENTAL HEALTH FIELD

The environmental health field is composed of individuals whose efforts are directed toward controlling, preserving, or improving the environment in order that people may have optimum health, safety, comfort, and well-being now and in future generations. Given the complexity of the environment, a variety of disciplines and career categories have developed, including engineering, comprised of sanitary, mechanical, and chemical; education; science, comprised of generalists or specialists in biology, chemistry, medicine, etc.; environmental health, comprised of generalists or specialists in three basic types of categories: (1) sampling and analysis, routine inspections, and distribution of public information, (2) investigation consultation, planning, and education; and (3) supervision, administration planning, enforcement, and public relations.

The Environmental Health Practitioner

The environmental health practitioner is an applied scientist and educator who uses the knowledge and skills of the natural, behavioral, and environmental sciences to prevent disease and to promote human well-being.

The environmental health practitioner makes inspections; conducts special studies; samples air, water, soil, and food; reviews plans; acts as an educator, public relations officer, and community organizer; plans programs; acts as a consultant to civic groups, business, industry, and individuals; and enforces environmental and public health laws. The environmental health practitioner is involved in a multitude of program areas, including accident prevention, air pollution control, communicable diseases, environmental emergencies, food processing and food handling, outbreaks of food-borne infections and poisonings, hazardous substances control, housing, indoor environment, insect and rodent control, institutional environment, meat, milk, migrant labor camps, noise control, nuisance abatement, occupational health, planning, product safety, radiation control, recreational sanitation, public and private sewage, solid and hazardous waste management, swimming pool sanitation, water pollution, and public and private water supply.

The environmental specialist of today and the future must be a highly skilled, well-educated and trained generalist possessing a multitude of competencies as an effective member of the health team. These competencies are divided into 6 general and 12 specific areas as follows:

General Competencies

A. General Science

1. knowledge and comprehension of general inorganic and organic chemistry
2. knowledge and comprehension of general biology
3. knowledge and comprehension of general microbiology
4. knowledge and comprehension of general college math, including algebra, trigonometry, and basic statistics
5. knowledge and comprehension of physics (mechanics and fluids)
6. knowledge and comprehension of epidemiological principles
7. knowledge and comprehension of risk assessment techniques

B. Communications and Education

1. knowledge and comprehension of various communications, verbal, written, and computerized

2. knowledge and comprehension of how to work with people
3. knowledge and comprehension of the use of audiovisual aids
4. knowledge and comprehension of group dynamic techniques and group processes
5. knowledge and comprehension of interviewing techniques
6. knowledge and comprehension of teaching and learning principles
7. understanding of the need for public information and proper relationships with the news media
8. understanding of how to work with and motivate community organizations including the use of small group dynamic techniques
9. knowledge and comprehension of use of databases

C. Planning and Management

1. knowledge and comprehension of techniques needed to establish a program in any of the environmental health areas
2. knowledge and comprehension of electronic data processing techniques and their application
3. knowledge and comprehension of techniques used to establish priorities
4. ability to design surveys and survey forms
5. ability to use survey techniques to determine the extent of given environmental health problems
6. ability to interpret survey findings
7. ability to determine the advisability of legal action and when to initiate it

D. General Technical Skills

1. knowledge and comprehension of the principles of learning and possession of skills in training, testing, evaluation, and use of aids in various areas of environmental health
2. knowledge and comprehension of how to use inspectional techniques to determine environmental health problems
3. knowledge and comprehension of a variety of sampling techniques related to air, water, food, hazardous chemicals, etc.
4. ability to accurately collect samples, complete record forms, and interpret result of lab samples in light of surveys made
5. ability to use survey and field sampling and test instruments

E. Administrative and Supervisory Skills

1. knowledge and comprehension of environmental and public health laws, regulations, ordinances, codes, and their application
2. knowledge and comprehension of supervisory techniques used in environmental health programs
3. knowledge and comprehension of administrative techniques used in the management of environmental health programs

4. knowledge and comprehension of the significance of the application of existing environmental and public health laws
5. knowledge and comprehension of the systems approach to the analysis of environmental health problems
6. knowledge and comprehension of the vital role of continued maintenance in the permanent resolution of environmental control problems
7. knowledge and comprehension of the relationship between health departments, other public agencies, voluntary agencies, business, and industry
8. knowledge and comprehension of basic principles of economics and how they relate to the existence of environmental health problems and the potential for successful environmental health programs
9. knowledge and comprehension of overall health problems and health priorities
10. knowledge and comprehension of risk management techniques

F. Professional Attitudes

1. desire to work with people and to use basic environmental health sciences to resolve environmental health problems
2. sense of obligation to fulfill the requirements of the job and to carry out assigned duties in a professional manner
3. approach recipients of services with a cooperative attitude
4. display courtesy in personal relationships with fellow employees
5. accept constructive criticism from employees, peers, and the public
6. sense of dedication to the environmental health profession reflected in participation in continuing education
7. control emotions and perform in a mature manner during periods of stress
8. desire to communicate public health principles

Specific Competencies

A. Environmental Chemical Agents

1. knowledge and comprehension of potential chemical contaminants of food
2. knowledge and comprehension of potential chemical contaminants of potable water supplies
3. knowledge and comprehension of transport requirements for hazardous chemicals
4. knowledge and comprehension of the techniques and procedures for identifying environmental chemicals
5. knowledge and comprehension of the means of disposal for environmental chemicals
6. understanding of decontamination of objects or substances that have been contaminated with environmental chemicals
7. knowledge and comprehension of field tests used to determine the presence and concentration of environmental chemicals
8. knowledge and comprehension of detergent and disinfectant chemistry

9. ability to evaluate detergents in an "in-use" situation
10. knowledge and comprehension of economic poisons and how they affect the human ecology of the region
11. knowledge and comprehension of principles and practices in the formulation and application of economic poisons
12. knowledge and comprehension of bait formulations used in pest control
13. knowledge and comprehension of the safety features needed to prevent accidents with environmental chemicals
14. knowledge and comprehension of disinfectant detergents and their use

B. Environmental Biological Agents

1. knowledge and comprehension of the epidemiology of vector-borne diseases
2. knowledge and comprehension of the natural habitat and control of common microorganisms and insects of public health and economic significance
3. knowledge and comprehension of basic life cycles of microorganisms, insects, and rodents of public health significance
4. ability to identify a variety of microorganisms and insects of public health or economic significance in the field
5. knowledge and comprehension of environmental factors related to vector control
6. ability to identify scope of field problems and to determine control activities required
7. knowledge and comprehension of advantages and limitations of microbicides and insecticides and their effect on the ecology of the region
8. knowledge and comprehension of the operation of sprayers and other pest control equipment
9. knowledge and comprehension of the epidemiology of microbial, insect, and rodent-borne diseases
10. knowledge and comprehension of environmental procedures used in microbial, insect, and rodent control
11. knowledge and comprehension of biological control of microbes, insects, and rodents

C. Environmental Physical Agents

1. knowledge and comprehension of public health and ecological effects of noise on the individual and community
2. knowledge and comprehension of instrumentation and procedures involved in noise measurements
3. knowledge and comprehension of existing laws pertaining to nuisances and noise abatement
4. knowledge and comprehension of practical applications of control measures
5. ability to implement surveys designed to define the extent of noise problem
6. ability to evaluate results of surveys and to establish long-range and short-range goals for control

7. knowledge and comprehension of work-related noise stress
8. knowledge and comprehension of radiation theory and principles
9. knowledge and comprehension of dangers of radiation
10. knowledge and comprehension of use of radiation and radioisotopes
11. knowledge and comprehension of effects of radiation
12. knowledge and comprehension of safety precautions
13. knowledge and comprehension of monitoring techniques and instrumentation used in radiation detection
14. knowledge and comprehension of techniques of storage and disposal of radioactive materials
15. knowledge and comprehension of techniques of transportation of radioactive materials
16. knowledge and comprehension of techniques of decontamination
17. knowledge and comprehension of legal requirements of transportation, use, storage, and disposal of radioactive materials

D. Air

1. knowledge and comprehension of the different air pollutants and their sources
2. knowledge and comprehension of the relationship of weather conditions to air pollution
3. knowledge and comprehension of effects of air pollutants on the biosphere
4. understanding of the relationship of air pollution to topography
5. knowledge and comprehension of microflow of air
6. knowledge and comprehension of functional operation of air pollution control devices
7. knowledge and comprehension of preventive measures in air pollution control
8. knowledge and comprehension of corrective measures in air pollution control
9. knowledge and comprehension of the practical applications of air pollution control procedures and techniques
10. knowledge and comprehension of the principles of combustion engineering
11. knowledge and comprehension of air, air sampling techniques, and the ability to conduct air sampling
12. ability to implement surveys to clarify and identify the extent of problems
13. ability to evaluate results of surveys in light of long-range and short-range programs within the community
14. ability to design and implement cost-benefit analysis of control programs
15. knowledge and comprehension of air toxics

E. Water and Liquid Wastes

1. knowledge and comprehension of water sources
2. knowledge and comprehension of potable drinking water quality and standards (physical, chemical, biological, and radiological)
3. knowledge and comprehension of water-borne diseases and how they are transmitted

4. knowledge and comprehension of sampling and testing of potable water
5. interpretation of laboratory analysis of water samples
6. knowledge and comprehension of legal aspects of water quality control
7. knowledge and comprehension of different types of water usage
8. understanding of the protection and selection of individual water supplies
9. understanding principles of water treatment
10. knowledge and comprehension of physical and biological composition of sewage, including common and exotic industrial wastes
11. knowledge and comprehension of types of industrial wastes and their significance
12. knowledge and comprehension of the effects of sewage discharge on water quality
13. understanding of the epidemiology of sewage-associated diseases
14. knowledge and comprehension of the technology and basic engineering principles related to water flow
15. understanding the principles of individual sewage disposal
16. knowledge and comprehension of principles of municipal sewage treatment
17. knowledge and comprehension of small sewage treatment units
18. knowledge and comprehension of the measurement of absorptive quality of soils
19. knowledge and comprehension of the principles of nonwater sewage disposal
20. knowledge and comprehension of the techniques used in problems of emergency situations related to water and sewage
21. knowledge and comprehension of the techniques and potential hazards of sludge disposal

F. Food

1. knowledge and comprehension of food technology and its relationship to health
2. knowledge and comprehension of principles of food manufacturing, processing, and preservation
3. knowledge and comprehension of food-borne diseases and their control
4. knowledge and comprehension of epidemiological techniques and procedures
5. knowledge and comprehension of design, location, and construction of food establishments and their equipment
6. knowledge and comprehension of principles of food establishment operations, housekeeping, and maintenance
7. knowledge and comprehension of equipment design, operation, maintenance, and cleaning techniques
8. knowledge and comprehension of methods of motivating industrial management to understand, accept, and carry out its responsibilities in the food environment, personnel training, and personal supervision
9. knowledge and comprehension of legal requirements of food technology
10. knowledge and comprehension of inspection, survey techniques, and significance of data
11. knowledge and comprehension of the examination and licensure of food establishment managers
12. knowledge and comprehension of techniques used by different cultural and ethnic groups in food growing and preparation

13. knowledge and comprehension of institutional food-handling practices
14. ability to obtain public support for food programs
15. knowledge and comprehension of characteristic and properties of milk
16. knowledge and comprehension of dairy bacteriology
17. knowledge and comprehension of milk production and processing
18. knowledge and comprehension of legal standards of food and milk composition
19. knowledge and comprehension of techniques used to investigate dairy farms
20. knowledge and comprehension of milk processing operations and control
21. ability to inspect pasteurization plants

G. Solid Wastes

1. knowledge and comprehension of the types of solid waste generated in the community
2. knowledge and comprehension of the types of waste generated by common industrial processes
3. knowledge and comprehension of various methods of storage, collection, and disposal of solid waste
4. knowledge and comprehension of public health and ecological aspects of solid wastes
5. knowledge and comprehension of the use of systems analysis in waste disposal management
6. knowledge and comprehension of economics of solid-waste disposal
7. ability to evaluate the results of solid-waste surveys and to establish long-range and short-range goals
8. ability to implement surveys to determine the extent of the solid waste problems
9. ability to design, implement, and evaluate programs related to waste disposal vs. public health problems

H. Hazardous Waste

1. knowledge and comprehension of the health and safety concerns related to hazardous waste sites
2. knowledge and comprehension of the effects of exposure to toxic chemicals in a hazardous waste site
3. knowledge and comprehension of the route of entry to the body of hazardous chemicals, including inhalation, skin absorption, ingestion, and puncture wounds (injection)
4. understanding of the potential health effects due to acute and chronic exposure to various chemicals at the hazardous waste site
5. knowledge and comprehension of the symptoms of exposure to hazardous chemicals, such as burning, coughing, nausea, tearing eyes, rashes, unconsciousness, and death
6. knowledge and comprehension of the potential chemical reactions that may produce explosion, fire, or heat
7. understanding of the psychological effects of oxygen deficiency in humans related to an increase in specific chemicals in the immediate environment
8. understanding of the health effects of ionizing radiation related to alpha radiation, beta radiation, gamma radiation, and X-rays

9. knowledge and comprehension of techniques used to dispose of radioactive material
10. understanding of the potential types of hospital and research facility waste that may cause biological hazards for the individual and may be spread through the environment
11. knowledge and comprehension of the various safety hazards that may be found at hazardous waste sites
12. knowledge and comprehension of the potential electrical hazards that may occur from overhead power lines, downed electrical wires, and buried cables that have been subjected to potential damage from hazardous waste situations

I. Population and Space Utilization

1. knowledge and comprehension of the population explosion and its effect on the present and future needs of our society
2. knowledge and comprehension of the health hazards related to congestion
3. knowledge and comprehension of individual space needs
4. knowledge and comprehension of the effects of different cultures on population control
5. knowledge and comprehension of the use of community planning and zoning on space utilization
6. knowledge and comprehension of establishment of priorities for the proper use of existing space

J. Indoor Environment

1. knowledge and comprehension of cultural, economic, and sociological aspects of individual and multiple dwelling units
2. knowledge and comprehension of housing conditions needed for health, comfort, and well-being
3. knowledge and comprehension of impact of transportation on housing
4. knowledge and comprehension of real estate laws and prevailing practices
5. knowledge and comprehension of the various agencies involved in supervision and licensing of community shelters
6. knowledge and comprehension of techniques used to evaluate individual and multiple dwelling units
7. knowledge and comprehension of local, state, and federal housing programs
8. knowledge and comprehension of zoning laws and their effect on the use of individual and multiple dwelling units
9. knowledge and comprehension of the relationship of minority groups and poverty to housing use
10. knowledge and comprehension of indoor air pollution problems

K. Environmental Injuries

1. knowledge and comprehension of public health and ecological aspects of environmental injuries problems
2. knowledge and comprehension of the instrumentation, material, and procedures involved in determining the causes of accidents

3. knowledge and comprehension of epidemiological techniques used for studying accident problems
4. ability to motivate voluntary corrective action on the part of the public
5. ability to evaluate accidents and their causes

SUMMARY

Humans, who alone have the intellectual capacity to improve their life through science and engineering, share the Earth with numerous other biosystems, some of which affect health. Humans contribute to the destruction of biosystems and to their own health problems through degradation of the environment by means of air, water, and land pollutants. The pollutants are transported biologically or physically through the environment and cause a variety of environmental problems that adversely affect much of the Earth. People through use of skill, good sense, and planning can avoid this destruction and provide for a wholesome, safe environment for future generations.

The problems of the environment are numerous and complex, and can only be resolved by determining the cause, means of prevention, and necessary controls of specific hazards and combined hazards through the use of epidemiological research, special study techniques, risk assessment, and risk management techniques. Not only the ecosystem, the food chain, the growth of population, and the energy cycle must be understood, but also the use and abuse of energy and the impact of humans on their own environment. An economic determination as to the benefit-risk of each type of environmental impact must be made backed by reasonable decisions concerning these impacts. Further, the legal system and the trained environmental health practitioners, along with all levels of government, industry, and concerned citizenry, must be employed to voluntarily improve the environment of all living organisms.

It is certainly clear that individual or collective human decisions influence our environment to a considerable degree and that the quality of our life and the time and manner of our death are related to these decisions. It is not only important to eliminate specific environmental components that cause or contribute to disease or injury, but it is essential to develop a preventive strategy that will eliminate many of the environmental problems before they occur.

In this chapter, the general problems affecting the environment and some of the kinds of primary concerns have been discussed. In the succeeding chapters, each environmental health area will be treated separately, and the specific environmental problem will be approached from the background and status of the problem; necessary scientific, technological, and general information available; the source, scope, and potential for disease related to the problem; the potential for intervention; the resources that may be utilized; the standards, practices and technique available; the modes of surveillance and evaluation available; specific controls; and research needs. (See Preface for a description of how best to use this book and Volume II.)

THE FUTURE

The environmental health practitioner, specialist, and scientist of the future must have appropriate knowledge of health and the environment and recognize the people-made pressures that have been placed upon natural resources, both living and nonliving. These pressures are far more severe than have ever been previously suspected. There is substantial evidence that potentially health-threatening groundwater contamination is a problem of increasing concern in the United States and that toxic chemicals in hazardous waste dumps and other ground storage may pose serious health and environmental threats and create public health problems. Toxic chemicals are present in the air, water, and in the workplace and are growing in quantity and complexity. Many examples of potential damage to the ecosystem, as well as to people, have been cited and will be cited in the succeeding chapters.

Short-term research and short-term horizons have been the techniques used for trying to resolve long-term problems. There is concern, not only in government, but also in the private sector. It is, therefore, necessary to have a better understanding of the interdisciplinary activities that will be needed to carry out the long-term research that will determine the kinds of risks that are occurring and the kinds of risk management that will be necessary in the future.

Research

New research techniques include molecular epidemiology, which is based upon the measurement in the exposed individual of the interaction of a toxic chemical or its derivative with a tissue constituent or a tissue alteration resulting from exposure to the chemical. It provides an indirect measure of individual exposure. Recent research has determined means for detecting and measuring the interaction of a foreign chemical with easily accessible normal human constituents, such as chemical carcinogen interactions with deoxyribonucleic acid (DNA). It is now possible, at times, to detect a few altered DNA molecules out of millions of cells. Altered chromosomes can be determined, and important advances will be occurring in detecting and quantitating human exposure to foreign chemicals.

Susceptibility to chemical toxicants varies widely in the human host population. The host factors are genetic diversity, current and prior disease, sex, and age. There are approximately 2000 genetically identifiable human diseases. Genetic conditions are likely to enhance the risk to individuals of developing environmentally or occupationally associated adverse health effects. The extent of the risk is unknown. The extent of the risk of enhancement is unknown. Considerable study is needed in genetic diversity-susceptibility and biological mechanisms.

Exposure to humans, other animals, and the environment of pollutants or mixtures of substances is not clearly understood. There are no valid general rules for determining the presence of synergistic activity in the mixtures of chemicals to which these individuals or the environment is subjected to.

The emerging technology of biotechnology and microelectronics are of considerable interest and potential concern. Biotechnology deals with genetic engineering, which hopefully will produce new medical products, agricultural products, chemicals, and other products. Microelectronics involves the production of microelectronic shifts with the use of a variety of virtually unstudied chemicals, such as gallium arsenide, silicon, and halogenated hydrocarbon solvents.

There is inadequate data available for the physical, chemical, and ecological variations in fresh waters, oceans, and the atmosphere over extended periods of time. It is, therefore, difficult to determine what is a natural change over a long period of time and what is caused by human activity. There is a need to understand a normal range of variations in ecosystems and how they can deal with the pollutants that are currently flowing into them and that will flow into them in the future.

As pollutants move from one environmental medium to another, the rate of transfer is not understood. Quantity assessments are necessary, and the results of these movements of pollutants need to be understood. At what rates are aerosols formed from chemically reactive organic pollutants and what effect do these aerosols have on the air-water interface, the precipitation of the aerosols, and the temperature and moisture content that may affect them?

Study is needed to determine the behavior and biologic effects of chemicals in various environmental media and what impact they may have on the overall global ecosystem.

The improvement in quantitative risk assessment is most likely to come from a better understanding of biologic processes. Little is known about basic pharmacokinetic dynamics and the environmental mechanisms related to toxicity, other than cancer. Very little is known about actual exposure patterns, the potential of short-term biologic screening to provide early prediction of effects, and a validation of risk assessments.

Environmental Changes and Their Consequences

In order to understand environmental changes and their consequences, it is necessary to know the movement from the point of exposure to the end point of disease. There are four rough stages in environmental toxicology. They are

1. Exposure, where the ambient condition brings the organism into contact with the hazard. For example, a toxic material that may be found in the air or water.

2. The dose where the internalized quantity of the hazard creates a specific body burden that may be toxic to an organ or cell.
3. Effects seen that are markers of intermediate biologic effects that are either a step in the toxicologic process or a parallel manifestation of effect. These indicators of exposure or dose are predictors of toxicity. They may include chromosome aberrations, mutations, and cell or cell killing enzyme activity.
4. The endpoint is the ultimate toxicological effect, which may be cancer, heart disease, or other diseases or injuries.

Perturbations are complex mixtures of substances or impacts, and equally complex patterns of human activities that influence exposure. Examples include hazardous waste dumps and related groundwater problems, as well as the dynamics of the ecosystems that may be affected by these. The perturbations, which also include the introduction of new technologies or changes in resource use patterns, affect ecosystems physically and biologically. The physical changes include nutrient and energy flow, physical structure of the system, and transport and transformation of substances that have been introduced. The biological changes include direct increases in populations due to secondary effects and resulting changes in linkages and functioning of the system as a whole. The ecosystems, then, are related to exposures in humans or other animals and plants, of concerns to humans, and finally ends up with health effects that are either improper or proper. The improper health effects may be changes in target organs of the body or death to the total organism.

Recommendations for the Future

The research areas that have been previously discussed are summarized in some specific types of studies that need to be carried forward. They are

1. The dynamics of stressed ecosystems, that is, the changes that occur in ecosystems that result from people-made stresses.
2. Ecologically significant end points, that is, the stressing of the ecosystems should become a basis for potential regulatory attention.
3. Ecological markers and sentinel events, that is, a study should be done to identify early indicators of potential ecological change.
4. Total toxicity of complex mixtures, that is, research should be devoted not only to the effects of individual substances, but to the complex of common mixtures, such as solvents, agriculture runoff, and wastewater sludges, as well as airborne chemical mixtures.
5. Movement of pollutants out of sinks. Sinks are ecological areas, such as wetlands or underwater sediment and landfills, that have been used for the disposal of toxic pollutants for many years. The pollutants may move across the media. Research is needed to identify under what conditions these materials or their conversion products will move out of the sinks and into ecosystems and human environments.

6. Human exposure, that is, data should be developed on human exposure from all environmental sources, ranging from inhalation in the ambient air, in the workplace, and other indoor environments, to smoking and the ingestion of consumer products, as well as food and water, and the absorption of substances by skin contact.
7. Effects on multiple organs and physiologic systems, that is, research should be expanded to study the neurobehavioral and immune systems.
8. Plant perturbations, that is, an analysis should be made of the effect of fossil fuel combustion and the potential for global temperature increases, changes in the carbon cycles, and other possible concerns.
9. Biotechnologies, that is, a better understanding is needed for the genetic manipulation of biological organisms to determine whether potentially serious consequences may occur.
10. Linkages among ecological systems, that is, an understanding needs to be found of the transport and fate of materials between heterogenous ecosystem units.

Chapter 2

FOOD PROTECTION

BACKGROUND AND STATUS

Effective food protection is carried out on a daily basis by large numbers of personnel who attempt to make this essential component of human life safe, attractive, appetizing, nutritious, and free of disease or poison. The food and beverage industry is extremely large.

Many changes have occurred in the food industry in the last 50 years. Today we use many worldwide food sources instead of depending primarily on home-grown food or food grown nearby. Our food is mass produced and nationally distributed. Food establishments have increased in size, number, type, and complexity. The short-order food establishment has replaced, in many cases, the "greasy hamburger joints" of the past. However, problems continue to increase in the new establishments, as well as in the problem establishments of the past, which still do exist, with management being the key to the problem.

Food is contaminated by microorganisms from soil, water, air, surface, animals, insects, rodents, and people. Food is also contaminated by chemicals from soil, water, air, pesticides, herbicides, fertilizers, and radionuclides. Contamination may be introduced during processing, as demonstrated in outbreaks of botulism from vacuum-packed fish and tuna fish, or during any stage of production, transportation, storage, preparation, and serving. Raw foods are shown to contain disease-producing organisms. Salmonella is contained in 15 to 30% of raw dressed poultry. *Clostridium perfringens* is cultured from 50% of red meat in stores. *Salmonella aureus* is present in most pooled milk supplies and in roughly 20% of cheddar cheeses. Salmonella are found in 15–30% of commercial egg products, with *S. enteritidis* being a serious problem currently. *C. botulinum* type

E is found on raw fish coming from fresh or salt water. Shellfish are contaminated with organisms causing infectious hepatitis and cholera. *Escherichia Coli* 0157:H7 has been associated with outbreaks of disease from hamburger. More than 10,000 food items are sold in stores today. Each one has its own unique potential for causing the spread of food-borne disease.

Many people carry pathogenic organisms that can be spread through food. *S. aureus* may be found on the skin or mucous membranes of 3–50% of the population. Salmonella may be found in the feces of 0.2% of all people. *C. perfringens* is carried by 80% of the population, and 6.4% of the fecal specimens from food handlers contain enteropathogenic *E. coli*.

Eighty to 90% of total human exposure to most chemicals, including heavy metals, pesticides, and radionuclides, comes from food consumption. Food is presently contaminated heavily by chemicals. These chemicals, such as pesticides, may be deliberately applied to crops and stored commodities to kill insects. Other chemicals, such as food additives, are added to food to achieve some desirable effect, such as improved taste, color, and longevity. Drugs used in the care of animals and chemicals used in food-packaging materials may accumulate within food. Another group of chemicals enter by means of environmental pollutants or as fungal toxins. At present the most important chemicals and chemical contaminants of food for which measurements can be made include toxic metals, such as mercury, selenium, lead, arsenic, cadmium, and zinc; industrial chemicals, such as polychlorinated biphenyls, chlorinated dibenzo-*p*-dioxins, chlorinated dibenzofurans, and pesticides; fungal toxins, or mycotoxins, such as aflatoxin; and miscellaneous chemicals, such as the polynuclear aromatic hydrocarbons and nitrosamines. In addition, the level of food-borne disease due to microorganisms is increasing and will continue to increase because of the tremendous amount of food processed and sold in a variety of different types of operations.

This chapter is concerned with food-borne disease; food microbiology; plans review; physical facilities; storage, preparation, serving, and protection of food; housekeeping procedures; cleaning and sanitizing; solid waste disposal; insect and rodent control; personal health of employees; inspectional procedures; and public health laws.

SCIENTIFIC AND TECHNOLOGICAL BACKGROUND

Food Microbiology

Microorganisms are unicellular and microscopic, and perform positive or negative functions in our lives. Certain microorganisms are important in the production of vinegar (*Acetobacter*), sauerkraut (*Lactobacillus*), and bread (Saccharomyces). However, our major concern will be with those organisms that can cause food-borne infections and food poisoning. Microorganisms are found in

Bacterial Growth Curve Growth Phases

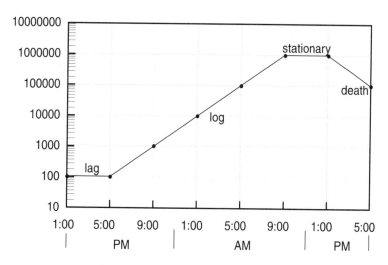

Figure 2.1 Bacterial growth curve. Source: HACCP, *Regulatory Applications in Retail Food Establishments,* U.S. Food and Drug Administration, Rockville, MD, 20857, page 23, 1991.

soil, water, air, on animals, in body openings, and on body surfaces. These organisms can be transmitted directly to food and water by hands or indirectly by fomites, insects, and rodents, and by aerosols due to coughing and sneezing. Bacteria can exist in the vegetative or spore form. The spore is resistant to heat and chemical destruction. Aerobic bacteria, such as *Pseudomonas*, grow in the presence of oxygen. Anerobic bacteria, such as *Clostridium*, grow in the absence of air. Facultative bacteria, such as *Escherichia coli* and *Salmonella*, can grow under either condition.

Reproduction and Growth of Microorganisms

In order for bacteria to grow, they need an adequate temperature, moisture, presence or absence of oxygen (depending on the organisms), and nutrients. Bacterial cells reproduce by a single cell split to form two identical cells (transverse fission). This takes about 20–30 min for those bacteria that cause food-borne infections and poisonings. Bacterial growth follows four basic phases: the lag phase lasting about 2 hr; the logarithmic or growth phase, where growth is at a maximum rate; the stationary or resting phase, where bacteria die at the same rate as they are being produced (spores are produced here); the period of decline and death, or death phase, which may occur in 18–24 hr. (Figure 2.1)

Environmental Effects on Bacteria

Temperature

The temperature of the medium in which bacteria live determines the rate of growth, multiplication, and death of the organism. Bacteria can generally survive very low temperatures, even though they are dormant; high temperatures will generally kill bacteria. The consistency of the medium, the number of organisms initially present, and the species also affects the rate of bacterial death. Generally the resistance of an organism to heat is greatest at a pH of 7.0, which is favorable for growth. As the acidity or alkalinity increases, the death of organisms due to heat increases. Psychrophiles are organisms found on uncultivated soil, lakes and streams, on meats, and in ice cream, and live best at temperatures of 0–5°C. They frequently are the cause of spoilage in refrigerated foods. Mesophiles are parasites found on plants, or may be harmless parasites on animals and humans. They live best at temperatures between 20 and 45°C. Thermophiles are organisms causing off flavors in pasteurized milk. They live best at temperatures exceeding 55°C.

Radiation

Ultraviolet (UV) rays in sunlight are lethal to bacteria. UV wavelengths of 210 nm (1 nm = 1×10^{-9} m) to 310 nm contain bactericidal power. The most effective ultraviolet wavelength against the majority of bacteria and some molds and viruses is 265 nm. Some of the shorter X-rays and gamma rays are more effective but are dangerous to people. However, under controlled conditions, these rays, along with high-energy electrons, are being used for the irridation of food. Radiation kills bacteria logarithmically and causes the remaining cells to form mutants susceptible to attack by bacteriophages.

Pressure

Bacteria are resistant to applied pressure.

Sound Waves

Supersonic vibrations rupture bacterial cell walls and destroy bacteria. Sound waves of low intensity but high frequency are most effective.

Moisture

Moisture is needed by microorganisms for growth and metabolism. Water activity or Aw is the ratio of vapor pressure of food to that of pure water, which has a value of one. For growth and metabolism, most bacteria need an Aw of 0.91, most yeast need an Aw of 0.88, and most molds need an Aw of 0.82. Since the osmotic pressure within the cell is usually greater than the surrounding media, water flows through the semipermeable cell membrane into the cell. By adding 10–15% sodium chloride or 50–70% sugar to the media, the water flows out of the cell, causing cell disruption and death of the bacteria. Drying the media stops the growth of bacteria and some die. As long as the product stays dry, spoilage or growth of disease-producing organisms will not occur.

Chemicals

A vast variety of chemicals will kill bacteria. These chemicals include antibiotics, alcohol, chlorine, bromine, iodine, iodophores, phenols, and quaternary ammonium chloride compounds. The effectiveness of the chemical is based on the organic load present, time, temperature, pH, specificity for organism, and mechanical action.

Detergents and Disinfectants

The detergent and/or disinfectant used in cleaning is determined by the soil load present, the type of soil, the type of water used, the cleaning operation, and the kind of disinfecting agent used for final sanitization. The degree of cleaning, which is a process of soil, tarnish, and stain removal from objects, is determined by the concentration of detergent, type of water, temperature, time, velocity, and type of cleaning procedure.

Soil varies in composition. It may contain water-soluble sugar, salts, and some organic acids, or it may contain water-insoluble mineral deposits and/or animal and vegetable fats, such as grease or carbon. Food particles may be bound to surfaces through absorption or by grease. Scale deposits may be formed by the interaction of alkaline cleaners and chemicals in the water. These chemicals include soluble iron and manganese salts, calcium and magnesium carbonate and bicarbonate, calcium sulfate, calcium chloride, magnesium sulfate, and magnesium chloride.

Adequate cleaning is enhanced through increasing the temperature of the cleaning liquid, since the rise in temperature decreases the bond between the soil and the surface, increases the solubility of materials, and increases the rate of the chemical reaction. Force or energy applied during cleaning helps to mechanically remove soil from surfaces. The cleaning procedure brings the detergent solution into intimate contact with the soil, removes the soil from the surface, disperses the soil into the solvent, and prevents redeposition of the soil on the clean surface.

Basic Detergent Mechanisms

The basic detergent mechanisms are as follows:

1. Establish intimate contact between detergent and soil through "wetting" and "penetration." Wetting reduces the surface tension of the soil and allows the detergent solution to penetrate and spread out. Penetration is the action of a liquid entering into porous materials through cracks and pinholes.

2. Displace the soil from the surface through dissolving, peptizing, and saponification. Dissolving is a chemical action of detergents that liquifies water-soluble materials or soils. Peptizing is a chemical action that brings materials into a colloidal solution. Saponification is the chemical reaction between an alkali and animal or vegetable fat to produce soap.

3. Disperse the soil in the solvent through suspension, dispsersion, or emulsification. Suspension is the action holding insoluble particles in solution. Dispersion is the action of breaking up clumps of particles and thereby suspending them. Emulsification is the action of breaking up fats and oils into small particles that are suspended in the cleaning solution.

4. Prevent redeposition of dispsersed soil through rinsibility. Rinsibility is the condition of a solution, permitting it to be flushed easily and completely from a surface.

5. Soften water through precipitation, sequestration, and chelation. Precipitation is the physical settling of particles. Sequestration is the "surrounding" or "tying up" of magnesium and calcium compounds by sequestering agents. A sequestering agent is a substance that removes a metal ion from a solution system by forming a complex ion that does not have the chemical reactions of the ion that is removed. Chelation is the same action as sequestration, except with organic materials.

Detergent Formulation

All detergents differ and should be formulated to take care of specific types of soils and surfaces. The ideal detergent should act quickly, be completely soluble, have good wetting and penetrating action, dissolve food solids easily, emulsify fats, suspend particles in solution, rinse off easily, have water-softening properties, be noncorrosive to metal surfaces, and be economical in use dilution.

Types of Detergents

Manual immersion cleaners are used in hand dishwashing or equipment-washing procedures. This cleaner is usually a blend of anionic and nonionic organic liquids or powders containing special additives for corrosion inhibition and skin protection.

Housekeeping cleaners are both liquid and powder and are used for cleaning a variety of surfaces, both by hand and machine.

Liquid all-purpose cleaners are composed of anionic and nonionic synthetic detergents that are low sudsing and will not damage or dull synthetic or wax floor finishes when used on floors.

Heavy-duty cleaners or wax scrubbers are liquid products formulated to emulsify and soften wax and synthetic floor finishes so that they may be easily removed prior to refinishing.

Low-sudsing cleaners are used in battery-operation automatic machines that both scrub and vacuum the floor. This operation is most useful on large floor areas.

Wall cleaners may be either all-purpose cleaners or heavy-duty cleaners. Occasionally, where it is necessary to remove deteriorating paint residues, a special powdered wall cleaner including trisodium phosphate may be used.

Concrete cleaners are a powder type of general cleaner similar to the powdered wall cleaner. The concrete cleaner must be more highly alkaline and specifically formulated for cleaning soiled concrete surfaces.

Machine products are either powder or liquid in form and are specially designed for cleaning dishes and utensils. These products are highly alkaline, causing skin irritation if used in manual cleaning.

Acid detergents may be made of synthetic detergents and hydrochloric acid for use in toilet bowls and urinals. They effectively remove rust and other mineral stains from toilet bowls. These strong acid detergents should never be used in bathtubs and sinks, which are generally made of porcelain. Another milder type of organic acid detergent is composed of organic acids or phosphoric acids and nonionic detergent. These products are very useful for removing mineral films from stainless steel surfaces and also from the arms of dishwashers and the milk stone found in milk vats and containers.

Detergent Evaluation

The best type of detergent evaluation is an in-use test. The manager must determine if the product is accomplishing the cleaning job required. Obviously the food service manager must take into consideration the temperature of the cleaning solution, the time it is used, the concentration, the amount of friction used, and the initial amount of soil on the utensils or equipment.

Disinfection Evaluation Methods

The recommended disinfection evaluation method is the AOAC Use Dilution Confirmation Test. The test is carried out by placing sterile, stainless-steel ring carriers in a broth culture of the test organisms and drying the rings for 20–60 min. The contaminated carriers are immersed in a tub of use solution of the germicide for 10 min at 20°C. The carriers are immersed in tubes of appropriate nutrient media and incubated for 48 hr. There must not be any growth in at least 10 separate tubes from ring carriers. In practice, the USDA often requires perfect results in 60 or more ring carriers.

Definitions

Detergent disinfectants are disinfectants with cleaning ability. They are also called germicidal detergents, detergent germicides, disinfectant detergents, and germicidal cleaners.

A disinfectant is a product that kills all vegetative bacteria but does not kill spores. Germicide and bactericide are synonymous with disinfectant.

Sanitizer is a product that kills some, but not all, bacteria. It usually reduces bacterial counts to a generally acceptable level for public health standards.

Sterilization is a process that kills all living organisms, including spores. Sterilization may be achieved by the use of saturated steam at 121–123°C for 30 min, through dry heat at 160°C for 1 hr, or by the use of certain chemicals such as ethylene oxide gas for 2 hr.

A bacteriostat is a product that prevents bacteria from multiplying without actually killing the bacteria. Bacteriostats are ineffective on hard surfaces.

An antiseptic is an antibacterial agent used on the skin.

A bactericide is a substance that kills bacteria.

Sanitizing Agents

Sanitizing agents include various forms of chlorine, iodophors, quaternary ammonium compounds, and heat. Chlorine may be available as calcium hypochlorite at 70% available chlorine, or sodium hypochlorite at 2–15% available chlorine. Hypochlorites are effective sanitizers at a minimum concentration of 50 ppm of free available chlorine and if applied for 1 min if the surfaces to be sanitized are clean. Organic material, low temperatures, and high pH influence the bactericidal effectiveness. The temperature for the sanitizer should be at least 75°F.

Chlorine may also be available as chloramine-T at 25% available chlorine, or dichlorocyanuric and tricyanuric acids at 70–90% available chlorine. These organic chlorine products having a slower bactericidal action are used at a minimum concentration of 200 ppm for 1 min. The temperature for the sanitizer should be at least 75°F.

Iodophores are soluble complexes of iodine usually combined with nonionic surface-active agents. The iodophores have a rapid bactericidal action in an acid pH range in cold or hot water. They are less affected by organic matter than hypochlorites, are nontoxic in ordinary use, noncorrosive, and nonirritating to the skin. The yellow or amber color of the solution is proportional to its concentration. Iodophores are used at a minimum concentration of 12.5 ppm for 1 min. Quaternary ammonium compounds (QAC) are effective sanitizers at a minimum concentration of 200 ppm for 1 min. They are more stable in the presence of organic matter than chlorine or iodine compounds. They are noncorrosive and nonirritating to the skin. Hot water can be used as a sanitizer if the temperature in the final rinse at the entrance of the manifold is at least 170°F for 30 sec or at least 170°F for 30 sec when equipment is immersed in hot water.

Field Tests for Sanitizers

The concentration of chlorine in rinse water can be determined by use of indicator test paper impregnated with starch iodine or by the use of an orthotolidine colorimetric comparison unit. Quaternary compounds can be tested by certain test papers that use color comparison. Iodophore concentrations can be determined by comparing the color of the solution to a color comparison kit. Hot-water sanitizing can be determined by bacteriological sampling with Rodac contact plates or swab samples. An average plate count per utensil surface should not exceed 100 colonies.

PROBLEMS IN FOOD PROTECTION

Although our scientific technology is so advanced that we can put humans on the moon and give them protected life-support systems, including food, our food programs on Earth, in many cases, are the same as they were 50 years ago. The excellent programs are scientifically oriented and conducted by trained environmentalists; in our poor programs, investigators inspect food establishments to license them; in our nonexistent programs, the public is deprived of any type of protection from food-borne disease. Technological developments in food processing are occurring so rapidly that their public health implications are not being properly evaluated. Industry is investing more than 100 million dollars a year in these new processes. New recently discovered production hazards include botulism and typhoid organisms in commercial products, and salmonella in liquid and powdered eggs and egg products, in instant nonfat dry milk, and in drugs and diagnostic reagents. Governmental agencies have been unable to adequately supervise industry because of insufficient resources, limited authority, lack of knowledge, inadequate workforce, preoccupation with legislative response, and inadequate channels of communication between the agency and industry. Public health agencies can be satisfied no longer with expending most of their efforts in supervising the food-service industry; they must now concern themselves with the source of the raw food and the means of processing it. Because of widespread distribution of food, the potential for large outbreaks of disease has increased immensely. There is an increasing concern over long-term health effects of chemical residues, pesticides, additives, and radioactive fallout. Heavy metals, such as mercury, are found in the fish food chain consumed by humans.

Source of Exposure

It is estimated that there are between 24 and 81 million cases of acute gastroenteritis in the U.S. each year with an estimated cost of 7.7 to 23 billon dollars per year. An estimated 10,000 people die from foodborne illness. Over 80% of the outbreaks of foodborne disease, and over 60% of the cases reported, are bacterial in nature. Salmonella are the most frequent agents. However, since less than 1% of the severe gastroenteritis cases are investigated thoroughly enough to determine the causative agents and routes of transmission, the true scope of this disease is not known. In fact, in known outbreaks of disease, about 50% of the causes are unknown.

Reporting is a major problem, since patients are either indifferent or reluctant to cause trouble, physicians usually only report fatal gastroenteritis, and industry is afraid to lose money. Investigations are hampered by nonreporting, delayed or inadequate reporting, inadequate professional staffs and laboratories, inability to obtain food samples rapidly or total absence of food samples, and lack of cooperation on the part of the institutions or industry. About 40% of the states report no outbreaks, and about 40% more report one or two outbreaks per year. California reports 30–50% of all food-borne outbreaks reported annually in the U.S. Ninety-three percent of all reported cases of disease due to water, milk, and food, which involves 10,000 people annually, are caused by food products. Many outbreaks of disease are tracked back to comminuted food such as fish or meat products which have been chopped, flaked, ground, or minced. Improper holding temperatures compound the problem.

Food-Borne Infections

Food-borne infection is direct infection in which the organisms are ingested into the body through food or water and then continue to multiply. In some infections the organisms die off after the clinical disease is over. In other infections, such as the salmonella group, the individuals may continue to be carriers for varying periods of time. They pass off live organisms in their feces.

Streptococcus. Streptococcal infections are usually due to *Streptococcus fecalis* or *Streptococcus pyogenes*. The source of these organisms are the feces, nose, or throat. *Streptococcus pyogenes* may also come from infected wounds or dirty bandages. Almost any foods may become infected, but turkey, other meats, and milk have been most frequently incriminated. Prior to pasteurization of milk, and even on occasion subsequent to pasteurization, individuals with beta-hemolytic streptococci have infected the udders of cows or the milk itself. The incubation period is 1–3 days. The reservoir of infection is people. It is communicable for 10–21 days.

Salmonella. Salmonellosis is a costly, highly communicable disease in the U.S. An estimated two to four million human cases occur annually. Since the diseases are primarily food or water borne, they are a potential threat to every person in the country. Although the true magnitude of the problem is not known, it has been determined that the highest incidents of reported cases occurs among the very young and elderly. The interrelationships between people, animals, and fomites make salmonellosis a difficult problem to resolve.

Salmonella are divided into three groups based on their host preferences. Group 1 is primarily adapted to people and includes *Salmonella typhi*, which causes typhoid fever, and *S. paratyphi* A, which causes paratyphoid fever. The organisms are transmitted by milk, shellfish, other food products, or water contaminated with feces from human cases or carriers. Group 2 is primarily adapted to animal hosts and includes several important pathogens of domestic animals, such as *S. cholerae-suis*, and serotypes of *S. enteritidis*. These organisms, especially serotype Dublin, cause gastroenteritis in humans. Children may be severely affected by these infections. Group 3 includes organisms adapted to either animals or humans showing no distinct preference.

In recent years, there have been numerous outbreaks of *S. typhimurium, S. newport, S. montevideo, S. pullorum, S. derby, S. infantis,* and *S. enteritidis. S. derby* was a particularly serious problem in many hospitals in the eastern U.S. during the 1950s and early 1960s. The problem became so severe that maternity wards and nurseries in certain hospitals were forced to close. An outbreak of *S. infantis* occurred in a major institution in the East, resulting in over 700 cases within a 30-day period. The organisms are usually spread through meat, poultry, eggs and egg products contaminated directly or indirectly through contact with human or animal carriers. From 1985 to 1990, there have been 293 outbreaks of *S. enteritidis* in the U.S. with 10,253 cases and 46 deaths.

In 1985, a salmonellosis outbreak involving 16,000 confirmed cases in 6 states was caused by low-fat and whole milk from one Chicago dairy where the pasteurization equipment had been modified to facilitate the running-off of raw milk, resulting in the pasteurized milk, under certain conditions, becoming contaminated with raw milk input.

In August and September of 1985, *S. enteritedis* was isolated from employees and patrons of three restaurants of a chain in Maryland where at least 71 people became ill, resulting in 17 hospitalizations. Scrambled eggs from a breakfast bar were epidemiologically implicated.

In 1984, 186 cases of salmonellosis (*S. enteritedis*) were reported on 29 flights to the U.S. on a single international airline where 2,747 passengers were affected overall.

A cycle of infection includes feed and fertilizer, livestock, transportation, processing of food, processed foods, and animals or humans. Processed foods, in addition to those already mentioned, that are most commonly contaminated with salmonella include poultry, salads, meat pies, pressed beef, sausage, cold cuts, smoked fish, reheated meats, gravies, dried egg products, frozen eggs, synthetic creams, custard, cream puffs, prepared fish dishes, and shellfish, especially oysters.

Outbreaks of salmonellosis have also been traced to dirty cutting boards and other utensils and equipment. The incubation period is 6–72 hr. The reservoir of infection includes poultry, rodents, turtles, cats, dogs, and people. Its communicability varies from days to weeks.

Shigella. Bacillary dysentery or shigellosis is caused by different members of the genus *Shigella.* The disease is spread through moist-prepared foods, milk, other dairy products, or water contaminated directly or indirectly with feces from actual cases or from carriers. The incubation period is 1–7 days, commonly 4 days. The reservoir of infection is people. It is communicable for about 4 weeks.

Cholera vibrio. Cholera is a severe gastrointestinal disturbance caused by the *Cholera vibrio.* The vibrio enters the body through food and water. The incubation period is from a few hours to 5 days, usually 2–3 days. Communicability usually lasts for a few days, but a carrier state could last for months. The reservoir of infection is people. In Africa in 1994, thousands of people died from cholera when they were displaced because of the war.

Listeriosis. Listeriosis is caused by the organism *Listeria monocytogens.* It is most frequently isolated from people, although the reservoir of infection can also be wild or domestic mammals and fowl. The disease is found at the extremes of age, during pregnancy, and in immunocompromised people. It may cause septicemia, meningoencephalitis, nausea, vomiting, delirium, coma, and death. The disease may be spread by people, food, or be a nosocomial infection. The incubation period is unknown, but is probably a few days to 3 weeks. Fetuses and newborn infants are highly susceptible, with the mother shedding the organism in urine for 7–10 days after delivery. An outbreak in 1985, in California, from contaminated soft cheese caused 90–100 deaths.

Norwalk-type disease (epidemic viral gastroenteritis). Diarrhea caused by Norwalk virus, a 27 nm particle, which is a parvovirus-like agent, is a limited mild disease with symptoms of nausea, vomiting, abdominal pain, malaise, and low-grade fever. The occurrence of the disease is worldwide, with people being the only reservoir of infection. Mode of transmission is probably the oral-fecal route through food and water. The incubation period is 24–48 hours with communicability occurring during the acute stage of disease and shortly thereafter. From 1983 to 1987, ten outbreaks of disease caused by Norwalk virus were reported involving food such as fruits, vegetables, salads, eggs, clams, bakery items, and ice.

Viral Hepatitis. Hepatitis virus A is spread through human feces to food, shellfish, and water, and is in turn consumed by people. The disease may also be transmitted by means of fomites. The incubation period is 15–50 days, usually 28–30 days. The reservoir of infection is people. Most cases are noncommunicable after the first week of jaundice. Viral hepatitis is endemic worldwide. Food has been implicated in over thirty outbreaks of Hepatitis A since 1983. In 1987, in Louisville, Kentucky, imported lettuce was the suspected food. In 1988, in Alaska, ice-slush beverage was suspected, in North Carolina, iced tea was suspected, and in Florida, raw oysters from an unapproved source was suspected. In 1989, in Washington, no food was identified. In 1990, in Georgia and Montana, frozen strawberries were suspected, while in Baltimore, shellfish appeared to be the cause of the outbreak. The HBV infection is common in certain high-risk groups, such as parenteral drug abusers, homosexual men, people in hemodialysis centers, and certain health professions where there is routine exposure to blood or serous fluids. The incubation period is usually 45–180 days, although symptoms may occur in as little as 2 weeks. The reservoir of infection is people. People may be infective for years.

Astrovirus. Unclassified virus containing a single positive strand of RNA of about 7.5 kilobase, surrounded by a protein capsid of 28–30 nanometers in diameter, which may cause sporadic viral gastroenteritis in children under four years of age and about 4% of the hospitalized cases of diarrhea, with additional symptoms of nausea, vomiting, malaise, abdominal pain, and fever. The mode of transmission is through the oral-fecal route from person to person, or ingestion of contaminated food and/or water. The incubation period is 10–70 hours. The disease has been reported in England, Japan, and California. The reservoir of infection is people. Susceptibility is found in the very young and the elderly in an institutional setting.

Calicivirus. Virus classified in the family Caliciviridae, containing a single strand of RNA surrounded by a protein capsid of 31–40 nanometers in diameter, which may cause viral gastroenteritis in children 6 to 24 months of age and accounts for about 3% of the hospitalized cases of diarrhea, with additional symptoms of nausea, vomiting, malaise, abdominal pain, and fever. The mode of transmission is the oral-fecal route from person to person, or the ingestion of contaminated food and water. The incubation period is 10–70 hours. The disease has been reported in England and Japan. The reservoir of infection is people. Susceptibility is found in the very young.

Enteric adenovirus. Virus representing serotypes 40 and 41 of the family *Adenoviridae* containing a double-stranded DNA surrounded by a distinctive protein capsid of about 70 nanometers in diameter, which may cause 5–20% of gastroenteritis in young children, with symptoms of diarrhea, vomiting, malaise, abdominal pain, and fever. The mode of transmission is the oral-fecal route from person to person, ingestion of contaminated food and/or water, or it may be transported by the respiratory route. The incubation period is 10–70 hours. The disease has been reported in England and Japan in children who are in hospitals or day care centers. The reservoir of infection is people. Susceptibility is found in young children.

Parvovirus. Virus belonging to the family *Parvoviridae* containing linear single-stranded DNA surrounded by a protein capsid of about 22 nanometers in diameter, which may cause viral gastroenteritis in all age groups, with symptoms of diarrhea, nausea, vomiting, malaise, abdominal pain, and fever. The mode of transmission is the oral-fecal route from person to person or ingestion of contaminated food, especially shellfish and water. The incubation period is 10–70 hours. The disease has been reported in England, Australia, and the U.S. The reservoir of infection is people. There is a general susceptibility.

Rotavirus. Virus belonging to the family *Reoviridae* containing spherical RNA 65–75 nanometers in diameter, with a wheel-like appearance, which may cause gastroenteritis with vomiting followed by diarrhea, malaise, fever, abdominal pain, and dehydration. Death may occur from dehydration or aspiration of vomitus. The mode of transmission is person to person or may be through contaminated food. Identical or similar viruses are found in various animals. The disease has been reported in the U.S. and is the most common cause of diarrhea in infants and young children. The reservoir of infection is people and possibly animals. There is a general susceptibility.

Brucellosis. Brucellosis, also known as undulant fever or Bang's disease, is caused by *Brucella abortus*, *B. melitensis*, and *B. suis*. The organisms are transmitted through raw contaminated milk, dairy products made from raw milk, and contact with the tissues or discharges of infected animals. The incubation period is highly variable, usually 5–30 days. Reservoir of infection is animals. There is no evidence of communicability from person to person.

Diphtheria. Diphtheria is caused by *Corynebacterium diphtheriae*. The organisms are transmitted from humans through milk or directly from human to human through discharges from the nose or throat. The incubation period is 2–5 days. The reservoir of infection is people. Communicability is typically 2–4 weeks.

Tuberculosis. Tuberculosis of the extra pulmonary type is caused by *Mycobacterium tuberculosis*. The organisms, both of the human and bovine type, are spread through raw contaminated milk or dairy products, and contact with infected animals and humans. The incubation period from infection to the primary lesion is about 4–12 weeks. The reservoir of infection is people and diseased cattle. Communicability lasts as long as there are infectious tubercle bacilli in the sputum.

Tularemia. Tularemia is caused by *Pasturella tularensis*. The organisms are spread through the meat of wild rabbits, squirrels, other animals, and bites of infected ticks and deer flies. The incubation period is 2–10 days, usually 3 days. The reservoir of infection is wild animals and various hard ticks. The disease is not directly transmitted from person to person.

Diarrhea Caused by Escherichia coli. Strains of *E. coli* that cause diarrhea are of at least three different types. They are invasive, enterotoxigenic (toxin producing), and enteropathogenic. The invasive strains cause a disease that is found in the colon and that produces a fever and occasionally bloody diarrhea. The pathologic changes are similar to those seen in shigellosis. The enterotoxigenic strains act more like *Vibrio cholerae* in producing profuse, watery diarrhea without blood or mucus, abdominal cramping, vomiting, and dehydration. The enteropathogenic strains belong to the group that is associated with outbreaks of acute diarrheal disease in newborn nurseries. The diseases are typically food borne or water borne, especially in areas where there is poor protection for the food and water, and endemic diarrhea is present. Traveler's diarrhea is most usually due to an enterotoxigenic *E. coli*. The disease is spread by the fecal-oral route. Contaminated hands, as well as improper protection of food and water, may lead to the outbreak of the disease. The incubation period is 12–72 hr. The reservoir of the infection is infected people. The disease is readily transmitted from person to person or from water, food, or fomites to people. An enterotoxigenic *E. coli* outbreak occured in 1975 aboard two cruise ships due to contaminated crab meat. In 1980 an outbreak occured in a Mexican restaurant and in a Texas hospital. Enteropatogenic *E. coli* has occurred for over 50 years in infant nurseries.

Hemorrhagic colitis. An acute disease, caused by *E. coli* 015:H7 (called enterohemorrhagic strain). Generally has symptoms of severe abdominal cramping, watery diarrhea, which may become grossly bloody, occasional vomiting, and possible low-grade fever. The incubation time is 2–9 days. The disease has been found in the Pacific Northwest, Canada, and Michigan. The reservoir of infection is cattle, and it is transmitted by raw or improperly cooked ground beef or raw milk. It is not communicable from person to person. There is a general susceptibility, but the very young may develop renal failure and the very old may develop fever and neurologic symptoms which can lead to death. It is controlled by proper refrigeration and cooking of ground beef, and using only pasteurized milk.

Three outbreaks occurred in 1982 in Michigan and Oregon involving hamburgers and in Canada involving club sandwiches served to the elderly in a nursing home. The Canadian outbreak was implicated in nineteen deaths. Two large outbreaks occurred in 1984, one in 1985, and three in 1986. Thirty-seven people were involved in the 1986 Wala Wala, Washington outbreak, with seventeen people hospitalized and two deaths. Ground beef was implicated.

Campylobacter Enteritis. Diarrhea caused by *Campylobacter* is an acute enteric disease of varying severity. It is caused by *Campylobacter jejuni* and *Campylobacter coli*. The organisms are an important cause of diarrheal disease in all parts of the world and in all age groups. *Campylobacter* may be responsible for more enteritis than either salmonella or shigella. Most outbreaks have been associated with foods, unpasteurized milk, and unchlorinated water. These organisms are an important cause of traveler's diarrhea. The incubation period is 3–5 days, with a range of 1–10 days. The reservoirs of infection are farm animals, cats, dogs, rodents, and birds, including poultry. The disease may be transmitted for several days to several weeks. Chronic carrier states may occur in animals and poultry, and they may become the primary source of infection.

Yersiniosis. An acute, enteric disease with acute watery diarrhea, enterocolitis, false appendicitis, fever, headache, pharyngitis, anorexia, and vomiting.. Incubation time is 3–7 days, usually under 10 days; caused by *Yersinia pseudotuberculosis* and *Y. enterocolitica,* with numerous serotypes and biotypes; found worldwide, especially in infants and children. It is associated with household pets, especially sick puppies and kittens. Reservoirs of infection include domestic animals, birds, and mammals. It is transmitted by oral-fecal route from contact with infected persons or animals, or eating or drinking food or water contaminated with feces; communicable as long as symptoms last, which may be 2–3 months or a chronic carrier state; general susceptibility.

Plesiomonas shigelloides. Gastroenteritis caused by *Plesiomonas shigelloides,* a gram-negative, rod-shaped bacterium which has been isolated from freshwater, freshwater fish, shellfish, and many animals, is usually a mild, self-limiting disease with fever, chills, abdominal pain, nausea, diarrhea, or vomiting. In severe cases, diarrhea may be greenish-yellow, foamy, and blood tinged. Although this condition cannot yet be considered a definite cause of human disease, the organism has been isolated in the stools of patients with diarrhea, but also is sometimes isolated from healthy individuals. The disease occurs primarily in tropical or subtropical areas with rare infections reported in the U.S. or Europe. The mode of transmission is probably from contaminated water, drinking and recreational, or water used to rinse foods which are uncooked or unheated. The mode of transmission is probably through the oral-fecal route. The incubation period is 20–24 hours. The reservoir of infection is not clear. It may be people and/or animals. There is a general susceptibility with children under fifteen. People who are immunocompromised or seriously ill with cancer and blood disorders are at greatest risk for severe symptoms.

Vibrio vulnificus. A bacterium that causes gastroenteritis, "primary septicemia," or wound infections; generally with symptoms of diarrhea and bullous skin lesions; consumption of the microorganisms in raw seafood by people with chronic diseases, especially liver disease, may result in septic shock rapidly followed by death. Incubation time is 16 hours for gastroenteritis; the disease is sporadically found in the U.S. The reservoir of infection is humans and other primates. It is transmitted through water, sediment, plankton, and consuming raw shellfish; people with open wounds or lacerated by coral or fish can become contaminated in sea water; not communicable from person to person; general susceptibility. It is controlled by avoiding sea water if one has open cuts and treating them properly if they do occur, the proper disposal of feces, and cooking all shellfish thoroughly.

Bacterial Food Poisoning

Bacterial food poisoning or food intoxication is due to the consumption of food containing a toxin created by bacterial growth in the food.

Staphylococcus. Outbreaks of staphylococcus food poisoning, probably the most frequent kind of food poisoning, usually affects large groups of people at picnics, church suppers, in hospitals, cafeterias, and other mass feeding operations. The organisms, which come from humans, are present in boils, carbuncles, pimples, hang nails, postnasal drip after colds, and wound infections. Humans then infect such foods as creams, hams, potato salad, chicken salad, ham salad, egg salad, meat and meat products, poultry, turkeys, custards, eggnogs, cream pies, eclairs, casseroles, and warmed-over foods. The organisms grow best at 50–120°F for a minimum of 5 hr. The bacterial growth does not alter the appearance or flavor of the food and does not produce off odors. The outbreak of the food poisoning occurs when individuals consume the enterotoxin produced by the staphylococcus. The enterotoxin causes inflammation and irritation of the stomach and intestine, resulting in vomiting and diarrhea. Since individuals vary in their susceptibility to the toxin, some become quite ill and others are not affected at all.

In July 1976 in Oakland County, Michigan, 87 senior citizens who were receiving hot meals on a special program fell ill with staphylococcus food poisoning. The beef base used in the preparation of the gravy was found to be the cause of the outbreak. Staphylococcal intoxication is due to exo-enterotoxins from *Staphylococcus aureus*, which may be present in the nose, on the skin, and in lesions of infected people and animals. The incubation period is 1–8 hr, usually 2–4 hr. The reservoir of infection is people or animals. The organisms are readily transmitted to foods to cause an intoxication.

An outbreak of Staph food poisoning occured in Texas when over 1300 children became ill from chicken salad improperly prepared, stored, and held at improper temperatures. In 1989, there were multiple outbreaks associated with eating canned mushrooms.

Clostridium perfringens. C. perfringens food poisoning, which has increased sharply in the last few years, is caused by the consumption of large quantities of organisms in food. Although no free toxin has been found by researchers, the consumption of small quantities of *C. perfringens* does not result in illness. This indicates that the organisms do not grow in the individual and, therefore, must cause intoxication rather than food infection. Most reported outbreaks of this food poisoning have been associated with mass feeding, such as banquets at schools or hospitals and dining rooms of college residence halls.

The organisms are present everywhere, but come primarily from soil, human or animal intestinal tracts, fecal material, and sewage. Although the vegetative forms are killed by cooking the food, spore forms are not. The organisms, which are circulated through the air or by the hands of food-service personnel, are found on meat, particularly large roast beefs and pork loins, on turkeys, and in gravies, dressings, or prepared dishes containing meat, poultry, fish, vegetables, or macaroni products. The organisms require 13–14 amino acids, 5–6 vitamins, temperatures of 60–125°F (the optimum temperature is 110–117°F), and anaerobic conditions for growth. Heating of foods such as gravies, use of warmers, and slow cooling in ambient air of large pieces of meat causes the air to escape and provides excellent incubation temperatures for rapid growth of bacteria and toxin production. The incubation period is 6–24 hr, usually 10–12 hr. The reservoir of infection is soil, the gastrointestinal tract of healthy people, and animals.

Clostridium welchii. Clostridium welchii type A food poisoning, similar to *C. perfringens* food poisoning, is usually caused by infected food handlers. Boiled, braised, steamed, stewed, or inadequately roasted meat allowed to cool slowly and then served warmed or cold the following day is the vehicle in the spread of this organism. (See *C. perfringens* for incubation period and reservoir of infection.)

Clostridium botulinum. Clostridium botulinum produces in food the most deadly toxin known to humans. A small amount of the pure toxin could kill thousands of people.

There are between 10 and 30 outbreaks of botulism each year in the U.S., mostly associated with home canned foods and occasionally commercially prepared foods. Sausages, meat products, canned vegetables, and seafood products have most frequently been involved. In 1987, eight cases of type E botulism occurred in New York and Israel from the consumption of an uneviscerated, dry-salted, air-dried whole whitefish called Kapchunka. One person died. A bottle of chopped garlic-in-oil mix caused three cases of botulism in Kingston, New York.

Infant botulism, first recognized in 1976, affects infants under twelve months of age. It is thought to be caused by the ingestion of *C. botulinum* spores which colonize and produce toxin in the intestinal tract of infants. The only food source implicated is honey.

The organism is found throughout the world in the spore form in soil. The spore is resistant to heat, chemicals, and physical stress. Since home-canned, preserved, or processed foods may not be adequately cooked or processed, and since the spore is so resistant to environmental changes, most outbreaks of botulism have been traced to home-cooked, rather than commercially processed, foods. However, outbreaks of botulism, as well as food containing the toxin have been traced to commercially prepared foods. As the spores vegetate, bacteria grow and multiply, and produce the toxin in anaerobic conditions during storage. Boiling the food for a few minutes will destroy the toxin.

There are six known types of *C. botulinum*. Type A, associated with human illness, is the common cause of botulism in the U.S. Type B, also associated with human illness, is most frequently found in soils in the world. Type C is associated with outbreaks of botulism in cattle, mink, waterfowl, and other animals. Type D, responsible for forage poisoning of cattle, is most commonly found in Africa. Type E, associated with human illness, is usually found in outbreaks associated with fish and fish products. Type F, associated with human illness, has only recently been isolated and is relatively rare. Type A and B toxins are found mainly in canned vegetables and fruits, although beef, pork, fish and fish products, milk and milk products, and condiments are also incriminated.

The key factors in the growth of *C. botulinum* and toxin production from the disrupted cells are pH, availability of oxygen, salt content, time, and temperature. A pH close to neutral favors growth, whereas a pH of 4.5, as is found in tomatoes, pears, and red cabbage, inhibits growth. Types A and B grow best at 95°F, but the temperature can range from 50–118°F. Type E grows best at 86°F, but can grow and produce toxin at 38–113°F. Although a food mass is aerobic, the conditions next to the bacterial cell may be anaerobic. Smoked fish can develop anaerobic conditions in the visceral cavity and under the skin. The interior of sausage can become anerobic. While some foods become foul and rancid due to growth of the organism, others show only minor changes in odor or appearance. In some foods no change occurs at all, despite the fact that the food is lethal. Vacuum-packed smoked fish is an example of a food causing serious outbreaks in recent years, yet showing no physical changes. The incubation period is 12–36 hr, when neurologic symptoms usually occur, although it may last several days. The shorter the time, the more severe the disease and the higher the death rate. The reservoir of infection is soil marine sediment and the intestinal tract of animals and fish. Botulism is not communicable from people.

Vibrio parahaemolyticus. Vibrio parahaemolyticus is associated almost exclusively with seafood and is found in nearly all seafood products. The number of recorded cases in Japan range from 10,000 to 14,000 annually. Several outbreaks occurred in the U.S. when steamed crabs were eaten. The organisms can live in salt water separate from the host. The incubation period is usually between 12 and 24 hr, but it can range from 4 to 96 hr. The reservoir of infection is in marine areas during the cold season, where the bacteria are found free in water or in fish and shellfish in the warm season. It is not spread by people.

Bacillus cereus, Food Poisoning. Bacillus cereus food poisoning is a gastrointestinal disorder in which there may be a sudden onset of nausea and vomiting, and in some cases diarrhea. The symptoms typically last about 24 hr. The infectious agent is *Bacillus cereus,* which is an aerobic spore former. The spore, when becoming a vegetated cell, produces two enterotoxins that are heat stable. These cause vomiting. One other enterotoxin is heat labile. This causes diarrhea. The incubation period is 1–6 hr where vomiting occurs, and 6–16 hr where diarrhea is most prominent. It is not communicable from person to person.

Chemical Poisoning

Chemicals accidentally introduced into food or leached into food from a variety of containers cause rapid illness and at times death. The incubation period is less than 30 min. The cause of the outbreak varies with the chemical and its origin.

Antimony. Antimony is leached from chipped gray enamelware by acidic foods. A typical antimony poisoning is caused by storage of a lemon punch in a large, chipped gray enamel pot.

Arsenic. Arsenic is found in ant, roach, or rodent baits, insecticidal fruit sprays, and herbicides. These poisons accidentally contaminate foods. Since arsenic is an accumulative poison, small doses over extended periods of time lead to poisoning. Arsenic is also a fairly frequent contaminant of drinking water.

Cadmium. Cadmium, present in plating materials of containers and trays, is leached into food by acidic substances, such as fruit juices. Cadmium might also seep from industrial operations into a water supply.

Chlorinated Hydrocarbons. Chlorinated hydrocarbons are synthetic chemical pesticides, such as DDT, lindane, and endrin. Since these pesticides have been used extensively and are fat soluble, they can create long-term hazards. Excess quantities of these poisons cause nervous system disorders. The main modes of entry into the environment in an agricultural setting are through the air, soil, or water. This may occur through air spraying of chemicals, ground spraying, exposure from industrial sources, and spreading of sludge on soil. The chemicals may enter the groundwater supply, directly contaminate the food supply, or be bioconcentrated in the food chain.

Copper. Copper poisoning is caused by the leaching of copper from food contact surface, such as carbonated water-machine tubes, into food or drink.

Cyanide. Cyanide, found in silver polishes, can accidentally contaminate foods.

Lead. Lead is present in lead arsenate, a pesticide used on apples. Lead may also be leached by acidic beverages from improperly glazed pottery vessels and utensils. Automobile radiators used as condensers for illegally distilled whiskey add lead to the final product.

Mercury. Mercury, commonly found in many industrial processes, commercial products, and homes, is an insidious chronic poison, producing symptoms that resemble emotional and psychological disorders. Mercury poisoning also appears in an acute form. Methyl mercury, the principle form of mercury found in food, penetrates the placenta, causing birth defects. It is also found in mother's milk. Fungicides containing mercury are used on grain consumed by farm animals, who are ultimately consumed by humans, leading to severe mercury poisoning and eventual death in humans. Fish and shellfish that have consumed mercury and are later used as food also cause blindness, paralysis, and death in humans.

Organic Phosphate Compounds. Organic phosphates, such as DDVP, parathion, and TEPP, are used as insecticides, fungicides, and herbicides. Although this group of poisons are most dangerous when inhaled, they also constitute a hazard when ingested with food.

Polychlorinated Biphenyls (PCBs). Polychlorinated biphenyls are found accumulated in fat tissues of wildlife in much of the world. PCBs are also found in fish; animal feed, which leads to contaminated meat, milk, and eggs; paints that have leached into feed; heat exchange fluids; and cardboard cartons from recycled paper. The potential for food poisoning will increase with the increased availability of this family of over 200 chemicals resembling DDT that are stored in the fatty tissues. PCB poisoning has occurred in Japan. In one reported case in 1968, 1000 people became ill, resulting in five deaths from ingesting rice oil contaminated with PCBs.

Zinc. Zinc poisoning is caused by zinc leaching into acidic foods stored in galvanized containers.

Poisonous Plants and Fungi

Certain plants and fungi are poisonous: castor beans contain a toxin called ricin; ergotism is caused by a parasitic fungus of rye (*Claviceps purpurea*); favism is due to the bean, *Vicia fava*; poisonous mushrooms are often fatal; shellfish poisoning is due to eating shellfish that have consumed a plankton called *Gonyaulax*; and cyanide is produced in green or sunburned potatoes and wild celery.

Aflatoxins. Aflatoxins are formed by molds on foods at harvest time, in storage, or in conditions of water damage. Aflatoxins are found in peanuts, brazil nuts, pecans, copra, corn, and cottonseed. Aflatoxins, known as very potent carcinogens for certain animal species, can be produced whenever proper temperature, time, humidity, and the proper strain of *Aspergillus flavus* are present. It is thought that some aflatoxins in Africa might be associated with liver cancer.

Fungi. Fungi may invade plants, causing them to undergo metabolic changes that produce toxic substances. The phytoalexins are examples of this. Genetic manipulation also produces new plant varieties having a stress mechanism that causes problems to the human being. It is known that compounds inhibiting the proteolytic activity of certain enzymes are found throughout the plant kingdom, particularly among the legumes. One of the better known inhibitor reactions is the inhibition of trypsin. Protease inhibitors may also be found in peanuts, oats, chickpeas, field beans, buckwheat, barley, sweet potatoes, rice, lyntol, lima beans, navy beans, garden peas, white potatoes, wheat, and corn. Trypsin is essential for adequate digestion and utilization of food. The presence of the inhibitors decreases the digestion and absorption of the proteins, causing reduced growth and decreased efficiency in utilizing food. Since protein is the most costly part of the diet, the trypsin inhibitors create an economic as well as public health effect.

Lectins. Certain plants contain substances that agglutinate red blood cells. These substances are called phytohemagglutinins or lectins. They are present in seeds and to a lesser extent in leaves, barks, and roots. They produce important environmental toxins, such as ricin, from the castor bean.

Plants. Certain substances, known as hepatoxins, that are toxic to the liver are found in plants. These plants are poisonous to livestock and humans. The seeds of these plants, which are of the genus *Senecio*, contaminate wheat and corn harvested from the same land.

Goitrogens. Goitrogens, natural products found in plant foods eaten by humans and animals, cause hypothyroidism, which is an enlargement of the thyroid gland. Consumption of additional iodine controls this problem. The plants in this group of the genus *Brassica* include cabbage, turnip, mustard greens, radish, and horseradish.

Saponins. Saponins, comprising at least 400 species and 80 families of plants, produce the toxin known as glycoside that hemolyze red blood cells. Saponins are found in soybeans and alfalfa, and are of particular concern, since soybeans and alfalfa are highly nutritious plants that are essential to our society.

Mycotoxins. Mycotoxins, produced by fungi, produce acute and chronic effects. Many humans and animals suffer with these fungal metabolites. Cases that come to the attention of public health officials are those in which large amounts of the toxin have been consumed and result in serious illness or death. Mycotoxins are secondary metabolites of molds that may be toxic, carcinogenic, or hepatoxic.

Species of *Aspergillius, Penicillium, Rhizopus,* and *Streptomyces* produce aflatoxins. Species of *Fusarium* produce trichothecenes. Species of *Penicillium* and *Aspergillus* produce penicillic acid.

There are three major classes of molds that invade agricultural products. They are field fungi, storage fungi, and advanced decay fungi. The physical factors that bring about mycotoxin production include moisture, temperature, mechanical injury, blending of grains, hot spots, and time. There must be a variety of conditions present that are favorable for the mycotoxin in order for the mycotoxins to be produced. Aflatoxins can grow on peanuts, cotton seed, seed oil, corn, legumes, dried fruits, wines, and dairy products. Zearalenone is found almost totally on corn. Ochratoxin is found on food grains such as wheat and barley. Citrinin is a co-contaminate with ochratoxins. Patulin is produced on apple rot and is found in apple juice. Penicillic acid is found on tobacco and storage grain, such as wheat, corn, peanuts, and cotton seed. Trichothecenes are found on a variety of different foods. Aflatoxins not only cause toxic effects, but also may be mutagenic and carcinogenic. When two or more toxins are combined, there may be a variety of toxic effects on experimental animals. Mycotoxins cause toxic effects in the liver, digestive tract, urinary system, the skin, hematopoietic system, reproductive organs, and nervous systems.

Allergic Reactions

Individuals may have peculiar allergic reactions to any kind of food, ranging from simple headaches to unusual central nervous system disorders, including convulsions and death. Infants and children often complain of abdominal distress and sometimes of genitourinary tract problems. The cardiovascular system and skin can be involved in an allergic reaction.

Parasitic Infections

Most parasitic infections are due to the ingestion of parasites in food or water.

Amoebic Dysentery. Amoebic dysentery, or amebiasis, is caused by *Entamoeba histolytica.* Water or food is contaminated with sewage or human feces containing the amoeba. Disease rates are higher in areas of poor sanitation, mental institutions, and among homosexuals. It may cluster in households or institutions where sanitation is good. The incubation period varies from a few days to several months or years. Commonly it is 2–4 weeks. The reservoir of infection is a person who is usually a chronically ill or asymptomatic cyst passer. Communicability may continue for years. The most dramatic incident in the U.S. of an outbreak of amoebic dysentery occurred in 1933 at the Chicago World Fair where 1000 people became infected and 56 died due to defective plumbing.

Trichinosis. Trichinosis, which is caused by *Trichinella spiralis*, has an incubation period of about 9 days, although it varies from 2 to 28 days. In heavy infections, the incubation period may be 24 hr. The live larvae are found in raw or improperly cooked pork, pork products, whale, seal, bear, or walrus meat.

Tapeworms. Tapeworms are caused by the ingestion of live larvae in raw or insufficiently cooked beef, fish, or pork. The beef tapeworm is *Taeniasis saginata*; the fish tapeworm is *Diphyllobothrium latum*; the pork tapeworm is *Taenia solium*. The beef tapeworm causes intestinal infections in people, whereas the pork tapeworm may cause intestinal infections and somatic infections by the larva. The larval disease, which is called cysticercus, is a tissue infection that may affect vital organs, and therefore, may cause fatality to occur. The beef tapeworm causes the intestinal infection, whereas the pork tapeworm may cause both diseases to occur. The fish tapeworm generally causes few symptoms. In some individuals, a vitamin B_{12} deficiency anemia may occur along with diarrhea. The incubation period for the pork tapeworm is 8–12 weeks. The incubation period for the fish tapeworm is 3–6 weeks. The reservoir of infection for beef and pork tapeworms is people. The intermediate host for the beef tapeworm is cattle, whereas the intermediate host for the pork tapeworm is pigs. The reservoir of infection for fish tapeworms is people and other hosts, include dogs, bears, and fish-eating mammals. The beef tapeworm may be communicable for as long as 30 years. The fish tapeworm is not communicable from person to person. An outbreak of *D. latum* occurred in Los Angles in 1980 when 39 people consumed sushi (a raw fish dish made of tuna, red snapper, and salmon). The salmon appeared to be the fish incriminated.

Giardiasis. Giardiasis is a protozoan infection, usually of the upper small intestine, which is caused by *Giardia lamblia*. The symptoms of the disease, when they are there, include chronic diarrhea, abdominal cramps, fatigue, and weight loss. The disease is transmitted by contaminated water and less frequently by contaminated food. Person to person contact may also occur and may also contribute to the spread of the disease when the hands of a contaminated person transfers the cysts from the feces of the individual to the mouth of another individual. This may occur in a variety of institutions, including day-care centers. The incubation period is 5–25 days or longer, with a possible 7- to 10-day period being most frequent. The reservoir of infection is people. The period of communicability lasts during the entire period of the infection.

Angiostrongyliasis. Angiostrongyliasis is a disease of the central nervous system that is caused by a nematode, *Angiostrongylus cantonensis.* The disease is spread through the ingestion of raw or improperly cooked snails, slugs, prawns, fish, and land crabs. The incubation period is usually 1–3 weeks, although it may be longer or shorter. The reservoir of infection is the rat. The disease is not transmitted from person to person.

Toxoplasmosis. Toxoplasmosis is a protozoal disease that may be spread from the mother to the child through the placenta, if she is infected, but may also be acquired by eating raw or undercooked infected pork or mutton. The disease may also be spread through water or dust contaminated by cat feces. The infectious agent is *Toxoplasma gondii.* The incubation period may be 10–23 days when consuming contaminated food. The reservoir of infection includes rats, pigs, cattle, sheep, goats, chickens, birds, and cats. The disease is not transmitted from person to person.

Scombroid Poisoning. Scombroid poisoning is caused by eating scrombroid fish or eating fish of the family Mahi-mahi, which is also called dolphin fish. The poisoning is due to a histamine-like substance produced by several species of *Proteus* bacteria or other bacteria. The histamine is produced from histidine in the flesh of the fish. The individual person has headaches, dizziness, nausea, vomiting, a peppery taste, a burning throat, facial swelling and flushing, stomach pain, and itching of the skin. The disease may also be caused by tuna or mackerel. The cause of the disease is inadequate refrigeration of scromboid fish. The incubation time is 10 min to 1 hr. The disease is not transmitted from person to person, and the reservoirs of infection are the fish that have been mentioned.

SPECIAL FOOD PROBLEMS

Church dinners are usually prepared by volunteers in their homes and then transported at varying times and temperatures to the place of consumption of the food. Time, temperature, personal health, methods of preparation, and initial ingredients are uncontrolled. This has led to outbreaks of food-borne disease and intoxication.

Although fairs and special events are limited timewise, they provide an environment where tens of thousands of people congregate and eat. The food operations should be similar to a permanent food operation but usually are not. Problems include lack of hot and cold running water, inadequate refrigeration, unwholesome food, poor handling of food, flies and other insects, lack of screening, and improper waste disposal. These food operations must be licensed and inspected by the health department.

Vendors that sell food to the public from vehicles of any type must meet the same criteria as permanent food establishments. Problems related to vendors include lack of hot and cold running water, improper waste and liquid waste disposal, lack of refrigeration, improper cleaning and sanitizing, and the inability of the health department to make inspections because of the high degree of mobility of the vendors. Sanitary conditions deteriorate very quickly in this type of operation. All vendors must be licensed.

Food stores must adhere to all of the regulations of other preparation and serving establishments. Problems usually include unwholesome food or damaged and rusted cans, improper refrigeration, stacking above freezer line, inadequate cleaning of equipment and utensils, flies, roaches, rodents, and inadequate solid-waste disposal.

The remnants of consumed food create solid-waste problems. If garbage is improperly wrapped or stored, it becomes a fly and rodent attraction. If it is contaminated by highly infectious individuals, it becomes an infection problem. When garbage is removed to a landfill, it becomes a part of the overall landfill solid-waste disposal problem. When burned it creates air pollutants.

The individual infected with food-borne disease organisms may become a carrier and transmit the infection to others. Although the initial spread of disease may be due to a food-borne organism, the secondary occurrences may well be traced to other environmental factors. An example of this would be the serious and prolonged outbreaks of *Salmonella derby* that occurred in U.S. hospitals in the 1960s. The initial outbreak was thought to be caused by contaminated eggs. The secondary outbreaks were caused by the carrier spreading the organism in a variety of ways.

Individuals who become ill from food-borne disease require treatment either at home or in hospitals. The estimated cost to society of food-borne illness is between 7.7 and 23 billion dollars.

POTENTIAL FOR INTERVENTION

The potential for intervention in food problems related to food protection and disease outbreaks consists of the techniques of isolation, substitution, shielding, treatment, and prevention. In the area of isolation, all prepared, stored, or served food must be kept from sources of hazardous materials, especially chemicals. The food also must be isolated from insects and rodents, and from contaminating sewage and water. In the area of substitution, mechanical processes should be substituted for hand operations in the preparation of food when possible. In the area of shielding, food on a buffet table must be protected from the customers.

Treatment is used for food-handlers with food-borne diseases. Prevention is the major technique utilized for intervention in disease processes. Prevention includes everything from refrigeration and freezing, adequate cooking, and proper storage, to removal of sick food-service workers from food preparation and service.

The potential for intervention generally is very good; specifically, care should be given to the selection of food and special techniques used for the handling, processing, and serving of specific types of hazardous foods.

RESOURCES

Scientific and technological resources in the area of food include the Institute of Food Technology, the Canning Trade Inc., the International Association of Environmental, Milk, and Food Sanitarians, the National Environmental Health Association, the American Public Health Association, various land grant colleges such as the University of Iowa and Purdue University, schools of public health such as the University of Michigan, the University of California, and the University of Minnesota. Civic associations concerned with food are basically in the area of nutrition and provision of food for individuals who are starving or who have inadequate food supplies. The governmental organizations involved in food include the Federal Food and Drug Administration, various local and state health departments, and many of the state departments of agriculture throughout the country. The U.S. Department of Agriculture is another fine resource. County extension agencies are an immediate source of help. The U.S. Senate Agricultural Committee is an important source for new legislation. The "Food Code," 1993 reccomendations of the U.S. Public Health Service Food and Drug Administration, is excellent. Shellfish and Sanitation Branch, U.S. Food and Drug Administration, 200 "C" Street S.W., Washington, D.C. 20204.

STANDARDS, PRACTICES, AND TECHNIQUES

Plans Review

A successful food-service operation is dependent on good facilities, proper arrangement and utilization of equipment, good-quality raw food, and well-trained personnel. State health departments and, where designated, local health departments are responsible for reviewing the plans of all new food establishments prior to construction, the conversion of an existing structure for food use, and the remodeling or change of a food establishment. Plans and specifications, including architectural, structural, mechanical, plumbing drawings, intended menu, and volume of food prepared, stored, served, and sold must be submitted to the health department, which gives preliminary approval for construction and/or remodeling and makes an inspection prior to licensing. The inspection includes evaluating material and grouting of floors; material, finish, and color of walls and ceilings; presence of insect- and rodent-proofing devices; quantity and quality of lighting in various areas; amount and type of ventilation; type of toilet facilities for males and females; type of water supply; hand-washing facilities; construction of utensils and equipment; type of utensil washing and sanitizing; type of utensil storage; type of food-storage areas; type and quality of food refrigeration; type of locker or dressing rooms; type of sewage disposal; and type of solid-waste disposal. The food service operation is licensed and permitted to serve food after specifications set forth by the health department are met.

Food Service Facilities

Physical Facilities

Floors in food preparation, serving, and storage areas should be composed of smooth, nonabsorbent, easily cleaned materials to aid in cleaning and to prevent absorption of grease, organic material, and odors. Anti-slip materials or treatment may be used for safety purposes. When floor drains are used, they should be screened to prevent harborage of flies and roaches, and to enable easy cleaning. Nonrefrigerated, dry storage areas need not have nonabsorbent floors. Securely attached, tightly installed carpeting under coving or away from the wall may be used except in food preparation areas, storage areas, warewashing areas, hand washing areas, toilet room areas where urinals and toilets are located, and refuse storage rooms/areas.

Walls and ceilings should have light-colored, smooth, easily cleaned surfaces to facilitate good food preparation and food-service habits and to provide more adequate distribution of light. A minimum of 70 footcandles of light are needed, whereas 50 footcandles are required, on all food preparation surfaces and 20 footcandles on all other surfaces to provide an environment conducive to the elimination of dirt, grease, insects, and rodents, and to prevent accidents.

Plumbing and Water Supply

There must be an adequate quantity of bacteriologically and chemically safe hot and cold water under pressure from an approved source, since water can easily contaminate food directly or indirectly through equipment, utensils, and hands. Ice must be prepared from water from an approved source, and must be transported and stored in a sanitary manner, since ice is used in food and drink, and has been shown to contain numerous bacteriological contaminants. Ice used for cooling must not then be used as a food.

Plumbing and drainage systems carrying human wastes have been incriminated in outbreaks of typhoid fever, paratyphoid fever, dysentery, and other gastrointestinal diseases. Transmission of these diseases usually occurs through cross-connections, overhead leakage, stoppage in drainage systems, and submerged inlets. Backflow to the potable water supply system may occur directly in coffee urns, dishwashing machines, and double-jacketed kettles. Back-siphonage is possible where submerged inlets occur as in drinking fountains, flushometer valves, flush tanks, garbage-can washers, hose outlets, ice makers, lavatories, slop sinks, steam tables, and vegetable peelers. The backup of sewage is possible in enclosed equipment, such as ice makers, refrigerators, steam tables, and walk-in freezers. Sewage should be disposed of in a public sewage system or in a properly operated on-site sewage disposal system to prevent the spread of enteric disease and mosquito breeding.

Separate sanitary toilet facilities with self-closing doors are needed to prevent the spread of enteric disease. The hand-washing facilities must have hot and cold running water, soap, and disposable towels or air dryers to prevent the spread of disease through contaminated hands. (In-depth discussions on water supply, plumbing, and sewage disposal are found in Volume II.)

Hot water has a corrosive effect on pipes. Hard-water deposits are seven times greater at 180°F than at 140°F. Water softeners should be used where hardness exceeds 125 ppm. Since hot water is the sanitizing agent, in machine warewashing, the force and water pressure, volume of water, and temperature are very important. A water pressure of 15–25 lb/in.^{2} (which is 100 to 170 kilopascals) at the entrance to the rinse manifold is optimal. Water should be flowing at a rate of 9 gal/min in the rinse section. The gas or electric company should be called upon to determine the size of water heater necessary for any given establishment.

Ventilation Systems

Adequate ventilation must be provided for all rooms to reduce condensation, minimize soiling of walls and ceilings, and remove excessive heat and objectionable odors. All hoods should be equipped with noncombustible 2-in. commercial grease filters that are easy to install, remove, and clean. The surface of hoods should have sealed joints and seams, be accessible, and be easily cleaned. The function of the hood is to collect vapors, mists, particulate matter, fumes, smoke, steam, or heat. The velocity of the air capturing these items is called the capture velocity. The canopy hood is an overhead hood completely covering the equipment it is designed to serve. The filter in the hood should be a minimum of 2.5 ft above an exposed cooking flame, 4.5 ft above charcoal fires, and 3.5 ft above other exposed fires. Hoods should be so designed as to not interfere with normal combustion and/or exhaust of combustion products from commercial cooking equipment, other processes, and heat equipment.

To determine the total quantity of air exhausted from a hood, the following formula should be used:

$$Q = VPD$$

where Q is the flow rate of air exhausted in ft³/min, V is capture velocity in ft/min, P is the perimeter of the open sides of the hood in feet, and D is the distance between the cooking surface and the face of the hood in feet.

Make-up air, composed of outside air equal to 100% of the quantity of the air exhausted, must be introduced into the building in such a way as to not interfere with the exhaust system.

Storage of Food

Wholesome food can be readily contaminated if stored improperly. Food must be protected against unsanitary conditions, insects and rodents, poisons, cleaners, other chemicals, and bacterial growth. Temperature, time of storage, initial contamination and moisture content are also vital factors to be considered in food storage.

Dry Storage

Unsanitary conditions result from dirt, dust, contamination by overhanging sewage pipes, sneezing, coughing, unclean utensils and work surfaces, and unnecessary handling.

Food-product insects can easily contaminate a large amount of grains, flour, and rice. These supplies should be stored in a clean, dry area, on racks at least 6 in. off the floor, and 1 ft from walls, and should be rotated at least once every 2 weeks during the summer months and once every 3–4 weeks during the winter months. The keeping quality of the food is improved because of adequate air circulation and because it is possible to prevent insects and rodent harborage under or behind the food. Mice and rats contaminate large quantities of food through defecation and urination on the food and through nibbling. This hazard can be detected by using a blacklight to determine if urine is present and a flashlight to determine if fecal material is present.

All poisons, cleaning materials, and other nonfood items must always be clearly labeled, stored in their original containers, and kept apart from all foods. However, only those poisons and cleaning materials needed for the operation of the establishment can be kept here.

Refrigerated Food

Bacterial growth can be controlled by proper cooking and by storing food at or below 41°F or above 140°F in shallow containers. The temperature at which food is stored is probably one of the most important of all environmental conditions, since it affects the rate of decomposition and the growth of microorganisms. The time of storage is equally important in preventing disease, undesirable changes in taste, odor, and quality of the raw products, and therefore, it is essential that the initial contamination of raw food be minimized. Chemicals added to food limit food spoilage, but add unusual tastes and odors if the food is stored for long periods of time. The presence or absence of air determine the amount of food decomposition due to different types of bacteria. Moisture, which aids in the growth of bacteria, has a considerable effect on food spoilage.

Refrigeration of food has been practiced for many centuries. Food has been placed in cool caves, wells, springs, and running streams. Ice has been used for at least 200 years. Today, the modern refrigerator and freezer is a major means of storage and preservation of food.

The rate of refrigeration is influenced by the heat transfer properties of the food, the volume of the food to be refrigerated, the kind of containers used, the heat conductivity of the containers, the agitation of the food, and the temperature difference between the food and the refrigerator unit. (See Table 2.1 for refrigerated storage temperatures and shelf-life for cold perishable foods.)

Potentially hazardous ingredients for salads, sandwiches, filled pastry products, and reconstituted food need to be chilled to 41°F or below within 4 hours, or more rapidly if possible.

Frozen Food

The quality of frozen food is determined by the raw products, the method of preparation, the speed of freezing, the temperature and time involved in distribution and storage, and the method of packaging. The frozen food package must be strong, flexible, and prevent entrance or escape of liquid from the package. Desiccation of frozen foods is a physical change taking place when moisture escapes. Crystallization is also a physical change, usually occurring when frozen food is permitted to defrost and is then refrozen. Bacteria grow readily in the moisture film collected at the surface of the frozen foods when the temperature rises above 32°F. The general practice of defrosting frozen foods at room temperature is considered to be poor, since the outer surface becomes a fine bacterial medium while the inner core is still defrosting. Frozen foods should be defrosted in the refrigerator at temperatures not exceeding 41°F or should be cooked immediately. Frozen poultry, fish, shellfish, and frozen leftovers are particularly prone to contamination and bacterial growth, and may easily lead to outbreaks of food infection or food poisoning if not properly defrosted and adequately cooked. Frozen-food temperatures should be below 0°F. Fishery products should be at −4°F or below for 7 days in a freezer, or −31°F or below for 15 hours in a blast freezer to destroy parasites.

Protecting Wholesomeness and Detecting Spoilage of Foods

All raw food products and processed food should be clean, wholesome, free from spoilage, free from chemical contaminants, and free from organisms that cause disease or intoxication. Meat, milk, shellfish, and similar products should

Table 2.1. **Refrigerated Storage Temperatures and Shelf–Life**

Product	Temperature (°F)	Maximum Storage Period
Meats		
Bacon	28–30	15 days
Beef (dried)	62–40	6 months
Brined meats	31–32	6 months
Beef (fresh)	30–32	1 day
Fish (frozen)	–4	6 months
Fish (iced)	30–32	10 days
Hams and loins	28–30	21 days
Lamb	28–30	14 days
Livers	20–22	6 months
Oysters (shell)	32–38	15 days
Oysters (tub)	32–38	10 days
Pork (fresh)	30–32	15 days
Pork (smoked)	28–30	15 days
Poultry (fresh)	28–30	5 days
Poultry (frozen)	0–5	10 months
Sausage (franks)	35–40	2 days
Sausage (fresh)	31–27	15 days
Sausage (smoked)	32–40	6 months
Veal	28–30	15 days
Miscellaneous		
Butter	35	6 months
Cheese (American)	32–34	15 months
Cheese (Swiss)	38–42	60 days
Cream (40%)	5–10	4 months
Eggs (frozen)	0–5	18 months
Eggs (fresh)	38–45	2 months
Milk	35–40	5 days
Oleo	34–36	90 days

come from approved sources. All leftovers should be discarded, having been handled extensively and probably having been kept at improper temperatures. All food-processing and storage areas should be free of birds, animals, insects, and rodents. Frozen foods should be defrosted in such a way as to prevent rapid growth of bacteria. The best approach is to defrost under refrigeration at 41°F or below.

Visual Inspection of Foods

Decomposing meat may have an off odor, be slimy, and have an off-color. Beef spoils from the surface inward; pork spoils from the bone outward. Decomposing fish have an off-odor, sunken eyes, and gray or greenish gills. An indentation will remain on the flesh of the decomposing fish, which is easily pulled away from the bones. Decomposing poultry will be sticky or slimy, have off odors, and exhibit darkening of wing tips. Canned foods can be swollen, dented, rusted, without labels, and have off-odors and off-color. With the exception of large pieces of beef, all of the previously mentioned foods should be discarded. Large pieces of beef can be trimmed by the butcher and the trimmings discarded. Ground hamburger and ground pork decompose rapidly; discoloration and off-odors result.

Thawing of Frozen Foods

To protect the wholesomeness of frozen foods and to prevent potential outbreaks of disease, the following thawing procedures should be followed: Cook frozen vegetables, small cuts of meat, chicken, fish, and prepared foods in the frozen state; thaw frozen fruits, juices, large cuts of meat, poultry, and shellfish in the refrigerator at or below 41°F.

Potentially Hazardous Foods

Fresh meats may contain *C. perfringens*. Since growth of the bacteria and development of the toxin occur best at room temperatures, refrigerate meat immediately after cooking and until used.

Fecal streptococci and *Staphylococcus aureus* are also found on meat. Partially cooked hams may contain fecal streptococci and staphylococci. Smoked hams that are cooked to an internal temperature of 150°F are not sterile and may contain organisms. Since cold cuts, including hot dogs, are handled by personnel, there is a potential hazard of staphylococcal food poisoning or salmonella food infection.

Salmonella, the primary contaminant of poultry, causes many outbreaks of salmonellosis; *C. perfringens* and fecal streptococci may also be contaminants.

Fish and shellfish coming from sewage-polluted waters may contain enteric organisms. *Clostridium botulinum* Type E is found on fish. Shellfish have been incriminated in outbreaks of infectious hepatitis and cholera.

Canned foods, such as tuna fish and mushrooms, and prepared frozen foods, such as some pizza, have been shown to contain botulism toxin.

Chicken salad, tuna salad, ham salad, potato salad, custard-filled pastries, dairy products, smoked fishes, egg salad, and other mayonnaise or cream-based salads have led to substantial outbreaks of food poisoning and food infection. The lack of proper refrigeration, improper handling of the food, and saving of leftovers are the contributing factors.

Food contaminated with water and smoke during fires, defrosted frozen foods, and unrefrigerated perishables are considered unwholesome and should be discarded.

Preparation and Serving of Food

Since food becomes contaminated easily during preparation or serving, and since organisms already present in the food can cause disease and poisoning, the following rules of food protection should be followed:

1. Use only good quality, wholesome food.

2. Keep the food clean and free of insects and rodents.
3. Clean preparation and serving areas carefully.
4. Use only well-constructed, easily cleaned equipment.
5. Keep equipment very clean.
6. Handle food as little as possible.
7. Use good personal hygiene at all times.
8. Refrigerate perishables as quickly as possible.
9. Cook foods long enough to kill organisms.
10. Keep food below 41°F or above 140°F in serving areas.
11. Provide proper sneezeguards in cafeterias and smorgasbords.
12. Use indicating thermometers accurate to +2 or −2°F to determine the temperature of the refrigerator and steam table.

Be particularly careful of ground and chopped foods, foods made of several raw materials, rich or nutritious foods, foods with high moisture content, and foods that require considerable handling during preparation, such as salads and ground meats.

Utilize the proper precleaned equipment and tools for the job. Use forks, knives, tongs, spoons, scoops, or single-use gloves on hands. Remove small quantities of prechilled perishable ingredients from the refrigerator immediately prior to use. Wash all raw fruit and vegetables thoroughly before cooking or serving. Cook all raw animal foods such as eggs, fish, poultry, pork, and game animals to an internal temperature of at least 155°F for 15 seconds. Cook all products which are stuffed, including pasta, to an internal temperature of at least 165°F for 15 seconds. See Table 2 for roast beef and corned beef. Place all custard, cream filling, and puddings below 41°F.

When food is served, the server must be careful not to place his or her fingers on the eating surface of the dishware or silverware. Plates of food should not be stacked on each other, since the undersurface of plates could contaminate the food beneath it.

Raw animal food cooked in a microwave oven must be:

1. Rotated or stirred throughout or midway during cooking because of the uneven distribution of heat.
2. Covered to retain surface moisture.
3. Heated an additional 25°F above previously stated temperatures for raw animal food products, because of the shorter cooking times.
4. Allowed to stand covered for two minutes after cooking to obtain temperature equilibrium.

When food is served in smorgasbords, buffets, and cafeterias, it should be kept behind protective guards below the speaking level of people. The food-holding equipment should be made of nonabsorbent, smooth, easily cleaned, corrosion-resistant material. Glass used must have a safety edge. Since hot or steam tables

Oven Parameters Required for Destruction of Pathogens on the Surface of Roasts of Beef and Corned Beef.

Oven Type	Oven Temperature Roast Weight	
	Less than or equal to 4.5 kg (10 lbs)	Greater than 4.5 kg (10 lbs)
Still Dry	177°C (350°F)	121°C (250°F)
Convection	163°C (325°F)	163°C (325°F)
High Humidity[1]	less than 121°C (250°F)	less than 121°C (250°F)

[1]Relative humidity greater than 90% for at least 1 hour as measured in the cooking chamber or exit of the oven; or in a moisture-impermeable bag that provides 100% humidity.

Minimum Holding Times Required at Specified Temperatures for Cooking All Parts of Roasts of Beef and Corned Beef.

Temperature °C (°F)	Time	Temperature °C (°F)	Time	Temperature °C (°F)	Time
54 (130)	121 minutes	58 (136)	32 minutes	61 (142)	8 minutes
56 (132)	77 minutes	59 (138)	19 minutes	62 (144)	5 minutes
57 (134)	47 minutes	60 (140)	12 minutes	63 (145)	3 minutes

[1]Holding time may include postoven heat rise.

Table 2.2.   Food Code 1993, U.S. Public Health Service, U.S. Department of Health and Human Services, Public Health Service, Washington, D.C. 20204, p. 52.

and cold tables are not intended to alter the temperature of food, but rather maintain it at the proper temperature, the food has to be preheated to 140°F internal temperatures or precooled to 41°F internal temperature. Hot tables should be maintained at 160°F in order to achieve the 140°F needed in the food. Where ice is used for cooling, the ice should be $3/_4$ in. in diameter or less, should be clean, and should come from an approved source. Food containers should be a maximum of 6 in. deep, and food should never be stacked above the ice level. Ice storage units need drains, which have an air gap, in order to prevent backflow of sewage. Tongs, forks, spoons, picks, spatulas, and scoops should be used by food-service workers, if possible. If the customer uses these implements, they should be changed frequently. Ice cream scoops, dippers, and spoons should be stored either in a cold running water well, which is frequently cleaned, or in a clean, dry manner. Sugar should be kept in closed containers or preferably in single-service packets. The remnants of all food served to the customer should be discarded.

Where food is transported in hot and cold food trucks to various parts of an institution, special care must be taken to preheat and precool the unit. The food containers and the truck must be carefully cleaned immediately upon completion of the serving process. The set-up trays must be covered to avoid contamination by dust, aerosols, and microorganisms.

Single-service articles are an excellent method of serving food, since the articles are thrown away after each use and a premeasured portion of food can be set up at the processing plant or kitchen. Contamination by hands and improperly washed utensils is avoided. Caution must be used in the storage of single-service materials to prevent contamination by dust, dirt, sewage, insects, and rodents. Items should be discarded in lined containers with self-closing lids.

Design and Installation of Food-Service Equipment

The design of food-service equipment should be based on good research, sound engineering, good environmental practices, and the practical knowledge of experts from industry and the field of public health. The standard for equipment design is now basically set by the National Sanitation Foundation (NSF) in Ann Arbor, Michigan. This nonprofit organization develops standards based on the aforementioned criteria.

Food-service equipment should contain the following design features:

1. easily disassembled, cleaned, and maintained equipment with as few parts as possible
2. smooth, nonabsorbent, nontoxic, odorless, and easily cleaned food-contact surfaces
3. food-contact surfaces containing no toxic materials, such as cadmium, lead, or copper
4. nontoxic, nonabsorbent, easily cleaned gaskets, packing, and sealing materials that are unaffected by food or cleaning products
5. easily cleaned splash zone areas
6. food product surfaces that will not chip, crack, or rust

There are use limitations for cast iron, ceramic, and crystal; utensils, galvanized metal, linens, napkins, and sponges; pewter, solder, flux, and wood.

Existing NSF standards or criteria should be reviewed at least every 3 years to ensure that the food-service equipment keeps pace with the public health and industry advances. The competent food-service manager should obtain copies of NSF standards and review them carefully before purchasing new equipment.

The installation of equipment is almost as important as its design and construction. If equipment is installed into the wall but is not flush with the wall, roaches may easily hide behind it. Overhead equipment such as hoods must be within easy reach, enabling food-service personnel to remove the filters regularly and to clean the hoods and filters frequently. Large equipment should be mobile or installed on legs to facilitate cleaning behind and underneath. Drains should be located in areas in the floor where spillage is likely or where equipment is scrubbed. Hand-washing facilities and service sinks should be conveniently located to ensure their use when needed. Clean metal storage shelves must be provided for the storage of cleaning equipment and utensils. See Figure 2.2 for a proper layout of kitchen equipment.

Figure 2.2.   Layout of kitchen equipment

Housekeeping and Cleaning in Food-Service Facilities

In a satisfactory environment in a food-service facility, soil is removed, bacteria is destroyed, and an aesthetically pleasing picture is provided through good physical facilities, proper color, adequate light, and well-managed cleaning practices. Soils are divided into four basic types: (1) fresh soil found immediately after use of equipment or facilities; (2) a thin film of soil caused by ineffective cleaning within which microbes live; (3) built-up deposits of soil caused by consistent ineffective cleaning and composed of soap films, minerals, food materials, and grease; (4) dried deposits of soil caused by drying of heavy, crusty deposits or baking of organic material onto dishes or trays by improper warewashing.

Fresh soils and thin films of soil on floors and walls are removed by the use of a good detergent and spraying technique, and plenty of physical action. Built-up deposits are best removed by the use of detergent spraying techniques plus mechanical scrubbing machines and special heavy-duty detergents. Dried deposits on dishes are first soaked in a good detergent and hot water, and then scrubbed with plastic or metal pads. Dried deposits on baking trays are best removed by cleaning with live steam. The best approach to cleaning, however, is prevention of the accumulation of deposits and soil through good daily cleaning. (Live steam is dangerous! Be careful of its use!)

General Daily Cleaning

At the conclusion of each work day, an assigned food-service worker should wash all kitchen, storage, and serving area floors in the following manner:

1. Starting at the far end of the room, spray an area of the floor measuring approximately 100 ft^2.
2. Spray an adjacent area of approximately the same size.
3. Pick up the detergent with a very clean mop and rinse in a clean pail of water.
4. Repeat steps 1, 2, and 3 until the entire floor is washed.
5. Clean floor drains.

All dry-food storage areas should be swept with a treated mop daily to decrease contamination by dust and dirt, and also to determine if an insect or rodent infestation exists. All outside areas, especially where solid waste is stored, must be checked frequently during the day and swept and washed daily. This prevents an unsightly mess and also helps destroy potential insect and rodent harborage.

Bathroom floors should be cleaned at least twice a day using the same procedure outlined for kitchen floors. In addition, the trash in the bathrooms should be emptied a minimum of twice daily, the sinks washed, and the soap and single-service towels refilled. Toilet bowls and urinals must be cleaned thoroughly. Deodorants are not a substitute for good cleaning procedures. Under no condition should food be stored in restrooms. The food-service attendant must thoroughly scrub his/her hands for at least twenty seconds with a cleaning compound and rinse thoroughly with clean water. Special attention must be paid to fingernails and between the fingers. After defecating or handling waste, it is recommended that handwashing take at least two minutes to be thorough. Finally, an alcohol based, instant hand sanitizer solution should be used. The sanitizer should be the strength equivalent of 100 mg/L chlorine or above.

Periodic Cleaning

Walls and ceilings should be washed down with a good detergent periodically, but not less than once a month. Lighting fixtures should be taken apart and cleaned and light bulbs replaced at least once every 2 months or when the bulbs burn out.

Cleaning Schedules

The establishment of definite work assignments in writing, and the development of a cleaning manual and schedule is essential to good cleaning practices. A food-service worker must be taught to disassemble and clean pieces of equipment and to utilize cleaning equipment and materials. A good food-service manager supervises this part of the food-service operation closely. The food-service manager should inspect his or her own establishment on a weekly basis using a flashlight to see behind and under equipment. A self-inspection sheet can be obtained from the health department or industry, or may be found in the chapter entitled Institutional Environment.

Cleaning and Sanitizing Equipment and Utensils

Good cleaning consists of prescraping and prerinsing with cold or warm water as soon after usage of equipment and utensils as possible; soaking in hot detergent solution, without overloading or improper stacking; power scrubbing manually or mechanically; power rinsing; sanitizing; and air drying.

All equipment, including counters, tables, carts, display cases, steam tables, shelves, storage racks, salad and vegetable bins, drain boards, meat blocks, stoves, stove hoods, coffee urns, meat and vegetable choppers, meat tenderizers, food mixers, griddles, cutting boards, ice-cream counter freezers, steam-jacketed kettles, vegetable peelers, and garbage cans, must be cleaned thoroughly each day. In all cases, empty the contents of the equipment and rinse and scrape thoroughly. For large, immovable equipment, spray all portions with a good QAC compound, scrub with clean brushes, and rinse with QAC. Scrape meat blocks and cutting boards clean and wash. Turn meat blocks regularly. Where food spillage occurs in the oven, sprinkle salt on food and heat the oven to 500°F. When the spillage is carbonized, cool oven and scrape food deposit. Ice-cream counter freezers must be emptied each night and the contents discarded. Rinse the unit thoroughly with water while still assembled, add dishwashing detergent, and operate for 1 min, rinse and drain all detergent, disassemble all parts, scrub thoroughly with a brush and detergent, rinse and reassemble, and use a final rinse of chlorine. Garbage cans should contain plastic liners, which are removed and replaced frequently. The cans should be scrubbed mechanically with a garbage-can washer or with a brush at least once every 3 days.

Hand Warewashing

Prescrape and prerinse all dishes and utensils, and place in the first compartment of a three-compartment sink containing detergent and hot water at 95°F or above. Scrub dishes well with plastic brushes. Do not use wash cloths or sponges, since they spread microorganisms. Rinse dishes thoroughly in the second sink at a temperature of 110–120°F. In the third sink, use a chemical sanitizer or hot water at a minimum of 170°F for 1 min. Remove and allow to air dry. An accurate temperature measuring device must be kept in the sink or be readily available. In all cases, predetermine the amount of water to be used to ensure that sufficient detergent and/or sanitizer is added. A test kit or other device needs to be provided to accurately measure the concentration of the sanitizer solution.

Machine Warewashing

Prescrape and prerinse dishes and utensils. Prewash heavily soiled items carefully. Place objects in racks so that all parts are exposed to the spray. Wash at 150°F for 40 sec. Rinse at 180°F at a pressure of 15–25 lb/in.^{2} (100 to 170 kilopascals) for 30 sec. Some variations in time and temperature will occur based on the type of dishwashing machine used. Suitable thermometers and pressure gauges should be attached to each machine. The accuracy of these gauges should be checked during an inspection.

A good mechanical warewashing operation depends on the following: (1) selection of the proper-size dishwashing machine; (2) proper-size hot-water boiler and booster heater; (3) proper layout of equipment and utilization of workforce; (4) adequate training for personnel; (5) proper supervisory control; (6) adequate storage of clean dishes and utensils; (7) discarding of chipped and cracked dishes, and rusted or corroded utensils; (8) adequate ventilation of dishwashing area to reduce strain and maximize personnel efficiency.

Soiled dishes are caused by improper scraping, prerinsing or prewashing, inadequate detergent, improper wash-water temperature, inadequate time for wash and rinse, and improperly cleaned dishwasher. Check all operations carefully and make corrections where needed. Unclog wash and rinse nozzles. Films are caused by water hardness: use adequate amounts of detergent and water softeners. Greasy films are due to low pH, insufficient detergent, low water temperature, and dirty dishwasher: check alkalinity, detergent, temperature, and wash and rinse nozzles. Streaking is due to high pH and improperly cleaned dishwasher: reduce alkalinity and clean dishwasher. Cooked-on egg or other proteins are caused by improper prerinsing and prewashing, and too high a washing temperature. Spotting is due to rinse-water hardness, too high or low a rinse temperature, inadequate time between rinsing and storage: soften water, check temperatures, and allow adequate time for air drying.

Dishwashers should be cleaned at the end of each work day and descaled at least once a week. Daily cleaning consists of removing, soaking, and scrubbing scrap traps, the suction strainer, and wash and spray arms. Return to the machine and run the machine without using a load of dishes. Once a week, after disassembling and scrubbing, reassemble, add 7 fl. oz of phosphoric acid or a 2% solution of acetic acid to the wash water, and operate the machine for 1 hr; drain off the solution, add baking soda, or 2 cups of detergent, and run the machine for 15 min; drain the machine and rinse it several times.

Solid Waste, Recyclables, and Returnables.

Solid wastes containing food are unsightly, odorous, and attract insects and rodents. Recyclables and returnables may also contain food. Unless a refuse can is in constant use in an establishment, it should have a tight-fitting lid. At night all receptacles should be tightly closed. The indoor storage area for refuse, recyclables, and returnables should be clean and effectively screened for flies. Self-closing mechanisms should be used for all exterior and bathroom doors. Garbage grinders should only be used when the liquid waste goes to a public sewer system.

Solid waste must be stored inside or outside the premises in receptacles with tight-fitting lids 12 inches off the ground. These containers should not absorb odors or leak water. The 10- or 20-gal cans should always have a plastic liner. Cans should be scrubbed at least every 3 days. All solid waste should be removed daily, or at the most every 3 days. Problems result from carelessness in dumping of waste, improper supervision by management, failure to empty cans at night, dirty cans, lack of a drain or clogged drain where cans are washed, failure to clean large metal storage boxes, failure to replace rusted or crushed cans and lids, and failure to use plastic bags. A complete discussion on solid-waste disposal may be found in the chapter on solid and hazardous waste in Volume II. Recyclables and returnables should be removed frequently enough to avoid unsightly conditions, odors, and prevent insect and rodent problems.

Insect and Rodent Control

Flies, roaches, and rodents spread disease by contaminating food. Food-product insects are a nuisance and a contaminant. Flies are present wherever there is highly organic material. They breed in floor drains or in other out-of-the-way areas. The best control for flies is removal of organic material, cleanliness, screening, and the use of safe pesticides. Roaches enter the premises through cracks; along pipes; in packages; through doors, windows, and other openings; in clothing; or through broken sewer lines. Cleanliness, sealing of roach entrances, and a good, safe poisoning program is required. Mice and rats enter establishments through holes in the walls, open doors and windows, and ruptured sewer lines. Control occurs through removal of harborage and food, poisoning, and rodent proofing. Food product insects infest dried fruit, cereal, grain, flour, rice, candy, and nuts. Control occurs through initial fumigation, rotation of stock, and destruction of highly infested foods. A detailed discussion on insects, rodents, and the use of pesticides occurs in another part of this book.

Personal Health of Employees

An employee's health, and cleanliness of person and clothing prevents the spread of many food-borne diseases. Hands are one of the greatest causes of the spread of disease. Hands must be scrupulously washed when starting work, after using the bathroom, after cleaning, and before preparing or serving food. Hands should be kept away from the head, mouth, nose, and other body parts. A 2-min hand and forearm scrub, including several soapings and rinsings, should be a routine practice for the worker. Food-service attendants should never smoke while preparing or serving food, should have clean and covered hair, and should have clean clothing changes daily. A food-service worker should not work if ill with an illness due to *Salmonella typhi*, *Shigella spp.*, *Escherichia coli* 0157:H7, or Hepatitis A virus infection. Employees with symptoms of intestinal illness such as abdominal cramps or discomfort, diarrhea, fever, loss of appetitie, or three or more consecutive days vomiting or jaundice; a boil or infected wound; past illness from an infectious agent which may create a potential carrier state; a record of being suspected of causing or being exposed to an outbreak of the above organisms, or family members sick from the above organisms should be restricted from work. All of these workers with infections or exposure to infections should be seen by a physician and a culture taken if needed. All individuals with diarrhea should have a stool specimen taken. If the specimen is positive for an enteric organism causing a food-borne illness, the worker should not be permitted to return to work until three consecutive negative samples are taken. In keeping with the requirements of the 1993 Food Code, the Applicant and Employee Interview form, (Figure 2.3), the Food Employee Reporting Agreement form (Figure 2.4), and the Applicant and Food Employee Medical Referral form (Figure 2.5), should be completed as needed.

A properly protected and stored First-Aid kit should be available for employee use.

MODES OF SURVEILLANCE AND EVALUATION

Flow Techniques and Inspectional Procedures

The department of health, or other designated departments at the state or local level, issue licenses for the operation of a food-service establishment based on laws approved by the appropriate legislative bodies. The health department formulates rules and regulations to be followed by the establishments, and conducts

---

**Form
1**

Applicant and Food Employee Interview

Preventing Transmission of Diseases through Food by Infected Food
Employees with Emphasis on *Salmonella typhi, Shigella* **spp.,**
Escherichia coli O157:H7, and Hepatitis A Virus.

---

*The purpose of this form is to assure that Applicants and Food Employees advise the
Person in Charge of past and current conditions described so that the Person In Charge
can take appropriate steps to preclude the transmission of foodborne illness.*

Applicant or Employee name (print)_____
Address _____

Telephone Daytime: _____ Evening: _____

TODAY:

Are you suffering from any of the following:

1. Symptoms
   Abdominal cramps?   **YES/NO**
   Diarrhea?   **YES/NO**
   Fever?   **YES/NO**
   Prolonged loss of appetite (more than 3 days)?   **YES/NO**
   Jaundice?   **YES/NO**
   Vomiting?   **YES/NO**
2. Pustular lesions
   Pustular lesion on the hand, wrist or an exposed body part?
   (such as boils and infected wounds, however small)   **YES/NO**

PAST:

Have you ever been diagnosed as being ill with typhoid or paratyphoid fever
(Salmonella typhi), shigellosis (*Shigella* spp.), *Escherichia coli* O157:H7
infection, or hepatitits (hepatitis A virus)?   **YES/NO**
If you have, what was the date of the diagnosis _____

HIGH–RISK CONDITIONS

1. Do you have a household member attending or working in a setting where
   there is a confirmed outbreak of typhoid fever, shigellosis, *Escherichia coli*
   O157:H7 infections, or hepatitis A?   **YES/NO**
2. Do you live in the same household as a person diagnosed with typhoid fever,
   shigellosis, hepatitis A, or illness due to *E. coli* O157:H7?   **YES/NO**
3. Have you been exposed to or suspected of causing a confirmed outbreak of
   typhoid fever, shigellosis, *E. coli* O157:H7 infections, or hepatitis A? **YES/NO**
4. Have you traveled outside the United States within the last 50 days? **YES/NO**

Name, Address, and Telephone Number of your Doctor:
Name _____
Address _____

Telephone Daytime _____ Evening_____

Signature of Applicant or Food Employee _____ Date _____

Signature of Permit Holder's Representative _____ Date _____

Figure 2.3   Food Code 1993, U.S. Public Health Service, U.S. Department of Health
and Human Services, Public Health Service, Washington, D.C. 20204, Annex 7, p. 1

**Form
2**

Food Employee Reporting Agreement

Preventing Transmission of Diseases through Food by Infected Food
Employees with Emphasis on *Salmonella typhi, Shigella* **spp.,**
Escherichia coli O157:H7, and Hepatitis A Virus.

---

*The purpose of this form is to assure that Food Employees notify the Person in
Charge when they experience any of the conditions listed so that the Person In
Charge can take appropriate steps to preclude the transmission of foodborne illness.*

I AGREE TO REPORT TO THE PERSON IN CHARGE:

FUTURE SYMPTOMS and PUSTULAR LESIONS:
1. Abdominal cramps
2. Diarrhea
3. Fever
4. Prolonged loss of appetitie (more than 3 days)
5. Jaundice
6. Vomiting
7. Pustular lesions:
   - Pustular lesion on the hand, wrist, or an exposed body part
     (such as boils and infected wounds, however small)

FUTURE MEDICAL DIAGNOSIS:

Whenever diagnosed as being ill with typhoid or paratyphoid fever *(Salmonella typhi),* shigellosis
(*Shigella* spp.), *Escherichia coli* O157:H7 infection, or hepatitis (hepatitis A virus).

FUTURE HIGH–RISK CONDITIONS:

1. A household member attending or working in a setting experiencing a confirmed outbreak of typhoid
   fever, shigellosis, *Escherichia coli* O157:H7 infections, or hepatitis A.
2. A household member diagnosed with typhoid fever, shigellosis, hepatitis A, or illness due to *E. coli*
   O157:H7.
3. Exposure to or suspicion of causing any confirmed outbreak of typhoid fever, shigellosis, *E. coli*
   O157:H7 infections, or hepatitis A.
4. Travel outside the United States within the last 50 days.

I have read (or had explained to me) and understand the requirements concerning my responsibilities
under the Food Code and this agreement to comply with:

   (1) Reporting requirements specified above involving symptoms, diagnoses, and high-risk
       conditions specified;
   (2) Work restrictions or exclusions that are imposed upon me; and
   (3) Good hygienic practices.

I understand that failure to comply with the terms of this agreement could lead to action by the food
establishment or the food regulatory authority that may jeopardize my employment and may involve legal
action against me.

Applicant or Food Employee Name (please print): _____

Signature of Applicant or Food Employee _____ Date _____

Signature of Permit Holder's Representative _____ Date _____

Figure 2.4.   Food Code 1993, U.S. Public Health Service, U.S. Department of Health
                     and Human Services, Public Health Service, Washington, D.C. 20204,
                     Annex 7, p.2.

**Form
3   Applicant and Food Employee Medical Referral**

Preventing Transmission of Diseases through Food by Infected Food
Employees with Emphasis on *Salmonella typhi, Shigella* **spp.,**
Escherichia coli **O157:H7, and Hepatitis A Virus.**

The Food Code specifies, under *Part 2-2 Employee Health Subpart 2-201 Disease or
Medical Condition,* that Applicants and Food employees obtain medical clearance from a
physician licensed to practice medicine whenever the individual:

1. Is chronically suffering from a symptom such as diarrhea; or
2. Meets one of the high-risk conditions specified in Paragraph 2-201.11(D) and is suffering
   from any symptoms specified in Subparagraph 2-201.11(B)(1).
3. Has a current illness involving *Salmonella typhi* (typhoid or paratyphoid fever), *Shigella* spp.
   (shigellosis), *Escherichia coli* O157:H7 infection, or hepatitis A virus (hepatitis), or
4. Reports past illness involving *Salmonella typhi* (typhoid or paratyphoid fever), *Shigella* spp.
   (shigellosis), *Escherichia coli* O157:H7 infection, or hepatitis A virus (hepatitis), if the
   establishment is serving a highly susceptible population such as preschool age children or
   immunocompromised, physically debilitated, or infirm elderly.

Name of Applicant or Food Employee being referred and population served:
(_____ (Name, please print) _____). **Highly susceptible population YES ☐ NO ☐**

REASON FOR MEDICAL REFERRAL:

The reason for this referral is checked below:
☐ Chronic diarrhea or other chronic symptom _____(specify)_____ .
☐ Meets a high-risk condition specified in Paragraph 2-201.11(D) _(specify)_ and suffers from a
   symptom specified in Subparagraph 2-201.11(B)(1) _(specify)_.
   diagnosed or suspected typhoid or paratyphoid fever, shigellosis, *E.coli* O157:H7 infection, or
   hepatitis.
☐ Reported past illness from typhoid or paratyphoid fever, shigellosis, *E.coli* O157:H7 infection, or
   hepatitis.
☐ Other medical condition of concern per the following description:

_____

PHYSICAN'S CONCLUSION:

☐ Applicant or food employee is free of *Salmonella typhi* , *Shigella* spp., *Escherichia coli* O157:H7, or
   hepatitis A and may work as a food employee without restrictions.
☐ Applicant is an asymptomatic carrier of *Salmonella typhi* and is restricted from direct food handling
   duties in establishments not serving highly susceptible populations.
☐ Applicant or food employee is not ill but continues as an asymptomatic carrier of _(pathogen)_ and
   should be excluded from working as a food employee in food establishments serving highly susceptible
   populations such as those who are preschool age, physically debilitated, immunocompromised, or
   infirm elderly.
☐ Applicant or food employee is suffering from Salmonella typhi, *Shigella* spp., *Escherichia coli* O157:H7
   or hepatitis A virus and should be excluded from working as a food employee.

COMMENTS:

Signature of Physician _____ Date _____

Figure 2.5   Food Code 1993, U.S. Public Health Service, U.S. Department of Health
and Human Services, Public Health Service, Washington, D.C. 20204, Annex 7, p. 13

special and periodic evaluations on site to determine compliance and whether licenses should be issued or renewed. All new food service establishments must submit plans for approval by the health department prior to construction and be evaluated on site prior to opening.

The environment health practitioner inspects the food-service establishment using the flow technique, whereby the flow of food from the point of delivery to the point of disposal is followed (see Figure 2.6) and problems are recorded. Simultaneously, cleanliness and the condition of floors, walls, ceilings, lighting, ventilation, hand-washing facilities, and bathrooms are checked, as well as plumbing problems, water supply, sewage disposal, housekeeping procedures, and insects and rodents inside and outside of the premises. The environmentalist should always carry a flashlight, clipboard, inspections sheets, pads, sanitizing testing equipment, light meter, and applicable rules and regulations of the health department.

If possible, the food-service manager or supervisor should accompany the environmental health practitioner on the on-site evaluation; otherwise, at the conclusion of the on-site evaluation, the environmental health practitioner should ask the food-service manager to go back through the establishment to see the major problems noted. The environmental health practitioner should complete the on-site evaluation form from his or her notes, and determine the length of time needed to correct each problem. If conditions are bad, another evaluation should be made within 1 week. If the health of the public is potentially endangered, conditions should be corrected at once. The environmental health practitioner should teach the food-service manager to set up an adequate cleaning schedule and should make recommendations on the carrying out of necessary corrections. It is important to discuss the public health reasons for compliance of the establishment with certain rules and regulations. Additional on-site evaluations are made as needed. When the food-service establishment will not comply, further action such as warning letters, administrative hearings, license suspension, license revocation, and court hearings are necessary. (See Figure 2.7 for an example of an inspection sheet.)

Food-Borne Disease Investigations

When the health department is notified that a possible food-borne disease outbreak has occurred, it must move quickly to obtain the necessary information from persons who ingested the suspected food. Surveys of the food preparation, storage, and serving operations must be made, and samples taken and processed rapidly. Unfortunately, many outbreaks of food-borne disease occur at large gatherings, such as church dinners, reunions, and picnics. This type of food operation complicates the investigation procedures.

Epidemiological Study Techniques

Environmental health personnel and epidemiologists should complete detailed questionnaires for all persons who may have eaten the food or drinks suspected

Preparation/Delivery	Storage	Preparation	Equipment	Serving	Cleaning and Storage	Waste Disposal
Raw foods	Refrigerator	Baking	Ovens	Milk containers	Removal of soiled utensils and dishes	Garbage disposal units
Finished products	Freezer	Frying	Stoves	Steam tables	Equipment cleaning	Garbage cans
	Dry storage	Cooking	Deep fryers	Cold tables	Dishwasher operation	Garbage rooms
		Salad making	Steam kettles	Counter freezers	Dishwashing techniques	Dempster dumpers
		Mixing	Slicers	Serving utensils	Use of detergents	Garbage can cleaning
		Grinding	Grinders	Dishware	Sanitizing	Cleaning of exterior premises
		Chopping	Toasters	Silverware	Storage of clean utensils, pots, pans, and dishware	
		Separating	Coffee urns	Single service utensils	Storage of single service articles	
		Combining	Pots and pans	Ice cream dippers		
		Food handling	Cutting boards	Ice handling		
			Meat blocks	Food handling		
			Griddles	Automatic vending equipment		
			Can openers			

Figure 2.6 Flow chart for food establishment inspection by type of operation. The environmental health worker checks construction and cleanliness of equipment, temperatures, and time, where applicable, during inspection.

FOOD SERVICE
INSPECTION REPORT_____

HEALTH JURISDICTION

KEY: S=Satisfactory
      U=Unsatisfactory
      R=Repeat Violation
      N=Not Apply

Name of Establishment	Owner or Operator's Name	Address

Person Interviewed and Title	Date	Time	am pm	Current License	Yes ( ) No ( )

Rating   *Rating*

Facilities

1. Floors–clean, good repair
2. Walls–clean, painted, good repair
3. Ceiling–clean, painted, good repair
4. Lighting–fixtures clean, good repair, adequate
5. Ventilation–adequate, proper design
6. Filters–duct work clean, adequate
7. Bathrooms–convenient, self-closing doors, clean, free from odors, adequate supplies
8. Handwashing facilities–adequate, good repair, hot & cold running wate, convenient, clean soap, single use towels or dryers
9. Plumbing–sewage & water supply, properly constructed, no cross-connections
10. Hot and cold running water
11. Screened doors & windows or other fly proofing devices

Food Protection

12. Ice–approved source
13. Approved source
14. Wholesomeness
15. Refrigeration of perishables below 45°F; Freezing below 0°F
16. Hazardous cooked foods above 140°F
17. Storage of foods
18. Protection of food from contamination
19. Preparation of food
20. Serving of food

Equipment and Utensils

21. Food Service Equipment–clean & good repair
22. Materials for cleaning & sanitizing
23. Equipment for cleaning–dishwashers, sinks, etc.
24. Cleaning Operation–temperatures, procedures, etc.
25. Sanitization–hot water or chemical
26. Cleanliness–equipment
27. Cleanliness–utensils, dishes, etc.
28. Storage of utensils, silverware, dishes, pots, pans, etc.
29. Construction of equipment, utensils, etc. (including thermometers)
30. Cleaning procedures for food contact surfaces
31. Cleanliness of food contact surfaces

Personal Health

32. Employee free from boils, diarrhea, etc.
33. Cleanliness of hands, garments, etc.

Solid Waste

34. Garbage storage & disposal
35. Rubbish storage & disposal
36. Outer premises

Insects and Rodents

37. Flies, roaches, rats, mice

Sampling Type

38. Sample number

Item	Recommendations	Date of Compliance

Figure 2.7   Inspection sheet

of having caused the food-borne disease. Once a pattern develops concerning the kinds of foods involved and the places in which the food was consumed, additional environmentalists should be detailed to make complete studies of all food operations. These studies must include a food history and the flow of preparation. The food, its source, method of preparation, storage, food temperatures, times, and other critical limits (such as the time needed at a certain temperature to prevent the growth of organisims which cause disease), and refrigeration must be determined. As part of this study, remnants of suspected foods should be collected aseptically in sterile containers or in their original containers and sent, refrigerated, immediately to the laboratory for analysis. It is necessary to properly identify the origin of the sample, the date and time of collection, the name of the environmentalist, and a brief statement about the symptoms of the patients and the suspected organisms or chemicals. Water samples should also be taken. All sewage systems and plumbing should be evaluated. The presence of insects and rodents must be determined. It is essential as part of this study to determine if any food handlers were ill within a period of 6–8 weeks prior to the onset of the disease. If an individual food handler had diarrhea or vomiting, a fecal sample should be taken. Food handlers should also be checked for any boils, carbuncles, and respiratory infections that could cause an outbreak of staphylococcus food poisoning.

Medical diagnosis of patients combined with good epidemiological data may determine the causative agent in the outbreak and assist the environmentalist in preventing further outbreaks of this type. If commercially prepared food is the cause of the food-borne disease, it is necessary to obtain data identifying the source of the food, when and where it was purchased, and any identifying codes or marks on the containers. This information should be immediately forwarded to the Food and Drug Administration. If the outbreak of disease is traced to a hospital or college cafeteria, intensive inspection, coupled with changes in procedures and good food-handler training, will help to prevent further outbreaks. If the outbreak of disease is traced to a church dinner or picnic, the environmentalist should provide information and suggest the presentation of a food-handling course to the group involved.

To gather data properly, an Individual Food-Borne Outbreak Report, a summary of Food-Borne Disease Reports, a Study Form for Food Preparation Establishments, and a History of Food Handlers Report should be used. These forms are obtained from local or state health departments (Figures 2.8, 2.9, and 2.10).

HAZARD ANALYSIS/CRITICAL CONTROL POINT INSPECTION

Hazard analysis/critical control point inspection (HACCP) is an in-depth inspection process used to resolve or prevent disease outbreaks by identifying the foods at greatest risk and the critical control points. These foods may be critical because they are naturally contaminated foods, large volume preparation, multi-step preparation, and temperature changes in the foods. The critical control points are those few steps in a process that are most important to bacterial contamination, survival, or growth. Menu analysis and food flow helps identify critical control points. They also are an operation or part of an operation where actual or potential risks are usually found and where preventive or control measures can be exercised that eliminate, prevent, or minimize a hazard that has occurred prior to this point. The critical limit is the maximum or minimum value to which a physical, biological, or chemical parameter must be controlled at a critical point to limit the risk that the identified food safety hazard may occur. The HACCP analysis helps determine the critical control points to prevent disease. The establishments are chosen for the program by use of a system similar to triage in mass casualties. The program may be very valuable in certain instances. Contact the Food and Drug Administration for further information.

1. Where did the outbreak occur?	2. Date of outbreak: (Date of onset 1st case)
State_____ (1.2) City or Town_____County_____	_____(3-8)

3. Indicate actual (a) or estimated (e) numbers	4. History of Exposed Persons	5. Incubation period (hours):
Persons exposed _____ (9-11)	No. histories obtained_____(18-20)	Shortest____(40-42) Longest____(43-45)
Persons ill_____(12-14)	No. persons with symptoms_____(21-23)	Approx. for majority_____(46-48)
Hospitalized_____(15-16)	Nausea____(24-26) Diarrhea____(33-35)	6. Duration of illnes (hours)
Fatal cases_____(17)	Vomiting____(27-29) Fever_____(36-38) Cramps____(30-32) Other, specify___(39)	Shortest_____(49-51) Longest____(52-54) Approx. for majority_____(55-57)

7. Food-specific attack rates: (58)

Food Items Served	Number of persons who ATE specified food				Number of persons who did NOT eat specified food			
	Ill	Not ill	Total	Percent ill	Ill	Not ill	Total	Percent ill

8. Vehicle responsible (food item incriminated by epidemiological evidence)_____(59-60)

9. Manner in which incriminated food was marketed:(Check all applicable)
(a) Food Industry (6-1)
  Raw ....☐ 1
  Processed ......☐ 2
  Home Produced
  Raw ....☐ 3
  Processed ......☐ 4
  (6-2)
(b) Vending Machine . ☐ 1

(c) Not wrapped ....☐ 1 (6-1)
  Ordinary Wrapping ....☐ 2
  Canned ....☐ 3
  Canned–Vacuum Sealed. ☐ 4
  Other (specify) ....☐ 5

(d) Room Temperature ......☐ 1 (6-2)
  Refrigerated ....☐ 2
  Frozen ....☐ 3
  Heated ....☐ 4

If a commercial product, indicate brand name and lot number

10. Place of Preparation of Contaminated Item (65)
  Restaurant ....☐ 1
  Delicatessen........☐ 2
  Cafeteria..............☐ 3
  Private Home.........☐ 4
  Caterer...............☐ 5
  Institution:
  School..............☐ 6
  Church.............☐ 7
  Camp.............☐ 8
  Other, specify........☐ 9

11. Place where eaten (66)
  Restaurant ....☐ 1
  Delicatessen........☐ 2
  Cafeteria..............☐ 3
  Private Home........☐ 4
  Caterer................☐ 5
  Institution:
  School..............☐ 6
  Church.............☐ 7
  Camp.............☐ 8
  Other, specify........☐ 9

DEPARTMENT OF HEALTH, EDUCATION AND WELFARE
PUBLIC HEALTH SERVICE
CENTER FOR DISEASE CONTROL
BUREAU OF EPIDEMIOLOGY
ATLANTA, GEORGIA 30333

Figure 2.8. Investigation of a food-borne outbreak.

12. Food specimens examined: (67)
Specify by "x" whether food examined was **original** (eaten at time of outbreak or **check up** (prepared in similar manner but not involved in outbreak)

Item	Orig.	Check up	Findings Qualitative	Quantitative
Example: beef	x		C. perfringens, Hobbs type 10	2×10^6/gm

13. Environmental specimens examined: (68)

Item	Findings
Example: meat grinder	C. perfringens, Hobbs Type 10

14. Specimens from patients examined (stool, vomitus, etc.): (69)

Item	No. Persons	Findings
Example: stool	11	C. perfringens, Hobbs Type 10

15. Specimens from food handlers (stools, lesions, etc.): (70)

Item	Findings
Example: lesion	C. perfringens, Hobbs Type 10

16. Factors contributing to outbreak (check all applicable)

	Yes	No
1. Improper storage or holding temperature	☐1	☐2 (71)
2. Inadequate cooking	☐1	☐2 (72)
3. Contaminated equipment or working surfaces	☐1	☐2 (73)
4. Food obtained from unsafe sources	☐1	☐2 (74)
5. Poor personal hygiene of food handler	☐1	☐2 (75)
6. Other, specify	☐1	☐2 (76)

17. Etiology: (77, 78)
Pathogen
Chemical _____
Other _____

Suspected .... ☐ 1   (79)
Confirmed .... ☐ 2
Unknown .... ☐ 3

18. Remarks: Briefly describe aspects of the investigation not covered above, such as unusual age or sex distribution; unusual circumstances leading to contamination of food, water; epidemic curve, etc. (Attach additional page if necessary)

Name of reporting agency: (80)	
Investigating official:	Date of investigation:

NOTE: Epidemic and Laboratory Assistance for the investigation of a foodborne outbreak is available upon request by the State Health Department to the Center for Disease Control, Atlanta, Georgia 30333.

To improve national surveillance, please send a copy of this report to:
Center for Disease Control
Attn: Enteric Diseases Section, Bacterial Diseases Branch
Bureau of Epidemiology
Atlanta, Georgia 30333

Submitted copies should include as much information as possible, but the completion of every item is not required.

Figure 2.9 Laboratory findings including negative results

FOOD PROTECTION CONTROLS

Until recently, the major controls used for food protection were control of temperature and time, and the proper preparation of food in a clean manner. Although these measures are still essential, there are new concerns due to the

DEPARTMENT OFHEALTH AND HUMAN SERVICE
PUBLIC HEALTH SERVICE
FOOD AND DRUG ADMINISTRATION

HACCP INSPECTION DATA

EST. NAME:	PERMIT NO.	INSPECTOR	
DATE:	TIME IN: :AM/PM	TIME OUT:	:AM/PM

Record all observations below–transfer violations to Inspection Report

FOOD TEMPERATURES/TIMES/OTHER CRITICAL LIMITS
Use Additional Forms If necessary

FOOD / STEP	1.	CRITICAL LIMIT	2.	CRITICAL LIMIT	3.	CRITICAL LIMIT	4.	CRITICAL LIMIT
A. SOURCE								
B. STORAGE								
C. PREP BEFORE COOK								
D. COOK								
E. PREP AFTER COOK								
F. HOT/COLD HOLD								
G. DISPLAY/ SERVICE								
H. COOL								
I. REHEAT								

OTHER FOOD TEMPERATURES OBSERVED   Use steps from above for location

FOOD	TEMP. °C/°F	STEP	FOOD	TEMP. °C/°F	STEP	FOOD	TEMP. °C/°F	STEP

Figure 2.10   Food Code 1993, U.S. Public Health Service, U.S. Department of Health and Human Services, Public Health Service, Washington, D.C. 20204, Annex 7, p. 5

MANAGEMENT/PERSONNEL OBSERVATIONS	

OTHER FOOD OBSERVATIONS	

EQUIPMENT, UTENSILS, AND LINEN OBSERVATIONS	

WATER, PLUMBING, AND WASTE OBSERVATIONS	

PHYSICAL FACILITIES	

Figure 2.10   continued

influx of chemicals into our food supply. Since specific controls do not exist at present, they are mentioned in a succeeding chapter under research needs.

Streptococcus infection control measures include pasteurization of milk and other dairy products, exclusion of persons with known streptococcal infections, and antibiotic treatment of known carriers or contacts.

Salmonella controls include pasteurization of milk and other dairy products, control and certification of shellfish areas, water treatment and chlorination, elimination of flies from human feces, and prohibition of food handling by individuals with a history of typhoid or paratyphoid fever.

Salmonellosis control measures include the use of good personal hygiene, removal of sick food handlers from duty, obtaining cultures from food handlers with diarrhea infections, and restricting them from work until three consecutive fecal samples are negative; storage of susceptible foods at temperatures below 41°F; avoidance of the danger zone for salmonella growth, which occurs between 60°F and 120°F (serve all hot foods above 140°F but precontrol to 165°F); never stuff chicken or turkey before cooking, cook in a separate pan; preparing small quantities of such things as egg salad, potato salad, chicken salad, or turkey salad; analyzing feed for salmonella contamination and maintaining disease-free animals; reporting immediately all suspected food infections to public health authorities and retaining samples of food; and making complete epidemiological studies.

Salmonella entridis. It appears that 75% of cases of disease are associated with raw or inadequately cooked Grade A whole shell eggs. Scrambled eggs should be cooked at 250°F for one minute and the egg temperature should be 165°F. Poached eggs should be cooked for 5 minutes in boiling water. Sunnyside fried eggs should be cooked at 250°F for 7 minutes uncovered and for four minutes covered. Eggs "over easy" should be cooked at 250°F for 3 minutes on one side and two minutes on the other. Boiled egges should be submerged in boiling water for 7 minutes.

Norwalk type diesase control measures are similar to salmonellosis control measures.

Shigellosis control measures include strict personal hygiene, elimination of carriers as food handlers, refrigeration of moist foods, cooking of foods at 165°F prior to serving, and elimination of flies.

Yersiniosis is controlled by proper disposal of human, cat, and dog feces; protection of water supplies; proper hand washing; proper preparation of food; pasteurization of milk.

Cholera control measures include filtration and chlorination of drinking water supply, growing of shellfish in certified sewage-free areas only, removal of known cases from food-handling operations, and vaccination when an outbreak of cholera occurs.

Hepatitis can be prevented by observing the same controls as stated under cholera.

Plasmodium Shigelloides can be controlled by purification of drinking water, protection of food from nonpotable water, and proper cooking.

Campylo bacterosis can be controlled by cooking meat thoroughly, pasteurizing milk, chilling foods rapidly in small quantities, avoiding contaminating prepared food with raw food.

Brucellosis, diphtheria, and bovine tuberculosis control measures include pasteurization of milk and dairy products and isolation of carriers. In addition, for brucellosis and bovine tuberculosis, sick animals should be eliminated.

Listeriosis can be controlled by cooking food thoroughly and pasteurizing milk.

Tularemia control measures include the use of protective gloves when handling wild animals and the elimination of arthropods.

Staphylococcus food poisoning control measures include removing food handlers who have nasal discharges or skin infections, cooking foods thoroughly, refrigerating foods immediately, using custard pastries within 2–4 hr after preparation and discarding the remainder, and washing hands thoroughly and frequently.

Bacillus cereus can be controlled by heating to 165°F or 133°F for 5 minutes for the toxin.

Clostridium welchii or *C. perfringens* control measures include the elimination of known carriers from food handling, cooking meat thoroughly before consumption, cooling meat rapidly in the refrigerator immediately after cooking, using drippings from meat for gravies only on the day of cooking, discarding all leftovers, and preparing chicken, turkey, or beef pot pies only from freshly cooked pieces of meat.

Botulism can be prevented by proper cooking of foods. If home-processed foods are used, they should be boiled for a minimum of 15 min before eating. Discard any canned or bottled foods that show signs of spoilage including off-odors, off-tastes, gas or foams, and off colors. If botulism is suspected, have a physician administer the proper antitoxin immediately.

Vibrio parahaemolyticus can be killed by thorough cooking.

Where substantial mold damage appears, food should not be consumed because of the potential for mycotoxins.

Amoebic dysentery control measures include protection of water supplies from human excreta, elimination of carriers from food preparation, proper sewage disposal, and filtration and chlorination of water supplies.

Giardiasis can be controlled by cooking food thoroughly, good personal hygiene, and control of sewage.

Beef and pork tapeworm control measures include proper inspections by trained veterinarians and thorough cooking of beef and pork. For fish tapeworms, cook fish thoroughly and avoid all raw smoked fish.

Trichinosis control measures include cooking garbage fed to pigs to an internal temperature of 137°F, eliminating rats from pig farms, and cooking all pork to an internal temperature of 165°F.

Toxoplasmosis can be controlled by cooking foods thoroughly, freezing food at or below 5°F for 24 hours, and washing hands after handling raw foods.

Angiostrongyliasis can be controlled by cooking foods thoroughly, freezing food at or below 5°F, not eating raw freshwater prawns, raw land mollusks, and raw crab.

Scrombroid poisoning can be controlled by ice or refrigerating fish right after capture, eating fish quickly after proper cooking, discarding sharp or peppery tasting fish.

Enteric adenovirus can be controlled by proper cooking of food, good personal hygiene, and proper disposal of sewage.

Escherichia Coli of all types can be controlled by chilling foods rapidly under 41°F in small quantities, cooking and heating to 165°F, practicing good personal hygiene, treating water, and proper disposal of sewage.

Astrovirus, Calcivirus, Parvovirus, and Rotavirus can be controlled by proper cooking, handling, and refrigeration of food if the virus is foodborne.

Chemical poisoning control measures include eliminating utensils and containers that may leach the chemical into solution and discontinuing the use of dangerous pesticides; protecting all food and food contact surfaces when using pesticides; and storing all chemicals in their original containers away from food storage, preparation, and serving areas.

Intoxication from poisonous plants or animals can be prevented by not eating these plants or animals.

Food-Service Training Programs

The key to a safe, sanitary food program is food-service supervisors and personnel properly trained in evaluating menus for potential hazards, good techniques of purchasing, storing, preparing, transporting, and serving wholesome food; preparation of HACCP Plan including flow diagrams, temperatures, times, and critical limits. Scheduling of cleaning operations, cleaning and sanitizing techniques, and proper use of equipment and materials must be understood and the necessary materials must be made available. Although all food-service personnel are important, the supervisor is the key. The supervisor schedules operations, teaches personnel, orders supplies, and inspects the establishment daily. Assistance is available from local and state health departments, who are happy to conduct necessary training programs. The supervisor should attend special training sessions held by the health department, and should be tested and certified on an annual basis by the health department. Renewal of the establishment's license should be contingent on this certification.

An educational program directed at homemakers and individuals who prepare food for mass feeding operations, such as church dinners, is useful in controlling the numerous outbreaks of disease that may occur in the home or at special functions.

Terminology

To better understand food-borne disease, it is necessary to have some knowledge of the following terminology.

- *Active immunity*. Antibodies produced in the person through contact with disease.
- *Carrier*. An individual harboring specific infectious agents in the absence of discernible clinical disease who serves as a potential source or reservoir of infection for other humans.
- *Chain of infection*. The spread of disease from the reservoir by means of a vehicle or vector to the host.
- *Contamination*. The presence of a pathogenic organism in or on a body surface or an inanimate object.
- *Communicable disease*. An illness due to an infectious agent or its toxic products being transmitted directly or indirectly to a well person from an infected person or animal, or transmitted through an intermediate animal host, vector, or inanimate environment.
- *Communicable period*. The time or times during which the etiologic agent may be transferred from an infected person or animal to humans.
- *Endemic*. The regular occurrence of a fairly constant number of human cases of a disease within an area.
- *Enterotoxin*. A toxin produced by staphylococcus aureus which specifically affects cells of the intestinal mucosa causing vomiting and diarrhea and arising in the intestine.
- *Epidemic*. The occurrence in a community or region of a group of illnesses of a similar nature, clearly in excess of normal expectancy, and derived from a common or propagated source.
- *Epidemiology*. The study of the causes, transmission, and incidence of diseases in communities or other population groups.
- *Etiological agent*. The pathogenic organism or chemical causing a specific disease in a living body.
- *Exotoxin*. Toxic substance produced by bacteria found outside of the bacterial cell.
- *Fomite*. An inanimate object not supporting bacterial growth but serving to transmit pathogenic organisms from human to human.
- *Foodborne Disease Outbreak*. An incidence of disease occuring after the ingestion of a food shown to cause the disease.
- *Hazard*. A biological, chemical, or physical property of food that may cause an unacceptable consumer risk.
- *Host*. The living body, human or animal, that provides food and shelter for the disease organisms.
- *Incidence*. The number of cases of disease occurring during a prescribed time period in relation to the unit of population in which they occur.
- *Incubation period*. The time interval between the infection of a susceptible person or animal and the appearance of signs or symptoms of the disease.
- *Infection*. The entry and development or multiplication of a particular pathogen in the body of humans or animals.

- *Parasite*. An organism living on the tissues and waste products of the host within the body of the host. All disease-causing organisms are parasites.
- *Passive immunity*. Antibodies produced in another host and then injected into the diseased person.
- *Report of disease*. An official report, usually by a doctor, of the occurrence of a communicable or other disease of humans or animals to departments of health or agriculture or both, as locally required. Reports should include those diseases requiring epidemiological investigation or initiation of special control measures.
- *Reservoir of infection*. Humans, animals, plants, soil, or inanimate organic matter in which an infectious agent lives and multiplies and then is transmitted to humans, who themselves are the most frequent reservoir of infectious pathogenic agents.
- *Resistance*. The sum total of body mechanisms placing barriers to the progress of invasion of pathogenic organisms.
- *Vector*. A living insect or animal (not human) that transmits infections and diseases from one person or animal to another.
- *Vehicle*. Water, food, milk, or any other substance or article serving as an intermediate means by which the pathogenic agent is transported from a reservoir and introduced into a susceptible host through ingestion, inhalation, inoculation, or by deposit on the skin or mucous membrane.

SUMMARY

Food protection is an essential part of the environmental health program, considering the vast variety of food produced, the enormous potential for disease, and the short- and long-range problems of chemicals entering our food supply. Food-borne disease outbreaks probably account for millions of illnesses each year. The various long-range problems related to food-borne disease cannot even be estimated.

Food protection consists of proper plan review, the development and use of adequate facilities in a correct manner; the procurement and use of correct equipment; necessary cleanliness; proper storage, preparation, and serving of food; and adequate disposal of food remnants. The cleaning and sanitizing techniques utilized are an essential facet of food protection. Important factors are also the health and potential for spreading disease of humans. A variety of inspectional techniques are utilized on a regular and special basis, depending on the nature of the food establishment and the types of problems likely to occur. Proper techniques and food-borne disease investigations eliminate outbreaks and prevent future outbreaks of disease. All states and many counties and localities have food protection programs.

RESEARCH NEEDS

Research needs will be discussed in the chapter on food technology.

Chapter 3

FOOD TECHNOLOGY

BACKGROUND AND STATUS

Food is a perishable product consisting of proteins, carbohydrates, fats, vitamins, minerals, water, and fibers. It is used by an organism in sustaining growth, repairing tissues, maintaining vital processes, and furnishing energy. Great pleasure is derived from the proper preparation of food and from its consumption. Food is altered favorably or unfavorably by microorganisms, enzymes, insects, and environmental changes. Microbes help produce sourkraut, bread, and cheese; ripen olives; and ferment milk. The action of the yeast *Saccharomyces cerevisiae* produces carbon dioxide in bread, making it porous and causing it to rise. The starches and proteins in bread are split by the action of the yeast, becoming more digestible, and characteristic flavors and aromas are produced. However, if certain spore-forming bacilli, principally *Bacillus mesentericus*, are introduced with the flour or yeast, discoloration and unpleasant odor results and the bread becomes unappetizing.

In this chapter, contamination, spoilage, and disease potential of certain types of foods will be discussed, including milk and milk products, poultry, eggs, meats and meat products, and fish and shellfish. The nature of enzymes, microbiological spoilage, factors affecting spoilage, techniques of preservation of foods, chemical preservatives, additives, pesticides, fertilizers, and antibiotics will also be discussed. The production, processing, testing, and environmental problems associated with each food group will be addressed. Applied on-site evaluation techniques and inspection forms will be introduced as needed. The foods were selected as the most frequent contributors to food-borne diseases, intoxication, or the kinds of food spoilage that the environmental health practitioner is likely to encounter.

During the early 1970s, famine with drought and starvation occurred in the Sahara Desert of Africa and in the Indian subcontinent. The United Nations estimated that at least 460 million people, excluding the communist countries of Asia, were suffering from serious food deficiency in 1970. It is estimated that 10–20 million young children are severely afflicted with the protein calorie malnutrition diseases of kwashiorkor and marasmus. Most of these children die without treatment. Another 200 million suffer from all other forms of malnutrition. The food crisis resulted from a decline in world food production and an increase in population. In the developing countries, the annual population growth averages nearly 2.5%. At this level the population doubles every 28 years. If this trend in population growth continues, there should be an estimated 1.2 billion people in India by the 21st century. In the 1990s starvation is occurring in many Third World countries. The situation is not anticipated to be improved in this century.

To combat hunger and death, and to improve the economy of many countries, more land is being cultivated and more environmental resources and people-made chemicals are being utilized. Water, which is a prime resource for increasing food output, is plentiful in some areas and essentially lacking in others. The changing of water patterns to increase productivity have also brought disease. In Egypt, for example, the Aswan Dam stimulated a large outbreak of schistosomiasis. Fertilizers when used properly are excellent for increasing crop yield, but when improperly used they become contaminants and potential cancer-causing substances.

Plant disease, insects, pests, and weeds caused an estimated 30% loss in potential food production throughout the world, resulting in a sharply increased use of pesticides. Unfortunately, some pesticides are inherently dangerous or are dangerous when misused. In the Canete Valley of Peru, the cotton crop was heavily treated with DDT and other hydrocarbons. The sprays killed not only the pests, but also their predators. The pests eventually developed resistance to DDT, whereas the predators did not, resulting in lost cotton crops. Numerous times pesticides contaminated raw food products and were concentrated as they were processed through the food or became a part of the runoff problem, resulting in varying levels of water pollution.

Additives are added to food to enhance flavor, taste, and keeping quality. Unfortunately, unintentional or intentional additives may be the cause of cancer or other related diseases. There is serious concern over the use of sodium in processed foods, since the sodium may produce adverse effects in hypertensive individuals. Other contaminants found in raw food before processing and remaining in the food until consumption include mercury, lead, cadmium, zinc, copper, manganese, selenium, and arsenic. In addition, a variety of pesticides, such as chlorinated hydrocarbons, polychlorinated biphenyls, and a vast group of chemicals are entering our food supply.

Animal feed may induce a toxic response in humans or animals. Unfortunately, there is a lack of research in this area, although scientists know the toxicants are found in animal tissue. Milk and dairy products are of particular importance, not only to the young but also to vegetarians. Milk and dairy products contain estrogens, nitrates, antibiotics, pesticides, radionuclides, and mycotoxins. Eggs may cause allergic disorders in children and adults due to a natural allergic response or to estrogens and antibiotics found in egg yolks. Aflatoxins were found in Iowa corn in 1990.

There has been a rapid increase in the use of frozen prepared foods and vacuum-packed foods. Improper processing of these foods has resulted in outbreaks of botulism. The situation in food technology can only worsen in the coming years, unless extreme methods are taken to control the many unwanted substances found in raw food products and unless research is conducted in the many unknown areas of contamination. Our understanding of acute responses of food poisoning is relatively good; our understanding of chronic responses, after years of consuming various types of products, is still very poor.

SCIENTIFIC AND TECHNOLOGICAL BACKGROUND

Chemistry of Foods

Enzymes produce desirable and undesirable changes in foods. As an example, enzymes may tenderize meat or cause off colors, off odors, and off tastes. An enzyme is an organic catalyst that speeds up a chemical reaction without being altered itself. It has the following characteristics: it acts most rapidly at body temperature, it has a high degree of specificity for a given test, it is produced by living cells including microorganisms, it continues to react after harvest of plants or slaughter of animals, it is retarded by low temperatures, it is destroyed by boiling, and it may react either within the cell (endozyme) in which it was produced or outside of the cell (exozyme) in the tissue.

The inactivation of the enzyme phosphomonoesterase by heat is a reliable means of measuring the efficiency of pasteurization of milk and ice cream in the phosphatase test. The enzyme is totally destroyed by proper pasteurization. However, false positives can occur due to certain bacteria present in milk, including streptococcus and aerobacter.

Food Additives and Preservatives

A food additive is any substance or mixture of substances other than basic foodstuffs added during the production, processing, storage, or packaging of foods. Additives are used to preserve, emulsify, flavor, color, and increase nutritive value. Intentional food additives include preservatives, antioxidants, sequestrants, surfactants, stabilizers, bleaching and maturing agents, buffers, acids, alkalines, colors, special sweeteners, nutrient supplements, flavoring compounds, and natural flavoring material. Unintentional food additives or contaminants include radionuclides, insect parts and excreta, insecticides and herbicides, fertilizer residue, other chemicals, dirt, microorganisms, and any other unintentional materials.

Antioxidants preserve freshness in meats by preventing rancidity of fats and preserve appearance and taste of fruits by preventing discoloration. Mold inhibitors, such as calcium or sodium propionate, keep bread fresh for longer periods. Emulsifiers, such as gum arabic, are used to maintain the consistency of French dressing. Stabilizers and thickeners, such as lecithin, are used in ice cream. Monosodium glutamate, nonnutritive sweeteners, and other agents enhance the flavor of food. Additives that add nutritive value to foods include iodized salt to prevent goiter; vitamin D in milk to prevent rickets; thiamine, niacin, riboflavin, or iron in bread and cereals to provide individuals with sources of these essential vitamins and minerals; and vitamin A in margarine to help prevent malnutrition. The major restrictions on additives are that they do not cause short-term or chronic poisoning or cancer, do not make decomposed food appear fresh, or do not cause any other unfavorable side reactions. For example, in curing meats, nitrites must not exceed 120 ppm combined with an activator of ascorbates or erythorbates at 550 ppm.

In addition to the nitrates or nitrites that are purposely added to food, considerable amounts of these substances are introduced into the food chain through the use of nitrogen fertilizers. Certain vegetables accumulate nitrogen compounds based on the amount of nitrate and molybdenum in the soil, the light intensity, and existing drought conditions. Further high concentrations exist from the accidental introduction of large amounts of nitrates or nitrites in the foods. When tests are conducted to meet the nitrate-nitrite standards, they should not only include levels added, but levels that already exist. Nitrates may also be introduced by groundwater usage.

Sodium chloride is added to diets in varying degrees in individuals. There is a suspected relationship between sodium chloride and human hypertension. The amount of salt added as an additive to food may well be a serious problem to the individual who has potential or existing hypertension.

Phosphates are introduced into food through the processing of poultry, in the soft-drink industry, and in the production of modified starches. Phosphates may also be introduced through the use of various fertilizers. Excessive daily intake of phosphorous causes a premature cessation of bone growth in children, subsequently affecting final adult height.

Additional knowledge is needed concerning the effects of the intentional addition of food additives to our processed foods. Cyclamates have been removed from the market. It is difficult to predict which additives previously thought to be safe will be evaluated and found potentially hazardous to humans. Even the widely used saccharin has been under suspicion. The Food and Drug Administration puts out a Food Additive Status List, which states the status of food additives that may be harmful to people.

Nitrates, Nitrites, and Nitrosamines

Nitrates and nitrites are used to preserve meats, including bacon, ham, hot dogs, pastrami, smoked fish, cured poultry, etc. Nitrite curing is more rapid than nitrate curing and, therefore, is used more often. Nitrites are added to the cured meat as a preservative. They are also effective microbial inhibitors. They retard the development of rancidity, enhance the flavor of certain meats, and maintain the bright red color of meats. The nitrite in the cured meat does not actually add color, but fixes the color pigment, myoglobin. Nitrites are not themselves carcinogens but rather are precursors. Nitrosamines are produced from nitrites and amines, which may be found in various places, including the decomposition of proteins. Amines are found in food, drugs, and pesticides. Nitrosation is the reaction between nitrous acid and secondary or tertiary amines. Nitrous acid comes from the nitrite added to the meat, and amines come from the meat protein. Over a hundred nitrosamines have been isolated and identified, with approximately 75% of these being carcinogens. Nitrosamines can cause cancer in animals after a single dose. Nitrosamines may also be found in imported and domestic beer.

Sulfites Used in Food

Sulfites have been used as preservatives in wine and other beverages, as well as in foods. Sulfiting agents retard the oxidation process of uncooked food, including vegetables, and extend the time of visual appeal. Fruits, green vegetables, especially lettuce, peeled or cut potatoes and apples, shrimp, seafood, and seafood salads are commonly sprayed or dipped into sulfiting agents. A sulfiting agent refers to sulfur dioxide, and several forms of inorganic sulfite that liberate sulfur dioxide under various conditions of use. Currently sulfur dioxide, potassium and sodium metabisulfite, potassium and sodium bisulfite, and sodium sulfite are considered to be safe for use in foods by the FDA. Sulfiting agents are added to foods because they control enzymatic and nonenzymatic browning, are antimicrobial, and act as an antioxidant, as well as a reducing agent. Sulfiting agents are used in a wide variety of food products.

Unfortunately it has been discovered that some individuals, especially asthmatics, may have a potentially severe adverse reaction to sulfites. The sulfites may induce an attack of asthma. Further, in some rare instances other types of hypersensitivity have occurred. Therefore, it is necessary that where sulfiting agents are permitted to be used, especially on salad bars, individuals be advised of this by the placement of a placard at the salad table.

Aspartame

Aspartame is marketed either as Equal® or Nutrasweet®. Aspartame is low in calories, has no odor or aftertaste, but because of its sweetness acts as a sugar substitute, and therefore, is useful in the diet of diabetics, heart patients, overweight patients, etc. There are some questions concerning the potential health effects of aspartame. Aspartame degrades into diketopiperazine at certain temperatures. The metabolism of aspartame, which is a peptide, proceeds in the same way as the metabolism of proteins. After metabolism, aspartame components become aspartic acid, phenylalanine, and methanol.

In some women of child-bearing age, the natural blood levels of phenylalanine fluctuate wildly. Therefore, small doses of aspartame might contribute to much higher levels of the phenylalanine of the individual. Some researchers have indicated that certain people had difficulty in metabolizing phenylalanine. High levels of phenylalanine in mothers apparently cause mental retardation in their babies. At lower levels, phenylalanine could affect the development of the fetus brain, thereby potentially decreasing IQ. Because aspartame is so useful in the diet of so many people, it is essential that further research be done in the area of the potential health effects of aspartame on people. If these health effects are proven to occur, certain restrictions or advisories should be printed on the aspartame containers.

F D and C Dyes

F D and C dyes in themselves are not carcinogenic, however, when the dyes are metabolized there is a possibility, depending on which dye is utilized, that a carcinogenic substance may be produced. As a result of this, at least 13 F D and C dyes have been banned from food use in the U.S. The gastrointestinal tract is an important metabolic organ. It contains mostly anaerobic bacteria, which aid in metabolizing F D and C dyes with the possible production of azo-reduction, which is the most important metabolic reaction of azo dyes that can contribute to carcinogenesis.

Anabolic Hormones

Anabolic hormones are used in cattle to increase weight. There is a potentially serious concern about the adverse affects to children who have been exposed to these hormones. In Puerto Rico, a study was done on young children that indicated premature development of the breasts, high incidence of ovarian cysts, very early development of vaginal bleeding, and uterus enlargement due to exposure to anabolic hormones.

PROBLEMS OF FOOD TECHNOLOGY

Contamination

Animal products contain microorganisms that are part of their normal flora, are present because of a disease process, or have been added at any point during the killing, processing, storage, preparation, and serving of the product. The organisms may be added by humans or from the environment. About 8% of fresh eggs contain microorganisms; dirty eggs are covered with organisms that easily penetrate the shell. The organism may be salmonella, shigella, and *Cholera vibrio*. Meat contains pseudomonas, proteus, *E. coli*, clostridium, and many other organisms. Although milk of a cow is sterile in a healthy udder, it frequently becomes contaminated by streptococci normally present in the milk ducts. Diseased cattle may be infected with pathogenic staphylococcus, streptococcus, tuberculosis, or brucellosis organisms.

Plant products, such as lettuce, cabbage, carrots, fruits, and vegetables, are contaminated with microorganisms from the air and soil, and by workers handling food.

Those bacteria causing food-borne disease or food-borne intoxication are of particular concern. The spores of *Clostridium botulinum* require temperatures well above the boiling point for destruction. Although intoxication is infrequent, it is of such a serious nature that particular care must be taken in preparing high-risk foods. This is especially true in home canning. Some molds will produce toxic materials known as mycotoxins. The best known of these are the aflatoxins, which have been found on peanuts, rye, wheat, millet, jellies, and jams.

Food Spoilage

The composition of food is important. Proteins are especially susceptible to spoilage by spore-forming gram-negative rods, such as pseudonomas and proteus, and by molds. Carbohydrates are particularly affected by yeasts, molds, streptococcus, and micrococcus. Fats undergo hydrolytic decomposition and become rancid. The pH of acidic foods, such as fruits, is low enough to prevent most bacterial spoilage; however, yeasts and molds grow well on these products. Nonacidic foods are subject to bacterial spoilage. Foods with a moisture content exceeding 10% will support the growth of microorganisms. Sugar and salt concentrations create an osmotic pressure that disrupts organisms: 5–15% salt will inhibit bacteria: 65–70% sugar inhibits molds. The presence or absence of oxygen determines the kinds of microorganisms that may or may not grow.

Food spoilage is due to two principal causes: chemical, including the enzymes just discussed, and biological, caused by bacteria, molds, and yeasts. Spoilage in most fruits is caused by molds and yeasts. Damage to the surface of the fruit increases the rate of deterioration. Spoilage of vegetables is due to bacteria and molds. Spoilage of properly refrigerated meat is usually confined to the surface. This is the reason why in the "aging" of beef, the surface is cut away and the interior portion, which is now more tender, is used. However, putrefactive decomposition occurs rapidly in ground meat and ground fish, because the bacteria are distributed throughout the food mass. Fish fillets become slimy and proteolysis occurs within several days because of the enormous number of microorganisms present initially and the handling techniques used. Shucked shellfish contains large numbers of bacteria that can lead to putrefactive decomposition if the shellfish are not refrigerated immediately. Milk, since it is drawn from the cow at temperatures favorable to rapid bacterial growth, must be immediately cooled to prevent bacterial decomposition.

Canning Spoilage

Underprocessing of canned foods results in microbial spoilage in the presence of high heat-resistant spores. This spoilage by microorganisms is a potential problem, since the raw food products and the processed foods may be contaminated by air, water, hands, equipment, and added ingredients. Flat sour spoilage is characterized by the production of acid without the production of gas. A slightly disagreeable odor and change of color may be present in the canned food. A swollen can of food may be caused by gas-forming thermophilic organisms. Swollen cans of meat are most likely caused by organisms producing putrefaction. Anaerobic bacterial growth in canned foods is of particular concern, since *Clostridium botulinum* may be causing the problem, rather than other members of the anaerobic family. Some characteristics of anaerobic activity are offensive odors, black sediment or residue, and reduction of oxygen. Spoilage in cans may also be due to a reaction between the food and the metal in the can, producing hydrogen gas; overfilling the can at too low a temperature; freezing the liquid portion of the food; inadequate removal of oxygen from the can before sealing; or sulfides present in the foods.

"Reduced oxygen packaging" is the reduction of the amount of oxygen in a package by mechanically evacuating the oxygen; displacing the oxygen with another gas or gases; or otherwise controlling the oxygen in a package to a level below 21% oxygen. The concern is the potential for supporting the growth of *Clostridium botulinium.*

Sources of Exposure

Extraneous Materials. Insect parts and rodent hairs are unwanted components of raw agricultural materials and processed food. Although large amounts of these materials are unacceptable, little is known about the relationship of the quantity of material to microbiological contamination. More research is needed in this area.

Pesticides and Fertilizer. Raw food products are contaminated by pesticides, herbicides, and fertilizers. As an example, in 1989, the pesticide Alar made national news when it was found to be contaminating apples. After washing, a residue of the chemical may be present. The Pesticide Chemical Act of 1954, as updated by The Federal Insecticide, Fungicide, and Rodenticide Act of 1988, prohibits interstate shipping of raw agricultural foodstuffs containing a residue of a pesticide unless it is safe or if the residue is within the tolerance level established by the Food and Drug Administration as safe. A further discussion on pesticides appears later in this book.

Radioactive Fallout. Radioactive material mixed with soil and rock are spread over a large area when a nuclear weapon is exploded or when a nuclear accident occurs. The fallout causes harm to people, animals, crops, food, water, structures, and fields. Radioactive strontium and iodine are most dangerous. Radioactive strontium is chemically similar to calcium; it enters the bones and may cause cancer. It has a half-life of 28 years. Radioactive iodine, which has a half-life of 8 days, may cause cancer of the thyroid gland. The major danger to people is that strontium-90 may be taken in with grass by cows and then passed through milk to humans. Milk sheds are sampled regularly for levels of strontium-90.

Antibiotics in Food. Such antibiotics as penicillin, streptomycin, bacitracin, Chloromycetin, Terramycin, and Aureomycin are used in the preservation of fish and meat, primarily to retard spoilage. Antibiotics and sulfa drugs are commonly used in treatment of bovine mastitis in cows. Antibiotics have also been used as food additives in uncooked ground beef and pork products to preserve keeping quality. Until recently, the Food and Drug Administration permitted the addition of chlorotetracycline and oxytetracycline to poultry water to extend shelf life. The major question for consumers is whether the antibiotics are toxic if ingested over extended periods of time. Another serious question is whether substantial usage of antibiotics in foods causes the development of resistant organisms that may be harmful to humans. Other concerns include the potential for human allergic reactions to the antibiotics.

Heavy Metals and the Food Chain. The heavy metals are considered to be atimony, arsenic, cadmium, chromium, lead, mercury, nickel, silver, thallium, and uranium. All individuals are exposed to heavy metals from a variety of sources, including natural sources, land pollution sources, water pollution sources, and aerial sources. The natural sources are due to chemical weathering and volcanic activities. In freshwater systems, chemical weathering of igneous and metamorphic rocks, as well as soil, causes these trace metals to enter surface waters. The decomposition of plants and animals also adds trace metals to the surface waters. Dust from volcanic activities, smoke from forest fires, aerosols, and particulates on the surface of oceans add additional amounts of trace metals.

Mining operations contribute higher levels of heavy metals, because the material brought from below the surface of the ground is now exposed to the air. Additional heavy metals are added by the corrosion of water pipes and the manufacturing of consumer products. Wastewater treatment by means of the activated sludge process usually removes less than 50% of the heavy metals present in the liquid. Sewage sludge, when placed on top of the ground or sprayed on the ground, becomes a source of cadmium, lead, mercury, and iron. Industrial sludge also adds a significant amount of heavy metals. Storm water runoff from urban areas contribute lead, cadmium, chromium, and zinc. The salt used on streets may increase the mobilization of metal ions, such as cadmium, mercury, and lead. Leachate from sanitary landfills can increase the levels of lead and mercury. Agricultural runoff, which includes soil erosion as well as animal and plant residues, fertilizers, herbicides, fungicides, and other pesticides, adds to the heavy metal problem.

Aerial sources of heavy metals come from the smelting of metallic ores. Other aerial sources come from fossil fuels and the addition of metals to fuels as anti-knock agents. These enter the body via these routes:

1.   through respiratory surfaces
2.   through adsorption from water onto body surfaces
3.   from ingested food particles or water through the digestive system

There are two main routes of entry into plants:

1.   through the plants surface above the ground
2.   through the root system

Terrestrial animals take up heavy metals through food and water. Further, a significant amount of absorption may occur through the lungs in some situations. Heavy metals can be bioconcentrated, as there is a progression upward through the food chain, and therefore, there may be a biomagnification of the heavy metals in larger animals.

Environmental Problems in Milk Processing

Environmental problems on dairy farms vary with the size of the operation and the age of the structures and equipment. The older, smaller operations may have problems with poor ventilation, poor lighting, and improper construction of the physical facility. Common to all farms are the increased problems of waste handling, flies, excessive use of pesticides, improper storage of pesticides, inadequate and improper cleaning of pipelines, milking equipment, and bulk storage tanks. A safe water supply is essential.

The basic problems in milk plants include cleaning of equipment and the constant use of properly operated, accurate instruments.

Environmental Problems in Poultry Processing

Basic problems in poultry plants include cleanliness of walls, floors, drains, equipment, conveyor belts, and interior and exterior surroundings. Blood is frequently splattered and mixed with dirt and feathers. This becomes very difficult to remove without constant, well-organized, efficient cleaning procedures. The amount of solid waste produced causes a serious removal problem. Sewage systems are frequently overloaded and the BOD (biological oxygen demand) is enormous. Rats frequently invade chicken-slaughtering plants because of the availability of food, blood, and other waste materials. Flies and roaches may become a serious problem. Water used in the chilling process is readily contaminated by birds, thereby continuing the cycle of contamination and bacterial growth. Improper supervision of cleaning, inadequate cleaning schedules, and a poor understanding of cleaning techniques increase problems. Quaternary ammonium chloride compounds cannot be used as sanitizers in poultry drinking water.

Environmental Problems in Egg Processing

Environmental problems include all those previously related to poultry. Further, because of the nature of the egg, bacteria may be introduced and grow rapidly due to improper handling, unhealthy workers, unclean equipment, utensils, facilities, contaminated uniforms, insects, rodents, and general debris. The water supply may be unsafe, and there may be an inadequate quantity of hot and cold running water under pressure for use in all preparation areas. The solid-waste problem is enormous because of the large quantity of shells and other materials that must be removed. A serious problem occurs when individuals, equipment, or air move from the raw eggs storage and breaking areas to the pasteurizing and packaging egg-processing areas. Improper removal of detergents and use of chemicals readily contaminates eggs. The use and storage of insecticides must be carefully controlled, since the eggs may readily absorb odors, as well as a variety of chemical contaminants.

Organisms penetrate the eggshell and grow quickly within the egg itself. Since eggs are used raw in eggnog and partially cooked in scrambled eggs, the chance of spreading food-borne disease is high. In addition, eggs are frequently used in specialized therapeutic diets for high-risk populations, such as the very young, sick, chronically ill, and aged. The feed ingredients are of concern since they may be contaminated. Clean, fresh eggs properly handled after production are not a public health problem. However, where the eggs are dirty and/or cracked, a potential condition arises in which pathogenic organisms could be introduced.

Environmental Problems in Meat Processing

The environmental problems relating to meat encompass all areas of stock-yards, slaughterhouses, meat-processing plants, retail meat-cutting operations, meat storage, and sales. Concern exists about the removal of large quantities of solid waste, intense fly problems, and the presence of rodents while the cattle are in stockyards. Disease may be spread from animal to animal and from human to animal.

At the slaughtering plant, organisms are added during many parts of the killing process. All equipment, appurtenances, and parts of the physical structure, such as walls, ceilings, and floors, contribute contaminants because of improper cleaning techniques. Adequate quantities of potable hot and cold running water may not be available for cleaning of equipment, physical structure, and for hand washing. Cutting instruments and food-contact surface may be improperly cleaned and sanitized. Meats may contain such pathogenic organisms as brucella, salmonella, streptococcus, mycobacterium, pseudomonas, and any other organisms that may inadvertently be introduced. Pork is hazardous since it may contain *Trichinella spiralis*. Refrigeration must be constantly monitored, since it is so essential to the control of bacterial growth. Employee health, personal cleanliness, and uniforms are also of concern.

At the retail market, all utensils, saws, and packaging equipment are readily contaminated and frequently improperly cleaned. Adequate quantities of hot and cold running water may not be available for equipment and hand washing. Employee health and personal hygiene practices, which are so essential, since the meat is frequently handled, may be poor. The walls, ceilings, cutting tables, and other pieces of equipment are frequently improperly cleaned. Meat may be overstocked in the display cases and not rotated properly. Once again refrigerator temperatures are of considerable concern. The number of microorganisms on meat varies from as few as 100 per gram of beef to several million per gram of beef, depending on slaughtering and processing techniques and the ultimate condition of the meat. Obviously, any chopped or ground meat may have bacterial counts in the millions.

Environmental Problems in Fish and Shellfish Processing

Fish and shellfish are potential vehicles for many organisms that cause disease in people. Fish acquire organisms from polluted water, harvesting equipment, processing equipment, or infected human beings. Shellfish are filter-feeders and concentrate microorganisms, natural toxins, and chemical contaminants from water. Many shellfish, such as oysters, clams, and mussels, are consumed raw, posing a serious public-health hazard. Infectious hepatitis, salmonellosis, typhoid fever, and cholera have been spread by shellfish. In recent years, outbreaks of botulism type E were caused by fish. The heat processing of fish may not create a high enough temperature to kill organisms. Coagulase-positive staphylococci are frequently isolated from crab and shrimp meat. Fish meal is of particular concern, since it is produced in the U.S. or abroad, and it may be contaminated with innumerable organisms plus the feces from rodents and birds. Shigella from imported shrimp have been involved in outbreaks of dysentery. Lately concern has increased about fish containing mercury and other chemicals found in polluted waters.

The major environmental problems in the harvesting of shellfish are due to the water in which the shellfish are cleaned, the harvesting boats, lack of proper prewashing of the shellfish to remove mud, the bacteriological quality of the washing water, and the disposal of human excretion.

The major environmental problems at the shucking and packing plant are cross-contamination from the shucking room, which is a dirty room, to the packing room, which must be kept immaculately clean; the presence of flies, which spread enteric organisms; improper plumbing, with the resultant possible spread of disease-causing organisms; inadequate hand-washing facilities; improper personal hygiene of workers, including inadequate hand cleaning; inadequate cleaning and bactericidal treatment; inadequate refrigeration of shell stock and shucked shellfish; and contamination of single-service containers.

Since employees can easily cut their fingers or hands with sharp knives or shells, it is essential that anyone with an infection be removed from the shellfish-processing areas.

Seafood implicated in histamine poisoning are the families *Scrombridae* and *Scromberesocidae* which includes tuna and mackeral. Non-scrombroid fish causing poisoning include mahi-mahi, bluefish, pink salmon, herring, anchovies, and sardines.

Impact of Food Processing on Other Environmental Problems

Food processing has an enormous impact on other environmental problem areas. Enormous quantities of solid waste are produced that must be processed in some manner, discharged to a receiving stream, burned, or removed to a landfill. The materials cause water pollution, air pollution, and soil pollution. For an in-depth discussion of these concerns, see Volume II.

Food Quality

Establishing standards or guidelines for food is a complex problem. Although microbiological standards have been established for water, milk, and dairy products, attempting to establish microbiological standards for solid foods is quite difficult. Virtually all foods, unless prepared in a sterile manner, will contain bacteria. The poorer the practices, the higher the level of bacteria. However, numbers of bacteria, although indicative of possible poor handling practices, may not necessarily indicate exposure to microorganisms that may cause disease. Frequently the number of microorganisms exceed 10 million per 1 g of food. Numbers of microorganisms may be increased sharply if there are roaches, flies, ants, rats, or mice present in the establishment.

Imported Foods

Imported foods may be of considerable concern to the population. The foods may come from processing plants that are run in a very unsanitary manner. Water used in processing may be contaminated. The raw food stuff may be contaminated with microorganisms, pesticides, or other chemicals. Leaking jars, dented cans, moldy food, and insect parts have been found in areas where imported foods have been stored.

Economics of Food Processing

Food production is controlled in many cases by the amount of fertilizers available and the energy that can be utilized. High-cost energy tends to decrease the amount of food produced in various parts of the world.

Inflationary forces in the U.S. and in other industrial countries have caused the cost of food to rise sharply. The farmers in certain areas are facing difficulties in getting adequate prices for the food they produce. The short-range effect may be to drive some of the smaller farmers out of business. The long-range effect will probably be further increases in the cost of food. The larger farmers tend, especially in the cattle-raising business, to concentrate the yields of food and therefore concentrate the byproducts, which become environmental contaminants.

POTENTIAL FOR INTERVENTION

Potential for intervention in the causation of disease or injury due to food processing varies with the type of processed food. With raw food products, a reduction in the use of pesticides under controlled conditions and the proper application of fertilizers and pesticides should help reduce the potential for disease due to these contaminants. A better understanding of the use of additives and preservatives, and the establishment of limitations based on this understanding, may be of further assistance in reducing disease potential. However, considerable research is needed in the entire area of food processing of raw foodstuffs to better understand the techniques for intervention in the disease process.

In the specific preparation of milk, meat, eggs, shellfish, and other highly perishable foods, the standards, practices, and techniques exist and the potential for intervention and disease prevention is excellent.

RESOURCES

Scientific and technical resources include the food industry, the National Environmental Health Association, the International Association of Milk, Food, and Environmental Sanitarians; the American Public Health Association; the National Sanitation Foundation; the American Dietetic Association; the University of Iowa; the University of Michigan School of Public Health; the University of Minnesota School of Public Health; the University of California School of Public Health; and many others.

Civic associations interested in food include the North American Vegetarian Society, the Nutrition Today Society, and the Society for Nutrition Education.

Governmental organizations include all state health departments and departments of agriculture, and many local health departments, the Environmental Protection Agency, the Department of Energy, and various sections of the Department of Health and Human Services, including the Center for Disease Control, the National Cancer Institute, the National Institute of Environmental Health Sciences, the National Institute of Occupational Safety and Health, the National Center for Toxological Research, and the Food and Drug Administration. In addition, the Department of Defense and the Consumer Product Safety Commission are concerned with food products. Departments of Agriculture of the various states and the federal government are important resources in a variety of areas. The U.S. Senate Agricultural Committee is an important source of legislation and information. The U.S. Fish and Wildlife Service Department of the Interior is also an important resource.

The Food and Drug Administration, Consumer Affairs and Small Business Staff, Department of Health and Human Services, 5600 Fishers Lane, Room 13-55, Rockville, MD 20857, is helpful to individuals trying to get additional information about food and food technology. The Food and Nutrition Service, Department of Agriculture, Room 512, 3101 Park Office Center Drive, Alexandria, VA 22302, is useful for those concerned with nutrition aspects related to food. The Food Safety and Inspection Service, Room 1163, South Building, Department of Agriculture, Washington, DC 20250, is useful for all areas of food and food technology. The National Agricultural Chemicals Association, 1155 15th Street, N.W., Washington, DC 20005, have experts in the areas of agricultural chemicals, including pesticides and their effects on food. The U.S. Department of Agriculture hot line for food safety is 1-800-535-4555.

STANDARDS, PRACTICES, AND TECHNIQUES

Food Preservation Techniques

The shelf life of food can be extended by inactivating or destroying enzymes or by inhibiting or destroying microorganisms. Food preservation techniques include the use of drying, low and high temperatures, irradiation, salt and sugar, and chemical preservatives. Various chemicals are also added during processing to enhance flavor and to promote better color and odor.

In all food preservation, it is essential to start with as clean a raw material as possible and to prevent contamination from hands, soil, water, air, surfaces, processing, and storage equipment.

Drying

Drying, also called dehydration or desiccation, is one of the oldest methods of food preservation still in use today. Dates, figs, and raisins are air dried. Forced hot air is used in drying potatoes, fruits, and vegetables. Milk and eggs are sprayed into heated cylinders. Enzymes are inactivated prior to drying by blanching, which is the removal of color by boiling, or by the use of sulfur dioxide or sulfites. Blanching removes sticky substances, checks browning, and makes vegetables easier to work with. Reconstitution of the dry food is aided by clustering the particles during drying, by adding a foaming agent such as nitrogen during concentration of the food prior to drying, or by causing the food to explode into particles. Drying will kill some organisms, but spores of bacteria, yeasts, and molds will survive and can later cause outbreaks of food-borne disease, or food spoilage.

Freeze Drying

Food is first frozen and water is then evaporated from the ice crystals without melting them. A small amount of heat is applied and pressure is reduced in this process, called sublimation.

Refrigeration and Freezing

Low temperatures retard food spoilage. The shelf life depends on the kind of food. Freezing is the removal of heat from a product. The formation of ice crystals in the food is due to free water or air being trapped in the package. During the freezing process, moisture from the warm food will form a vapor on the packaging material and will freeze more rapidly than the food mass. Ice crystals break down cell structure and cause a physical deterioration in food quality. This can be avoided by rapid freezing of high-quality raw materials. Freezing does not improve the quality of food, nor does it destroy many of the pathogenic organisms that may be present.

Heating

The rate of kill of microorganisms at high temperatures varies with the species of the organisms, presence of spores, quantity of organisms, consistency, pH and type of food medium, temperature, time, moisture, quantity of food, and ease of heat conduction and convection. To avoid overcooking a given food, which could affect food quality, the temperature and time of cooking should be adequate to kill pathogenic organisms, but not necessarily thermophilic or thermoduric organisms.

Pasteurization is a heat-treatment process well below the boiling point, which will destroy all harmful microorganisms and improve the keeping quality of food. It is used in the dairy industry and also in the preservation of dried fruits, syrups, honey, and juices.

Boiling is a common method of preparing and preserving food. Boiling does not kill all organisms, since at atmospheric pressure the temperature may not reach 212°F. Therefore, adequate time must be provided for a total bacterial kill. Although heating of food to boiling will destroy botulinum toxin, the spores may survive boiling temperatures for a considerable period of time. Baking in an oven at a temperature of 350–400°F may not cause the food temperature to rise to 212°F because of moisture present in the food and because baked goods have low heat conductivity. Frying kills organisms on the surface of the food, but may not affect the organisms within the food mass.

Canning

Canning is the preservation of food by subjecting selected prepared foods to high temperatures in a permanently sealed container (Table 3.1). Containers may be made of glass and sealed with tin, aluminum, paperboard, or plastic caps, which are either automatically vacuum sealed or screwed on. Metal cans made from tin-coated steel or aluminum are also used. Lacquered cans are used for highly colored foods to prevent color loss. Can sizes are as follows: no. 1, 10.94 oz; no. 2, 20.55 oz; no. 2$\frac{1}{2}$, 29.79 oz; no. 3, 35.08 oz; no. 10, 109.43 oz.

The entire canning process is important. However, certain key items should be noted:

1. Use high-quality food from which gross dirt and microorganisms have been removed.
2. Heat-process foods as soon as the air has been exhausted.
3. Cool promptly to avoid overcooking and undesirable changes in texture and flavor and the germination and multiplication of spores of thermophiles.
4. Since food is commercially sterile (all bacteria, but not spores are killed), the resistant spores of *Clostridium botulinum* may not be destroyed in nonacid foods.

Testing. Canned products should be tested by incubating samples at 100°F to test for swells, incubating samples at 130°F for 2 weeks to test for flat sours, inserting thermocouples in food center to check temperature, placing spore-forming test organisms in marked cans, and taste-sampling for quality.

Table 3.1.   **Canning Flow Diagram**

HARVESTING ➡ GRADING ➡ WASHING ➡
                Size;   Soaking, 20 sec.
                color   3 min.;
                quality   spraying, 60 lbs.
                                             pressure in drum

PEELING ➡ CAN WASHING ➡ BLANCHING ➡
                Invert cans;   Boiling
                wash   sulfur dioxide;
                                             sulfides

CAN FILLING ➡ ADD BRINE, SYRUP ➡
Dipping in hot lye: OR STOCK
machine or hand   1–3% salt solution;
peeling   preservatives;
                      additives (other);
                      sugar solution;
                      stock (for meat)

EXHAUSTING & ➡ COOKING ➡ COOLING ➡
CAN SEALING   By type of food & mass   Quickly, by water,
Remove air;   spray, water bath
draw vacuum

CODING ➡ LABELING ➡ INSPECTION ➡
Product letter;   Machine
grade letter, #; plant
letter, #; destination

CASING ➡ WAREHOUSE ➡ TRANSPORTATION ➡
Hand, machine   Dry, ventilated, cool,   Trucks, trains, planes
                       frostproof

Problems in Canning and Bottling. The following problems may occur in canning: microbial contamination, pinholes developing from chemical action of food on container, cans collapsing, glass containers breaking because of impact, thermal shock, internal pressure, and toxins developing.

Irradiation

Certain selected food, meat, poultry, fish products, vegetables, and spices have been preserved by irradiation. Organisms in food are destroyed by direct hits, causing ionization, fragmentation of DNA, and death. Some lethal effects are caused chemically by free radicals produced in the solvents. In irradiated water, hydroxyl ions, which are strong oxidizing agents, are formed. The hydrogen ions present are strong reducing agents. The oxidizing and reducing agents may produce off-flavors, and changes in taste, appearance, odor, and texture. On the beneficial side, radiation sterilizes very quickly, kills insects, parasites, helmints, pathogenic bacteria, molds, and yeasts effectively, keeps the temperature of the food from rising significantly, delays ripening, prevents sprouting, and extends the shelf-life of food by retarding spoilage.

Irradiated food is not radioactive, nor does it contain hazardous radiolytic products. The radiolytic products include glucose, formic acid, acetalehyde, and carbon dioxide, that are naturally present in foods or formed by heat processing. The change in nutritional value of the food caused by irradiation depends on radiation dose, type of food, packaging, processing, and storage time. Other food processing technologies create like problems. Vitamins A, E, C, K, and B-1 in foods are sensitive to radiation while riboflavin, niacin, and Vitamin D are more stable. No abnormal chromosomes have been shown to occur from eating irradiated food. A study conducted in the mid-1970s by the National Institute of Nutrition in India indicating the condition known as polyploidy or multiple sets of additional chromosomes has not been duplicated. Irradiation at levels up to 10 kilogray (KGy) does not increase risk from botulisim any more than other substerilizing techniques. Although irradiation suppresses microorganisms which cause spoilage, it cannot freshen-up spoiled food. If toxins or viruses are already formed in foods, irradiation will not destroy them. There is no scientific evidence to indicate any health hazard associated with irradiation of food containing pesticide residues and additives. Almost all commonly used food packing materials, including plastics, are unaffected by irradiated food.

The Joint Expert Committee on Food Irradiation concluded that irradiation to doses not exceeding 10 KGy is wholesome and safe for human consumption. The World Health Organization, in 1992, after a comprehensive review of data concluded that irradiation carried out under good manufacturing practices presented no toxicological hazard and caused no special or microbiological problems. In 1992, the Department of Agriculture, Food and Safety Inspection Service, issued a final rule on the irradiation of poultry products permitting the use of ionizing radiation. All poultry prepared from irradiation had to be clean, processed, and packaged in accordance with the Poultry Products Inspection Act and the poultry products inspection regulations.

Milk

Milk is the fluid secretion of the mammary glands, practically free of colostrum, containing a minimum of 8.25% nonfat milk solids and 3.25% milk fat. Milk contains the following average constituents: water, 87.3%; fat, 3.7%; protein, consisting of casein, 2.9%; albumin, 0.5%; lactose, 4.9%; and ash, 0.7%, which includes salts of calcium, copper, iron, magnesium, manganese, potassium, and zinc. Milk contains lactose in true solution, casein in permanent suspension, and butter fat in temporary colloidal suspension. Milk is an excellent source of vitamins A and B_2, and a fair source of vitamins B_1 and E. It also contains vitamins

D, K, niacin, pantothenic acid, B_6, B_{12}, folic acid, citrin, P-amino benzoic acid, biotin, choline, and inositrol. It also contains enzymes, epithelial cells, leukocytes, yeasts and molds, and extraneous foreign material and bacteria from within the udder from milking, people, processing equipment, and poor techniques. Microbial pathogens may include *Escherichia coli* 0157:H7, *Salmonella spp.*, *Listeria Monocytogenes*, and others.

Milk Production

Clean, safe, nutritious milk comes from healthy cows that live and are milked in an environment where clean, sanitized equipment, rapid cooling, and proper handling and storage techniques are utilized (Table 3.2 contains a diagram of the milking flow process).

Vat Pasteurization. The vat pasteurizer consists of a stainless-steel jacketed vat with a cover, agitator, inlet and outlet valves, airspace heater, and indicating, recording, and airspace thermometers. It is essential that the milk is protected during the entire process, that pasteurization temperatures of 145°F be maintained for a minimum of 30 min, and that the milk be agitated.

The vat may be operated in the following manner:

1. Heat milk to pasteurization temperatures and maintain these temperatures by spraying hot water on inside of the double-jacketed vat or by surrounding the inner jacket with heated coils.
2. Partially preheat milk by means of a heater and then bring milk up to pasteurization temperatures as in 1.
3. Preheat milk to pasteurization temperature and then maintain as in 1.
4. Remove milk from vat either partially cooled by turning off heating system or at pasteurization temperature.

The cover must have overlapping edges, have all of its openings protected by raised edges, and have all condensate diverted from it.

The inlet and outlet of the vat must have leak-protector valves with the following design:

1. seat in every closed position
2. grooves at least $^3/_{16}$ in. wide and $^3/_{32}$ in. deep at center
3. stops not reversible
4. close coupled valve seats flush with inner wall of pasteurizer or so close to the wall that milk in the valve inlet is no more than 1°F colder than pasteurized milk

To ensure that all milk is pasteurized, the inlet and outlet lines must be properly sloped to allow drainage of all milk; the inlet valve must be closed during the holding and draining periods; and the outlets valves must be closed during filling, heating, and holding.

Table 3.2.   **Milk Production Flow Diagram**

MILKING

COW ━━━▶
Wash udder with warm water and then with chlorine; strip, examine foremilk; rinse teat cups; sanitize teat cups

MILKING MACHINE ━▶
Prewash, sanitize; rinse in cold water; sanitize; attach cups to udder; set timer; strip udder

STRAINING ━━━▶
Stainless steel strainer with pad; material on pad indicates level of sanitation in milk house

COOLING AND STORAGE ━▶
Mechanical milk cooler; cool to below 41°F rapidly (within 4 hours)

TRANSPORTATION ━▶
Cans cooled below 41°F or below; bulk 41°F or below in stainless steel tank; wash, sanitize every 24 hours.

MILK PLANT

MILK PROCESSING

TRANSPORTATION ━━▶
Raw milk–cans; bulk

HOLDING TANK– RAW MILK ━━▶
Storage 41°F or below

BALANCE TANK ━━▶
Constant level of liquid

BOOSTER PUMP ━━▶
Moves milk

REGENERATION ━━▶
Hot pasteurized milk on positive pressure side of plate heats up cold, raw milk on negative pressure side

METERING PUMP
Opposite side regenerator from holding tank; regulates amount of milk passing

PASTEURIZATION

HOMOGENIZER ━━▶
Suspends milk, milk particles; diffuses equally

Vat
145°F for 30 min.; higher milk fat or added sweeteners– 150°F for 30 min
High temperature short-time
161° for 15 sec.; higher milk fat or added sweeteners– 166°F for 15 sec.
Ultra high temperature
by steam injection at 191°F for 1 sec.; 194°F for 0.5 sec. or 212°F for 0.01 sec.

REGENERATION ━━▶
Cold, raw milk on negative pressure side cools hot pasteurized milk

COOLER ━━━▶
Quickly lowers milk temperature

PIPES ━━━▶
To storage containers; to fillers

FILLERS ━━━▶
Bottles; cartons; cans; bulk containers

TRANSPORTATION ━▶
Below 41°F

RETAIL STORES ━▶
41°F

CONSUMER

Constant agitation of the milk during pasteurization ensures that all of the milk is uniformly heated and properly pasteurized. The airspace heater is used for the same purpose for the foam above the milk. The temperature of the air must be 5°F higher to ensure proper pasteurization.

The indicating thermometer and airspace thermometer show the temperature of the milk at any given time. The recording thermometer, which is attached by wires to a chart-and-pen arm, records the temperature at which the milk is being processed. The recording thermometer, which is checked daily against the indicating thermometer, must never read higher than the indicating thermometer. Charts should indicate the preheating time, pasteurization time, precooling time, airspace temperature, indicating thermometer temperature at a given time, record of unusual occurrences, name of milk plant, operator's name, location of recorder, and date.

High-Temperature Short-Time Pasteurization. High-temperature short-time (HTST) pasteurization (Table 3.3) consists of a system including:

1. Drawing of cold raw milk at 41°F or under from a constant-level tank.
2. Raw milk heated as it flows on the negative pressure side of thin stainless-steel plates in the regenerator, by hot pasteurized milk flowing in the opposite direction on the positive-pressure side of the plates.
3. Raw milk still under negative pressure passing through a positive displacement timing pump that pushes it under positive pressure through the rest of the system.
4. Heating raw milk to at least 161°F.
5. Raw milk under pressure at 161°F flowing through holding tube for a minimum of 15 sec.
6. Milk flowing past indicating thermometer and recorder controller at 161°F or higher and then through the forward-flow position of the flow-diversion valve. If the temperature is less than 161°F, the milk flows through the diverted-flow position into the diverted-flow line and back to the raw, constant-level tank.
7. Pasteurized milk flowing through the regenerator and cooling.
8. Milk passing through cooler where temperature is dropped to 41°F or under and being stored in a vat until packaging.

The recorder chart must have a scale with a span greater than 30°F, including 12°F from the diversion temperature. The temperature must be accurate to within 1°F at diversion temperatures. The chart must not rotate more frequently than once every 12 hr. The thermometric response must be 4 sec to travel 12°F within a 19° range. The date, number of location, name and address of plant, reading of indicating thermometer at a given point, amount and name of product, product temperatures at beginning and end, time in forward-flow position, unusual occurrences, and operator's name must be recorded on the chart.

It is always important to maintain proper pressure throughout the system.

Table 3.3.   **High–Temperature Short–Time Flow**

CONSTANT LEVEL TANK
Raw milk; reservoir for milk; tank for
cleaning system in place

REGENERATOR
Pasteurized milk always on positive
pressure side in case of leakage

METERING PUMP
Sealed at maximum speed; wired so it only
operates if FDV is fully operable

HEATER

HOLDING TUBE
¹/₄–in. slope per ft. of tubing upward in
downstream flow of milk preventing air
entrapment; permanently supported

FLOW DIVERSION VALVE (FDV)
3-way valve automatically controlling
product flow; forward flow position occurs
when recorder controller microswitch,
preset to operate at 161°F, energizes a
solenoid-operated air valve permitting
compressed air to flow across a
diaphragm, forcing it to seat securely on
diverted flow line; when diverted flow
position less than 161°F, the microswitch
does not activate the solenoid and the
diaphragm remains in place across the
forward flow line.

REGENERATOR

COOLER

STORAGE
41° or below

FILLER

CONTROLLER SENSOR
Automatically records pasteurization
temperatures; automatically controls FDV
position; sensor, maximum of 18 in.
upstream from FDV

INDICATING THERMOMETER
Mercury actuated, direct reading; scale not
less than 25°F including ±.5°F from
pasteurization temperature, graduated in
0.5°F divisions; accurate to ±0.5°F
throughout scale; thermometric lag
maximum 4 sec. to rise, 12 degrees in a
19 degree span.

Ultra-High-Temperature Pasteurization. Ultra-high-temperature pasteurization is
accomplished by steam injection or by the use of a vacuum. In the steam process,

automatically controlled injected steam is used to heat the milk to the following pasteurization temperatures: 191°F for 1 sec, 194°F for 0.5 sec, 201°F for 0.1 sec, 204°F for 0.05 sec, 212°F for 0.01 sec. The product is then rapidly cooled and passed through the flow-diversion valve before storage. The reason that the flow-diversion valve, which operates the same as in the HTST process, is located after the cooler is that a safety hazard, due to flashing steam, occurs if milk is diverted at 212°F; the milk would be diluted by steam if diverted at high temperatures; the response time of the thermal limit controllers would be too slow to prevent the forward flow of raw milk.

In the vacuum process, the milk must be heated to at least 194°F. Constant temperatures must be maintained in the pasteurizing zone by a constant uninterrupted, adequate quantity of steam. The steam pressure must be 35 lb/in.^{2} and the vacuum drawn in the chamber must be 8 in. Ultra-high-temperature pasteurization is used to increase the shelf life of milk, inactivate enzymes, and improve flavor.

Dry Milk. Milk is dried by spraying in a vacuum chamber. Salmonella surviving in the dry milk are a potential hazard.

Cleaning

Proper cleaning is absolutely essential during all phases of milk production and milk processing. One technique involves disassembling all equipment, rinsing thoroughly with cold water, scrubbing all equipment with brushes using hot water and detergent, rinsing thoroughly, reassembling, and sanitizing with 200 ppm chlorine. Sanitization may also be accomplished by the use of steam for 5 min or hot water at 180°F for 5 min. Acid cleaners should be used when necessary to aid in cleaning. Cleaning in place is a technique in which the circulating system is not disassembled. Instead, a cleaning solution is circulated through the system at a velocity of at least 5 ft/sec. The solution, which is alkaline and nondepositing, is heated to 120°F and circulated for at least 15 min. The solution is then drained and the system is thoroughly rinsed. Finally, chlorine or hot water is used for sanitization.

Ice Cream and Frozen Desserts

Frozen desserts are pure, clean, frozen, or semifrozen foods prepared by freezing while stirring a pasteurized mix composed of edible fats, nonfat milk solids, sugar sweeteners, flavors, and other ingredients such as eggs, fruits, and nuts. The pasteurization process is the same as used in milk, except that in the vats the pasteurization temperature must be 155°F for 30 min instead of 145°F for 30 min, and in the HTST system the pasteurization temperature is 175°F for 25 sec instead of 161°F for 15 sec. Contamination may enter the mix when fruits and nuts are blended into the mix after pasteurization and cooling, and before freezing.

Poor techniques, careless handling, or contaminated raw products can raise the bacterial count of the mix.

Counter Freezers or Soft Ice-Cream Freezers

Counter freezers or soft ice-cream freezers are a continuation of the frozen dessert process. Handling of the mix after delivery to the stores and improper cleaning of the equipment contributes to extremely high bacterial counts. Typically, the mix is delivered in plastic bags, packed in boxes, or packed in cartons similar to large milk containers. The mix is poured into the dispenser and then added to during the day as needed. At the end of the day, in many instances, the mix is drained into a bottle or jar and saved for the next day. The dispenser may or may not be rinsed or cleaned.

Proper technique consists of carrying out the following steps each evening: Drain mix at end of day to waste; rinse dispenser thoroughly with cold water; wash inside of machine with hot water and detergent for at least 5 min; rinse unit thoroughly; add to dispenser 200 ppm chlorine in water and allow to stand for 2 min; drain thoroughly and air dry. Before refilling the following morning, the dispenser should be once again sanitized with 200 ppm chlorine for 2 min and allowed to drain.

Poultry

Poultry are domesticated birds, such as chickens, turkeys, ducks, geese, and pigeons, that are raised for food consumption. The widespread use of poultry and poultry products, the substantial amount of handling and processing, the kinds of bacteria indigenous to the birds (salmonella), and the long history of poultry-associated food-borne diseases makes poultry processing an important area of concern in public health.

Poultry are mass produced in special housing with automated feed and water supplies. The feed ingredients may contain salmonella or other organisms transmitted to the chicks and eventually to humans. Salmonella may also be introduced into the environment of the poultry by dust, air, domestic animals, rodents, fomites, human waste material, flies, pigeons, and other birds. (See Table 3.4 for a diagram of the poultry process.)

Eggs

Eggs are produced from domesticated chickens, ducks, geese, turkeys, and other domestic fowl for human consumption. Eggs are mass produced in special units in which hens are automatically fed and provided water. Eggs have been implicated in outbreaks of disease. Even intact-shell eggs have been implicated in outbreaks of *Salmonella enteritidis*. (See Table 3.5 for a diagram of the egg process.) Egg pasteurization is usually performed in a high-temperature short-time pasteurizer and is similar to HTST milk pasteurization.

Table 3.4.   **Poultry Processing Flow Chart**

LIVE BIRD → **INSPECT AND WEIGH** → **SHACKLE** →
Healthy; vigorous

BLEED → **SCALD** → **DEFEATHER** →
Thoroughly; quickly   135° to 140°F;
helps remove feathers

WASH → **EVISCERATE** → **INSPECT** →
Spray

WASH → **CHILL** → **PACKAGE** →
Inside; outside   Quickly below 35°F; freeze   Cut up; whole; fresh, chilled,
frozen; in prepared meal,
either frozen, canned or
fresh (chicken roll)

Meat

Fresh, raw meats refer to the regular retail cuts of beef, veal, lamb, and pork. Processed meats are meats plus other ingredients, additives, and spices that have received special treatment, such as curing, smoking, or canning. Meats have been implicated in a number of outbreaks of disease including, most recently, ground beef-associated *Escherichia coli* 0157:H7

In order to simplify the slaughtering process, only the slaughtering of cattle will be discussed. (See Table 3.6 for a diagram of the slaughtering process.) In the slaughtering plant, the bones, feet, and condemned parts will be sent to rendering plants. Unusable portions of the meat are used for pet food. However, pet-food processing must be a totally separate operation from human-food processing.

Refrigerated beef, after being cut into prime or retail cuts, may be wrapped and frozen quickly at temperatures of –10 to 0°F. Quick freezing and proper wrapping are essential to preserve the quality of the food. Freezing may cause slight changes in flavor or color.

Specialized Processing of Meats

Curing, the addition to meat of sodium chloride mixed with sugar and spices, was originally developed to preserve meats without the use of refrigeration. Sodium nitrites and nitrate are added to stabilize the color, since most people prefer a pink look to their meat. Sugar is added to counteract the hardening effect of salt and to improve flavor. These ingredients can be dissolved in water to form a brine solution or brine cure. The meat can either be soaked in the brine, injected with brine, pumped with brine through the arteries, or rubbed with a salt and brine solution.

The smoking of meats is primarily for the addition of tastes, although some preservation takes place. Hardwood chips and sawdust are used in the smoke

Table 3.5.   **Egg Processing Flow Diagram**

WHOLE SHELL EGGS

HEN ⟹ **SHELL EGGS** ⟹

CANDLE ⟹
Classify; eliminate spoiled
eggs, blood spots; separate
cracked eggs

CLEAN ⟹
Spray wash with water at
same temperature as eggs

SHELL TREATMENT ⟹
Immerse eggs at 75°F in oil;
rotate for 10 min.

STORE ⟹
41°F or below; well-ventilated
area; clean area; relative
humidity 85–90%

SUPERMARKET
41°F or below

FLOW CHART FOR LIQUID OR FROZEN EGGS

HEN ⟹ **SHELL EGGS** ⟹ **CANDLE** ⟹

CLEAN ⟹
SANITIZE SHELLS ⟹
Spray
Do not use a quat
Rinse off excess sanitizer

BREAK SHELLS ⟹
Aseptically; separate white
from yolk

MIX THOROUGHLY ⟹ **CLARIFY OR FILTER** ⟹ **COOL** ⟹
40°F

FERMENT ⟹
To remove sugar

PASTEURIZE ⟹
140°F for 1.75 min. for
whole eggs; 134°F at pH9
for 1.75 min. for liquid egg
whites; 142°F for 1.75 min.
for liquid yolk

ADD ADDITIVES ⟹
Salt; sugar; adjust pH

PACKAGE ⟹
FREEZE
Rapidly to 0°F; defrost
when needed in refrigerator
at 41°F or below

FLOW CHART FOR DRIED EGGS

HEN ⟹ **SHELL EGG** ⟹ **CANDLE** ⟹

CLEAN ⟹
STORE ⟹
40°F for 72 hours before
sanitizing

SANITIZE ⟹

BREAK SHELLS ⟹ **MIX** ⟹ **CLARIFY** ⟹

HOLDING VAT ⟹
40°F for maximum of 48 hours

PASTEURIZE ⟹

SPRAY DRY ⟹
Under high pressure; drying
temperature 340°F

COOL ⟹
By coils to 90°F; final cooling
to 41°F or below

STORE
In sterile barrels or cartons

Table 3.6.   **Meat Processing Flow Chart**

SLAUGHTER ▬▬▶	ANTEMORTEM ▬▬▶	KILL ▬▬▶
Few animals at a time	**EXAMINATION** Inspectors and Veterinarians condemn or retain animals if the animal is in a coma, seriously crippled, dead or dying; has a disease or temperature of 105°F	Stunning by sharp blow on head or electrical prod; hung by hind legs on rail; jugular vein slit
DRESS ▬▬▶	**CARCASS WRAPPING** ▶	**PROCESS CARCASS** ▶
Blood completely drained; head removed; skin cut; belly opened; legs removed; viscera removed; viscera inspected by Vet or meat inspector (post-mortem inspection); carcass split; hide inspected	Wrapped in cloth for shaping	Chill within 1/2 hr. after slaughter by high velocity, cold air system or refrigeration; humidity in cooler 88–92%; hold in cooler 1–3 days or until rigor mortis is over; may hold longer for aging and tenderizing; temperature of 32°F
CUT FOR RETAIL ▬▬▶	**DISPLAY IN**	
At 32°F; primal cuts such as roast, steak, stew	**RETAIL MARKET** 32°F	

houses. Humidity and temperature are carefully regulated. Long-cured meats are smoked at an internal meat temperature of about 110°F. Short-cured meats are smoked at temperatures of 140–150°F and finally at temperatures of about 200°F. The length of smoking varies from a few hours to several months.

Cased meat products include sausages, frankfurters, and luncheon meats. Sausage is usually made from chopped meats, including beef, pork, veal, lamb, and mutton. It contains fillers, such as cereal, starch, soy flour, dry milk, and added water. Sausages may be cooked or smoked. Frankfurters, a type of smoked sausage, contain any of the above products and may also contain chicken; they are always fully cooked before being sold. Luncheon meats contain a variety of meat products, other ingredients, spices, and water; they are usually precooked before being sold.

Canned meats are usually cooked after packing. The storage of meats in cans depends upon the processing. In almost all cases, canned hams must be refrigerated.

Fish

Fish must come from safe and clean fishing waters. Fish may be preserved by icing, salting, smoking, freezing, or canning. The fish is harvested from the ocean and may be processed on the boat. If so, they are gutted, scaled, filleted, and iced. The fish may also be iced on the harvesting boat and then taken back to a fish-processing plant. Unfortunately, even artificially produced ice is not sterile because of handling, production, and storage, and may contaminate the fish.

In the smoking process, dehydration occurs, which helps in preservation. Smoking, however, is basically used for improving taste. The fish is first salted and then smoked with hardwood fumes in the same way that meat is smoked. Smoking may last a few hours or many days.

Fish preserved by canning are heat treated and commercially sterile. The keeping quality of the canned fish is based on the kind and amount of cooking.

Fish should be flash frozen. One of the problems is that during defrosting, a considerable amount of water is lost from the fish and the flavor is impaired. When salting, fish must first be cleaned and cut, immersed in a 20% salt brine for 30 min, and placed in water-tight containers.

Fresh fish, when inspected, will have the following characteristics: firm; no off odor; no discoloration or slime; eyes bright, full, and moist; red gills, not grey or brown; tight scales.

Shellfish Harvesting and Processing

Shellfish include all edible species of oysters, clams, and mussels. (See Table 3.7 for a diagram of the shellfish production process.)

MODES OF SURVEILLANCE AND EVALUATION

Milk Testing

Tests are performed on raw milk as it enters the plant to determine odor, appearance, off-color, blood, viscosity, foreign matter, sediment, and temperature. (See Figure 3.1 for an example of a milk farm inspection report.)

Laboratory tests conducted on raw milk include

1. direct microscopic count or breed count, which measures bacteria count, and identifies the type of bacteria and the source of contamination
2. methylene blue test, which gives a rough estimate of the bacterial load; decolorization in half an hour indicates a high bacterial load, whereas decolorization in 8 hr indicates a low bacterial load
3. resazurin test, which is similar to the methylene blue test
4. standard plate count
5. thermoduric bacteria test, which indicates use of unclean equipment on the farm or in transportation
6. coliform test, which indicates use of unclean equipment in production

Table 3.7.   **Shellfish Processing Flow Chart**

GROWING AREA ▰▰▰▷
Sanitary survey by industry and state authority under supervision of Food and Drug Administration (see NSSP program); evaluate sources of actual or potential pollution; bacteriological study of water and shellfish; chemical analysis of water and shellfish; graded as approved, conditional approval, restricted or prohibited

HARVEST ▰▰▰▷
From approved area only; from conditionally approved area only under supervision of health authorities, to ensure that shellfish are purified; from restricted areas only under controlled purification and supervision of health authorities; never from prohibited areas; human body waste must not be discharged from the harvest boat into the water

TRANSPORT ▰▰▰▷
Boats and truck, shellfish stored in thoroughly cleaned bins and away from bilge water; bins must be sanitized

WASH SHELLFISH ▰▰▰▷
Wash away mud; use potable water under pressure

SHUCK ▰▰▰▷
Separate from packing operation; not subject to flooding; fly-tight screening

PACK ▰▰▰▷
Clean room; surfaces washed, sanitized; impervious floors and walls; fly control measures; adequate safe, sanitary water supply; approved plumbing–avoid cross connections and submerged inlets; strict hand-washing procedures; use clean, sanitized single-service containers; label with packer's name, address, number, state, and date

REFRIGERATE ▰▰▰▷
Refrigerate quickly below 40°F; freeze quickly

CLEAN EQUIPMENT AND SANITIZE
Washing sinks, made of impervious, non-toxic materials; storage, shucking, and packing rooms cleaned within two hours after completing operations; sanitize with 100 ppm available chlorine

Laboratory tests conducted on pasteurized milk include

1. standard plate count, which determines the number of living bacteria and indicates the sanitary conditions under which milk was processed
2. phosphatase test, which determines the efficiency of pasteurization
3. coliform test, which determines the presence of coliform and indicates contamination after pasteurization
4. residual bacterial count from milk processing and storage equipment and containers
5. screening tests for abnormal milk

County	State	Date

Farm	Location	Date _____ Time _____	Ratings N–Not Apply U–Unsatisfactory S–Satisfactory

Item	Rating	Item	Rating
1. **Cow healthy & from disease free herd**		5. **Utensils & Equipment:**	

Item **Rating**   **Item** **Rating**

1. **Cow healthy & from disease free herd**
2. **Milk barn:**
   a. adequate lighting
   b. adequate ventilation
   c. floors, walls, ceiling clean
   d. floors, wall, ceilings, good repair
3. **Cow yard:**
   a. graded, drained, clean
   b. manure stored properly and removed
4. **Milk house:**
   a. floors, walls, ceilings, smooth & good repair
   b. adequate lighting
   c. adequate ventilation
   d. all openings screened
   e. facilities clean

5. **Utensils & Equipment:**
   a. smooth, non-absorbent non-toxic
   b. good repair
   c. easily cleanable
   d. clean
   e. sanitized
   f. proper storage
6. **Milk storage:**
   a. at or below 45°F within 2 hours
7. **Milking:**
   a. cleanliness of flanks and udders
8. **Fly control**
9. **Storage of pesticides**
10. **Safe water of adequate quantity**
    a. hot & cold water
11. **Proper sewage disposal**
12. **Adequate toilet facilities**
13. **Adequate handwashing facilities**

Recommendations   Time for Compliance

Environmentalist _____

Figure 3.1.   Milk farm inspection report form.

6. disc assay for presence of antibiotics
7. radioactive material testing

Vat Pasteurization Equipment Tests

All indicating thermometers, recording thermometers, and leak-protector valves used in vat pasteurization must be checked at installation and at least quarterly afterwards to determine if they are operating properly. The indicating thermometer, which must be accurate to +0.5 or –0.5°F within the pasteurization range or +0.5 or –0.5°F within the airspace range compared to a standard thermometer, is tested by inserting both thermometers into a well-agitated can of hot water at

145°F and making a direct reading. However, the indicating thermometers must read not less than the pasteurization temperature during the entire holding period. The error allowed in the thermometer's accuracy is taken into account when determining the indicating temperatures.

All recording thermometers and pen arms must be checked at installation, quarterly, and when needed, and must be able to adjust to a wide range of fluctuating temperatures. The recorder thermometer is adjusted to read exactly the same as the previously tested indicating thermometer while immersed in a can of agitated water at 145°F. The recorder sensor bulb is then immersed in a can of boiling water for 5 min and into a can of ice water for 5 min. Finally the recorder element is placed back into the can of agitated water at 145°F. The final temperature on the recorder chart must never read higher than the indicating thermometer.

The rotating time of the recorder chart is tested at installation and at least quarterly to make sure that the elapsed time is accurate. This is done by comparing the elapsed time on the recorder chart to an accurate watch.

The recording-pen arm temperature notation is compared to the indicating thermometer while the product is in the vat. The recorder must not be higher than the indicating temperature.

Proper functioning of the leak-protector valve is determined by disconnecting the outlet pipe and carefully checking for leakage around the valve while product pressure is being exerted on the valve.

High-Temperature Short-Time Pasteurization Equipment Tests

Indicating thermometers, including airspace thermometers, must be checked upon installation and at least once every 3 months against a standard thermometer to make sure that they are within +0.5 and –0.5°F accuracy of the test thermometer. This is done by inserting the thermometer in an agitated 10-gal can of water heated to within –3°F of pasteurization temperatures and taking a direct reading.

Measure the thermometric response of the indicating thermometer on the pipeline by determining the time that it takes to raise the temperature of the indicating thermometer, which has been placed in a water bath 19°F higher than the lowest reading on the thermometer and above pasteurization temperatures. It should rise 12°F within 4 sec.

The recording thermometer should be checked against the indicating thermometer at installation and quarterly by an environmental health worker and daily by the plant operator. The recording thermometer should never read higher than the indicating thermometer. The recording thermometer's temperature accuracy is tested in the same way that vat pasteurization temperature is tested, except that 161°F is used instead of 145°F. The time accuracy of the recorder controller is measured in the same way.

Cut-in temperatures of milk flow should be determined by raising the product temperature from 3°F below pasteurization temperatures to pasteurization at a rate of 1°F per 30 sec and determining the temperature at cut in. Cut-out temperature is determined by raising the pasteurized product flow slightly and then decreasing at a rate of 1°F per 30 sec. Determine the temperature at cut out. This ensures that milk is reaching full pasteurization temperatures.

The thermometric lag of the recorder controller must not exceed 5 sec. This is determined by inserting the recording thermometer in a can of hot water that is 7°F above pasteurization and determining the time it takes the mercury to rise from 12°F below pasteurization and the cut-in mechanism by the controller.

Check the flow-diversion valve for leakage. Measure the response time of the flow-diversion valve, which should not exceed 1 sec, by determining the elapsed time between the instant of activation of the cut-out mechanism and the fully diverted flow position.

Determine the time of product flow through the holding tube by electrodes and injection of 50 ml of saturated salt solution at the starting point. Determine the time it takes to reach the second electrode. This should be a minimum of 15 sec.

Poultry Testing

High bacterial counts indicate poor handling and processing techniques and improper refrigeration. Clean birds should be held at 32–34°F for a maximum of 4 days. If slime develops, the bacterial count per square centimeter of surface will probably be in the millions. A total count of organisms can be made using a swab technique or by pressing an agar plate against the surface of the bird. A rinse technique using sterile water may also be used.

The dye, resazurin, may be used to determine the condition of a bird, with swab samples in a solution noting the length of time the sample takes to decolorize to a fluorescent pink color. A fresh bird takes greater than 8 hr, a good bird takes greater than 5 hr but less than 8, a fair bird takes greater than $3^1/_2$ hr but less than 5 hours, and a poor bird takes less than $3^1/_2$ hr.

Egg Testing

The tests used with milk pasteurization equipment are also used with egg pasteurization equipment. In addition, total counts are taken from the whole eggs, egg white, egg yolk, salted yolk, sugared yolk, and blends. Tests are taken of liquid, frozen, or dried eggs. Yeasts and molds are also evaluated. Of particular concern are salmonella bacteria. Therefore, special tests are conducted to determine the kinds and quantities of salmonella present. Bacteria grow very rapidly in broken-out egg products when the temperature rises above 60°F. Although freezing and drying decreases the number of viable bacteria, these processes do not eliminate them, and therefore, frozen and dried eggs can and have caused outbreaks of food-borne disease.

Meat Testing

Programs monitor pesticide residues, cleaning agents, and antibiotics in meat tissue. Inspections are made at many stages during meat processing by the U.S. Department of Agriculture, the Food and Drug Administration, and state and local health authorities. Labels are required to show all ingredients used in any type of processed meat. Total plate counts may be taken from fresh meats, cured meats, canned meats, and a variety of meat products. Samples are usually taken of the surface and the interior of the meat. In addition, tests are conducted for yeasts and molds. Direct microscopic examinations are also used. Generally the test for contamination is not meaningful unless there is a population of at least one million organisms per gram of meat or a specific organism known to cause disease is present. The pH, salt, and nitrite levels are also determined.

Shellfish Testing

Shellfish are tested for coliform on a routine basis. However, where outbreaks of specific diseases, such as infectious hepatitis, cholera, typhoid fever, and bacillary dysentery occur, specific tests are conducted for these organisms. The bacterial content of shellfish can be sharply increased and added to during any phase of harvesting, processing, and storage. Shellfish may also be tested for various chemicals.

FOOD QUALITY CONTROLS

The federal government protects all food shipped in interstate commerce through the enforcement of the following acts: Pure Food and Drug Act, passed in 1906 and updated to the present; the Wholesome Meat Act of 1967; the Wholesome Poultry Products Act of 1968; and the Egg Products Inspection Act of 1970, last revised in 1992. In 1994, new regulations applied to meat processing were being added bimonthly. In the early 1990s, new regulations were being issued for poultry processing. In early 1995, new reform regulations were published for meat and poultry by the U.S. Department of Agriculture. It will take 3 years to implement them. Many federal agencies are involved in protecting the food supply. The National Shellfish Sanitation Program helps protect food supplies.

The National Shellfish Sanitation Program (NSSP) is a voluntary program composed of shellfish producing and receiving states, the shellfish industry, and federal agencies, coordinated and administered by the Food and Drug Administration. Each state supervises its local and imported fresh and fresh-frozen oysters, clams, and mussels, which must be certified by a certified dealer to have been grown, harvested, transported, processed, and shipped in accordance with the criteria. Plant inspections and record keeping are of primary importance. The FDA makes available for use the following: "Sanitation of Shellfish Growing Areas—Part I; Sanitation of the Harvesting, Processing, and Distribution of Shellfish—Part II; and Interstate Certified Shellfish Shippers List."

In April of 1994, the Clinton administration introduced pesticide/food safety legislative reform proposals which would require amendments to both the Federal Food, Drug, and Cosmetic Act and the Federal Insecticide, Fungicide, and Rodenticide Act. The major elements of the proposal's areas are as follows:

1. Tolerances for pesticide residues in all types of foods would be based on a strong health based standard, defined as a "reasonable certainty of no harm" to consumers of the food.
2. Special provisions would be made for setting tolerances for infants and children since these special populations have unique sensitivities, and on a mg of chemical/kg of body weight basis, would have greater exposure to the chemicals consumed. These provisions would be in response to the recommendations made in the 1993 report of the National Academy of Sciences entitled "Pesticides in the Diets of Infants and Children."
3. EPA would be required to review all existing tolerances and insure that pesticides meet the new standard within 7 years of enactment.
4. EPA could grant time-limited transitional tolerances of 5 years.
5. FDA would enforce levels of pesticides in food.
6. Every fifteen years, all ingredients in pesticides would be reviewed.
7. With adequate scientific information, EPA would phase out certain pesticides.
8. Uniform labeling would be used for pesticides.
9. The statute would promote integrated pest management.
10. There would be accurate uniform record keeping.

All states have health departments and departments of agriculture that provide additional protection through inspection services. Counties, cities, and multiple county health departments provide for local inspection services to protect the food supply. When food processing is poor or unsanitary, necessary legal action is taken.

SUMMARY

The production of adequate quantities of food to satisfy the needs of the world will continue to be an extremely serious problem because of the sharp increase in world population and the decrease in death rates brought about by better medical care. The production of food brings with it inherent difficulties, such as the use of scarce water supplies, energy supplies, and potentially hazardous fertilizers and pesticides. Additives to enhance color and taste and to preserve food are increasingly under suspicion of having the potential to cause a variety of diseases, including cancer. Current techniques of food technology have been incriminated in outbreaks of botulism from lack of adequate controls during food processing. As the quantity of prepared frozen foods and food shipped to varying parts of the country and the world increase sharply, a set of controls must be established to prevent the spread of disease.

RESEARCH NEEDS

Toxicants that may be part of plant foods require further study. Long-term feeding studies need to be conducted in experimental animals to determine special problems, such as sensitization and allergies. Analytical methods should be developed to measure the important toxicants found in foods. These toxicants include the pressor amines, goitrogens, anti-nutritional factors, safrole-related compounds, and allergic hemagglutinins. Sampling techniques must be utilized both on raw agricultural products and prepared products. Studies of plant breeding to reduce the level of toxicants found in certain selected food plants are needed. New plant varieties should be investigated to determine if genetic manipulation has caused an increase in the production of toxicants. Research is required to determine if existing plants produce abnormal toxic metabolites under certain stress conditions. A catalogue of toxic constituents of poisonous plants should be produced. A rapid, reliable, quantitative analytical method should be developed for paralytic shellfish poisoning. A complete study of the chemical nature of various poisons must be conducted on all paralytic shellfish poisoning. The mycotoxins must be studied in the laboratory and field situation to determine why they are produced. Methods must be developed for measuring the incidence of mycotoxin contamination in raw and finished foods. Studies are needed to determine the biotransformation of mycotoxins that occurs in food animals and to determine the potential for contamination of the food for human beings. Toxicological studies are needed to determine the amount of mycotoxin to which humans are exposed. A complete study of the biological activities of nitrates, nitrites, and nitrosamines in foods and a study of the effect of these chemicals on humans is needed. Research is needed on the effects of high intakes of sodium chloride and phosphates. Special emphasis should be placed on the relationship of these chemicals to hypertension and bone growth. Because of the large numbers and quantities of compounds found in foods, ongoing studies should be made to determine the interaction of food additives with major dietary constituents, the alterations of these dietary items, and possible resulting effects. Considerable research is needed in understanding the movement of mercury through the food chain. An understanding is needed of the adverse effect of lead and on the range of adverse effects of cadmium, copper, zinc, magnesium, selenium, and arsenic found in food. Research is needed on pesticides to determine the level of food contamination and their adverse effects in humans. Additional research should be carried out on the consumption of food additives, the interaction of food additives and other residues present in food, the chemical nature of many of the additives, and the chemical changes brought about by cooking. This fundamental research is essential to determine which additives should be eliminated from food and which should remain.

At present the FDA has listed 86 chemicals, including many of the coloring substances, as potential carcinogens. The use of risk assessment in some of these determinations is an excellent technique. The first uses of risk assessment appears to have been for aflatoxins in 1978 and for PCBs in 1979. In 1982, risk-assessment techniques were used to determine carcinogenic impurities in color additives that were not covered by the Delaney clause. Finally, risk assessment was extended to substances covered by the Delaney clause. Beside direct food additives, which may be potential carcinogens, indirect food additives may come from the packaging materials and food-processing equipment, color additives used for ingestion and for external purposes, potential contaminants of food or color additives, and unavoidable environmental contaminants in food and cosmetic ingredients. In certain substances, such as saccharin, although the FDA has found the substance to be potentially carcinogenic, the benefit that the substance gives to the population at large is more important than the potential risk that has been shown in some studies with laboratory animals. The FDA also allows the continued use of methylene chloride to decaffeinate coffee, although it is banned for use in cosmetics. The use of this decaffeination agent is justified on the grounds that through assessment the risk is minimal to people. The indirect food additives, such as the packaging materials that have been mentioned may come in contact with food, as well as processing equipment. The FDA has banned outright two indirect food additives. They are flectol-H and NBOCA. The FDA has also banned certain uses of bottles made from acrylonitrile copolymers and polyvinyl chloride because of the concern of leaching of these chemicals into the food. There may be numerous other chemicals that are part of packaging materials that should be studied as potentially hazardous to people through consumption of food.

Animal drug residues have been tested since 1962. The Congress in 1962 passed the DES (diethylstilbestrol) Proviso that allows the use of carcinogenic drugs in animals providing that the residues cannot be detected in edible portions of tissue or foods derived from the living animals. These substances need to be studied further, since there may be potential risk for carcinogenic activity in humans, over time, as was shown later with the drug DES. DES became the only animal drug successfully banned, since there was human evidence of potential cancer to humans. Other research concerns include the potential hazard to the food supply by groundwater or surface water contaminated by a variety of organic chemicals that may be toxic and hazardous. Additional study should be done on food crops to determine if airborne pollutants that may be hazardous to people can be carried forward during the processing of food and therefore create potential problems. The study of sludge-related contaminants put on land, and their potential effect in the food chain, need to be further researched.

Chapter 4

INSECT CONTROL

BACKGROUND AND STATUS

Insects and rodents are responsible for numerous outbreaks of disease in humans and animals. These pests or vectors also cause extreme annoyance because of bites. They contaminate huge quantities of food, which must be destroyed because of the potential spread of disease. Contamination can be caused by vectors coming into contact with disease-causing microorganisms and carrying the organisms on their bodies or by vectors depositing feces or urine on food.

Vector-borne disease varies in prevalence from state to state and from year to year according to weather and climatic conditions. In determining the potential incidence of these diseases, it is necessary to identify and determine the prevalence of the vectors that act as primary transmitters. For instance, in an outbreak of viral encephalitis in St. Louis, it was determined that mosquitoes spread the disease to the South and through Illinois, Indiana, and other states in the country. Other types of data include the incidence of a given vector-borne disease in an area; the type of climate, including maximum and minimum temperatures, rainfall, prevailing winds, and their daily and monthly occurrence; life cycles of the vectors; seasonal fluctuation in vector populations and their requirements for life; any unusual conditions that may help spread the given vector-borne disease, such as floods, earthquakes, and other natural disasters.

Insects are extremely adaptable to the human environment. They are found in the air, on and under the soil, and in fresh or brackish water. They live on or in plants and animals, and compete fiercely with other species as parasites. Insects have caused enormous problems for many centuries. With numerous insecticides available, insects continue to plague humans, for they are capable of developing a resistance to insecticides.

Rat fleas have transmitted plague organisms to millions of people over the last 15 centuries. An estimated 25 million people or 25% of the population of Europe died during the great plagues of the 15th, 16th, and 17th centuries. Today, in the

U.S., humans still contract plague from rodents. Sylvatic plague, which is contracted from wild rodents, particularly groundhogs, in the western part of the U.S., is seen most frequently today. In recent years, plague occurred in some of the southwestern states, including New Mexico. The potential for serious outbreaks of plague always exists as long as rats are present, Indian rat fleas are present, and the organism can be transmitted from rodent to rodent and from rodent to human. Pneumonic plague is a highly contagious, pneumonia-like disease related to bubonic plague. Fortunately, pneumonic plague has not appeared recently in this country.

The majority of the important human vector-borne diseases cannot be prevented by vaccines or by chemotherapy. Their control is based on the ability to reduce the source of vectors and the contact between vector and humans. Although in the last 25 years there has been a sharp reduction in many of the insect-borne diseases, they have not disappeared, and in some areas because of changes in environmental conditions or social conditions, the diseases are on the increase again. By 1972, malaria was eradicated from most of the world, yet in 1973 and 1974, during the period of conflict between Pakistan and India, malaria increased. Yellow fever was under control for many years, but there was a sharp increase in Nigeria in 1969. There was an increase in plague in South Vietnam from 1963 to 1973, and whenever wars have occurred in other areas, an increase in typhus and relapsing fevers followed. In 1994, typhus fever again occured in refugees from a war in Africa. As a result of movement from urban to rural areas, there has been a sharp decrease in available sewage systems and an increase in overflowing sewage and mosquito problems. In California, people moving into the woodlands in the suburbs have become exposed to the vectors of California encephalitis and dog-heart worm. As artificial lakes are created to enhance the aesthetic beauty of areas, mosquito problems increase. Schistosomiasis increased with the creation of major dams, especially the Aswan Dam in Egypt. There were devastating epidemics of malaria in Pakistan after the construction of the Sukkar Varrage. In India a severe epidemic of malaria occurred after the construction of the Mettur Dam. A detailed assessment of status is difficult to make, since there are limited statistics available that can be organized into some uniform analysis. In 1992, Lyme disease had become the leading tick-borne illness in the U.S., with 9677 reported cases.

SCIENTIFIC, TECHNOLOGICAL, AND GENERAL INFORMATION

Description of Insects

To identify insects, one must understand something about insect biology (Figure 4.1). Insects have a hard outer skin called an exoskeleton. The exoskeleton serves as a means of attachment of muscles and also protects the internal organs from injury. In most insects, the outer parts of the body wall are hardened into place. These hardened areas, or sclerites, are joined together by flexible body wall

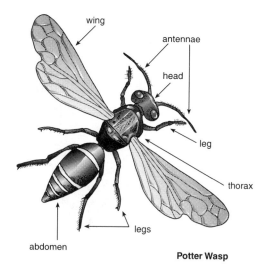

Figure 4.1.   Typical Insect.

segments called intersegmental segments. The scleritis may be covered with numerous small structures, such as hairs, scales, protuberance, and spines, which are useful for identification. Insect bodies are divided into three main regions: the head, thorax, and abdomen. The class arachnid, comprised of such insects as ticks, mites, scorpions, and spiders, has one or two main segments instead of three (see Figure 4.4 for adult arthropods of public health significance).

External Structure

The following discussion of the external structure (Figure 4.2) is not meant to be a detailed description of all insects. For this type of information, it would be well to obtain a book on entomology. The first body region, or head, contains mouth parts, antennae, large compound eyes and simple eyes (ocelli). The mouth parts are used in chewing, sponging, or piercing-sucking. Insects that have chewing mouth parts, e.g., roaches and silverfish, grind solid food. Insects that have sponging mouth parts, e.g., house flies, blow flies, and flesh flies, suck up liquid or readily soluble foods. They eat such solids as sugar and all liquids; they regurgitate a drop of saliva to dissolve the sugar. Insects such as mosquitoes, deer flies, sucking lice, and fleas have piercing sucking mouth parts. They easily pierce the skin of animals and humans, and suck their blood. The number of pairs of antennae vary with the type of insect, as do the eyes.

The second part of the insect, the thorax, contains the three sets of legs and may contain two sets of wings attached to the last two segments of the three-part

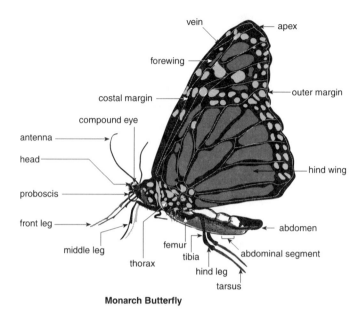

Monarch Butterfly

Figure 4.2.   Insect: External Structure

thorax. Insects are identified at times by observing the thorax. The wings, when present, contain a reinforcing structure called veins. The arrangement and number of wing veins aids in insect identification.

The third body part, or abdomen, contains the spiracles, which are the external openings of the respiratory system and the external reproductive organs.

Internal Structure and Physiology

The digestive system which is part of the internal structure (Figure 4.3) of an insect consists of an alimentary canal. This canal runs from the mouth to the anus and is divided into several sections. Digestion takes place in the canal, and undigested food plus waste is removed as feces. The digestive system is an important part of the transmission of disease, since the insect can pick up an organism from one host and pass it on to another during feeding and defecation.

The circulatory system is not enclosed in blood vessels but circulates freely throughout the body cavity. In most insects, blood is colorless or greenish-yellow. Blood does not carry oxygen or expel carbon dioxide. Its major function is in the removal of waste products from body cells.

The nervous system of the insect contains a brain located in the head, a double nerve cord extending backward along the ventral surface of the body cavity, and a nerve center or ganglia.

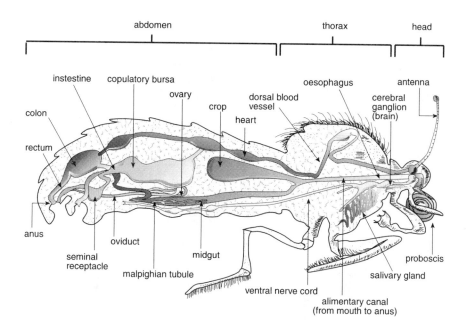

Figure 4.3.   Insect: Internal Structure.

Air enters the insect through spiracles into large tracheal trunks to tracheae and tracheoles.

Most insects have two sexes that mate to produce eggs. Insects that lay eggs are oviparous, and insects that deposit larvae are larviparous. Some insects have only one sex, the female, and reproduce without fertilization. Ants and bees are examples of unisexual reproducing insects. This occurs part of the time to produce workers, which are then sexually sterile.

Insect Development

The life cycle of the insect starts with the fertilization of the egg and is completed when the adult stage is reached. Some insects go through an incomplete metamorphosis (Figure 4.5). This means that their life cycle consists of egg, nymph, and adult stages. Examples of insects with incomplete metamorphosis are roaches and body lice.

Complete metamorphosis occurs in four stages, as egg, larvae, pupa, and adult. Examples of insects with complete metamorphosis are mosquitoes, flies, butterflies, and moths.

Insects molt or shed their protective exoskeleton while a new exoskeleton is being exposed. After the first day of molting, a new outer skin hardens and becomes colored.

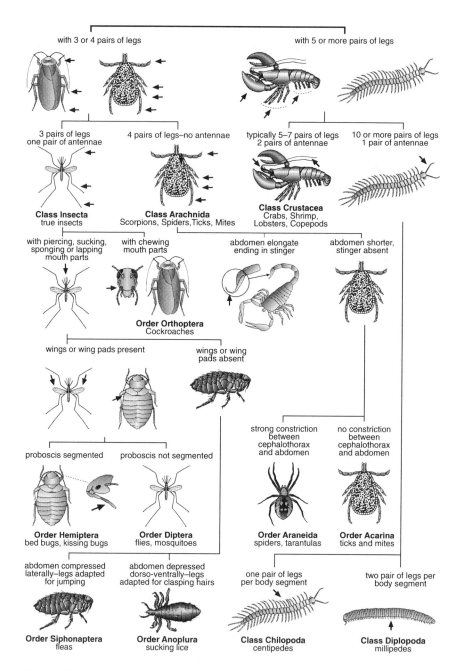

Figure 4.4.   Pictorial key to major classes and orders of adult arthropods of public
health importance. Source: *Pictorial Keys to Some Arthropods and
Mammals of Public Health Importance* (Washington, DC: U.S. Department
of Health, Education, and Welfare, Public Health Services,1964), p.3.

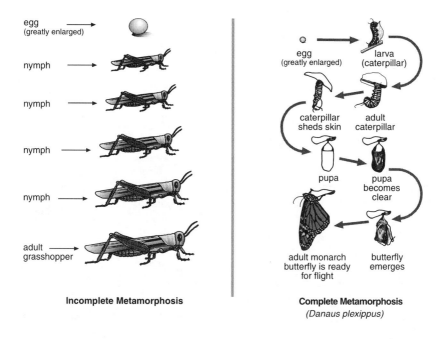

Figure 4.5.   Insect Life Cycle: Incomplete and Complete Metamorphosis

Insect Senses and Behavior

Insects have a sense of touch, taste, smell, hearing, sight, balance, and possibly orientation. Since insect skin is hard, the sense of touch is conducted through antennae or feelers, as well as other parts of the body that are sensitive to contact and pressure. The sense of smell, located primarily in the antennae, is highly developed to aid in locating food, mates, and a place to deposit eggs. Insects vary in their degree of hearing ability. Sound waves are picked up by fine sensory hairs or special organs appearing on the side of the abdomen or the lower part of the front legs. They also hear through cuplike organs on the antennae. Visual power varies considerably among insects.

Although there is no proof that insects reason, they do have a series of instinctive, highly complex patterns that they follow. Certain stimuli cause them to behave in certain ways.

TYPES OF INSECTS OF PUBLIC HEALTH SIGNIFICANCE

The insects to be discussed include flies, fleas, lice, mites, mosquitoes, roaches, ticks, and food-product insects. It is important to understand the biology, characteristics, breeding habits, and mechanisms of spread of disease of each of these insects in order to institute adequate chemical, physical, biological, or mechanical controls.

Fleas

Fleas breed in large numbers where pets and livestock are housed. They spread readily through homes, buildings, and yards, attacking pets, livestock, poultry, and people. Fleas are the vectors of many important diseases, such as bubonic plague, murine typhus, tapeworms, salmonellosis, and tularemia.

Fleas obtain food and spread disease by biting humans and animals. Some individuals have a simple swelling as a result of the flea bite; others have severe generalized rashes. The species of fleas that commonly bite humans include the cat flea (*Ctenocephalides felis*), the dog flea (*Ctenocephalides canis*), the human flea (*Pulex irritans*), and the oriental rat flea (*Xenopsylla cheopis*).

Female fleas lay their eggs after mating on the hair of animals. The eggs drop to the floor and may become enmeshed in mats, webs, overstuffed furniture, or cellar floors. When the eggs grow to fleas, they seek a blood meal from animals or humans. In the case of rat fleas, the adult life is usually spent on the Norway or roof rat. Fleas may leave their animal host and bite humans (see Figure 4.6).

Flea Biology

Fleas are small, wingless insects that range in length from 1 to 8.5 mm. They feed through a siphon or tube after they have bitten a warm-blooded animal. The female must have a blood meal before producing eggs. Fleas cannot tolerate extremes in temperature and humidity. This is why they live close to the animals or humans that they infest. They may live on the individual's body or in the individual's shelter.

Fleas go through complete metamorphosis (Figure 4.7). Under favorable conditions, this may occur in 2–3 weeks. The eggs are laid in small batches over a long time period, which is interrupted by blood meals. The adults are usually able to feed within 1 day after emerging from the cocoon. The adult flea can live several weeks without food. This is the reason why people can take their dogs or cats along on vacation and return to still have a flea infestation.

Flies

Flies annoy humans; they bite, infest human flesh and the flesh of domestic animals, attack and destroy crops, and cause numerous, serious diseases. Although biting flies do not contain any type of toxin, they can kill the victim when attacking in the thousands. The victim may die of anaphylactic shock as the result of all the foreign protein injected into the body. Eye gnats do not bite, they damage the delicate membrane of the eye.

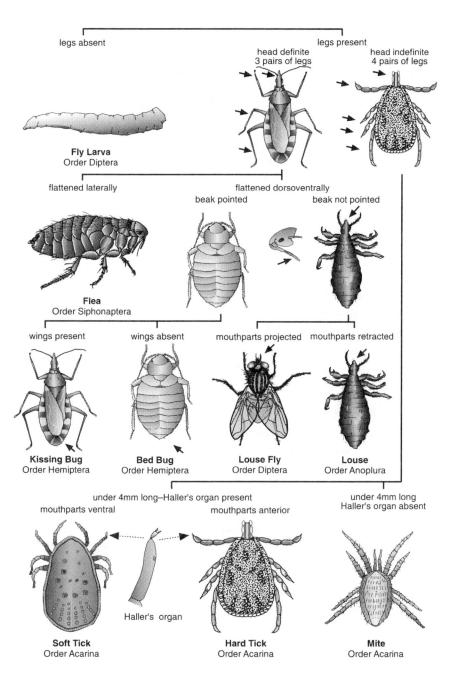

legs absent

legs present

head definite
3 pairs of legs

head indefinite
4 pairs of legs

Fly Larva
Order Diptera

flattened laterally

flattened dorsoventrally
beak pointed

beak not pointed

Flea
Order Siphonaptera

wings present

wings absent

mouthparts projected

mouthparts retracted

Kissing Bug
Order Hemiptera

Bed Bug
Order Hemiptera

Louse Fly
Order Diptera

Louse
Order Anoplura

under 4mm long–Haller's organ present

under 4mm long
Haller's organ absent

mouthparts ventral

mouthparts anterior

Haller's organ

Soft Tick
Order Acarina

Hard Tick
Order Acarina

Mite
Order Acarina

Figure 4.6.   Pictorial key to Groups of Human Ectoparasites. Source: *Pictorial Keys to Some Arthropods and Mammals of Public Health Importance* (Washington, DC: U.S. Department of Health, Education, and Welfare, Public Health Services,1964), p.4

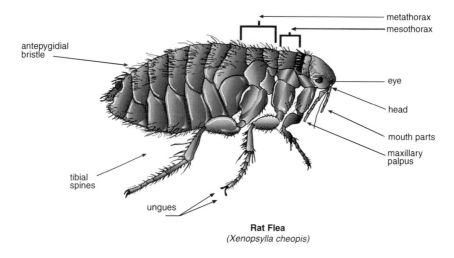

Rat Flea
(Xenopsylla cheopis)

Figure 4.7.   Flea Biology.

Many different species of flies lay eggs or larvae in the flesh of humans and animals. The larvae invade the flesh and produce a condition known as myiasis.

Thousands of species of flies exist. The ones that will be discussed here are those most common to humans and those that cause the most disease. These flies include the common house fly (*Musca domestica*), the stable fly (*Stomoxya calcitrans*), the face fly (*Musca autunnalis*), the blow fly or bottle fly (*Calliphora cynomyopsis, C. phaenicia, C. lucillia,* and *C. phormica*), the flesh flies (*Sarcophaga*), and the horse or deer fly (*Tavanus chrysops*). The house fly is the most important vector of microorganisms causing food-borne and infectious diseases. They are most active during daylight hours, when they move from food source to food source to obtain a balanced diet. They utilize sugar and starches for energy and to extend their life span. They utilize proteins for the production of eggs. Flies feed about three times a day. Between feedings, they rest on floors, walls, interior surfaces, ground, grass, bushes, fences, and other surfaces. During feeding on solid food, the fly regurgitates fluids to dissolve the solids, since they can only take up liquid food. During this feeding process, they regurgitate fluids containing microorganisms that cause disease and they also defecate. Flies are inactive at night. They usually rest on upper surfaces of rooms (see Figure 4.8).

Fly Biology

Flies belong to the order Diptera. This order has one or no pair of wings and halteres. Halteres are tiny knoblike structures located behind the wings and are considered to be the second pair of wings. The adult fly has three distinct body

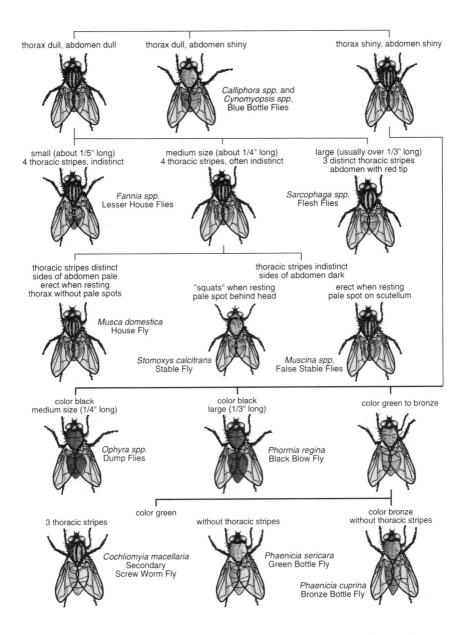

thorax dull, abdomen dull

thorax dull, abdomen shiny

Calliphora spp. and
Cynomyopsis spp.
Blue Bottle Flies

thorax shiny, abdomen shiny

small (about 1/5" long)
4 thoracic stripes, indistinct

Fannia spp.
Lesser House Flies

medium size (about 1/4" long)
4 thoracic stripes, often indistinct

large (usually over 1/3" long)
3 distinct thoracic stripes
abdomen with red tip

Sarcophaga spp.
Flesh Flies

thoracic stripes distinct
sides of abdomen pale
erect when resting
thorax without pale spots

Musca domestica
House Fly

thoracic stripes indistinct
sides of abdomen dark

"squats" when resting
pale spot behind head

Stomoxys calcitrans
Stable Fly

erect when resting
pale spot on scutellum

Muscina spp.
False Stable Flies

color black
medium size (1/4" long)

Ophyra spp.
Dump Flies

color black
large (1/3" long)

Phormia regina
Black Blow Fly

color green to bronze

color green

3 thoracic stripes

Cochliomyia macellaria
Secondary
Screw Worm Fly

without thoracic stripes

Phaenicia sericara
Green Bottle Fly

color bronze
without thoracic stripes

Phaenicia cuprina
Bronze Bottle Fly

Figure 4.8.   Pictorial key to common domestic flies. Source: *Pictorial Keys to Some
Arthropods and Mammals of Public Health Importance* (Washington,
D.C.: U.S. Department of Health, Education and Welfate, Public Health
Services, 1964), p. 37.

regions—the head, thorax, and abdomen. They have large compound eyes, one pair of antennae, sponging, and rasping or sucking mouth parts. The mesothorax, or middle section of the thorax, is much larger than the first and second sections. The size of the mesothorax is necessary for the powerful wing muscles used by the fly in flight.

Flies go through complete metamorphosis. The larvae feed differently and have a different habitat from the adult. The pupae are usually quiet and often are enclosed in a heavy puparium.

The house fly lives in almost all parts of the world, except the Arctic, Antarctic, and at extremely high altitudes. The normal life cycle runs from 8–20 days under average summer conditions. The female fly starts laying eggs within 4–20 days after it reaches adulthood. The eggs are deposited in batches of 75–150. The average female will lay five or six batches in cracks and crevices in the breeding medium away from direct light. The eggs hatch in about 12–24 hours during the summer months. The larvae grow quickly in the breeding medium. In warm weather the usual time for the larvae to develop through the three stages is 4–7 days. When the growth is completed, the larvae migrate from the growth medium and go into soil or under debris for the pupa stage. The pupa stage may be as short as 3 days or, at low temperatures, as long as several weeks. Following the pupa stage, the adult emerges. As an adult, the fly will mate and start the breeding process all over again. It is common to have two or more fly generations per month during warm weather. Adult flies can be kept alive for long periods of time at 50–60°F. This is why flies live through the winter and are ready to mate and start new fly populations during the warm months. Almost any kind of fresh, moist, organic matter is suitable for fly breeding. Flies can live in cereal, grain, animal manure, garbage, garbage-soaked soils, urine-soaked soils, or anywhere organic materials are present.

House flies move rapidly into new areas by flying. They travel as far as 6 mi within a 24-hr period, and possibly as far as 20 mi from their source of breeding. Flies are inactive below 45°F and are killed by temperatures slightly below 32°F. Their maximum activity occurs at 90°F. There is a rapid decline of activity at higher temperatures, and death occurs at 112°F. Humidity has a definite relationship to temperature. When the humidity is high and the temperature is low or very high, flies die rapidly. Above 60°F, flies live longest when the relative humidity is 42–55%. Flies tend to move toward light and are inactive at night. This is why the ordinary flytrap is successful. Flies are sensitive to strong air currents, and therefore, are quiet on extremely windy days. During high winds, house flies may be carried as far as 100 mi. The natural enemies of flies include fungi, bacteria, protozoa, roundworms, other arthropods, amphibians, reptiles, birds, and humans.

Face flies are barely distinguishable from house flies in physical characteristics. The larvae develop in fresh animal feces and the pupa burrow into the soil. Adults hibernate in houses and barns. They suck blood and other exudates from the surfaces of mammals, although they cannot pierce the skin. They are particularly annoying because they are found around the eyes, nostrils, and lips.

Stable flies have a piercing proboscis, which sucks blood from humans or animals. They are vicious biters, usually found around stables and horses. Stable flies are suspected of transmitting disease, although no proof exists at this point.

Blow flies or bottle flies have varying degrees of iridescence in their basic coloring. Their eggs are deposited on carcasses of animals, decaying animal matter, or in garbage containing animal matter. Eggs may also be deposited on fresh meat or decaying materials. On humans, the larvae of the flies can invade living tissue and cause a condition known as myiasis.

Flesh flies are comprised of a large number of species. They are usually light grey and have three, dark longitudinal stripes on the thorax. The abdomen has a checkered pattern. Most of the species breed in the flesh of animals or in their stools. They generally do not enter buildings, and therefore, are not very significant as vectors or nuisances.

Horseflies and deer flies are large flies that produce painful bites. They create a nuisance around swimming pools, streams, and sunny portions of damp woods.

Lice

Lice bite severely, causing skin irritation and sometimes a generalized toxic reaction. They inject an irritating saliva into the skin during feeding. The itching for some people is extremely annoying. Lice have the capacity to act as vectors for numerous diseases. They are particularly a problem during times of war and disaster. The three basic kinds of human lice are the body louse (*Pediculus humanus*), the head louse (*Pediculus humanus capitis*), and the crab louse (*Phthirus pubis*). Today lice infestations occur rarely, but they still do occur. In 1974, in a school system in Indiana, a lice infestation occurred that spread throughout the school system (see Figure 4.6).

Louse Biology

Lice go through an incomplete metamorphosis (Figure 4.9). The stages are egg; first, second, and third nymph; and adult. The egg, also called a nit, attaches itself to the scalp, skin, or area around the pubic hair, as well as to underclothes. Body heat causes the eggs to hatch. The nymph then goes through its three molting stages until it becomes a sexually mature adult. The nymphal stages require 2–4 weeks if the clothing is removed from the body periodically and 8–9 days if the lice remain in contact with the body. The total life cycle may take about 18 days. Lice move quickly from person to person by means of clothing, bedding, or close contact. Infested brushes and combs are also excellent means of transmission of lice.

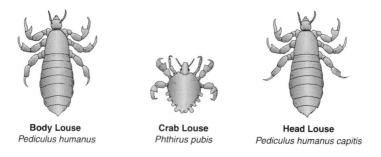

Body Louse
Pediculus humanus

Crab Louse
Phthirus pubis

Head Louse
Pediculus humanus capitis

Figure 4.9.   The three basic kinds of human lice

Mites

Mites are very small arthropods that are difficult to see with the naked eye. They do not have a distinct body segmentation. Their life cycle is very short, usually from 2 to 3 weeks. Mites increase in number very quickly under favorable conditions. They often infest food, stuffed furniture, and mattresses, and occur sometimes by the hundreds of thousands or millions. They enter premises with birds or rodents. The major problem is that they may cause several types of diseases, as well as annoyance or infestation of the body (see Figure 4.6).

Mite Biology

Since mites differ so greatly from one another, it is not possible to give a general picture of them. The mite lays eggs that hatch into larvae and then pass through two or more nymphal stages to the adult stage. When trying to identify mites, use a good entomology textbook.

Mosquitoes

Mosquitoes are small, long-legged, two-winged insects belonging to the order Diptera and the family Culicidae. The adults differ from other flies by having an elongated proboscis and scales on their wing veins and wing margin. Of the 2600 species of mosquitoes, approximately 150 are found in the U.S. The ones of most concern include anopheles, aedes, culex, culiseta, mansonia, and psorophora. The discussion that follows will be concerned with the mosquitoes most hazardous or annoying to humans.

Female mosquitoes have piercing and blood-sucking organs; males do not suck blood.

There are four stages in the life cycle of the mosquito. These include the egg, larvae, pupa, and winged adult. The eggs are laid and hatched in quiet, standing water. The female lay eggs in batches of 50–200 or more, and may lay several batches of eggs. Some species glue the eggs together into a floating mass. Other species deposit eggs singly on the water, on the side of containers above the waterline, or in moist depressions. The incubation period is 3 days. It then takes 7–10 days for the larvae to mature. The larvae become adults in 4–10 days, during which time the larvae skin is shed four times. Mosquito larvae feed on minute plants and animals or fragments of organic debris.

It is difficult to determine the life span of adult mosquitoes in natural settings. However, in most of the southern species, the life span is probably only a few weeks during the summer months. In the North, the female of culex, anopheles, and other species hibernate.

Mosquitoes inject saliva during feeding. It is the saliva that causes the itchy feeling (see Figure 4.10).

Mosquito Biology

Roughly equal numbers of male and female mosquitoes are produced. The male usually stays near the breeding place and mates with the female soon after reaching adulthood. Flight habits vary with the species of mosquito. Some stay within 100 yd of human habitation, and others travel as much as 10–20 mi. The preferred hosts, which include cattle, horses, other domestic animals, and humans, vary with the kind of mosquito. The female requires 2 days or more to digest a blood meal, lay eggs, and seek another blood meal. Mosquitoes may breed four or five times before they pass on disease to animals or humans. Since the spread of disease is intimately related to a given mosquito and its life cycle, each of the mosquitoes and the diseases that they transmit will be discussed.

Aedes. Of the 500 species of aedes in the world, about 40 are common to the U.S. They are of greatest concern in the northern areas. All aedes lay their eggs singly, on the ground, at or above the waterline, in tree holes, or in a variety of containers, including tires. The eggs hatch after the area has been flooded. Some species can survive long dry periods. Breeding places for *Aedes* vary tremendously. They may breed in small, wet depressions left by rains or melting snows. Some breed in coastal salt marshes; others in irrigation ditches. The *Aedes* are vicious biters and bloodsuckers. Some attack during the evening hours, some during the day, and some will bite at any time.

Aedes aegypti is a vector for yellow fever. It is found in the southeastern and southern U.S. It is a semidomesticated mosquito that lives very well in artificial containers in and around human habitation. Mosquitoes may breed in anything that will hold water. The life cycle can vary from 10 days to 3 weeks or more.

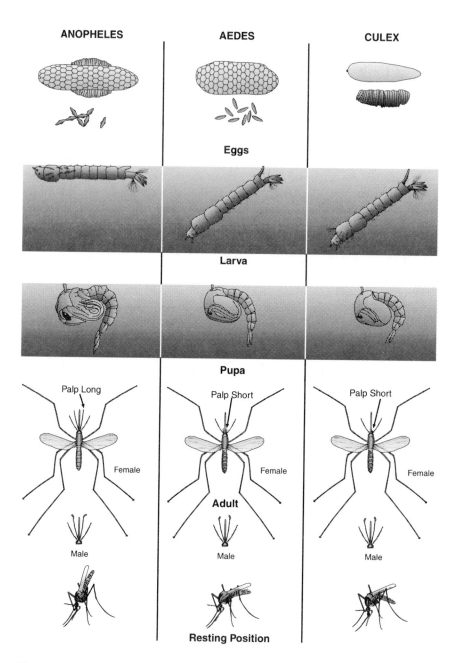

Figure 4.10.   Characteristics of anophelines and culicines. Source: *Pictorial Keys to Some Arthropods and Mammals of Public Health Importance* (Washington, DC: United States Department of Health, Education, and Welfare, Public Health Services, 1964).

Aedes aegypti are susceptible to cold and cannot survive the winter unless they are in the southern U.S. The adults prefer humans to other animals, and therefore, spread disease from human to human. They usually bite around the ankles, under coat sleeves, or on the back of the neck. The adult lives about 4 months or more. Their flight range is normally from about 100 ft to 100 yd.

Anopheles. Anopheles mosquitoes are found throughout the U.S. Most of the anophelines have spotted wings, while the culicines have clear wings. Anophelines rest in a position where their head, thorax, and abdomen are in a straight line, usually at an angle of 40–90°, while the culicines rest in a position almost parallel to the surface.

Anopheline eggs are laid singly on the water surface and float there. They are usually laid in batches of 100 or more, and hatch within 1–3 days.

The larvae are found either in fresh or brackish water. The larval stage takes from 4–5 days to several weeks, depending on the species of mosquito and the environmental conditions, especially the water temperature. The larvae eat microscopic plant life. Most adults are active only at night. They spend the daytime resting in dark, damp shelters. Their peak of activity occurs just after dark and just before daylight. Flight range is from less than one mile to several miles.

Anopheles quadrimaculatus is the most common species of anopheles that causes malaria. It is frequently found in houses and is likely to attack humans. The mosquito is distributed throughout the southeastern U.S. and into the Midwest and Canada. It breeds primarily in permanent freshwater pools, ponds, and swamps that have aquatic vegetation or floating debris. It is most frequently found in shallow waters, in sunlit or densely shaded areas. The larvae can withstand temperatures below 50–55°F, although development will not be completed. Development starts to progress at temperatures of 65–70°F and is most active at temperatures of 85–90°F. It takes 8–14 days at the 85–90°F range for the larvae to complete their cycle. The female, after mating, lays eggs for 2–3 days after the first blood meal. A single mosquito may lay as many as 12 batches, totaling over 3000 eggs. The adults are inactive and rest during the daytime in cool, damp, dark shelters, such as buildings, caves, and under bridges. Feeding and other activities occur almost completely at night. They feed readily in houses on humans and on other warm-blooded animals. Usually the adults do not fly more than one-half mile from the breeding place.

Anopheles albimanus are found in southern Texas and the Florida Keys. The larvae live in freshwater ponds or brackish water. The adults enter dwellings during the night and bite people, and leave at dawn for forested areas. They have not been known to cause malaria in the U.S.

Anopheles freeborni is the most important vector of malaria in the western U.S. These mosquitoes enter homes and animal shelters and bite avidly at dusk and dawn. They breed in any permanent or semipermanent water that is at least partially exposed to sunlight. The larvae are also found in slightly brackish water. This species is particularly adapted to rice fields and sunny arid regions. The mosquitoes leave their hibernating places in February, obtain blood meals, and lay their eggs. As the season progresses and the area gets dryer, the mosquitoes travel longer distances, sometimes 10–12 mi to seek shelter. During the winter, they are in semihibernation.

Culex. The culex include about 300 species, of which 26 are known to inhabit the U.S. Only about 12 of these species are common or pests or vectors of disease. The culex mosquitoes breed in quiet waters, artificial containers, or large bodies of permanent water, and often breed in areas where there are large quantities of sewage. The eggs are deposited in rafts of 100 or more. They float on the water surface until they hatch in 2–3 days. The adult females are usually inactive during the day, but bite during the night.

Roaches

Most of the 55 species of roaches known in the U.S. live outdoors and are not a problem to humans. Only five species are usually found indoors in the U.S. They mechanically carry dirt and organisms that may spread disease, and they destroy fabrics and book bindings. Their odors, which are quite offensive, ruin food. Roaches are broad, flattened, dark or light brown or black insects. They usually run at night to seek food and hide out during the daytime. When roaches are seen during the day, they are generally present in very large numbers. The females lay eggs in capsules called oothecae.

Roaches have been around for about 400 million years. Their fossil remains, which are similar to today's roaches, are abundant in the strata of certain types of soils. Roaches are present everywhere. It is important that moisture be present where they exist. They feed upon cereals, baked goods, small amounts of grease, glue, starch, wallpaper binding, fecal materials, dead animals, etc. During the feeding process, they regurgitate a brown liquid that may contaminate the food. They commonly rest in crevices, behind moldings of doors and window frames, in areas where pieces of equipment are joined together, under food preparation tables, in cabinets and closets, in the asbestos covering of hot-water lines, in cardboard boxes, in and under debris, and in almost any other place where a tiny crack may exist. They are also found in the backs of radios, television sets, paper bags from the supermarket, and in the pockets or cuffs of pants or other places in clothing (see Figure 4.11).

Roach Biology

The roach passes through incomplete metamorphosis. The stages include the egg, nymph, and adult. Roaches have an oval, flattened shape with the head

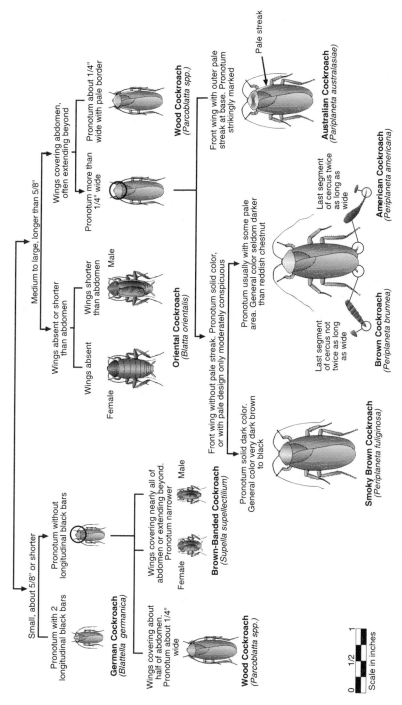

Figure 4.11. Pictorial key to some common adult cockroaches. Source: *Pictorial Keys to Some Arthropods and Mammals of Public Health Importance* (Washington, DC: United States Department of Health, Education and Welfare. Public Health services, 1964), p. 18.

hidden. They feed on almost anything. The eggs are enclosed in capsules called oothecae that are either deposited or carried at the end of the abdomen.

American Roach. The American roach, *Periplaneta americana*, is widely distributed throughout the world in mild climates, although it is a native of tropical and subtropical climates. The adult is reddish brown to dark brown. It usually forages for food on the first floor of buildings. Although it does not normally fly, it has the ability to do so. The female carries the egg capsule for a day or two, and then glues it to some object in a protected area. In 2 or 3 months, about 12 nymphs will hatch from each capsule. Adult females live for at least a year. During this period of time, they lay many capsules. The nymph, which is the same color as the roach, passes through 13 molts as it matures during the course of a year. It can grow very slowly where conditions are unfavorable. The length of the roach ranges from $1^1/_2$ to 2 in.

Brown-Banded Roach. The brown-banded roach, *Supelle supellectilium*, is found throughout the world, although its main habitat is the tropical areas. The adult is about $^1/_2$ in. in length, light brown, and has mottled wings. The female is reddish brown and lighter. The eggs are laid and hatched in the same way as the American roach. The color of the nymph is the same as the adult. It matures in 4–6 months. It lives in almost any part of buildings.

German Roach. The German roach, *Blattella germanica*, has a very wide range. It is most commonly found indoors and is the roach that readily becomes resistant to insecticides. The adult is light brown with longitudinal stripes on the back. The female carries oothecae for about a month and then drops it a day or so before the eggs are ready to hatch. There are about 30 nymphs in each capsule. The adult is about $^6/_{10}$ in. in size. The color of the nymph is the same as the adult. It undergoes six molts and matures in about 4–6 months. It lives in kitchens, bathrooms, and other areas.

Oriental Roach. The oriental roach, *Blatta orientalis*, is spread throughout the world. The adult is black to dark brown, with vestigial wings in the female. It is quite sluggish. It usually forages in the basement or first floor of buildings. The eggs are laid in an egg capsule, which is glued to some object in a protected place. About 12 nymphs hatch from each capsule in about 2–3 months. The roach is 1 to $1^1/_4$ in. in length, with the female being a little longer than the male. The color of the nymph is the same as the adult. It matures in about a year.

Ticks

Ticks belong to the class Arachnida. They are not true insects, but are important vectors of significant arthropod-borne disease (see Figure 4.6).

Tick Biology

Ticks belong to the same class as mites, spiders, and scorpions. They have three characteristics that distinguish them from insects: the head, thorax, and abdomen are all fused together in one body region; no antennae; and four pairs of legs in the nymph and adult stages instead of three. Ticks are divided into two groups, soft and hard.

The tick has a piercing organ that has barbs, allowing the tick to anchor itself to the host. It also has cutting organs to allow the piercing organ to enter the skin for a blood meal. The leathery body covering is capable of being greatly distended with blood. The female tick is somewhat flattened and tapered toward the anterior end before it takes its blood meal. Ticks are adaptable for blood sucking because they have a powerful, pumping pharynx. The blood is pumped backward into the esophagus and then into the stomach.

Ticks metamorphize completely. The female hard tick will feed once and lay large batches of up to 18,000 or more eggs. Most of the soft ticks take several blood meals and then deposit 20–50 eggs in a batch. The eggs hatch in 2 weeks to several months, depending on the environmental conditions. The larvae have difficulty attaching themselves to a host, and therefore, may go through long periods of starvation. After a blood meal, the engorged larvae usually will drop to the soil, shed its skin, and enter the nymph stage. The nymph may wait for long periods for a suitable host. After it has its blood meal, the nymph drops from the host, molts, and becomes the adult. The life cycle may take from less than a year to 2–3 years or longer. Ticks survive if they can get another blood meal. During this period the sexual organs are formed, and both the female and male tick become bloodsuckers and feed several times before copulation. The male hard tick will have a blood meal, usually copulate with one or more females, and then dies. The female drops to the ground and starts to develop eggs over a period of several days. Once the eggs are deposited, the hard tick female dies. Unfed hard ticks can survive frigid winter weather if they are sheltered. Engorged ticks and eggs are less resistant. Ticks are most active during the spring and summer. They usually die quickly if the area is dry. Soft ticks can exist in much dryer situations.

Food-Product Insects

Stored food-product insects become a serious problem in the kitchen, in the warehouse, pantries, or any other place where dried foods are stored. These pests eat or contaminate the food, making it unfit for human consumption. They are also annoying, since they leave the infested food and crawl or fly around the area. These food product insects include the confused flour beetle, saw-tooth grain beetle, flour grain beetle, larder beetle, black carpet beetle, cabinet beetle, grain weevil, rice weevil, Indian meal moth, Mediterranean flour moth, and bean and pea weevils. These insects may forage on dry foods, furs, skins, wool, silk, cheese, and almost any other product made from dried meats, cheese, or other organic materials.

ENVIRONMENTAL AND HEALTH PROBLEMS
CREATED BY INSECTS

Destruction of Food and Material

Insects infest cattle, causing weight loss, annoyance, and eventually a reduction in the supply of acceptable food. Further, they infest stored foods and contaminate them with feces, body parts, or live insects. Common stored food-product insects include grain and flour beetles; *Dermes pidae* beetles, which feed on a variety of plant and animal products, including birds, skins, live meat products, woolen and silk materials, cheese, and cereal grain products; grain weevils, which feed on stored whole grain; and bean and pea weevils, which infest dry beans and dry peas.

Transmission of Disease

Disease is transmitted from human to human, from human through vector to human, from human through vector to animal to human, or from human through vector through animal through vector to human. It is very important to understand these various types of disease transmission, since once the life cycle of the disease organism is known, it becomes much simpler to break the chain of infection and to stop the outbreak of the vector-borne disease.

Disease is transmitted mechanically, for example, flies pick up microorganisms of dysentery, typhoid fever, or cholera on their feet or body hair and transmit them to humans or their food; or biologically through various parts of the life cycle of the arthropod. In developmental biological transmission, the parasite grows or multiplies in the insect. For example, the viruses causing yellow fever or encephalitis increase in number within the body of the mosquito. In cyclical biological transmission, the parasite goes through an essential part of its life cycle in the arthropod, without increasing in number. For example, in filariasis, the mosquito sucks up a number of microfilaria into its gut and eventually these organisms enter the thoracic muscles, where they pass through several stages of their life cycle and then move to the proboscis of the mosquito. When the mosquito feeds, it infects the host. In cyclodevelopmental transmission, both an essential cycle occurs in the arthropod host and the parasites increase in number. An example of this would be the parasite *Plasmodium vivax*, which causes malaria.

In some cases, humans are accidentally infected with a disease that ordinarily would be an animal disease. This is called a zoonoses. An example of this is jungle yellow fever, where jungle mammals such as monkeys are normal hosts and wild mosquitoes transmit the disease from monkey to monkey. From time to time, the insect will bite a human, who then develops the disease. Another example of this is viral encephalitis transmitted by mosquitoes, such as St. Louis encephalitis, where the normal hosts would be wild birds.

Numerous tick- and mite-borne diseases are spread by eggs from the infected parents to the next generation. This is congenital transmission of the pathogen. The eggs are laid either on humans or animals, and when they hatch and become insects they transmit the disease to the host. Examples of this would be tick-borne typhus, tick-borne encephalitis, tularemia, relapsing fever, and scrub typhus.

People can also be poisoned by the bites and stings of insects. The victim may suffer an insignificant blister, anaphylactic shock, or death. Anaphylactic shock can result from an individual reaction to the protein or a large amount of protein injected into the body by the arthropods. Some insects, such as the black widow spider, cause death by their extremely poisonous toxin.

An infestation occurs when living insects burrow into the skin of the host. Lice or scabies burrow under the skin causing infestation. Table 4.1 lists some of the arthropods that affect human comfort.

Insect-Borne Diseases

Bubonic Plague

Bubonic plague is a highly fatal infectious disease, with toxemia, high fever, reduced blood pressure, rapid and irregular pulse, mental confusion, and prostration. Bubonic plague can become pneumonic plague when the disease spreads to the lungs. The incubation period in bubonic plague is 2–6 days. The incubation period in pneumonic plague is 3–4 days. In 1994 an extremely serious outbreak of pneumoic plague occured in India. The etiological agent is *Pasteurella pestis*. The insect involved is the oriental rat flea, *Xenopsylla cheopis*. The reservoir of infection is wild or domesticated rats, and possibly other rodents.

Colorado Tick Fever

Colorado tick fever is an acute fever lasting 2–3 days, with occasional encephalitis occurring. The incubation period is 4–5 days. The etiological agent is a virus. The insect involved is an infected tick, most frequently *Dermacentor andersoni*. The reservoir of infection is small mammals.

Dengue Fever

Dengue fever causes a sudden onset of fever, intense headache, joint and muscle pain, a rash, and prolonged fatigue, as well as depression. The incubation period is typically 3–15 days, but most often 5–6 days. The etiological agent is a virus of the group B Togaviruses. Various aedes, including *Aedes aegypti* and *Aedes albopictus* are responsible for causing this disease. The reservoir of infection is people, mosquitoes, and monkeys.

Table 4.1   **Some Arthropods Affecting Human Comfort**

Mites		
Chigger	*Trombicula alfreddugesi*	Intense itching; dermatitis
Rat mite	*Ornithonyssus bacoti*	Intense itching; dermatitis
Grain itch mite	*Pyemotes ventricosus*	Dermatitis and fever
Scabies mite	*Sarcoptes scabiei*	Burrows in skin, causing dermatitis
Ticks		
Hard ticks	*Amblyomma spp.*	Painful bite
	Dermacentor variabilis,	
	D. andersoni and	Tick paralysis; usually fatal
	Ixodes spp.	if ticks not removed
Soft ticks	*Ornithodoros spp.*	Some species very venomous
Spiders		
Black widow spider	*Latrodectus mactans*	Local swelling; intense pain and occasionally death
Brown spider	*Loxosceles reclusa*	Extensive loss of affected body tissue
Scorpions	Order: Scorpionida	Painful sting; sometimes death
Centipedes	Class: Chilopoda	Painful bite
Mayflies	Order: Ephemerida	Asthmatic symptoms from inhaling fragments
Caddis flies	Order: Trichopera	Asthmatic symptoms from inhaling hairs and scales
Lice	*Pediculus humanus and Pthirus pubis*	Intense irritation, reddish papules
Bugs		
Bed bugs	*Cimex lectularius*	Blood suckers; irritating to some
Kissing bugs	Family: Reduviidae	Painful bite: local inflammation
Beetles		
Blister beetles	Family: Meloidae	Severe blisters on skin from crushed beetles
Rove beetles	Family: Staphylinidae	Delayed blistering effect
Caterpillars	Order: Lepidoptera	Rash on contact with hairs or spines
Bees, wasps, and ants	Order: Hymenoptera	Painful sting; local swelling
Fleas	Order: Siphonaptera	Marked dermatitis frequent
Flies		
Punkies or biting midges	*Culicoides spp.*	Nodular swelling, inflamed
Black flies	*Simulium spp.*	Bleeding punctures; pain, swelling
Sand flies	Phlebotomus spp.	Stinging bite; itching; whitish weal
Mosquitoes	Family: Culicidae	Swelling; itching
Horse flies	*Tabanus spp.*	Painful bite
Deer flies	*Chrysops spp.*	Painful bite
Stable flies	*Stomoxys calcitrans*	Painful bite

Source: *Introduction to Arthropods of Public Health Importance*, H.D. Pratt and K.S. Littig, DHEW Publication No. (HSM) 72–8139 (1972)

Encephalitis

Arthropod-borne viral encephalitis includes eastern equine encephalitis, western equine encephalitis, California encephalitis, and St. Louis encephalitis. The symptoms of encephalitis include a short period inflammation, which involves parts of the brain, spinal cord, and meninges. Symptoms also include headaches, high fever, stupor, and disorientation. Fatality rates range from 5 to 60%. The incubation period is 5–15 days. The etiological agents vary with the type of the disease. Eastern and western equine encephalitis are caused by a virus in group A Togaviruses, whereas St. Louis encephalitis is caused by a virus in the group B Togaviruses. Various mosquitoes, including *Culex tarsalis, Culiseta melanura,* and other mosquitoes, transmit the disease. The reservoir of infection is suspected to be birds, rodents, bats, and reptiles. Surviving mosquito eggs or adults may keep the virus alive over the winter.

Lyme Disease

The early symptoms of Lyme disease includes a rash that starts out flat or raised and may develop into blistering or scabbing in the center. The lesions may have a bluish discoloration. Other symptoms include fatigue, headache, neck stiffness, jaw discomfort, pain or stiffness in muscles or joints, slight fever, swelling of glands, or conjunctivitis. If the symptoms of the disease are untreated, later symptoms occur, which include complications of the heart, nervous system, or joints. The incubation period ranges from 3 to 32 days after a tick bite. Typically a red ring-like lesion develops at the site of the tick bite. The etiological agent, which is a spirochete, is called *Dorrelia durgdorferi.* A tiny tick found in northeastern U.S., and as far south as Virginia, called *Lxodes dammini,* spreads the disease. In the south and southwest, *Lxodes scaturlaria* spreads the disease. On the west coast, *Lxodes pacificus* spreads the disease. Dogs, cats, horses, cows, and goats are domestic animals that may carry the ticks and the disease. Also, wild animals may carry the ticks and the disease. These are the reservoirs of infection.

Malaria

The symptoms of malaria include fever, chills, sweating, headache, shock, renal failure, acute encephalitis, and coma. In nontreated cases, there is a fatality rate of greater than 10%. The incubation period varies with the type of plasmodium causing the disease. *Plasmodium falciparum* has an incubation period of 12 days. *Plasmodium vivax* has an incubation period of 14 days. *Plasmodium ovale* has an incubation period of 14 days. *Plasmodium malaria* has an incubation period of 30 days. In certain strains of *Plasmodium vivax*, the incubation period is 8–10 months. The etiological agents have just been described. The anopheles mosquito transmits the disease. The reservoir of disease is people.

Relapsing Fever

The symptoms of relapsing fever include periods of normal temperature and then elevated temperature. In untreated cases, the fatality rate is usually 2–10%, although it may exceed 50% in the epidemic louse-borne disease form. The incubation period is 5–15 days but usually 8 days. The etiological agent is *Borrelia recurrentis*. The disease is transmitted either through lice or ticks. The reservoir of infection for the louse-borne disease is people. The reservoir of infection for the tick-borne disease is wild rodents or through transovarian transmission.

Rickettsial Pox

The symptoms of the disease include skin lesions, chills, fever, rash, headache, muscular pain, and malaise. The incubation period is probably 10–24 days. The etiological agent is *Rickettsia akari*. The rodent mite, *Liponyssoides sanguineus*, is the insect that spreads the disease through its bite. The reservoir of infection is the infected house mouse.

Rocky Mountain Spotted Fever

Rocky mountain spotted fever is also known as tick fever or tick-borne typhus fever. The symptoms of the disease include a sudden onset of a persistent fever, headache, and chills. A rash appears on extremities and spreads to most of the body. There is a 20% fatality rate in untreated cases. The incubation period is 3–10 days. The etiological agent is *Rickettsia rickettsii*. The disease is spread by the bite of an infected tick or by contamination of the skin with the crushed tissues or feces of the tick. The reservoir of infection is the tick, which may transmit the disease either through transovarian or transstadial passage within the tick. The tick may also be transferred from animals to people.

Scabies

The symptom of the disease is an intense itching, especially at night. The incubation period ranges from several days to several weeks. The etiological agent is *Sarcoptes scabiei*. The insect involved is the mite. The means of spread is through direct contact or by indirect contact by means of underclothing. The reservoir of infection is people.

Scrub Typhus

The symptoms are skin ulcers followed by fever, headache, profuse sweating, eruptions on various parts of the body, cough, and pneumonitis. The fatality rate ranges from 1 to 40%. The incubation period is from 6 to 21 days, but usually 10–12 days. The etiological agent is *Rickettsia tsutsuganushi*. A mite causes the disease. The reservoir of infection is the infected larval stages of mite.

Tularemia

Tularemia, which is also known as deerfly fever or rabbit fever, causes ulcers to form at the site of the innoculation by the insect. Headache, chills, and a rapid rise of temperature occurs. There is about a 5% fatality rate in untreated cases. The incubation period is 1–10 days, but usually 3 days. The etiological agent is *Francisella tularensis (Pasteurella tularensis)*. Although the disease may be transmitted by handling or ingesting insufficiently cooked rabbit or hare meat, or by the drinking of contaminated water or inhalation of dust from contaminated soil, it is also spread by a variety of insects. They include the deerfly; the mosquito, aedes; the wood tick, *Dermacentor andersoni*; the dog tick, *D. variabilis*; and the lone star tick, *Amblyomma americanum*. The reservoir of infection includes many wild animals, especially rabbits and beavers.

Typhus Fever — Epidemic (Louse-Borne)

Epidemic typhus fever causes symptoms of headaches, chills, prostration, fever, pain, and toxemia. In untreated cases, the rate of fatality varies from 10 to 40%. The incubation period is from 1 to 2 weeks, but usually 10 days. The etiological agent is *Rickettsia prowazeki*. The most common insect spreading the disease is the body louse, *Pediculus humanus*. The reservoir of infection is people.

Typhus Fever — Flea-Borne

Flea-borne typhus fever, also known as murine typhus or epidemic typhus fever, causes the same symptoms as the louse-borne epidemic typhus fever but they are milder. If the individual is untreated, the death rate is about 2%. The incubation period is 1–2 weeks but usually 12 days. The etiological agent is *Rickettsia typhi*. The insect involved is the flea, primarily the rat flea, which is *Xenopsylla cheopis*. Rats are the reservoir of infection, but cats and other wild or domestic animals may spread the disease.

Yellow Fever

The symptoms of the disease are fever, headache, backache, prostration, nausea, vomiting, and jaundice. The fatality rate in areas where yellow fever is endemic is less than 5%. In other areas, it may be as high as 50%. The incubation period is 3–6 days. The etiological agent is a virus of the B Togaviruses. The disease is spread by the bite of the infected *Aedes aegypti* mosquitoes.

Fleas

There are many important diseases transmitted by fleas beside bubonic plague. Murine typhus fever is a disease caused by *Rickettsia typhi*. The organism is transmitted by fleas to humans. The important vector in the U.S. is the oriental rat flea. The disease is spread from rodent to rodent and occasionally to humans. Infection occurs when the site of the flea bite is contaminated by the flea's feces. Although murine typhus fever has largely disappeared, it still could be the cause of a serious epidemic.

Fleas spread tapeworm infestations. They act as an intermediate host for several tapeworms that usually infest dogs and cats or rodents, and occasionally infest humans. The dwarf tapeworm (*Hymenolepis mana*) frequently infests children.

Salmonella enteritidis, which often causes outbreaks of food-borne disease, may be transmitted from rats and mice to humans via fleas. The flea contaminates food with its feces or it contaminates humans directly.

Flies

Flies transmit disease mechanically or biologically. Since flies breed and feed in human feces, they are capable of carrying organisms mechanically to food or to people on their mouthpart, in their vomitus, on their hairs, which are on their body and legs, on the sticky pads of their feet, or in their feces. Mechanically transmitted diseases include typhoid fever, paratyphoid fever, cholera, bacterial dysentery, amoebic dysentery, pinworm, roundworm, whipworm, hookworm, tapeworm, and salmonella other than those that are the causative agents of typhoid and paratyphoid fever. Biologically transmitted diseases include African sleeping sickness, onchocerciasis (blinding filariasis), loiasis (African eyeworm disease), bartonellosis, and sandfly fever.

Lice

Lice are responsible for the spread of epidemic typhus fever, trench fever, and relapsing fever. Typhus fever has resulted in tens of thousands of deaths and hundreds of thousands of illnesses. Typhus fever is caused by a rickettsia. The rickettsia goes through part of its life cycle within the louse. It may change from the organism, causing murine typhus fever, which is rather mild, or to the mutant causing epidemic typhus fever, which is extremely serious. The rickettsia multiplies rapidly within the midgut of the louse after it has ingested blood with the organism *R. prowazeki*. The epithelial cells of the midgut become so loaded with rickettsiae that they rupture and are released into the contents of the gut and feces of the insect. The insect may die as the result of this. The disease is transmitted through bites or feces deposited on human skin.

Trench fever caused over a million cases of illness in soldiers on the western front in World War I. In World War II, the organism and disease once again appeared on the German-Russian front. The organism responsible for this disease is *Rickettsia quintana*.

Relapsing fevers are a group of closely related diseases in which the patient's temperature rises and falls at regular intervals. Louse-borne relapsing fever caused epidemics in parts of Europe, Africa, and Asia. When the louse bites the patient with relapsing fever, it takes in about 1 mg of blood which may contain many spirochetes. The spirochetes leave the gut of the louse and reappear in great numbers in the blood of the louse in about 6 days. The louse becomes infected for life. The disease is not transferred to human beings by bites or feces, but only by crushing the louse or damaging it in some way so that the blood contaminates the skin or mucous membrane of the human being.

Mites

Scabies or mange-like conditions are produced primarily by the mange or itch mites. Sometimes the mites cause only mild infections, but often they can cause serious infections, serious skin irritations, and severe allergic reactions. Outbreaks of scabies have occurred, especially during wartime and other emergency periods when many people are together in close quarters. The mites burrow under the skin and leave open sores to become the source of the secondary infection. The victim becomes very pale and loses considerable sleep. The individual may have intense itching, redness, or rashes. The primary means of transmission of the scabies mite is through physical contact with an infected person. Itch and mange mites are also found on domestic animals.

Chiggers transmit the rickettsia causing scrub typhus, a disease that has afflicted Americans, especially during wartime while in the Orient. Scrub typhus occurs most frequently where there are tall grass fields and neglected coconut plantations. The rickettsia are transmitted from infected victims through the egg to the larvae, which then feeds on rodents and humans. In past outbreaks of scrub typhus, mortality rates varied from 3 to 50%, depending on the virulence of the strain of rickettsia. Chigger bites are thought to transmit hemorrhagic fever. The disease, which is probably caused by a virus, affected American troops in Korea. The disease is also found in Siberia and Manchuria. It is fatal in about 5% of the cases.

The house-mouse mite transmits *Rickettsia akari* from the house mouse to humans. This causes outbreaks of rickettsial pox. Another possible disease transmitted by mites is encephalitis, which is usually found in fowl or wild birds. A number of different mites, especially the chiggers and rodent- and bird-mites, may cause skin irritation or dermatitis. These mites are found in tropical or subtropical areas, including the subtropical areas of the U.S. They are most abundant in wooded areas, swamps, roadsides, or any place inhabited by birds or wild rodents. Occasionally mites have been found to invade the respiratory tract of laboratory animals such as dogs, monkeys, and birds. Mites may serve as the intermediate host for tapeworms. The tapeworm is transmitted from one host to another. Of course, the possibility always exists that humans are the inadvertent host.

Mosquitoes-Aedes

Aedes aegypti is the principal mosquito vector of yellow fever. The virus lives in the victim's blood for 3–4 days at the onset of illness. During this time, a mosquito can bite the infected human and start the process of spreading yellow fever once again. The mosquito becomes infectious after a period of 10–14 days and can harbor the virus for life. There is no known means of infecting human beings other than by the mosquito bite, which transmits the filterable virus to the person. Destroying the breeding places of *Aedes aegypti* is the principal means of controlling the spread of disease.

Aedes albopictus (Asian tiger mosquito) was first found in the U.S. breeding in imported tires in Houston, Texas in August 1985. It was probably imported from Northern Asia, more specifically, Japan. The mosquito is exceedingly adaptable. It is a container-breeder where rain water is stored and can be found in artificial containers including old tires, tree holes, and other artificial depressions containing water. It is a link between forest, rural, and urban areas, since it can live in both, especially in houses. The eggs will survive cold and can live through a Midwestern winter. *Aedes albopictus* is an established vector of dengue fever, dengue hemorrhagic fever, and is two to three times more susceptible to infection with the dengue virus than *Aedes aegypti*. It transmits all four dengue serotypes transovarially and transstadially. In laboratory studies it has been shown to be a highly efficient vector of California encephalitis group viruses (La Crosse virus). It can also transmit yellow fever, Japanese encephalitis, eastern and western equine encephalitis, and possibly St. Louis encephalitis. It is a vicious biter and seeks out people to bite.

Other problem-causing aedes are discussed briefly in the following section. *Aedes atlanticus-tormentor-infirmatus* are vicious biters and have been known to drive cattle from woodlands. The virus of California encephalitis was isolated from these mosquitoes. *Aedes canavensis* are serious pests in woodlands and also transmit the virus of California encephalitis. *Aedes cantator* live in salt marshes and migrate to shore town and resorts, and can be a nuisance on the east coast.

Aedes cinerus occasionally pester the northern states. *Aedes borsalis* is a severe pest for humans and cattle throughout the arid and semiarid parts of the western U.S. It is a vicious biter and attacks during the day or night. *Aedes nigromis* inhabits the western plains from Minnesota to Washington and south to Texas and Mexico. This mosquito is a serious pest that bites severely in the daytime. *Aedes punctor* and other related species are important pests in woodlands and recreational areas in northeastern U.S. and mountains of the West. *Aedes sollicitans* live in salt marshes and are severe pests along the Atlantic and Gulf Coasts and are also found in the Midwest. They are fierce biters that migrate just before dark. *Aedes seencerii* are found on the prairies of Minnesota, North Dakota, and Montana, and in several northwestern states. They are fierce biters that attack during the day. *Aedes sticticus* are found mostly in the northern states. They are usually quite abundant after a flood. They are severe biters during the evenings and also bite in cloudy or shady areas during the day. The *Aedes stimulans* are found in the northern states. They bite during the daytime. *Aedes taeniorhynchus* are found on coastal plains along the Pacific Coast. They are most abundant in salt marshes and are severe pests and fierce biters, primarily at night. *Aedes triseriatus* are fierce biters found in tree holes, old tires, tin cans, and other artificial containers. *Aedes trivittatus* are fierce biters and annoying pests found in the northern U.S. They usually rest in grass and vegetation during the day. *Aedes vexans* are found in rain pools or flood waters in the northern states from New England to the Pacific Coast. The adults travel 5–10 mi from breeding places. They viciously bite at dusk or after dark. There are many other species of *Aedes* too numerous to discuss in this volume; the principal species have been covered.

Anopheles

The major vectors of malaria include *Anopheles albimanus* in the Caribbean region, *Anopheles freeborni* in the western part of the U.S., and *Anopheles quadrimaculatus* in most parts of the world and east of the Rocky Mountains.

Humans are the only important reservoir of human malaria. The disease is transmitted when the anopheles ingest human blood containing the plasmodium in the gametocyte stage. The parasite develops into sporozoites in 8–35 days. The sporozoites concentrate in the salivary glands of the mosquitoes and are injected into humans as the mosquito takes its blood meals. The gametocyte appears in the blood of a host within 3–14 days after the onset of the symptoms. Malaria is also transmitted from person to person through blood transfusions or contaminated syringes.

The incubation period varies, depending on the organism, from 12 days to 10 months. The infection may be transmitted as long as the gametocytes are present in the patient's blood. The disease is carried by a patient for as little as 1 year or as long as a lifetime depending on the organism. The *Plasmodium vivax* is usually carried from 1 to 3 years. *Plasmodium falciparum, Plasmodium ovale,* and *Plasmodium malariae* are also malaria parasites.

Other troublesome anopheles include *Anopheles punctipennis,* which is a vicious biter out of doors and apparently does not enter homes readily. It breeds in rain barrels, hog wallows, grassy bogs, swamps, and margins of streams. *Anopheles walkeri* breeds in sunny marshes along the edges of lakes and in saw grass. It readily bites humans and is a good laboratory vector of malaria.

Culex

Culex pipiens pipiens, the northern house mosquito, lives in the northern U.S.; *Culex pipiens quinquexasciatus,* the southern house mosquito, lives in the southern U.S. Both of these mosquitoes are severe pests. They are extremely annoying, bite fiercely, and feed continuously on humans. Both of these mosquitoes breed in huge quantities in rain barrels, tanks, tin cans, storm-sewer catch basins, poorly drained street gutters, polluted ground pools, cesspools, open septic tanks, effluent drains from sewage disposal plants, and in any other highly unsanitary condition. The mosquitoes do not migrate far from their area of harborage. They are capable of transmitting innumerable diseases.

Culex tarsalis are naturally infected with the viruses of St. Louis and Western encephalitis, and have the capability to transmit either of these diseases. The mosquito bites birds and then transmits the virus to other birds, horses, or humans. It is the most important vector of encephalitis in humans and horses in the western U.S. The insect is active soon after dusk and enters buildings for its blood meal. It is widely distributed west of the Mississippi River, and also in southern Canada and northern Mexico. The larvae develop in a variety of aquatic settings. They develop in arid and semiarid regions, wherever water is trapped, in effluent from cesspools and other organic materials, and in artificial containers. The females deposit 100 –150 eggs in a raft. Hatching occurs within 18 hr. Adults are active from dusk to dawn and remain at rest in secluded areas during the day. They are found on porches, shaded sides of buildings, under bridges, or in other protected areas. The majority of these mosquitoes rest in grass and shrubs or along the banks of streams. They fly up to 11 mi, even though they normally stay within 1 mi of their breeding place.

Encephalitis, which is caused by an arthropod-borne virus, is a group of acute, inflammatory diseases of short duration that involve parts of the brain, spinal cord, and meninges. The disease is not transmitted directly from human to human, but from mosquito to bird to mosquito to human.

Culiseta

There are 10 species of culiseta in the U.S., of which five are fairly widespread. Although they are relatively unimportant as pests, two of the species are naturally infected with encephalitis viruses. It is not known, however, whether they spread encephalitis to humans. These two species are *Culiseta inornata*, which are frequently found in cold water, and *Culiseta melanura*, which are found in most of the eastern U.S. from the Gulf states to Canada.

Mansonia

Of the three species of Mansonia found in the U.S., one is widespread and common. *Mansonia perturbans* is a serious biter and pest found in the southern and eastern states, and also in the Great Plains, Rocky Mountain, and Pacific Coast states. In Georgia, this mosquito was found to be naturally infected with the virus of Eastern encephalitis. It breeds in marshes, ponds, and lakes that have a thick growth of aquatic vegetation. The larvae develop slowly. The difficulty with this mosquito is that it cannot be killed in its larvae stage by an ordinary surface larvacide, since its breathing equipment is inserted into plants through which it breathes. The females bite during the daytime in shady humid places, but they are most active in the evening and the early part of the night. They enter houses and bite viciously.

Psorophora

Thirteen species of *Psorophora* are found in the U.S., 10 of which are rather widely distributed in the southern and eastern states. Their breeding habits are similar to the *Aedes*. *Psorophora ciliata* which breeds in temporary pools, is a vicious biter and a serious nuisance in the south, midwest, and eastern U.S. It attacks during the day or evening. *Psorophora confinnis* is a fierce biter that attacks in the day or night. Their numbers are often so great that they kill livestock. They are principally found through the southern U.S. and in Nebraska, Iowa, New York, Massachusetts, and southern California. They breed in temporary rain pools, irrigation waters, or seepage pits. They are a great problem in rice fields. *Psorophora cyanescens* is a severe pest that attacks in the day or night. It is particularly abundant in Oklahoma, Arkansas, Alabama, Mississippi, Louisiana, and Texas. *Psorophora ferox* is a vicious, painful biter found in wooded areas of the south and east.

Roaches

Although there are no records of outbreaks of disease caused by roaches, it is possible that they act as vectors. Roaches cross areas containing sewage or fecal material. They also cross food and could mechanically transmit enteric organisms from their breeding places to food. It is also possible that they transmit organisms through their feces and through the small amount of material regurgitated while they are feeding. Roaches have been shown to carry salmonella in their gut.

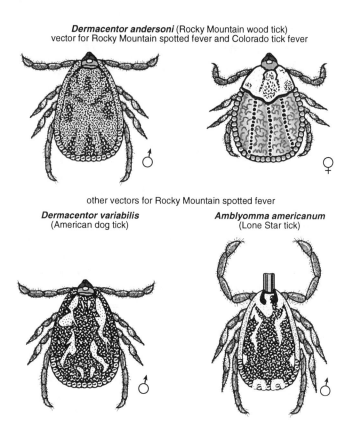

Dermacentor andersoni (Rocky Mountain wood tick)
vector for Rocky Mountain spotted fever and Colorado tick fever

other vectors for Rocky Mountain spotted fever

Dermacentor variabilis
(American dog tick)

Amblyomma americanum
(Lone Star tick)

Figure 4.12. Tick-vectors for Rocky Mountain spotted fever and Colorado tick fever.

Ticks

Ticks can transmit pathogenic organisms mechanically or biologically (Figure 4.12). Their mouthparts become contaminated when feeding on an infected animal and may in turn contaminate another animal or human while feeding. This is mechanical transmission. Further, ticks serve as reservoirs of viruses, rickettsiae, spirochetes, bacteria, and protozoa. These organisms are transmitted from the infected adults through the eggs to the larval, nymphal, and adult stages. This is called transovarial or transstadial transmission. This mechanism is of considerable concern because the tick-borne disease can survive adverse weather conditions, since the organisms are being transmitted through the eggs.

Tularemia or rabbit fever is caused by the bacillus *Pasteurella tularensis*. This bacterial disease infects numerous hosts, including rabbits, rodents, and humans.

Ticks transmit this disease organism to the genera *Dermacentor* and *Haenaphysalis*. In addition, the deer fly, which is also a blood-sucking arthropod, may be an important vector of the spread of tularemia. Ticks transmit the disease organisms to their own eggs, keeping the organism alive from generation to generation.

There are two peaks of tularemia in the U.S., occurring during the fall-winter season and also during the spring-summer season when the ticks and deer flies are most prevalent.

Protozoal diseases spread by ticks include cattle tick fever, Texas cattle fever, and anapromosis. The diseases are spread primarily from cattle to cattle. Rickettsial diseases include Rocky Mountain spotted fever and Q fever. Rocky Mountain spotted fever is caused by the organisms *Rickettsia rickettsia*. The wood tick, *Dermacentor andersoni*, and the rabbit tick, *Haenaphysalis leporistalustris,* spread the disease from animal to animal and from animal to human in the western part of the country. The American dog tick, *Dermacentor variabilis*, is the most important vector in the eastern part of the country. The Lone Star tick, *Amblyomma americanum*, is probably a vector of Rocky Mountain spotted fever in Texas, Oklahoma, and Arkansas. These *Dermacentor* ticks easily spread spotted fever, because the larvae and nymphs feed on infected animals and the adults attack humans or other large animals. Spotted fever has been reported in 46 of the 50 states in the U.S. The western strain of the disease is much more virulent than the eastern strain.

Rocky Mountain spotted fever resembles typhus fever, but it is a more severe infection. The mortality rate may be as high as 25%. The rickettsiae invade the endothelial cells and smooth muscle of the arterioles, which may cause obstruction and destruction of the vessels. Although the disease is not transmitted directly from human to human, the tick remains infected for life and continues to transmit the disease to humans for periods of up to 18 months.

Q fever is a rickettsial disease caused by the organism *Coxiella burnetii*. Outbreaks of the disease often occur in stockyard workers. The wood tick and Lone Star tick, among others, spread rickettsiae to humans. The disease is also airborne or is passed through raw milk from infected cows. The tissues and feces of ticks become massively infected. It is suspected that the most common means of transmission occurs when the organisms are inhaled in dust or droplet form. About 1% of untreated victims die of Q fever. Relapsing fevers, which are tick or louse borne, are caused by spirochetes. The diseases occur on all continents, with the possible exception of Australia. The spirochetes causing relapsing fever most frequently in the U.S. belong to the genus *Borrelia*. The louse-borne type of the disease is usually due to *Borrelia recurrentis*. The tick-borne disease is spread through soft ticks in the genus *Ornithodoros* found in 13 western states. The spirochetes are transmitted through the offspring of the ticks or through their rodent hosts, which then serve as reservoirs for the disease.

The viral disease spread by ticks is Colorado tick fever. New cases occur in the Rocky Mountain areas of the west. The wood tick, or *Dermacentor andersoni*, carries the disease. Although dog ticks are shown to carry the virus, they have never been incriminated in any outbreak of the disease. The reservoirs of infection include ground squirrels, porcupines, and chipmunks.

Ticks, in addition to causing the above-mentioned diseases, may also cause a paralysis of considerable concern to both physicians and veterinarians.

Food-Product Insects

Stored food-product insects are a nuisance and contaminate food in large quantities. They must be destroyed and removed from the food supply. Their role as agents in the spread of food-borne diseases has not been proven.

Impact of Insects on Other Environmental and Health Problems

Pesticides are the major means of control of pests today. Unfortunately, the World Health Organization estimates that approximately 500,000 cases world-wide of pesticide intoxication occur annually and that about 1% of the victims die as a direct result of contact with the pesticide. Pesticides also cause resistance in a disease-bearing vector, thereby exacerbating disease production.

Economics

It is difficult to make a good economic analysis of a pest control program. Public health specialists are unable to evaluate the associated costs and benefits of large or small programs, and economists fail to deal adequately with the benefits or values of health vs illness, the value of human life, or even the value of a reduction in the number of pests. There also is uncertainty about the technical feasibility of various types of controls. Even the standard techniques used, such as source reduction, larvaciding, and adulticiding, vary tremendously with region and country. Nowhere has a major effort been made in specific data collection to justify or to establish cost-benefit ratios.

POTENTIAL FOR INTERVENTION

The potential for intervention is excellent in the general area of insect control and for specific pests. The techniques of intervention include removal of adult harborage, adulticiding, larvaciding, and biological control. (For additional information, see the section on controls.)

Intervention by isolation techniques is difficult to accomplish, since insects move rapidly through a given area. Substitution is another inadequate technique. Shielding can be accomplished by use of proper-sized screens or blow-down air devices. Treatment is used for disease carriers but not effectively for the insects. Prevention is the strongest potential intervention technique in the spread of particular insects.

RESOURCES

Some scientific and technical resources available include Tulane University School of Public Health, Colorado State University, various land grant colleges such as Pennsylvania State University and Purdue University, Notre Dame University, the American Public Health Association, the National Environmental Health Association, and the Entomological Society of America.

The civic and professional associations in pest control include the American Mosquito Control Association, the International Association of Game-Fish and Conservation Commissioners. Pest control operators associations and chemical manufacturers associations are other important resources.

Government resources available include state health departments, many local health departments, the U.S. Department of Agriculture, the National Bureau of Standards, the Department of Defense, the Environmental Protection Agency, the Food and Drug Administration, the National Science Foundation, and Department of Health and Human Services, and county extension agents.

STANDARDS, PRACTICES, AND TECHNIQUES

Mosquitoes are by far the largest group of insects controlled by humans. The standards, practices, and techniques utilized vary tremendously from area to area. Generally, surveys are made to determine the extent and location of the mosquito populations and then extensive larvaciding or adulticiding programs are put into effect. In the event of an outbreak of disease, extensive larvaciding and adulticiding programs are again carried on in those areas where mosquitoes are most prevalent. In many areas there are unorganized mosquito control programs. These programs are put into effect when citizens demand action because of the annoyance due to mosquitoes. Generally public works employees or contracted commercial firms will do the actual control work. The work consists of the use of larvacides and fogging units, or mist units for adulticiding. Control may be done in a haphazard way or it may be carried out weekly on a regular basis where specific high-volume mosquito areas exist. Where organized mosquito programs are in existence, the community is taught first how to reduce mosquitoes through environmental controls and adequate prevention; then proper engineering approaches are used, as well as larvacides and adulticides. Although the individuals carrying out this work must be licensed by their states, and adhere to the principles of the Environmental Protection Agency, there still is a substantial need to establish proper standards, practices, and techniques for insect control.

MODES OF SURVEILLANCE AND EVALUATION

Inspections and Surveys

Fly Surveys

The major value of a fly survey is to determine those areas in which the flies are in greatest concentration, to determine the kinds of flies in order to use effective controls, and to determine the potential spread of disease. There are several different types of fly surveys. They include fly traps, fly grills, and reconnaissance surveys. The selection of the sample area should be based on a clear understanding of what information is needed, the kinds of problems antici- pated, the areas where the greatest number of flies are found, and the areas where the greatest refuse and other solid-waste problems exist.

Fly traps consist of baited traps, fly paper, or fly cones. The fly traps have the advantage of capturing a cross-section of the kinds of flies in a given area. They also provide an approximate count of the various species. The baited fly trap may be lined with animal feces, sugar, fish heads, or some other decaying matter. The flypaper strip gives a rapid count of flies, but this only will represent a small portion of the fly population. Flies are drawn to fly cones by the attractant, and then as they attempt to escape, they move upward toward a light that is in the fly cage.

Fly grills are the most widely used means of surveying for flies. The grill depends upon the natural tendency of flies to rest on edges. The grill is placed over a natural attractant, such as garbage, manure, or other decaying materials. The number of flies landing on the grill during a 30-sec period is tabulated. It is possible to physically count the number of flies and make some judgement of the different kinds of flies on the grill. If there is a huge number of flies, the grill should be subdivided and a count made on the subdivided portion. A minimum of 10 counts should be made in each subdivision sample, and the five highest counts should be recorded.

It is possible to make a rapid survey by driving through an area or walking through an area in which you would expect to find large numbers of flies. A rough estimate can be determined by checking the usual fly-breeding areas and making a physical count.

In an attempt to determine if a fly survey or a fly control program is effective, it is wise to make a fly survey initially in the usual breeding areas and then return after the breeding areas have been eliminated and resurvey. Obviously, chemical control procedures should be used prior to the cleanup procedures so that the flies do not move on to other areas.

Mosquito Surveys

Surveys are necessary for the planning, operation, and evaluation of mosquito control programs. The basic survey is used to determine the species of mosquitoes, sources, breeding places, locations, quantities, flight distance, larval habits, adult resting places, and any other information that may be needed in the actual mosquito control program. It is also necessary to continually evaluate the ongoing program. This is done by conducting surveys on a periodic basis during each season of the year, especially the mosquito-breeding season, to determine the effectiveness of chemicals and other controls. The surveys consist of making field evaluations of the quantity of mosquitoes, the types of mosquitoes, the kinds of mosquito breeding areas and potential mosquito breeding areas, the numbers and types of larvae, and the development of accurate, comprehensive maps that indicate where all the larval and adult problems may be found.

A larval mosquito study may be carried out by an environmentalist who goes to areas where there are artificial breeding places and standing bodies of water and removes water containing larvae by means of a white enamel dipper, which is about 4 in. in diameter. The dipper has an extended handle so that it may be lowered into the body of water. Some environmentalists prefer to use white enamel pans that are about 14 in. long, 9 in. wide, and 2 in. deep. The pan is used to sweep an area of water until the pan is half full. The water is then examined for larvae. Another technique is to examine cisterns, artificial containers, or other areas with a flashlight. A large bulb pipette or siphon is used to draw out water with larvae into a pan or onto a slide, and then observations are made. The environmentalist should always, after observing the habits of the larvae in the water, record the number of dips he or she makes and the count of larvae. A rough estimate can be made over time of larvae increases or decreases by going to the same area and using the same dipping procedures.

Adult mosquito surveys can be carried out by setting up bait traps, window traps, carbon dioxide traps, insect nets, light traps, or checking the daytime resting places. Bait traps are put in areas where mosquitoes are expected to be found. The mosquitoes are attracted to a bait animal that is placed in a trap. When they enter the trap, they cannot bite the animal and they cannot be released. The mosquitoes are taken alive and evaluated. Bait animals have been such animals as horses, calves, mules, donkeys, and sheep. Window traps use the same principle as the animal bait trap, except a human being is the attractant. The trap is set on the outside of the window. The mosquitoes are attracted to the building by the presence of the human being and then they enter the trap and cannot leave. Insect nets are used for collecting mosquitoes in grass and other vegetation. Light traps are traps that contain a light that acts as an attractant to the mosquitoes. Once the mosquitoes enter, they are incapable of leaving the trap area. An evaluation can be made of the adult resting places. The type of resting place depends on the kind of mosquito. This material has been previously covered.

Once all the information is gathered and a determination is made of the kinds of mosquitoes that are present in an area, comprehensive maps are drawn, and the control program is established and put into action.

Food-Product Insect Surveys

The easiest means of determining the presence of food-product insects is searching through dry-food products and looking for live insects. The insects may also be present along the seam of the food bag. At times the insects can be seen crawling near the food or flying about. In any of these cases, a food-product insect problem exists and preventive measures should be taken.

CONTROLS

Before DDT was used for mosquito problems, the principal control techniques consisted of identifying the species of mosquito and its breeding habits, and then designing a method that would eliminate the larva, either by means of engineering techniques or by means of chemical larvacides. Engineering techniques consist of improving drainage areas, adjusting the slope and depth of streams, and using methods and techniques for the removal of any obstructions hampering the flow of water, which in turn increases the potential for mosquito breeding. Biological control is the technique utilizing parasites, such as *Bacillus thuringiensis*, predators, and pathogens of insects to eliminate them. It also includes the use of sterilization techniques, genetic manipulation, management of the habitat, insect growth regulators, and biologically produced chemical compounds. In recent years there has been considerable interest in this area. Currently mosquito biolarvicides have become an important part of integrated mosquito control management programs. Biological control is basically the fostering and manipulation of the natural enemies of insects. Fish are an excellent biological control for mosquito larvae. There are over 1000 species of freshwater fish that consume mosquito larva and pupa. However, the one that is most frequently used is the minnow *Gambusia affinis*. This minnow is a surface feeder and inhabits permanent waters. It has been used in the control of anopheline mosquito larvae to reduce malaria, and *Aedes aegypti* to reduce yellow fever. The technique of building insects out by removing harborage along with chemical control is now called integrated pest management.

Insect growth regulators are used as a means of control. Attractants will lure the insect to trap or poison. Repellents are used on the skin and body as a means of control to prevent the insect from alighting and biting, or otherwise contaminating the individual. Since certain pesticides are being banned as a result of new research, it is essential to reevaluate the chemicals being used very frequently. Read the labels for proper usage.

Specific Scientific and Technical Controls

Flea Control

In flea control, infested animals and their habitats are treated. Animals are treated by shaking a dust containing 10% methoxychlor or 5% carbaryl. These chemicals have a slight toxicity and a long residual period. Particular attention should be paid to the back, the neck, and the top of the head. The powder should be rubbed thoroughly into the hair of the animal. The animal's quarters or the house that is infested with fleas should be treated with either carbaryl, methoxychlor, or ronnel. Each of these chemicals are slightly toxic and have long residual periods. If there is concern about a rodent flea infestation, it is important to dust the rat run with any of the previously mentioned chemicals. The dust will then get onto the rats or mice and will kill the fleas that are present. It is essential that the fleas be killed before a rat-control program starts. Otherwise, the rats will die and the fleas will leave the dead animals and possibly attack humans. In addition, infested premises may be treated with 2% malathion or 0.5% diazinon. In barns and yards, you may use 2.5% diazinon or 2.5% ronnel, malathion as a 4% dust, or carbaryl as a 5% dust. Insect growth regulators, such as methoprene, are juvenile flea hormone analogues which disrupt egg development and sterilize adult cat fleas.

Fly Control

The four basic fly-control techniques include proper solid-waste disposal, chemical control, mechanical and physical control, and biological control.

Since house flies breed in garbage, decaying matter, feces, dead animals, and any kind of organic material, it is essential that this solid waste be removed and stored properly. The storage should take place in plastic bags within metal cans with tight-fitting lids. The refuse should also be stored, depending on the kind, in concrete containers (for animal manure), in large metal containers with tightly closed lids, or in refuse rooms where the doors and windows are flyproof. The maximum size of the can should be not greater than 30-gal capacity, since it becomes difficult to remove heavier cans. Where garbage, heavy refuse, or ashes are stored, the maximum size of the cans should be 20 gal. Collection should take place during the summer at least twice a week, since flies go from the eggs to the pupa stage within 5 days and then move into the ground. By removing the cans at least once every 4 days, the fly eggs, larvae, or pupa are removed and disposed of. During the winter, the solid waste should be removed at least once a week. In food operations and in other situations where there are large quantities of organic waste material, this material should be removed on a daily basis. Solid-waste disposal should occur in either an incinerator or a properly operated sanitary landfill. Garbage that is used for hog feeding must be cooked to an internal temperature of 137°F before it is served to the hogs. One other technique utilized for disposal is through the home garbage grinder.

Animal feed and feces become breeding areas for flies. Animal manure should be adequately stored, and dog and cat feces should be picked up and removed in order to prevent fly breeding. It is also essential to check the animal feed periodically to make sure flies are not breeding in it. In a home situation, dog food or cat food left outside becomes an attractant.

Large fly populations can breed in high weeds. It is therefore necessary that weeds be controlled either chemically or physically. The type of chemical chosen for control of flies is determined by the effect that is desired. Residual sprays are put on surfaces on which flies will alight. These sprays should last for considerable periods of time. The chemicals include 2–5% malathion mixed with 6–12% sugar emulsion or a 1% diazinon emulsion. Space sprays are used in areas where quick knockdown of flies is desired, and where the aim is to kill the adult fly in large quantities. The space spray should contain 0.1% synergized pyrethrum plus 1% malathion. Fly cords should be impregnated with 2.5% diazinon. The cord should be made of strong cotton, $^3/_{32}$ in. in diameter and strung at a rate of 30 linear ft of cord per 100 ft^2 of floor area. Cords are accepted because flies will rest on them absorbing the insecticide. A larvacidal treatment should consist of 2.5% diazinon emulsion, 1% malathion emulsion, 2.5% ronnel emulsion, or 2% DDVP emulsion. Dry baits can consist of diazinon, malathion, ronnel, or DDVP plus sugar. For wet bait add water.

It is important to realize that insects become quite tolerant and resistant to insecticides. As a result of this, the insecticides that are now recommended and probably the safest to use for humans at this time may not be those recommended for future use. The environmental health specialist should contact the local county extension agent if he or she has any questions concerning the problems of fly control and chemical safety.

It is known that flies not only are resistant to certain insecticides, but apparently develop an inherited resistance over several generations. As a result of this, flies are able to either absorb a lower rate of insecticide, store the insecticide without being killed, excrete the insecticide, detoxify it, or use an alternate mode of accomplishing its body functions when a chemical blocks its normal route.

There are several techniques used in applying insecticides. They include space spraying for the adults; mist spraying through large blowers or insecticide bombs for adults; fog generators, which produce an aerosol or smoke for adults; and residual spraying with hand sprayers for adults and larvae.

Flies can also be controlled chemically by treating the animal by the use of fly repellents or by the use of fly attractants to draw flies into traps or onto poison.

One of the most effective physical controls is fly screens, which must be 16-mesh in size. The screens have to fit tightly to the windows or door frames. Fly traps, which have been discussed under survey techniques, can be utilized for some fly control. Flies can be electrocuted if they cross an electrically charged field or alight on a screen that is electrically charged. Electronic fly traps, although effective in certain instances, can increase the number of repirable fly particles by 10–500 times over the background levels. Air shields can be installed so that air blows down and out from the doorway, thereby preventing flies from entering a building.

The biological control of flies is based on the use of sterile flies and on the dissemination of organisms that are pathogenic to the insect. These techniques are basically in experimental stages and are not generally used. Predatory animals will consume a certain quantity of flies; however, it is necessary to be cautious that the predators do not themselves become pests.

The most important means for control of flies is through the elimination of food and garbage. Therefore, it is extremely important that all exterior areas and all interior areas are thoroughly cleaned and all solid waste is removed.

Within the home, flies are best chemically controlled by the use of synergized pyrethrum plus malathion, diazinon, or baygon. Each of these chemicals must be used very carefully. DDVP strips may be used to kill flies. These resin strips will last for about 3 months. One strip should be used per 1000 ft^3 of space. Care must be taken in the use of DDVP strips around food or food services and where individuals have respiratory difficulties.

Louse Control

Body louse control is accomplished by washing the clothes of the individual in very hot water, drying, and then ironing. Wool clothing, which is especially good for louse habitation, should be thoroughly dry cleaned and then pressed. Clothing may also be put in a plastic bag and allowed to remain for a period of 30 days. This will kill the existing lice as well as the eggs. The individual can be treated with a 1% malathion powder, which is effective against eggs and adult lice.

Head lice are controlled by cutting the hair very short and then washing it thoroughly. It is essential to kill both the lice and the eggs. The best technique for head louse control is to shampoo, dry the hair thoroughly, tilt the head back, cover the eyes, apply a 25% benzyl benzoate emulsion, which is worked into the hair and scalp, and combed through the hair. Shampoo the hair again after 24 hr, dry, comb, and brush it to remove the dead lice and loosen the eggs.

In crab louse control, the pubic hair should be washed very thoroughly and the hair shaved off. A vaseline ointment containing pyrethrum or a 1% lindane cream or lotion may be used.

It is essential to complete louse control by eliminating all the bedding and clothes used by the lice-infested individuals. These garments and materials should be washed very thoroughly or dry cleaned. It is also essential that hair is washed frequently and thoroughly.

Mite Control

Mites are controlled through the use of chemicals or proper cleanup. In trying to control mites within a house, or even within areas where animals live, pyrethrum bombs can be used as knockdown agents. Obviously the pyrethrum has to be synergized. As soon as the mites are knocked down, they must be vacuumed up and destroyed. For residual treatment use 1% malathion. This insecticide should be applied onto the top of foundations, around the plate and ends of joists, baseboards, edges of floor areas, and around windows and doors. On the outside, mites can be controlled with 0.5–1% malathion at a rate of 2.5 gal of 40% emulsifiable concentrate per acre.

Human infestation can be controlled by use of 68% benzyl benzoate or 12% benzocaine. The chemicals should be left on the body for 24 hr before washing. The second treatment is needed in 10–14 days. Scabies may also be treated with eurax, which is a salve containing 10% N-ethyl-O-crotontoluide in a vanishing cream. Two applications should be given at 24-hr intervals.

Mites are controlled through elimination of rats, house mice, birds, and so forth. With chigger mites, the environment should be modified to permit plenty of sunlight and air to circulate freely.

Larval and Adult Mosquito Control

Mosquito larvae are controlled mechanically, biologically, or chemically. Mechanical control involves emptying and removing all temporary containers, such as old tires and cans. If possible, depressions in the ground should be smoothed out. High weeds, which trap water, must be cut on a regular basis. It is also important to ensure that all sanitary landfills or any other type of landfill has a final slope of 0.1–0.5 ft/100 ft for drainage purposes. Clean, straighten, and drain all ditches so that water runs freely. Any growth in bodies of water, particularly around the shoreline, must be removed so that the water flows readily rather than sits stagnant. Of particular importance is the closing or sealing of all seepage pits, septic tanks, and other areas where sewage may be trapped in a stagnant situation. Impounded waters can be a source of mosquito larvae breeding. This can be remedied by cleaning the major vegetation off the shoreline, filling, and removing the water from low places. The water level should be so controlled that a reduction in level of about $1/_{10}$ ft per week will occur during the mosquito-breeding season. Additional research should be done in the areas of mechanical mosquito larvae control, since there are huge numbers of mosquitoes in marshy areas and in areas where certain types of food, such as rice, are grown. It would be detrimental, at present, to change the ecology of these areas. On irrigated lands, the mosquito problems can be considerable. They generally find habitation in stored water, the irrigation conveyance and distribution system, the irrigated land, and in drainage systems. The key to control is to use only the amount of water necessary and to have all systems freely moving so that stagnation of water cannot occur. The biological technique for eliminating larvae would be to: use the toxins of *Bacillus thuringiensis*, use insect growth regulators, or stock the lakes and other impounded areas with top-feeding minnows of the gambusia variety. In residential areas, it is important to change the water in the bird baths, stock garden ponds with goldfish, and remove all containers holding stagnant water.

Chemical control is accomplished by applying #2 fuel oil, #2 diesel oil, kerosene, or petroleum distillates. Petroleum distillates tend to be more environmentally friendly than the other petroleum compounds. The petroleum is toxic to the eggs, larvae, and pupa. Pyrethrum larvacides have been used for many years. Methoxychlor may act as a systemic poison and as a contact poison, which penetrates the body wall or respiratory tract. In addition, baytex and malathion have been used effectively in mosquito larvae control.

At dark, mosquitoes are best controlled by use of screens, repellents, and by space spraying. Screens should be 16×16 or 14×14 mesh to the inch to keep mosquitoes out. In areas where very small mosquitoes exist, such as the *Aedes aegypti*, it is necessary to have a screen with a mesh of 16×20 or 16×23. Bed nets are useful in temporary camp areas where mosquitoes are quite prevalent. Bed nets are made of cotton or nylon cloth with 23–26 meshes/in.

There are several types of mosquito repellents on the market today. They include bimethyl thalate, Indalone, and biethyl toluamide. The repellents are applied to the neck, face, hands, and arms. They will prevent mosquito bites for a period of from 2 to 12 hr. This protection depends on the person, the species of mosquito attacking, and the abundance of mosquitoes available.

Space spraying for adult mosquitoes is accomplished by the use of aerosols, fogging, misting, dusting, or by airplane application. The chemicals generally used include carbaryl and malathion. Care must be taken to avoid large concentrations of pesticides in any type of space spraying in residential areas. Where residential areas are space sprayed, personnel must ensure that the trucks have blinking red or orange lights as signals and that children are not driving their bicycles in and out of the mists. Serious accidents have been caused when automobile drivers were blinded by the fog or mist and drove into children on bicycles.

Residual spraying is another technique for the destruction of adult mosquitoes, where 2.5% malathion is used. However, the asian tiger mosquito in Houston appeared to be resistant to malathion. The chemicals are applied to surfaces, and they continue to kill mosquitoes for a period of 10 to as much as 32 weeks, depending on the type of insecticide and mosquito. Another valuable technique is the use of strips of DDVP in enclosed basins. These strips continue to give off vapona insecticide over a period of time and will kill the mosquitoes within the catch basins.

It is important to realize that mosquito-borne outbreaks of disease have not been eliminated from this country or other countries. In the years to come, all health departments should develop adequate, comprehensive mosquito-control programs in conjunction with the general public and the federal agencies.

Computerized programs can be utilized to aid in finding reoccuring mosquito breeding areas and determining level of control. Such an excellent program is being used by the Marion County Health Department in Indianapolis, Indiana.

Roach Control

Roaches are controlled by keeping the premises extremely clean and by discarding paper bags, food cartons, and other materials from food stores and warehouses. Roaches are controlled chemically by the following insecticides: diazinon, 2% spray or 2% dust; malathion, 3% spray or 4% dust; baygon 1% spray or 2% dust. Borax is also used to control roaches by placing the powder in out of the way areas, such as wall openings, under cabinets, under bathtubs, in attics, and in other hard to get at places during construction. Roaches that repeatedly cross these areas will be killed by the chemical. German roaches may also be controlled by the use of an insect growth regulator called hydroprene, which augments the level of juvenile hormones. The insect develops a hormonal imbalance that neither allows it to remain an adolescent or progress to an adult, causing the death of the insect.

Tick Control

An individual is protected against ticks by keeping clothing buttoned, trouser legs tucked into socks, and by wearing the type of clothing that will prevent ticks from penetrating and getting to exposed body parts. Individuals should always inspect their clothing and their bodies after they have been through a tick-infested area. The ticks should be removed immediately and destroyed. If ticks become attached, the simplest technique for removal is a slow straight pull that will not break off the mouth part and leave it in the wound. A drop of chloroform, ether, vaseline, or fingernail polish rubbed over the tick and the area will help the removal. Tick repellents can also be used. However, there is no one repellent that appears to be perfect for all ticks. Clothing may be treated with indalone, dimethyl tareate, or benzyl benzoate. Dogs and cats may be dusted with 0.75–1% rotenone or 3–5% malathion.

Ticks inside of buildings are controlled by the use of baygon or 0.5% diazinon. Sprays should contain 0.2–0.5% DDVP as a fumigant to drive the ticks out from behind baseboards and out of the cracks in walls. Tick-infested areas on the outside may be treated with a 4% malathion dust or DDVP. The control program must go on for many months so that the infected eggs can be destroyed as they become adults. Another technique for the removal or destruction of ticks is to remove the host. Where dogs are infested with ticks in a certain area, the area and the dog should be carefully treated, and then the dog should be kept away from the usual sleeping quarters, which prevents the ticks from getting a blood meal.

Food-Product Insect Control

Control of food-product insects include the following: (1) locate the source of infestation by examining the foods and areas where infestation is most likely; (2) clean all cabinets and storage areas thoroughly by vacuuming and washing with hot water; and (3) keep all shelving dry and spray with appropriate insecticides. These insecticides may include diazinon of 0.5% concentration or 1% baygon; (4) after drying and spraying, put only fresh food back into the area; and (5) make sure that foods are rotated frequently. During the hot months, rotate the dry foods at least once every 2 weeks. During the cold months, the foods should be rotated at least once every month.

Public Health Laws

A typical public health law gives the state board of health the power to develop rules and regulations to control nuisances that are dangerous to public health, to control fly and mosquito breeding places and to control the spread of rodents. These provisions are typically found in the powers and duties of the state board of health. In addition, the food codes that come under the general public health laws have specific provisions that food establishments shall be protected by all reasonable means against flies, other insects, and rodents. Further, housing codes stipulate that dwelling units, school units, or any other type of dwelling or shelter must be free of insects and rodents. A health code may also contain a provision making it unlawful for any individual to maintain any lands, places, buildings, structures, vessels, or watercraft that are infested by insects and rodents. It may also stipulate that any person owning, leasing, occupying, possessing, or having charge of any land, place, building, structure, stacks, or quantities of wood, hay, corn, wheat, other grains, vessels, or watercraft must eliminate infestations of insects or rodents or be prosecuted under the pest eradication sections of the Health Code.

SUMMARY

Vector-borne disease will continue to be a major deterrent to human settlement and agricultural development in certain areas of the country and the world. The insects discussed were of public health importance, including fleas, flies, mites, mosquitoes, roaches, ticks, and food-product insects. These insects destroy food material and cause disease. The potential for intervention is excellent, providing that humans are willing to use a comprehensive program of habitat removal, biological control, chemical control, and good public health education. The pesticides used in chemical control, while effective, have caused potential hazards to humans, either through the short-range problem of intoxication or the long-range potential hazard of carcinogenesis or mutagenesis. The destruction of the vector should be supplemented by chemotherapy and vaccines to limit disease potential. In working toward a good pest-control program, the potential hazards to the environment should be considered. The use of pesticides for all purposes should be coordinated to reduce the environmental degradation while increasing the health potential for society.

RESEARCH NEEDS

Additional study is needed to determine the best technique to be used for the control of arthropod vectors of disease or other public health pests, and the significance of this control on the ecology and behavior of the various ecosystems. A reporting system where information from health departments of cities, counties, and states is sent to the Center for Disease Control to help quickly identify the existing levels of insect or other pest problems, is needed. Additional study is needed to determine the problems related to environmental manipulation. Tolerance levels for biting nuisances should be established. It is necessary to have a combined study by agricultural experts, medical experts, and environmental experts to determine the problems of poison persistence and resistance as it relates to pesticide management.

Chapter 5

RODENT CONTROL

BACKGROUND AND STATUS

Rodents are members of the order Rodentia. These mammals have teeth and jaws adapted for gnawing. The order Rodentia includes squirrels, beavers, mice, rats, lemmings, porcupines, and chinchillas. For the purpose of this chapter, four rodents — the Norway rat, the roof rat, house mouse, and the deer mouse — will be discussed (Figure 5.1).

The first three rodents, known as murine rodents (they are actually of the subfamily Murinae), are important because they are found everywhere that humans are found. Murine rodents live in buildings, destroy food and property, endanger health, and compete with humans for existence. They are found in all parts of the country and are dangerous whether they are out of doors or within the home.

In rural and semi-rural areas rodents pose the greatest problems by destroying and contaminating food. They inhabit outside areas, improperly managed solid-waste disposal areas and attics, or other parts of the home, depending on the species. Further, the deer mouse (family crlcetidoe) is a primary reservoir host for hantavirus infection.

In urban areas, rodent infestations occur in luxurious suburbs where mice inhabit houses and rats inhabit lawns, undersides of doghouses, or any area where construction has taken place or sewers have been disrupted. The Norway rat typically lives in sewers or along creek or river banks and migrates to areas where food may be found.

Rat infestations become extremely significant within the inner city. To simply count rat bites or determine the amount of food loss or property damage is inappropriate. Human misery increases as the rat continues to spread. The rat is

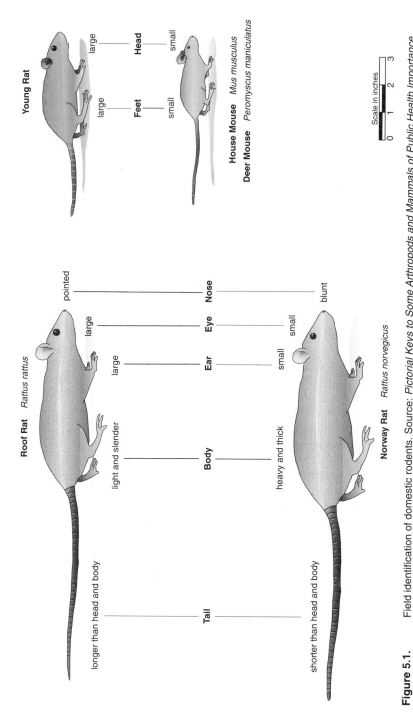

Figure 5.1.   Field identification of domestic rodents. Source: *Pictorial Keys to Some Arthropods and Mammals of Public Health Importance* (Atlanta, GA: U.S. Department of Health, Education, and Welfare, Public Health Service, Communicable Diseases Center, 1964), p. 55, as adapted by Herman Koren, 1994.

an important part of the inner city community. It is a symptom of the breakdown of the community, not only from an environmental standpoint, but also from a social and economic standpoint.

Rat control within the city is an extremely complex problem that must be resolved by involving not only the various municipal and state agencies, but also by involving the community itself. Behavioral patterns must be altered, lifestyles changed, social and economic factors improved, and serious concern with the rat problem shown if the rodent problem, including disease and community degradation, is to be eliminated.

More than 60 million people live in 300 cities where rats are a major environmental problem in the U.S.. The Urban Rat Control Program was started in 1969 to eliminate the conditions that allowed rats to grow and multiply. The program emphasized the control and improvement of both the physical and social environments. As a result of this program, intensive efforts in rodent control realized considerable solid-waste removal and some reduction in rat populations. However, although the federal government by 1979 had spent many millions of dollars in attempting to remove or reduce rat populations, the results were insubstantial. The program was temporarily effective, but as soon as it was removed from an area, the rat problem generally returned. However, the City of Philadelphia, Marion county (Indianapolis), and some other communities are maintaining effective rodent control efforts as of 1995.

Rodents are developing immunity to anticoagulants in both the U.S. and Europe. As their resistance increases, so too will their numbers.

SCIENTIFIC, TECHNOLOGICAL, AND GENERAL INFORMATION

Characteristics and Behavior of Murine Rodents and Deer Mice

Murine rodents are capable of almost continual reproduction and produce large quantities of newborn mice and rats. Some of the young are killed or eaten. The newborn rodent is born blind, but gains sight in about 12–14 days. For about 21 days it depends on its mother for food. If forced to, it can start to live by itself within 3–4 weeks of birth. Sexual activity and fighting occurs in about 2–3 months. The young learn through experience and imitation. At 3 months, they are very active and completely independent (see Table 5.1 for detailed description of murine rodents).

Daily life patterns rarely vary, unless there is a sharp increase in the rodent population or a decrease in the food supply. When food is abundant, rodents are most active during the first half of the night; when food is in short supply, they are active at any time. Generally, however, if rodents are seen during the day, there are large numbers present. Rodents tend to use regular paths and runways. They live as close to their food source as possible and travel to it with the least amount of exposure. This is why runways tend to be along the edges of walls or through covered areas.

Table 5.1   **Description of Murine Rodents and Deer Mouse**

Characteristics	Norway Rat (Rattus norvegicus)	Roof Rat (Rattus rattus)	House Mouse (Mus musculus)	Deer Mouse (Peromyscus maniculatus)
Weight	16 ozs.	8–12 ozs.	$1/2$–$3/4$ ozs.	$2/5$–1 $1/4$ ozs.
Total length including the tail	12 $3/4$–18 in.	8 $3/4$–17 3/4 in.	6–7 $1/2$ in.	4 $4/5$–9 in.
Head and body	Blunt muzzle; heavy, thick body	Pointed muzzle; slender body	Small muzzle and body	Same
Tail	Shorter than head and body	Longer than head and body	Equal to or longer than head and body	Same
Ears	Small, close set	Large, prominent	Large, prominent	Same
Hind foot	$1^1/2$ in. or longer	Less than $1^1/2$ in. in length	$3/4$ in. in length	
Teeth	Strong, well-developed; single pair of incisors in upper and lower jaw	Same	Same	
Color	Great range	Black, tawny, gray	Gray, brown, white	White feet; usually white underside; brownish upper surfaces
Distribution	US and Southern Canada	Southern Gulf states and Pacific Coast	US and Southern Canada	Throughout most of North America
Environmental Distribution	Share human habitat	Less dependent on humans; live away from human habitation	Least dependent on humans; live anywhere	Same; usually forest and grasslands
Gestation	22 days	22 days	19 days	21–23 days
Mating	Female can mate again wtihin 48 hours after giving birth	Same	Same	
Litter size	4–6 litters of 8–12 young each	4–6 litters of 6–8 young each	8 litters of 5–6 young each	2–4 litters or more, depending on weather; with 3–5 young each
Number weaned	20 per female annually	Same	30–35 per female annually	
Length of life	One year	One year	Same	2 yrs. in wild, 5–8 years in captivity
Harborage	Outdoors–in buildings under foundations, in waste disposal areas; indoors–between floors and walls, under solid waste, in any concealed area	Outdoors–trees and dense vines; indoors–attics, between walls, enclosed spaces	Nest anywhere	Nest anywhere, but primarily out of doors; in Great Plains region they enter homes and cabins
Range	100–150 ft	100–150 ft	10–30 ft	.13–1.6/ha or greater
Droppings	To $3/4$ in.; capsule-shaped	To $1/2$ in.; spindle shaped	To $1/8$ in.; rod shaped	
Food	Omnivorous– garbage, meat, fish, cereal	Omnivorous– usually vegetables, fruits, cereal	Omnivorous– usually cereals	Primarily seed eaters; also nuts, acorns, fruits, insects, larvae, fungi and some green vegetation
Daily food requirements	$3/4$–1 oz. dry food; $1/2$–1 oz. water	$1/2$–1 oz dry food; 1 oz. water	Nibbles; $1/10$ oz. dry food; $3/10$ oz. water	

Rodents have an avoidance instinct to alien objects that aids survival. Rodents will sense and avoid unusual changes within the environment. They may avoid new food for several days. When poisoning it is wise to prebait the area with unpoisoned bait for several days before poison is added. This is necessary when the poison kills swiftly; when the poison is an anticoagulant and death is slow, it is possible to start poison baiting immediately. In environments where there are constantly new strange objects, rodents are less prone to avoidance. Rodents will also avoid areas containing several dead rodents, around poisoned bait, out of a sense of danger.

Rodents have tremendous versatility in climbing, jumping, reaching, and swimming. Mice and roof rats are excellent climbers; Norway rats vary from fair to good. Rodents can climb the vertical walls of most brick buildings as long as they can get a toenail hold. They can also climb or cross-wires easily. Rats can reach as much as 13 in. along smooth vertical walls. They can make a standing high jump of about 2 ft and with a running start, they can jump a little over 3 ft. Mice can high jump more than 2 ft with a running start. Rats can jump outward 8 ft horizontally when jumping or dropping at least 15 ft downward. With a running start, the distance is increased. Rats can swim as much as a half mile in open water. If they are placed in a tank of water, they will repeatedly dive to the bottom to look for exit pipes. This is why they survive so well in sewer lines and use them as a transportation route.

As mentioned, rodents live close to food sources if they can maintain their safety. Rats usually live within 100–150 ft of their food source, while the house mouse lives within 10–30 ft. Deer mice usually live within 1/3 of 4 acres. Rodents will build their nests in any relatively quiet hiding place. Outside, they burrow into the ground. Norway rats burrow 5–6 ft; roof rats will only burrow in the absence of Norway rats; house mice will burrow extremely well if there is no other harborage present. Rodents also nest in piled-up trash, lumber piles, discarded appliances, around stables, under animal houses, in old garages and sheds, and on the banks of streams, creeks, and rivers. Indoors, rodents nest between double walls, under floors, above ceilings, in attics and basements, and in any closed-in spaces behind counters, equipment, or stairwells. Their nests are usually bowl-shaped, and are about 8 in. in diameter for rats and 5 in. in diameter for mice.

Rodent Senses

Rodents have the senses of touch, vision, smell, taste, and balance. In addition to the normal sense of touch, rats have highly sensitive whiskers with a complex nerve net at the base of each. Their bodies are covered by guard hairs, facilitating night travel.

Rodents are apparently color blind. Their sense of vision is underdeveloped, although they are able to detect objects as far away as 45 ft.

Rodents have a keen sense of smell. They are attracted to the entrance of new bait boxes by the urine of other rodents, if the odors are strong enough to travel some distance. Human odor is commonplace, since rodents live with humans constantly.

The sense of taste is not as well developed in rodents as it is in humans. They probably cannot taste poisonous material in baits with the possible exception of red squill, which has a very bitter taste. Bait shyness is developed in rodents because the rodents become sick, rather than because they are able to taste the poison within the bait. In the case of anticoagulants, the poisoning is done so slowly, over a period of 7–10 days, that the rodent does not connect sickness with the anticoagulant.

Rodents have an excellent sense of balance. They usually land on their feet if tossed into the air. Rats and mice can sometimes fall two to three stories without injury. Mice have fallen as much as four stories without injury. This sense of balance enables rodents to cross from one building to another over wires and cables, or cross from a tree into the open window of a house.

Rodent Signs

Signs of rodent habitation apart from the sight of live or dead rodents include sounds, droppings, runways, tracks, rub marks, and burrows. The most frequently seen outdoor rodent sign is a burrow or gnawed material. The most frequently seen indoor rodent sign is rodent droppings. Obviously, the most positive proof of an infestation would be to sight live or dead rodents. As mentioned, rodents travel at night; rodents seen during the day suggest a sizable infestation. A technique used in determining whether rodents are present in a building at night is going into the building and suddenly turning on the lights or flashing a large-beam flashlight. Either the rodents will be seen or the noise of their flight will be heard.

Sounds of rodents are very unique. You can hear them running, gnawing, or scratching within double walls or floors. You may also hear squeaking noises from young rodents in the nest or the noises of adults fighting.

Droppings are a sure sign of the presence of rats and mice. The Norway rat's droppings are $3/_4$ in. long by $1/_4$ in. in diameter. The roof rat's droppings are generally smaller than those of the Norway rat. The house mouse droppings are usually about $1/_8$ in. long. The age of the dropping can be determined by color and texture. If the droppings are fresh they are usually black, soft, glistening, and appear moist. If they are old, they are brittle, graying, and dry (see Figure 5.2).

Since rodents follow the same path frequently, they establish runways, tracks, and rub marks. The runways are easily identified because the area is smooth from the constant movement of the rodents. The area is also relatively free of dust, in contrast to surrounding areas. Tracks are detected by shining a light obliquely with the ground. The tracks are marks in dusty or muddy areas. Dark smears or rub

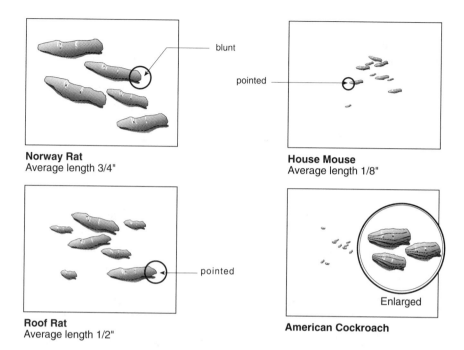

Norway Rat
Average length 3/4"

House Mouse
Average length 1/8"

Roof Rat
Average length 1/2"

American Cockroach

FIGURE 5.2.   Recognizing rat and mouse signs. Source: *Control of Domestic Rats and Mice* (Atlanta, GA: U.S. Department of Health, Education, and Welfate, 1972), p.7.

marks are found along the walls, pipes, or rafters from the grease on the coats of the rodents. Generally the runways are along walls, under boards, behind stored objects, and in, around, or under accumulated solid waste. Tracing the runways, harborage, food and water supplies, and the methods of entry of the rodents into the building facilitates proper control measures.

Rodent Population Characteristics

A rodent population behaves like any other society. It goes through the characteristic stages of birth, growth, maturity, and death. Each population area has the capacity to support rodent populations by the environment. This includes the amount of food, harborage, living space, and population in the area. The population size, affected by reproduction patterns, tends to increase in the spring and fall. Mortality rates counterbalance population increases. Female rats and mice live longer than males; however, accurate mortality data is difficult to obtain.

Population density is also affected by movement. Rats and mice generally spend their lifetimes in a limited area. If within the home area there is a problem of sharply changing temperature or population increase, portions of the population will move. Rodents frequently migrate to additional food sources. This behavior is important in a successful rodent control project, since if the food source or harborage is removed prior to poisoning, the rodents will scatter and sharply increase the rodent population of other areas. The major food source for rodents is garbage, stored food products, crops, and grains. Harborage is a limited problem since rodents are so adaptable. Rodents are predators and will, in the absence of a food source, feed upon each other; they are, of course, the victims of predators as well. Where the population of rodents is high, more are killed by predatory animals. Cats, with the exception of Siamese cats, are poor rodent predators. Although cats are good mouse catchers, they generally play with the mice rather than kill large quantities of them. Dogs, while good rat catchers, are not used for rodent control work since they are often mutilated by the rodents. The presence of domestic animals also provides a source of food for the rodents since owners leave pet food and water out for their pets, providing an excellent fresh food source for the rodents on a continuing basis. Other rodent predators are snakes and birds. Rodents will compete fiercely for food and water, nesting sites, harborage, and females in heat. When Norway and roof rats compete, the fiercer Norway rats generally are successful. Norway rats will compete with each other — many times over females. This competition among members of the same species is associated closely with their social organization. The aggressive animals live closest to the food source, whereas the least aggressive animals live furthest away from the food source and are in most danger of attack by predators.

RODENT PROBLEMS AND THEIR EFFECT ON HUMANS

Rodents cause psychological, sociological, and economical human suffering and a variety of diseases. It is estimated that each rat damages between one and ten dollars worth of food and other materials each year by gnawing or feeding. They contaminate 5–10 times more food and materials than they utilize. Many fires of unknown origin are attributed to rats gnawing through the insulation of electrical wires. They also gnaw through walls, floors, and doors to either obtain food or to grind down their teeth, which grow at a rate of 5 in. a year. Americans lose an estimated 500 million to one billion dollars annually from rodent damage.

It is difficult to properly determine the number of rat bites occurring each year, since many are unreported. Many years ago the author, while involved in a community rodent control program, conducted a survey of several city blocks within the inner city. At least 1500 rat bites had occurred during a year in the target rat control area. It was difficult to get good statistics, since many of the individuals interviewed failed to report the rat bites to the health department or other medical authorities. An estimated 45,000 individuals are bitten by rats each year. Apparently these figures are low. In order to get a proper figure, it would be necessary to conduct extensive surveys of many inner city areas throughout the country.

Rats bite helpless infants and defenseless aged or invalid adults. Rat bites are disfiguring; they may become infected and in rare cases lead to death. The psychological problem associated with rat bites is enormous. The individual is extremely frightened by the occurrence and may bear emotional scars for years.

During the course of one of the rat-control projects participated in by the author, a mother came running out of the door screaming that her child was being bitten by a rat. When the health department extermination crew ran into the building, they found a large Norway rat clinging to a 6- or 7-year-old girl. The rat had to be beaten into unconsciousness before it would let go of the child. Rats that bite once will probably bite again. That is why it is so important to report rat bites and then to exterminate the premises where the bite has occurred.

Sixty million Americans live in dilapidated and dirty inner city sections within the nation's larger cities. To these individuals the rat is the uninvited king that bites, destroys, stalks its prey, makes irritating noises, defecates, urinates, and spreads disease. It is a symbol of urban decay. Empty lots piled 3 and 4 ft high with garbage, trash, and debris or houses with the plaster falling down, the electrical wires showing, the floor boards ripped up, the staircases collapsing, and overpowering odors are home to rats.

Today the sewer systems of our major cities are infested with rats. Sewers provide easy entrance and exit to areas, food, liquid, shelter, and a constant temperature. The food found in sewers includes food wastes, undigested food in fecal material, ground garbage from garbage disposal units, and food that is washed or thrown into sewers. Rats enter sewers by burrowing into the ground, and then through a broken pipe into the sewer, or they swim through storm drains. Combined sewers have a far greater rate of rat infestation than the modern sanitary sewer, since the opportunity for entrance is increased. During the summer months, when the level of liquid decreases within the sewer, the rats may take to the exterior areas and search for food in yards and other areas. Rats generally will not leave the sewer area unless the competition for space and food becomes acute. Then they move to a new section of the sewer or leave the system altogether. Obviously, since rats are constantly in contact with all manner of infectious organisms, the opportunity for spreading disease is multiplied. When rodent problems occur in new neighborhoods, they are frequently traced to broken or improperly connected sewer lines.

Diseases

Rats and mice may be the reservoir of infection for many human diseases, including murine typhus fever (see insect chapter), plague (see insect chapter), rat bite fever, rickettsial pox (see insect chapter), salmonellosis (see food chapter), trichinosis (see food chapter), and Weil's disease.

Rat Bite Fever

Rat bite fever or Haverhill fever is transmitted through the teeth and gums of a rat infected with the organisms *Streptobacillus moniliformis*. This disease, which may then also be spread by milk, is highly infectious and causes acute febrile symptoms. If untreated the fatality rate is 7–10%. The incubation period is 5–15 days, usually 8 days. The reservoir is infected rats.

Weil's Disease

Weil's disease or leptospirosis is caused by the organisms *Leptospira incetrohaemorrahagiae*. The individual becomes infected by direct or indirect contact with infected rodent urine. The organism enters the body through the mucous membranes or through minute cuts or abrasions of the skin. The disease ranges from mild to severe, but is seldom fatal. The incubation period is about 10 days, with a range of 4–19 days. The reservoir of infection is rats or farm and pet animals.

Hantavirus Infection

Hantavirus infection, also known as hantavirus pulmonary syndrome, is caused by the hantavirus, a genus in the family *Bunyaviridae,* which are lipid-enveloped viruses with a negative-stranded RNA genome composed of three unique segments. Humans may be infected when they encounter and inhale aerosolized microscopic particles that contain dried rodent urine, feces, or saliva from infected deer mice. The disease might also be transmitted by rodent bites. It is not transmitted from person to person. The disease is extremely severe with a fatality rate of over 50%. The incubation period is about 1–2 weeks, but ranges from a few days to up to six weeks. The reservoir of infection is the deer mouse.

POTENTIAL FOR INTERVENTION

Isolation, substitution, shielding, treatment, and prevention techniques are all forms of intervention in controlling rat problems. The potential of each technique varies from excellent to poor, depending on the individual's involved, the nature of the housing areas, the concern of citizen's organizations, and government. Isolation is ineffective since it is almost impossible to isolate rodents. Substitution is another unacceptable procedure. Proper rodent-proofing techniques in all areas are used in the shielding technique. Treatment is the application of rodent-specific pesticides. Commercial applicators use the anticoagulants and other chemicals, whereas government programs use red squill, raticate, and anticoagulants. Prevention is by far the most effective technique for reducing rodent problems. Prevention is the removal of all food and harborage, and proper maintenance of property to eliminate rodent entry.

RESOURCES

Scientific and technical or industry resources include, among others, the National Pest Control Association, the area schools of agriculture, the National Environmental Health Association, and the American Public Health Association.

Civic and professional resources consist of the community civic associations responsible for neighborhood beautification, neighborhood housing improvement, and specific rodent control programs. Pest control operators associations and chemical manufacturers associations are important resources.

Governmental resources include local and state health departments, the federal urban rat-control program, and county extension agents. The City of Philadelphia Department of Public Health Division of Environmental Health is an important resource. They prepare a report annually of their activities, entitled Philadelphia Environmental Improvement Program.

STANDARDS, PRACTICES, AND TECHNIQUES

Rodent Poisoning and Trapping

The safest rodent poisons include the anticoagulants, such as warfarin, pival, fumarin, diphacinone, and red squill, (a safe single-action poison). Somewhat less safe, but still effective, is the anticoagulant brodifacoum. Bromethalin is quite toxic to humans and pets, but is very effective. Although many other rodent poisons are used for rat and mouse control, they are too toxic to be used in the home. Some of the more toxic poisons include zinc phosphide and ANTU. Rodent poisoning is discussed in detail in the chapter on pesticides.

Rodent trapping is only fairly successful for rats, since rats drag traps, if necessary chew off a leg caught in a trap, or even cause a trap to malfunction. Mouse trapping is successful if single mouse traps, rather than four-unit traps, are used if the trap is placed perpendicular to the wall and within the runway and if fresh bait is used on a routine basis. The bait should be changed each day or every other day. Mice are attracted by peanut butter, bacon, nuts, and other fresh foods. The bait should always be tied to the trigger so that the mouse will not steal it. Another technique of baiting is to take cornmeal or some other tasty food and make a tiny path up to the trigger and cover the trigger with the food.

Glue boards, which are similar to fly paper but sturdier, have been used but are not very successful. The rodent sticks while crossing the glue board and must then be killed and disposed of.

Food and Harborage Removal

The potential for rodent infestation is directly related to the amount of food and harborage present. Rodents find food in exposed garbage, pet food, or stored food, such as grains, fruits, meats, and vegetables. Rodents also find food sources in gardens, on farms, in horse manure, and any other type of organic material. Although they prefer fresh food, rodents will eat anything.

Most harborage is provided by poorly stored rubbish and trash, lumber, abandoned vehicles, discarded appliances, old shacks and sheds, high weeds, hay, double walls, false ceilings, drawers, attics of houses, and any other area where they may be able to conceal themselves and nest. Frequently, rodents are found in cellars or closets within the home behind piles of dirty, discarded clothing.

An effective rat-control program requires removal of available food supplies and shelter for the rodents. As many rats as possible must first be exterminated to avoid scattering during a comprehensive cleanup program. Reduced food supply decreases the rodent population because of competition for food and harborage. The cleanup technique is extremely important, not only for the reduction and final elimination of most of the rodents, but also for the elimination of flies, roaches, mosquito-breeding areas, and for the reduction of potential disease.

All refuse should be stored in 30-gal metal cans with tight-fitting lids 12 in. off of the ground. Only heavy-duty plastic is acceptable because rodents chew it and also plastic cans tend to tip easily. Heavy duty rubber cans may be used providing the lids screw on tightly. A good refuse container is rust-resistant, watertight, tightly covered, easy to clean and handle, and of heavy-duty construction. Fifty-five gallon drums, such as oil drums, should never be used as a refuse storage container, since the lids do not fit properly on the container and when full they are too heavy to lift.

Remove pet food immediately after mealtime, making certain to dispose of all spilled pet food as well. Pet food has been a vital source of rat problems in numerous instances.

All dry food products should be stacked on pallets 12 in. above the floor and 12 in. from walls. Rotate food at least once every 2 weeks during the summer and once a month during the winter. Clean behind and under dry food on a daily basis. All spilled foods should be removed immediately. It is important to use a bright light or flashlight when sweeping storage areas to locate all spilled food and to determine if rodent signs are present.

Materials such as lumber and pipes should be stored in columns 100 feet from the house and 12 in. above the ground. Since high weeds contribute not only to rodent problems but also to fly and mosquito breeding, it is essential that they be trimmed regularly and destroyed. Cut grass, brush, and dense shrubbery within 100 feet of the home.

Abandoned automobiles are rodent harborages and should be removed and disposed of promptly in keeping with air pollution or solid-waste ordinances.

Rickety old sheds are excellent rodent shelters. Materials usually found in these sheds should be removed and such sheds either refurbished or torn down. Numerous rodent problems are traced to old sheds, which provide excellent harborage for rats.

Solid-Waste Disposal and Storage

Solid waste should be collected at least twice a week during the summer and at least once a week during the winter. All business areas should have solid waste collected on a daily basis. Many of the large metal containers found behind businesses are open invitations for rats to come, eat, and live. If these containers are improperly covered and maintained, food is spilled, and units grow odorous and attract flies, roaches, and rodents. It is essential that these containers are perfectly maintained, that the surrounding areas are policed several times a day, and that the lids are placed firmly on the units. Compactor trucks are best for collection of garbage and other types of solid waste. Compactors cannot be used for the collection of large items, such as refrigerators and lumber. Compactor trucks prevent blowing and spilling of materials, are leakproof, and are easy to load and unload. Solid waste taken to properly operated sanitary landfills sharply reduces potential rodent problems. Properly operated incinerators will destroy solid-waste materials; however, when improperly operated, partially burnt organic material is removed from the incinerator and becomes an excellent rodent attractant.

Garbage may be ground up and sent through the sewers to sewage treatment plants. This technique is fine for the home but provides enormous quantities of fresh garbage for the rodents living in the sewers. Composting can be utilized; however, if the compost heap is improperly maintained, it also becomes a rodent attractant and supplies rodent food. Further information on solid-waste disposal will be found in the chapter on solid and hazardous waste found in Volume II.

Safety Precautions for Hantavirus Infection

Residents of hantavirus affected areas should follow these general household precautions to reduce the risk of being exposed to this most serious disease.

- Eliminate potential insect parasites of rodents.
- Eliminate rodents through poisoning and trapping with sping loaded traps and peanut butter as an attractant.
- Reduce rodent shelter and food sources within 100 ft of the home.
- Before rodent elimination, ventilate buildings thoroughly for at least 30 min.
- Wear rubber or plastic gloves when removing rodents or cleaning.
- Before removing gloves, wash gloved hands in a general household disinfectant and then soap and water (three tablespoons of bleach in one gallon of water can be used). Then wash and disinfect hands thoroughly.

- Leave several baited spring-loaded traps inside of the house at all times.
- Clean up rodent contaminated areas by spraying everything that has come in touch with the rodents with a general purpose household disinfectant, bag the materials and/or dead rodents and bury in a hole 2–3 feet deep or incinerate. Then spray and mop floors, counter tops, and surfaces, clean carpets, launder clothes.

In homes with hantavirus or heavy rodent infestation, wear disposable coveralls, rubber boots or disposable shoe coves, rubber or plastic gloves, protective goggles, half-mask air-purifying (or negative pressure) respirator with a high efficiency particulate air (HEPA) filter or a powered air-purifying respirator with HEPA filters. Decontaminate on site, launder on site, or immerse launderable items in a disenfectant until laundry can be done.

All potentially infective waste material, including respirator filters which cannot be incinerated or buried deeply on site should be double-bagged, marked as infectious waste, and disposed of in accordance with infectious hazardous waste regulations.

Workers in affected areas exposed to rodents hould have baseline serum samples taken, understand symptoms, nature, and severity of the disease, and seek immediate attention if ill.

Campers and hikers in affected areas should: avoid contact with rodents or rodent nests, avoid use of empty cabins or enclosed shelters unless thoroughly cleaned, disinfected, and ventilated; avoid sleeping on the bare ground or near rodent shelters, feces, or burrows; use only chlorinated, iodinated, boiled, or bottled water.

Rodent Proofing

Rodent proofing is the elimination of all holes $\frac{1}{2}$ in. or larger for rats and $\frac{1}{4}$ in. or larger for mice to prevent these rodents from entering an establishment. Rodent proof all doors, windows, gratings, vents, pipe openings, and foundation walls within easy access of rats. Techniques include cuffing and channeling, screening, use of metal guards, steel wool, concrete, and curtain walls. Channeling and cuffing is the application of metal bent at right angles to the bottom and sides of doors. Kickplates are also used on doors. The door must be lowered so that the opening beneath it is less than $\frac{1}{2}$ in. from the floor. Vents and windows must be screened against rodents and flies. Existing fire-control ordinances or rules and regulations should be checked to avoid violation. The screening should be 17-gauge galvanized hardware cloth for protection against rats, and 19-gauge ($4 \times 4\frac{1}{2}$ in. mesh) for protection against mice. Where sheet metal is used, it must be 24-gauge galvanized sheet metal.

Metal guards are protection against rats entering a building by coming across wires and pipes. The metal guard is conical in shape, with the cone facing outward. The openings around pipes or conduits have to be covered with either sheet metal patches or with concrete bricking mortar. The difficulty with the use of concrete is that if it is in an area, especially around radiators, where expansion and contraction takes place, the concrete will expand and contract at a different rate than the wall and may easily crack and crumble, thereby creating a new rat opening. Curtain walls, which are L-shaped walls, are made of concrete. They are poured along the foundation walls to keep rats from burrowing under the foundation. The curtain has to be at least 2 ft below ground level, 1 ft in length, and 4 in. in height. (See Figure 5.3 for these types of rodent proofing.) Place metal roof flashing around the base of wooden, earthen, or adobe dwelling to a height of 12 in. above the ground and 6 in. below the ground. Also, place 3 in. of gravel below the base of homes or under mobile homes to discourage burrowing. Rodent proofing also calls for the elimination of dead spaces, such as double walls, double floors, or enclosed areas of stairways, whenever possible. If elimination is impossible, then these areas should be checked frequently for rodent infestation and measures taken as are needed for control. Once a building has been rodent proofed, adequate rat and mouse control work should take place within the building before it is occupied.

Continued Maintenance

Continued maintenance is a necessary part of any rodent control activity. Once food and harborage are removed, the last rodent is killed, and rodent proofing is complete, many assume the job is finished. It is not, since there are innumerable ways in which rodents may still invade a premises. New pipes may be put into a building. New construction may take place. Doors or windows may be left open. Walls may crumble. Screens may be broken. Garbage and trash may start to accumulate again, and new harborage may be created. It is essential that the individuals responsible, whether they be tenants, landlords, or homeowners, thoroughly survey the premises at least once every 2 weeks to determine if a rodent problem is redeveloping. Flashlights should always be used indoors to check for rodent signs. Continued maintenance is the key to any private or public rodent control program. Without it the conditions will not only reoccur, but the rat problem will be substantially increased, since it takes a period of time before a rat population will find some sense of equilibrium within its environment.

MODES OF SURVEILLANCE AND EVALUATION

Rodent surveys are conducted by professional personnel, technical personnel, members of the community, or businessmen who are trained to look for rat signs. There are four basic types of rodent surveys. The first one is a survey in a single building, business, or industry where rodents are suspected. The surveyor looks

Figure 5.3.   Rodent-proofing devices. Source: *Control of Domestic Rats and Mice* (Atlanta, GA: U.S. Department of Health, Education, and Welfare Public Health Service, Communicable Disease Center, 1964), p. 22.

for previously discussed rodent signs, for food and water sources, for improper solid waste storage and removal, for rodent openings, and for any other potential rodent harborage. This information is noted on a survey form and the exact place where each of the problems occur is marked on a rough map on the back of the form. It is wise to use a system of symbols with explanations to avoid cluttering the map.

The second type of survey is one in which the environmental health practitioner or technician makes a study of an area and determines the types of problems in a given city block. This study includes such information as the address of a building, information concerning the building (i.e., residential, business, vacant lot, etc.), potential rodent food and harborage, active rodent signs, potential rodent entries, and so forth. A map of the area is then drawn. It is also wise, at this time, to make a comprehensive housing survey to determine the rodent problem within houses and the types of housing problems needing correction.

A third type of survey is one in which a professional health educator or environmentalist goes from house to house with a questionnaire to determine from residents the types of rodent problems encountered. Again this information could be put on a checksheet. The information should be indicated as a questionnaire survey, in contrast to a physical survey. Together the two surveys are of considerable value in a rodent-control program.

A fourth type of survey is one in which the environmentalist goes out to a vacant area, farmland, park, or other outdoors area away from homes to determine where rodent conditions exist. A detailed map is drawn of the area to indicate precisely where rodent problems are found and the extent of the problems.

It is wise to utilize a simple complaint form to record rodent or other problems on an individual basis.

RODENT CONTROL

Many of the techniques of rodent control include poisoning, trapping, removal of food sources, removal of harborage, storage of solid waste, alternate methods of disposal of solid waste, and continued maintenance. Additional control measures are used in large communities, single homes, sewers, and outside areas.

Community Rodent Control

Although rodent problems are not unique to any part of the urban environment, those that are of greatest concern exist within the inner city. Slum tenements and poorly kept houses, coupled with debris, litter, and other solid waste, contribute to the large rat populations. Where single-family, three-story, or four-story dwellings once existed, now dwellings have been divided into six- to eight-family apartments. The population density of a given dwelling can rise from 4 or 5 to 50 or 60. Obviously, bathroom accommodations are lacking and there is insufficient space for adequate living. Many of these dwellings have numerous rat entrances, usually bad sanitary conditions, and severe rat infestations. The human population is generally highly transient. There seems to be a lack of cohesiveness between members of the same building.

Even where civic organizations and block councils exist, only a few residents may have participated in electing the officers and boards of directors. At times, individuals may simply appoint themselves as officials. Generally these associations start out to improve their neighborhood by keeping taprooms out, obtaining stoplights or stop signs, improving recreational facilities and housing conditions, and possibly by getting involved in some health matters. The typical single-block organization consists of a group of people who live near to each other on a given street. These individual block organizations range from totally inactive to highly active. Usually within each block one finds one or more civic-minded people who are willing to work together on a project to improve the neighborhood. In many cases, the larger civic association does not truly represent the wishes of the individual block groups, since the more vocal and better-educated individuals tend to come to civic association meetings, whereas the less vocal tend to stay home and be concerned with individual problems. Unfortunately, if you only use these organizations, rat-control projects will be unsuccessful, since the people who need the most help are generally those who do not attend the civic association, home and school association, or health and welfare meetings. To stimulate this group of individuals to work together, it is important to go into each individual block and work on a house-to-house basis to attempt to set up small meetings in the homes. A good approach is to knock on doors, introduce yourself officially as a member of the health department (make sure you always show your identification card), and explain the severity of the rat problems in the area, as well as in other areas. The environmentalist should also give some indication as to the nature of the programs that can be brought into the area to help improve it. These individuals within the blocks are then encouraged to come to a mass meeting at a local school to further discuss the rodent control problems and the kind of things that might be done to improve the environment. Each individual should be encouraged to bring friends, neighbors, and relatives to the meeting. Announcements should be distributed the day prior to the meeting and on the day of the meeting. Sound equipment should be used on a truck to announce the community meeting for rodent control at the local school. Explanations of the rodent problem given over the loudspeaker should be simple. The meeting place and time should be spelled out clearly.

It is essential at this mass meeting to have representatives from all the civic associations and the various official agencies to discuss informally the needs of the community as they see them, to listen to citizen problems, and then to explain some proposed techniques for conducting a community rodent-control program. It is important that the individuals at this meeting vote to participate in each of the phases of the rat program. During such a meeting each of the citizen complaints should be recorded and submitted to the appropriate official agency for correction, where possible. In one such meeting in which the author was involved, the community's major concern was the fact that the young people were constantly getting into trouble with the police. There was an apparent language barrier. When this problem was identified, the police department sent community relations people who spoke the language of the community into the area and worked with them to improve community-police relations. This helped not only to satisfy a need of the community, but it also indicated that the official agencies were interested in the public and that it would be a good idea to work together on community rodent control.

Another technique used to organize blocks is to pull together members of the community by the use of sound equipment at the time that rodent poison is placed by the health department. Here the members of any given block have an opportunity to meet face to face. This is an important time to hold an election of officers for that given block. Once officers are elected, the health educator in conjunction with the larger community organization should assist the block in developing its own block organization.

Another technique is to ask individuals, when you are going door to door prior to the first mass meeting, about the individual who they turn to when they have problems. Frequently, key people in a block will be identified. These individuals are natural leaders and could be utilized as such in a block organization. There are innumerable block organization techniques. The key is not necessarily which technique to use, but rather to help organize a block, stimulate the people to work together, and try to maintain the organization so that a ready supply of willing workforce exists for continued maintenance to keep the program from falling apart. These block organizations may be used later in helping bring satellite health clinics into an area to help increase immunizations, and so forth.

An important person in a block organization can be the health aide. This individual is a member of the community hired by the health department who receives specific health training. As a member of the community and therefore recognized by the community, and as a member of the health team, this individual can greatly assist in developing a comprehensive rodent-control program, as well as other comprehensive health programs.

Presurveys

Three types of presurveys may be used. First is the horseback type of survey, where the environmentalist rides up and down alleyways or walks the city blocks to get some approximate idea of the kinds of existing housing conditions, the amount and type of solid waste, and the kind of rat signs readily visible. This survey gives a rough idea of the needs of the community.

The second kind of survey, the opinion attitude survey, should contain several basic points. (1) What is the extent of the community organization, including the size and scope of the group, potential leaders, and desire to organize? (2) What are the needs and attitudes of the people toward elimination of roaches, rats, and poor housing? (3) What is the extent of the pest infestation, including mice, rats, and roaches? (4) What action has the citizen taken against roaches, mice, and rats? (5) How many complaints has the individual made about pests, and to whom have they been made? (6) How many rat bites have occurred in the person's house in the last year? How many have been reported to the health department? How many have been reported to others? (7) Has the individual heard about rat bites in the neighborhood within the last year? If so how many?

The information in this opinion attitude survey, when compiled, gives the environmentalist some understanding of the community's self-concept and whether the individuals are anxious to help correct the rodent situation.

The technical survey consist of a professional evaluation of rodent signs inside and outside of the property, of potential rodent entrances, of the availability of food and harborage, and of the rat-proofing problem. The data should be compiled and combined with the information in the opinion attitude survey to make a good evaluation of the existing rodent problems within the community.

The survey data should be kept and used as a means of comparison after completion of the program. This will give the environmentalist a means of comparing the results of the actions taken and also the results of continued maintenance.

In addition, it is well to add the material from the files of the health department relating to rodent complaints, solid-waste complaints, and rat bites that were reported to the health department within the areas under study. It is recommended that a comprehensive map with proper coding be drawn to indicate the degree of severity of rodent infestations and also the other environmental problems associated with rodent infestation.

Area Selection

The area selected for a community rodent-control program should be based on the extent of community organization, the willingness of groups to form and work together, the number of complaints for a 3-year period relating to rats and solid waste, the number of rat bites for the past 7 years, and the various surveys that have been conducted. These proposed areas should be stated clearly on a map and in writing. The number of houses involved, the severity of infestation, and the extent of cleanup should then be presented to a committee composed of regulatory agencies and civic leaders. The ultimate decision of where to work should be made jointly by these representatives of the communities and the official agencies.

Planning

In order to plan an effective community rodent-control program, it is necessary to establish a working committee structure. The steering committee can be composed of representatives of the health department, housing department, police sanitation unit, solid-waste department, and community relations department. The functions of a steering committee are to plan details of operation, resolve day-to-day problems, supply general knowledge of existing community resources, determine on a departmental basis the availability of workforce, act as liaisons between the project and the individual official departments or agencies, and help evaluate and redirect the program when needed.

A second committee, called a citizens advisory committee, should be established and can be composed of representatives of the health and welfare council, civic associations, block councils, block leaders, unofficial agencies, a local religious leader, a local school principal, a representative of the landlords, a business person, and others as needed. The functions of this committee are to review the suggested programs developed jointly by official agencies and citizens and make recommendations to the health director or environmental health director who is the coordinator of the rodent-control program regarding changes or additions. These recommendations may include selecting certain city blocks for intensive rat-control activities, techniques to assist in stimulating community interest, and practical methods of carrying out effective community rodent-control programs. The steering committee and citizens advisory committee should define the problem clearly in writing and establish goals, as well as develop the program and help in its implementation. The written definition of the problem is based on the previously mentioned surveys. The goals should be of two types: (1) long range, to establish and maintain dwelling units, yard, building, and other areas that are free of rats; to help provide a community with a spirit of self-help and pride, leading to neighborhood betterment; to improve the total dwelling unit and neighborhood; to stimulate interest to undertake similar programs in other infested areas. (2) immediate goals, to remove garbage, trash and other solid waste from yards, homes, buildings; to remove rodent harborage; to poison and otherwise destroy the rats in the neighborhood; to remove abandoned automobiles; to cut down weeds in vacant areas; to rodent-proof homes; to train community leaders in the various phases of rat control; and to encourage block organizations to promote successful programs and continued maintenance.

The Program

The community program consists of a series of meetings and actual rat control activities. The first meetings, which are for stimulation and interest, have already been discussed. The major meeting called at a school or other similar facility has also been discussed. It might be added that, during the course of this major meeting, the responsibilities of government and each of its branches and the responsibility of each of the groups of tenants, landlords, businesses, and industry should be explained. The next meeting called should be a training meeting. Personal safety, safety of the child and family, removal of dead rodents, and the covering up of rodent odors are taught at this time. It is important to note that only certified environmentalists or pest-control operators can place rodent poisons. The training meeting is also used to present a total picture of how to rid an area of rats. Therefore, further discussions, demonstrations, and visual aids in the removal and storage of solid-waste material, in rodent control through rodent proofing, and in the necessity for constantly policing potential rodent problems and continually maintaining a rodent-free environment are conducted. At times, the general meeting may be combined with the training meeting. The individuals within the project will decide which is better for their community.

During a discussion of cleanup techniques, it is important to mention the team approach, the use of teenagers, and the use of volunteers. Everyone can help, from a small child to an aged person. All that is necessary is to remove the solid waste to the curb in the best manner possible so that the solid-waste agency workers can pick up the material and remove it. Additional meetings are held as needed during the course of the project to explain various aspects of the project. It is important to establish at the first meeting a timetable for the project and then to reinforce the dates of the timetable at each meeting.

In addition to meetings, other techniques used to stimulate community help are handouts, and especially, mobile sound equipment, since the spoken message, especially in the early evening, is received by all, including the working people within the family, who are then given the opportunity to participate in the rodent control program.

The second phase of the program consists of the actual implementation of activities. Implementation of activities are as follows. (1) First poisoning — red squill used preferably and applied by the health department. (2) Second clean up. The latter should be carried out by the community and any volunteers such as school children and Y-teen groups. All solid waste, including junk, garbage, refuse, and debris, have to be removed from basements, cellars, and vacant lots. The community supplies the workforce to bring this material to the curb. The city supplies the necessary personnel to pick up the material and truck it to the landfills, recycling center, or incinerators. (3) Additional poisonings, using anticoagulants. (4) Total cleanup carried out by the community on a special cleanup day. (5) The city should go into a direct enforcement program to get the individuals who do not comply with the wishes of the community to remove all their solid waste and destroy all their rats. Inspections should be made of all of the areas. Owners of houses in violation should be cited and if necessary brought to court. Vacant houses should be cleaned and sealed, and the cost should be assessed on the owners. Anti-litter, -garbage, and -refuse storage ordinances should be enforced strictly. All vacant lots and alleyways should be cleaned by the community, and the waste should be removed by the solid-waste agency. (6) The program should be evaluated by the environmentalist by making careful postsurveys of all areas. A comparison should be made between the presurvey and the postsurvey to determine the effectiveness of the program. (7) The community, with the assistance of the environmentalist, should now stimulate their block organizations to work on continued maintenance and rodent proofing, where rodent proofing is possible. Each individual should attempt to rodent proof his or her own structure.

The schedule for the program is generally as follows: first week—first poisoning; second week — first cleanup; third week — second poisoning; fourth week — second cleanup; fifth week — additional poisonings; sixth week — survey; seventh week forward — continued maintenance.

The area-baiting concept is extremely important since rat pressures are reduced initially and during cleanup, rats will not scatter. Eventually, as more and more solid waste is removed, the rats will die from lack of food or will be poisoned by the additional rat poison available to them.

Individual Rodent Control

When an individual is faced with rodent problems, the premises should be surveyed to determine where the rats are coming from and appropriate measures of poisoning, cleanup, and rodent proofing should be taken. It is important that the individual contact the health department and inform them of the problem. In this way, the health department can determine if this is more than a local situation. An individual should never use rodent poisons other than anticoagulants. If the situation is beyond the individual's control, a trained exterminator should be called to carry out the necessary rodent-control functions.

In businesses, trained exterminators should be utilized for rodent control. However, owners should be alert to the use of highly toxic substances. Rat problems are the responsibility of the individual business owner.

Sewers

Rodents can be controlled in sewers by use of a poison, usually red squill, raticate, or an anticoagulant in paraffin blocks placed along the edges of the sewers. The rats consume the poison and usually die in the sewer.

Outside Areas

It is important to use adequate rodent poisoning. If the area is completely away from homes, carbon monoxide or chloropicrin can be pumped into the burrows to destroy the rats. Other rodenticides, such as red squill and anticoagulants, may be used.

Government Programs

Local and state governments have been involved in rodent control through the enforcement of nuisance laws, food laws, and other types of public health laws. The federal government became involved in rat control in 1967, when Congress authorized 40 million dollars for a 2-year period for rat prevention. In the ensuing years, additional millions of dollars were allocated to the Department of Health and Human Services for rat-control projects in various communities. Since 1969, the Partnership for Health Amendment (PL 90-174) has initiated new rodent-control programs and strengthened existing ones in cities with serious rat problems. Since 1969, programs have been initiated in New York, Atlanta, Baltimore, Charlotte, Chicago, Cleveland, Milwaukee, Nashville, Norfolk, Philadelphia, Pittsburgh, St. Louis, and Washington, DC These rodent-control programs generally consist of three phases: (1) preparatory phase, during which the concept was sold to community officials and mass media on how to control rats; (2) the attack phase, consisting of surveys, education, cleanup, poisoning, and code enforcement; and (3) follow-up phase, which included continued maintenance. As part of these federal efforts, plans were drawn up for comprehensive rodent-control programs. These included citizen participation, community information, and education; effective local administration; adequate municipal service, including garbage, trash, and junk pickup; enforcement of codes and ordinances; removal of dilapidated buildings; systematic poisoning; training and employment of local residents in rodent-control efforts.

In 1979, the federally funded rat-control projects were once again renewed. In the 1980s and 1990s, rodent control became solely a state and local effort. Philadelphia and New Orleans used some funds from Preventive Health Block Grants for rodent control. Ultimately local citizens working with local government must solve the problem.

SUMMARY

Rodents live close to humans and can readily cause disease or other health-related problems. Although at present the level of disease caused by rodents is low, the potential is always there. Rodents not only destroy property and cause fires and disease, they also bite helpless individuals and are the symbol of urban decay. Rodent control can be carried out if a given community is willing to work together to remove all the food and harborage, apply the necessary rodent poisons, and carry out proper rodent proofing. Rodents will continue to be a serious problem in our society because of a lack of adequate control programs.

RESEARCH NEEDS

It is necessary to develop a better understanding of the resistance that rats are starting to acquire toward anticoagulants. The mechanism of resistance must be understood in order to prevent the elimination of this relatively safe type of rodent poison. New rat-control methods are needed. Research should be conducted for the development of single-dose chemosterilants and repellents. It is also necessary to determine how to motivate citizens to keep their homes and neighborhoods rat free.

Chapter 6

PESTICIDES

BACKGROUND AND STATUS

Pests cause a reduction in size, yield, storage, and market quality of crops and food and serve as vectors in the spread of diseases. To control pests, a series of chemicals, called pesticides, have been developed. Pesticides include acaricides or miticides, used against mites; algaecides, used against algae; attractants, used to attract insects, birds, and other animals; chemosterilants, used to interfere with reproduction; defoliants, used to remove leaves from plants prior to harvest or to eliminate unwanted plants; fungicides, used against fungi; herbicides, used against weeds; insecticides, used against insects; molluscides, used against slugs and snails; ovicides, used against insect eggs; repellents, used to drive animals or insects away; rodenticides, used against rats, mice, and other rodents.

Pesticides are biologically active chemicals that kill or modify the behavior of problem insects, animals, microorganisms, weeds, and other pests. Pesticides are used as aerosols, sprays, dust, in granular form, or as baits. They may be effective on contact, be taken up by the plant, enter the lungs or trachea of animals, or be eaten. The quantity of the pesticide used, the type of pesticide, and how it is used are very important in the control of troublesome pests.

Pesticides produce useful and harmful effects, depending on the type and quantity used and the method of application. About half of all pesticides are used in farming. Roughly 5% are used by governmental agencies and the balance by residential and industrial users. Currently, we are using approximately one billion pounds of pesticides annually.

Pesticides are categorized by their lifetime of effectiveness as follows: nonpersistent, lasting several days to about 12 weeks; moderately persistent, lasting

1–18 months; persistent, which include most of the chlorinated hydrocarbons such as DDT, aldrin, and dieldrin, lasting many months to 20 years; permanent, including mercury, lead, and arsenic, lasting indefinitely. PCBs or polychlorinated biphenyls used in asphalt, ink, and paper behave very much like the persistent pesticides and require close control to avoid contamination of the environment.

Pesticides that degrade or deteriorate rapidly are also of great concern because of their extreme toxicity and because they are nonselective in their action on animals, humans, and pests. Organophosphates would be an example of this type of pesticide.

A pesticide moves through an ecosystem in numerous ways. It is introduced by surface application, spraying, or other techniques and may stay in the air or be washed down by rain. The concentrations of the pesticide continue to increase in the soil over time, and where leaching occurs, the pesticide can move into surface or underground water supplies. Some pesticides become tightly bound to soil particles, polluting the surface waters when the surface particles are washed into them by the force of heavy rains. Some pesticides are ingested by minute, aquatic organisms and scavengers and become concentrated as they move up through the food chain. It is known that oysters, for instance, will concentrate DDT in their tissues 70,000 times greater than amounts found in the surrounding waters. Fish also concentrate pesticides as part of the food chain. Eventually, the pesticides may reach humans and, at least in the case of DDT, are stored in the fatty tissues.

The major pesticide laws in effect in the U.S. were totally rewritten in 1972 and updated in 1978 and in 1988 as amendments to the 1947 Federal Insecticide, Fungicide, and Rodenticide Act (FIFRA). This law forbids anyone, including the federal government, from using a pesticide contrary to label instructions and gives the Environmental Protection Agency (EPA) the authority to restrict the use of pesticides to trained persons. The law applies to interstate and intrastate use and sale of the product. It provides screening procedures for new pesticides suspected of causing cancer, birth defects, or mutations. Based on this law, the EPA has taken action against the use of kepone, DDT, aldrin, dieldrin, heptachlor, chlordane, mirex, and mercury-based pesticides. At present, the only exception to these actions is made by the EPA if the agency feels that the benefits outweigh the potential adverse effects and no alternatives are available, if a significant health problem occurs without its use, or if an emergency exists.

Under present EPA orders, all individuals, including public health workers, who are involved in the use of pesticides, must take comprehensive examinations and be registered in the use of pesticides by category of employment. Since the use of pesticides is increasing, the dangers will increase and the current status in succeeding years will deteriorate unless further action is taken.

Current Issues

The first pesticide act was enacted in 1947. When the U.S. EPA was founded in 1970, the FIFRA authority was transferred from the U.S. Department of

Agriculture to the Environmental Protection Agency. In 1972 the Federal Environmental Pesticide Control Act passed by Congress was an amendment to the original Pesticide Act mentioned earlier. It provided for direct controls on the use of pesticides, for classification of certain pesticides into a restricted category, for registration of the manufacturing plants, and for a national monitoring program for pesticide residues. Environmental effects and risks were added to the pesticide registration process.

The 1972 FIFRA amendments required a review of all of the registered products being used. The review was to be completed by 1975. Unfortunately because of the large amount of data that was needed and the large number of products to be assessed, the General Accounting Office (GAO) determined it would take until the year 2024 for this work to be done. As a result of this, the entire reregistration process simply broke down.

From the very beginning, the EPA had problems with the pesticide regulation program and with the process of implementing the 1972 amendments to FIFRA. However, this act still did not resolve the many problems related to pesticide regulation.

The issues were registration, tolerances (which are standards) for pesticide residues, federal preemption of state tolerances, reregistration, inert ingredients of pesticide formulations, regulatory options, and pesticides in groundwater.

Under FIFRA the regulation of pesticides was done through the registration of the individual pesticide products. The products were not permitted to present unreasonable adverse effects to people or the environment if the pesticide was used on a food crop or animal feed. The regulations also required that a maximum acceptable level of pesticide residues remaining on a treated crop be determined by the EPA, and be monitored and enforced by the Food and Drug Administration. The data required for registration of pesticide products included health and environmental data, environmental fate, carcinogenicity, chemistry of the product, toxicity to fishlife, and mutagenicity.

The EPA established tolerances for pesticide residues in foods. Tolerances determined the maximum amount of pesticide residue that could be permitted in food or animal feed without being considered to be an adulteration of the product. Tolerance-setting procedures to protect human health include: the anticipated amount of pesticide residues found on food; the toxic effects of these residues; estimates of the types and amounts of food that make up our diet; field trials of pesticide use and residues; toxicity studies; product chemistry data; and plant and animal metabolism studies including metabolites.

The EPA recognizes that the diet of infants and children may differ substantially from those of adults and that they may be exposed to pesticied residues in food at levels higher than adults receive. Using a computerized data base known as the Dietary Risk Evaluation System (DRES), EPA combines survey information on food consumption and data on the pesticide residues to estimate dietary exposure. DRES breaks out a number of subgroups in the population, including infants, children and other age groups, several different ethnic groups, and regional populations. EPA appropriately identifies childhood or infant exposures for special consideration when looking at the risks for discrete periods of time. EPA also calculates a cumulative lifetime exposure that integrates the exposure rates experienced in infancy and childhood and takes the exposures experienced in adulthood. If risks are at an unacceptable level, then the EPA takes action to reduce those risks.

The federal government, under the 1988 law, preempted the tolerances established by state law. The problem was that the EPA standards were currently considered to be a floor and not a ceiling for standards for any given pesticide. Therefore, the states could have more stringent standards. However, if this occurred, it could have interfered or hindered the flow of products through interstate commerce and hindered the marketing of pesticide products, pest treatment services, and the treated commodities. The states argued that the data that had been used to determine the allowable amount of pesticides in the food or feed product may not have been accurate, and over a long term it was the right of the states to protect their citizens.

The FIFRA amendments authorized the EPA to conduct a "generic" review of the active ingredient's safety. The EPA had identified some 600 active ingredients considered to be commercially important among the 1500 active ingredients officially registered with them. There were an estimated 40,000 studies about these pesticides in the EPA's files. Only two pesticide active ingredients had been reregistered under this process until this point. Also under the 1988 law, all existing pesticides must be reviewed and reregistered, with industry being required to provide the test data for the review. By 1993, the EPA has issued 31 reregistration documents.

The EPA also investigates inert ingredients. An *inert* ingredient is that part of the pesticide formulation that is not intended to have any pesticidal activity. It is used either to dilute the pesticide or to propel it or deliver it in some manner. Unfortunately, some of the *inert* ingredients have potentially adverse effects on people. Vinyl chloride gas is a human carcinogen that has been used as an aerosol propellent. Of the approximately 1200 compounds used as *inert* ingredients in pesticides, the EPA has determined that 55 are known toxics that may cause animal cancer or nerve damage. Fifty-one compounds are structurally related to the known toxic compounds. Nine hundred compounds are of unknown toxicity.

The EPA can revoke the registration for a pesticide if the information indicates that the product presents an unreasonable risk to human health or the environment. This process is called *deregistration*. If the chemical is cancelled, the stockholders, owners, and other individuals can demand payment of the EPA for their losses. In the last 23 years the EPA has cancelled the registration of 36 potentially hazardous pesticides and has eliminated the use of 60 toxic inert ingredients in pesticide products.

Pesticides and Groundwater

Pesticides can reach the groundwater supply through misuse or mismanagement related to waste disposal, spills, leaching, etc. There are four main issues related to pesticides in groundwater. They are

1. Pesticides need to be detected in groundwater.
2. The EPA needs to determine at what level pesticide residues in groundwater should trigger action.
3. If a pesticide is detected in groundwater as a result of normal use and if the groundwater pesticide limits will be exceeded, the EPA needs to decide what is an appropriate remedy.
4. The EPA needs to decide if a pesticide should be immediately prohibited if it reaches substantial levels.

The National Pesticide Survey was conducted by the EPA in all 50 states between 1988–1990. Preliminary results indicate that 10% of the nation's community drinking water wells and about 4% of rural domestic wells have detectable residues of at least one pesticide. One percent of community wells and 0.8% of rural wells have one or more pesticides exceeding health advisory or maximum containment levels. More than half of the nation's wells contain nitrates, but fewer than 3% have concentrations of health concern.

Pesticides and Groundwater Strategy

"Pesticides and Groundwater Strategy" was released by the EPA in 1991. It includes the following six federal policies:

1. Encourage, where appropriate, less burdensome environmental agricultrual practices concerning use of pesticides and fertilizers.
2. Determine appropriate regulatory approach to cancel chemicals that may threaten groundwater.
3. Confine legal sale and use of cancelled pesticides to states that have EPA approved state plan.
4. Provide research and technical assistance supported by EPA to Department of Agriculture and U.S. Geological Survey to assess groundwater problems and vulnerable groundwater systems.
5. Provide clear instructions for the field use of pesticides by improving training and certification programs.
6. Promote and encourage companies to conduct more monitoring studies, develop safer pesticides, and prevent degradation of groundwater.

Risk/Benefit Balancing Under FIFRA

There are four steps in the EPA's risk assessment process: hazard identification, dose/response assessment, exposure assessment, and risk characterization. There are two steps in benefit assessment: biological analysis and economic analysis.

In hazard identification, the EPA evaluates a pesticide's inherent toxicity, that is, the types and degrees of harmful effects a pesticide may cause by determining the effects on animals, especially cancer, in laboratory studies. In dose/response assessment, laboratory animals are exposed to various doses of the chemical during various time periods. The acute or chronic effects are determined and a NOEL (No Observed Effect Level) is established where non-cancer effects were found. The NOEL is divided by an uncertainty factor of 100 or more to determine the Reference Dose (RFD). At or below this level it is assumed that a lifetime exposure would not cause harmful effects. A "negligible risk" standard is used for cancer, where an individual's chance of cancer is one in one million if exposed for a lifetime. Exposure assessment refers to the level, duration, and route of exposure of people to chemicals identified in laboratory tests as causing harmful effects. Risk characterization is the estimate of the risk from exposure to pesticides to people by integrating the above factors and extrapolating exposure in animals to humans.

In benefit assessment, the EPA decides whether to cancel or approve a new pesticide by determining the effectiveness and economic value of a pesticide compared to alternative chemical and nonchemical controls.

Other Issues

Additional issues in the use of pesticides include food contamination, air pollution, potential indoor air pollution, preparation of professional pesticide applicators, and pesticides related to fish and wildlife. Pesticides in foods or on foods have become a major issue today. The scare related to the purposeful contamination of two grapes in 1990 brought an entire food supply to a total halt and created untold problems for the country providing the food, the agencies evaluating the food, and the public, who in some way end up paying for the food through taxes or through the disposal of the rotted fruit. There are some questions concerning the amount of pesticides that are introduced into the food chain and the concentration of these pesticides in the finished product because of the use of raw products that have been exposed to the chemicals. The consumer must be protected, but the food supply must not be destroyed.

Pesticide application can cause an air pollution problem. Air currents may carry pesticides to the wrong area or the chemical may be dumped on the wrong area and potentially affect people and animals.

A variety of pesticides are used on lawns and on pets. The individuals applying these pesticides are untrained in the storage, mixing, application, and disposal of the pesticides. Pesticides may be stored in garages or other enclosed areas, and therefore, constitute a potential indoor air pollution problem.

Professional pesticide applicators need to be certified and recertified by their state agencies. Many states have more stringent standards than the minimum standards established by the U.S. EPA. It is therefore necessary to determine the standards of a given state and then decide how best to train the applicators and test them.

Pesticides may have a long-range effect on fish and wildlife, depending on the type of accumulation that may occur. It has been known for years that DDT causes the thinning of egg shells, which in turn prevents the successful hatching of the chick in a variety of birds. Even a pesticide that is as valuable as diazinon has been found to be harmful to fish and wildlife and, therefore, can no longer be used on golf courses and sod farms.

SCIENTIFIC, TECHNOLOGICAL, AND GENERAL INFORMATION

Types of Pesticides of Public Health Importance

The chapter will basically be concerned with the types of pesticides used to control insects, ticks, mites, spiders, rodents, and plants of public health significance. Pesticides should have certain qualities to be acceptable for use. They should be specifically toxic to harmful insects, etc., harmless to humans, inexpensive and easily used, rapidly degradable to nontoxic substances, nonflammable, noncorrosive, nonexplosive, and nonstaining.

Insecticides may be used as stomach poisons, contact poisons that penetrate the body wall, fumigants that enter through the insect's breathing pores, desiccants that scratch or break the body wall or absorb into the waxy protective outer coating. Pesticides may also be listed as larvicides that kill larvae, ovicides that kill the insect's eggs, or adulticides that kill the adults.

Inorganic Insecticides and Petroleum Compounds

Prior to 1945, numerous inorganic pesticides were used widely. These included the arsenical, Paris Green, used against the potato beetle; hydrogen cyanide, used against red scale; lead arsenate, used against the gypsy moth; and sodium arsenite, an insecticide, and a weed killer. All arsenicals are now banned. Compounds of copper, zinc, and chromium were also used as pesticides. Chlorine and sulfur made extremely toxic compounds and were used along with salts of arsenic, lead, mercury, and selenium. Unfortunately, many of these compounds were quite toxic to humans. Also, some of the insects developed resistance to certain inorganic pesticides.

Petroleum oils, such as kerosene, diesel oil, and #2 fuel oil, were used as mosquito larvicides. These oils, which are still in use, have certain toxic properties, because they penetrate the tracheae of larvae and pupa of mosquitoes and anesthetize them. A fraction of these oils mechanically interfere with the breathing process of insects, causing suffocation. Sulfur acts as a repellent against chiggers. Borax is still used in buildings for roach control with varying results.

Botanicals

Probably the earliest pesticides used were pyrethrum, which is extracted from the flowers of *Chrysanthemum cinerarifolium*, rotenone, which is derived from Peruvian cuve, and red squill, which is derived from the inner bulb of the plant *Urginea maritima* belonging to the lily family. These chemicals are highly specific to the pests and have a very low toxicity to humans. Today pyrethrum is used primarily in combination with other insecticides. However, permethrin synergized by piperonyl butoxide seems to be effective against mosquitoes which are organophosphate resistant. Further, sumithrin is effective against mosquitoes and mites. It is especially used to disinsect aircraft coming from foreign countries prior to landing. Methoprene is directly ovicidal to cat fleas and it also sterilizes the adult. Pyrethrum continues to be a quick insect knockdown agent. Synthetic compounds similar to pyrethrum, called allethrin, resmethrin, sumithrin, and permethrin, have been developed and are utilized in the same manner as natural pyrethrum. Rotenone is used to kill fish without leaving toxic byproducts for human beings. Natives of some tropical countries crush and throw plants such as Derris and cuve into the water, and the chemicals present in the plants paralyze the fish. Rotenone is also utilized for killing of fleas and other ectoparasites on domestic pets. Red squill in its fortified state is used effectively in killing Norway rats. Because of the difficulty of obtaining these pesticides from abroad during World War II, and the military need for chemicals to kill disease-producing insects, the U.S. developed a series of organic pesticides in the early 1940s. In addition, such insecticides as DDT were recognized in Switzerland in 1939. Benzene hexachloride was recognized as an insecticide in 1940 in France and England.

Chlorinated Hydrocarbons

The chlorinated hydrocarbon insecticides are combinations of chlorine, hydrogen, and carbon, They act primarily as central nervous system poisons. The insect goes through a series of convulsions and finally dies. The first major chlorinated hydrocarbon was dichlorodiphenyl trichloroethane (DDT). DDT has been highly useful for the control of mosquitoes, flies, fleas, lice, ticks, and mites, reducing considerably the level of malaria, plague, typhus fever, yellow fever, encephalitis, and so forth.

Although DDT has been banned for general use in the U.S., with the exception of a serious uncontrollable emergency, it is still an effective chemical for the control of mosquitoes, which may cause malaria or other diseases. DDT enters the ecosystem and is stored in animal fat. However, in public health a decision must be frequently made as to the relative importance of one hazard vs another. Therefore, DDT is still used abroad in the interior of homes in areas where malaria is prevalent. The DDT, if applied carefully, will leave a residue on the structures that will last from 6 to 12 months. It also will not readily escape into the environment. The dosage in the residual spray should be 100–200 mg/ft^2.

Methoxychlor and DDD are part of the DDT group. Methoxychlor is safer than DDT, since it is less toxic to mammals. It is utilized in many household sprays and aerosols. It is readily metabolized and eliminated in the urine of vertebrates. It also does not remain within the environment for more than a short period of time. It is used to kill mosquito larvae and control flies and insects that attack livestock; it is also used in agricultural areas. As a larvicide, methoxychlor is applied at a rate of 0.05–0.20 lb/acre. DDD is now banned.

Benzene hexachloride (BHC) was widely used in public health and in agricultural programs. It has a musty odor and a short residual life. The gamma isomer of BHC has significant insecticidal activity. BHC is used currently abroad as a residual spray at a dosage rate of 25 or 50 mg/ft^2 for the control of mosquitoes causing malaria or other diseases. It will have a residual effect for about 3 months.

Lindane is the pure gamma isomer of BHC. It is highly effective as an ingested poison and as a residual contact insecticide. It has been used for control of lice, ticks, and other insects. Lindane vaporizers, used to control flies in food establishments, are dangerous and should never be used. In Southeast Asia, lindane was used in irrigation waters to control rice stem-borers. Unfortunately, the chemical killed the fish that were the protein sources for the local population. In overseas areas where the vector of plague, the flea *Xenopsylla cheopis,* still exists and where DDT is not effective, it is recommended that a 1% lindane solution be applied to ensure adequate control of this rodent flea. Lindane may also be used by professional personnel for the control of fleas on household pets, provided the pets are not under 2 months of age. Lindane in a 1% emulsion may be used for treating infested household sites within the house and in the yard. One percent lindane powders are used to control body lice. The brown dog tick can be controlled with a 0.5% lindane spray, with spot treatments on baseboards, floors, wall crevices, and areas where the animal sleeps. Lindane can still be used for dogs. It is banned for any type of fumigation.

Chlordane is dissolvable in many solvents but not in water. It is used to produce oil solutions, emulsifiable concentrates, wettable powders, and dusts. It acts as a stomach poison, contact insecticide, and fumigant. It is effective in spot control of ants, American roaches, silver fish, and has also been used extensively on soil insects, particularly termites. Chlordane has been extremely effective over the last several years, although German roaches have built up a resistance to it. It is probably the least toxic of the chlordane series, which includes chlordane, heptachlor, aldrin, dieldrin, endrin, isodrin, and toxaphene. In July 1975, the EPA suspended the use of chlordane because it is suspected of causing cancer in animals and it readily contaminates the environment.

Heptachlor has been used effectively for mosquito larvicide control. Heptachlor production was also suspended by the EPA as a suspected link to cancer in animals and because it is highly toxic to humans. It remains for long periods of time in the environment and readily contaminates the water, soil, and air. Aldrin and dieldrin, considered to be effective chemicals for insect control, were found to be highly toxic when misused, killing fish, birds, mammals, and even human beings, and is also suspected of carcinogenic activity; these too have been suspended from use since 1974. However, dieldrin still is used overseas at a rate of 25 or 50 mg/ft^2 as a residual spray in those areas where malaria or other mosquito-borne diseases are prevalent.

Endrin is one of the most poisonous of the chlorinated hydrocarbons. It is highly toxic, persists in the environment, and is a hazard to animals and humans. It should be used only when necessary and under strict supervision. It is not recommended for general use. It is now banned for general use.

Isodrin and toxaphene are both highly toxic chemicals, persisting for long periods of time in the environment. They are not recommended for use as insecticides. They are now banned for general use.

Chlordecone, better known as kepone, is a very effective chlorinated hydrocarbon when used as insect bait. It may last as long as 1 year without being altered. When used in proper dosages, that is, 0.125% peanut butter bait, it is effective against both roaches and ants, and produces a high kill. However, kepone, when used improperly or produced improperly, can be very toxic to humans. Therefore, the production of the chemical should be closely monitored, the disposal of waste in the chemical process controlled, and the bait carefully handled. Kepone can be placed in a paraffin bait and still be very effective for the control of American roaches. Kepone has been banned.

Organophosphates

The organophosphates are derived from phosphoric acid and inhibit the enzyme cholinesterase. In many cases, they have replaced the chlorinated hydrocarbons, because they are effective against insects that have become resistant to the chlorinated hydrocarbons; they are biodegradable; they do not contaminate the environment for long periods; and they have less long-lasting effects on organisms that are not meant to be treated with these chemicals. However, they vary tremendously in toxicity. The organic phosphorous insecticides include TEPP, chlorpyrifos, dichlorvos, phosdrin, and parathion, which are highly toxic; bayer 29493, baytex, dimethoate, fenthion, DDVP, and diazinon, which are moderately toxic; abate, gardona, dipterex, malathion, and ronnel, which are slightly toxic.

TEPP is used in greenhouses and on fruits and vegetables. It is highly toxic when mishandled and causes severe poisoning. It contaminates the environment for short periods of time.

Phosdrin and parathion are highly toxic insecticides that are fatal to humans if only one drop is placed in the eye. They should only be used by highly experienced, licensed, commercial operators. They are used as larvicides for mosquitoes at a rate of 0.1 lb/acre in rural areas away from children and animals. Phosdrin is now banned by the EPA. Most uses of parathion has been voluntarily cancelled.

DDVP, dichlorvos, and vapona are all the same compound. This pesticide is useful as a fumigant because it is highly volatile. It is quite toxic, but breaks down quickly. It is generally used in fly control as a spray or fog or in impregnated strips. In strip form it is effective for mosquito control for 2.5–3.5 months if used at a rate of one strip per 1000 ft^3. It may also be used in catch basins with one strip suspended 12 in. below the catch-basin cover, per basin. Dichlorvos is used in a sugar solution as a bait for fly control. It is mixed with water and used as an outdoor space spray for flies or mixed with water and used as a fly larvicide. Dichlorvos, since it has only a short residual life, is most effective rapidly and presents a short residual hazard to the environment. However, vapona strips should not be suspended over food, for a drop of the chemical could collect at the bottom of the strip and fall into the food, creating the potential chemical food poisoning. It would not be wise to hang vapona strips in areas where individuals suffer from upper respiratory ailments, since the chemical is discharged slowly over a long period of time and could become either an irritation or hazardous to the individual. Dichlorvos should not be used around food.

Diazinon is utilized in fly and roach control and other insect control problems related to vegetables and fruits. The residual period in the environment is fairly short, varying from 1 week to at most 2 months. Diazinon should not be used in poultry farms, since it is toxic to birds. In fly control, diazinon is mixed with petroleum compounds or water and used as a space spray or mixed with water and used as a larvicide. In roach control, diazinon is effective in reducing or eliminating all roaches, including German roaches. Only the German roach appears to be developing some resistance in some areas of the country to the compound at this time. Solutions usually contain the following concentrations of diazinon: spray, 0.5%; dust, 1%. However, pest-control operators are permitted to use 1% spray and 2–5% dust. Diazinon cannot be used on golf courses or sod farms.

Dipterex or trichlorfon is used in sugar and water as a bait for flies. In areas where garbage, organic materials, and manure are controlled, the fly bait works effectively and rapidly. Dipterex is also used as a bait with some success in roach control.

Abate is used at a rate of 0.05–0.1 lb/acre for larvicidal control of mosquitoes. Chlorpyrifous is used in mosquito control for ground-applied outdoor space, spraying at a rate of 0.0125 lb/acre. It is also used as a mosquito larvicide at a rate of 0.05–0.125 lb/acre. It is effective in control of roaches at a 0.5% concentration. When painted on a surface over which German roaches crawl, it has a strong residual effect, causing a kill of 90% of roaches for up to 1 year. The chemical persists in the environment from several days to as long as 1 year. Dimetholate is used for outdoor space spraying for flies and also for larvicidal control outdoors for flies. Fenthion must be used by trained mosquito-control personnel only. It is used in ground-applied outdoor space spraying for mosquitoes at a rate of 0.001– 0.1 lb/acre. It is used as a larvicide at a level of 0.05–0.1 lb/acre, but must be carefully handled by trained personnel. Also, it is effective in roach control, but must only be used by pest-control operators as a spray at a rate of 2.0% concentration.

Gardona is a relatively safe, nonsystemic, broad-spectrum organophosphate used for fly and tick control. However, it is highly toxic to bees. It persists in the environment for only short periods of time. Naled is moderately toxic to animals. It has a very short residual period in the environment. It is used as an outdoor ground-applied space spray at a rate of 0.02–0.1 mg/f^2 for mosquito control. It is also used as an outdoor space spray for flies in liquid form, and as a bait in a sugar solution. Ronnel or korlan is used to control flies in an agricultural area. Its toxicity is slight and its residual period in the environment is very short. It is hazardous to livestock or dairy food.

Carbamates

The carbamates are derived from carbonic acid. Most of the carbamates are contact insecticides. They inhibit the cholinesterase activity and act as nerve poisons, similar to the organic phosphorous compounds, but inhibition is reversible without antidotal treatment. Several of them produce a rapid knockdown, as produced by pyrethrum. The carbamates include sevin, which is also called carbaryl and dimetitan; baygon, which also called propoxur; and landrin.

Carbaryl is widely used in public health and agriculture. It is one of the safer insecticides for animals, but is highly toxic to bees. It is formulated only as a solid, which is then used as a wettable powder, slurry, or dust. It remains for a relatively short period of time in the environment. Carbaryl dusts are used in a 2–5% concentration to kill fleas on dogs and cats older than 4 weeks. It is also used in the United Sates to kill the oriental rat flea in murine typhus control programs. The sprays and dusts have been used in adult mosquito control. Carbaryl, when used as an outdoor space spray, is concentrated at a level of 0.2–1.0 lb/acre for the control of mosquitoes.

Dimetitan is highly toxic when ingested and moderately toxic when absorbed through the skin. It is impregnated into plastic bands and suspended near the ceilings of farm buildings for use in fly control.

Propoxur (baygon) acts as a stomach poison and contact poison in roach control and also in tick control. The spray has a long-lasting residual contact. This insecticide also differs from others in that it has a flushing or irritating action that forces insects out of hiding areas, and it has a rapid knockdown action. Baygon is used to control mosquitoes, flies, sandflies, ants, other insects, and the resistant brown dog tick. It has some toxicity for animals. Apparently, it lasts for short periods of time in the environment as a contaminant. Baygon is used in a dosage of 100–200 mg/ft^2 for residual spraying in mosquito control. For roach control, it is used either as a 1% spray or a 2% bait.

Landrin has been tested by the World Health Organization as a residual spray in anopheles mosquito control.

Biolarvicides

Biolarvicides are naturally occuring crystaline delta-endotoxins produced by *Bacillus thuringiensis*, which are lethal to the larva of mosquitoes. The mosquito larvae ingest the delta-endotoxin which reacts with gut secretions and causes gut paralysis and disruption of the ionic regulation capacity of the mid-gut epithelium resulting in death within minutes.

Insect Growth Regulators

Insect growth regulators are chemicals which cause mosquitoes to die in the pupa state by interfering with the normal metamorphosis, thereby preventing the immature insect from emerging as an adult. Methoprene, which is an insect growth regulator, is a true analog of the mosquito's own juvenile hormone. During the third and fourth larval instar stages the juvenile hormone levels normally drop to a very low level or zero. The ingestion of the methoprene by the larva maintains the high level of juvenile hormone and therefore the insect does not develop the physical features for adult emergence and it dies.

Fumigants

Fumigants are gases that kill body cells and tissues after penetrating the body wall and respiratory tract of insects. They are purchased in either solid, liquid, or gaseous form. Fumigants are of limited use in public health work due to special hazards. They are flammable, toxic, highly reactive, costly, tend to corrode metals or damage dyes in fabrics, and lack chemical stability. The fumigants include hydrogen cyanide, which is extremely hazardous to animals and humans; methyl bromide, which has little or no warning odor; carbon disulfide, which is highly flammable and explosive: chloropicrin (tear gas), which is highly irritating; ethylene dibromide, which is now banned, desorbs very slowly from certain products; ethylene oxide, which is highly flammable and explosive; phosphine, which may be a fire hazard; and sulfuryl fluoride, which is not recommended for food fumigation.

The fumigants are important because they do provide a means of destroying large quantities of insects that infest food and may also be utilized in homes where there are severe insect infestations. It is extremely important that all fumigants be handled very carefully and that they be applied only by trained, licensed pest-control operators who understand the nature of the chemicals and their hazards.

Desiccants or Sorptive Dusts

Certain desiccants, which in effect damage the outer waterproof layer of the arthopod's exoskeleton by either absorbing the fatty or waxy material or by abrasion, are used in insect control. These desiccants include finely powdered silica gels and silica aerosols.

Other Types of Insecticides

Attractants are materials used to lure insects into traps or to make poison baits more inviting. Attractants include such things as sugar, peanut butter, and fish. Sex hormones have also been tried. Some chemical attractants include methyl eugenol, ethyl acetate, and octyl butrate.

Repellents are substances that produce a reaction in insects that makes them avoid animals or humans. A good repellent works for several hours; is nontoxic, nonirritating, and nonallergenic; has a pleasant odor; is harmless to clothing and accessories; is effective against many insects; and is stable in sunlight. Some repellents include oil of citronella, sulfur, dimethyl phthalate, and indalone.

Piperonyl butoxide is a compound that acts as a synergist when added to insecticides. It is a low hazard compound in itself, is not known to create environmental problems, and is most effective when used in combination with insecticides that require a booster to do an effective killing job.

Red Squill

Red squill is an extremely effective rodenticide against Norway rats. It may be used fresh with live bait or water bait. Fortified red squill should be used, since the effectiveness of this natural chemical, which comes from the inner bulb scales *Urginea maritima*, varies in effectiveness with production techniques and the time of storage of the pesticide. Red squill kills very rapidly, and although a single rat may develop shyness to bait, rats in general will continue to come back and feed upon it for periods of time if the bait is mixed properly. In the past, it had always been recommended that red squill be mixed with fresh baits. This is an effective means of control of rats for a short period of time. However, if the red squill is mixed with cracked corn and rolled oats and bound with peanut oil, the peanut oil will not only help preserve the bait for weeks under all types of conditions, but also will act as an attractant to the rats, who will come back and feed upon the red squill bait. The poison is of considerable usefulness in public health, since it is one of the least hazardous rodenticides to humans and domestic animals. Red squill causes animals to regurgitate; since Norway rats cannot regurgitate, the poison produces cardiac arrest, convulsions, and respiratory failure. It is used at a ratio of 3.5–10.0% in baits.

Anticoagulants

The anticoagulant poisons include fumarin; diphacinone; indandione; PMP, also known as valone; warfarin; warfarin plus, also known as sulfoquinoxalin; chlorophacinone; and brodifacoum. Each of these poisons can be used in liquid or dry bait, with the exception of brodifacoum, which is used in dry bait. They provide excellent control of Norway rats, roof rats, and also mice. Since the anticoagulant poison depends on an accumulative action, it is necessary for rodents to feed upon poison bait for a period of several days. Generally, it takes from 1–2 weeks to get an effective kill of the rodents present. Brodifacoum can cause death in 4 or 5 days and present a lethal dose in one feeding. Some resistance has been noted to warfarin in Europe and also in some parts of the U.S. However, the anticoagulants are still extremely useful and safe. It would take a large dose of anticoagulant poison bait to cause any harmful effects in humans or animals with the exception of brodifacoum, where smaller doses may be harmful. The chances of getting such large doses are apparently insignificant.

Fumarin is odorous, nonflammable, and soluble in water and oil. It is highly toxic to rats and mice. It does not deteriorate in baits. It does present a slight problem of secondary poisoning to cats, dogs, and to the individual applying the poison. Diphacinone is highly toxic to rats, cats, dogs, and rabbits. Again, the quantity of poison necessary to harm cats, dogs, and rabbits tends to protect them. It is only hazardous to wildlife and fish if they feed on it continuously for a period of days. No deterioration is found in the bait. There is a slight possibility of hazard to the individual who is using the poison.

Indandione is an odorless compound that is also known as pival. It is soluble in water and oil, is moderately toxic to dogs, and is hazardous to fish and wildlife if eaten continuously over a period of time. It is a slight hazard to the person applying the poison.

PMP is odorless, insoluble in water but soluble in oil, does not deteriorate in baits, has a slight chance of causing secondary poisoning to cats and dogs, and creates a slight problem for the applicator.

Warfarin is relatively insoluble in water and should not be added to baits that contain much vitamin K. It is highly toxic to cats, moderately toxic to dogs, and relatively nontoxic to humans. There is no deterioration in the baits. There is a slight hazard to the applicator. Warfarin plus is warfarin containing hydroxycoumarin and sulfaquinoxalin. This compound is available in ready-to-use baits. Sulfaquinoxalin inhibits vitamin K-producing bacteria in rodents and therefore increases the effectiveness of the warfarin. Smaller amounts are needed for control. Chlorophacinone in laboratory tests has shown good results against Norway rats.

Brodifacoum is an effective anti-coagulant which kills warfarin-resistant Norway rats, house mice, and can also be used for roof rats. It may be harmful or fatal if swallowed and must be kept away from humans, domestic animals, and pets. The product reduces the clotting ability of blood and causes hemorrhaging. The bait should be in permanent bait stations resistant to destruction and kept away from children, pets, domestic animals, and non-target wildlife. It can be used around homes, industrial, commercial, agricultural, and public buildings, and in transportation facilities. It should not be used in sewers.

Other Rodenticides

Bromethalin is an effective rodenticide for rats and mice who are anticoagulant-resistant. It may be harmful or fatal is swallowed and must be kept away from humans, domestic animals, and pets. It also causes eye irritation and should not come in contact with skin. The product causes an uncoupling of oxidative phosphoxlation in mitochondria with ATP depletion and leads to fluid buildup between the myelin sheath, pressure on the nerve axons, and increased spinal pressure. Acute poisoning has symptoms of headache, confusion, personality change, seizures, coma, and possible death. The bait should be in tamperproof bait boxes or in areas inaccessible to children, pets, domestic animals, or wildlife. It can be used for control of Norway rats, roof rats, and house mice in and around homes, industrial and agricultural buildings, and similar structures. It may also be used in alleyways, located in urban areas, and in and around port or terminal buildings, but should not be used in sewers.

There are a variety of other rodenticides available. They will each be discussed. ANTU, or alpha naphthyl thiourea, is a compound that causes death by inhibiting the clotting of blood and causing internal hemorrhaging. It is very toxic to Norway rats, but less effective on other species. There is a medium degree of hazard in its use, since there is no known antidote for it. Tartar emetic is the best substance used when the compound is ingested accidentally. ANTU should not be used more than once a year, because the rat population will refuse the bait. The poison has also been used as a 20–25% tracking powder. ANTU is quite toxic to dogs, cats, and hogs. It is ineffective against roof rats.

Phosphorus bait is a fast-acting poison effective on Norway rats, roof rats, and roaches. It is highly toxic to humans, especially children. Phosphorus baits should only be used when absolutely necessary, in the absence of children, away from food, and under the strict supervision of pest-control operators.

Zinc phosphide is usually used to kill rats and mice. It is generally prepared as a 1% bait with meat or diced fruit. Tartar emetic can be added to this product to make it less hazardous to humans, however, it should never be placed in any area where children, dogs, or cats could consume it. It is an extremely hazardous poison and therefore should be used with the greatest of care.

Sodium fluoroacetate, or 1080, is an extremely effective poison against rats and mice. However, the poison is extremely hazardous to people, and therefore should only be used with the greatest of care and by highly skilled, licensed pest-control operators only. It causes death by paralyzing the heart and central nervous system. The degree of hazard of this chemical is so great that it is recommended that it either never or rarely be used. The residue of the poisons must be destroyed by burning in an open field away from any possible human or animal activity. The operator must be extremely cautious in the destruction of this residue. This chemical has now been banned.

Fluoracetamide, or 1081, is very effective for rat control in sewers, in either dry or watered baits. It should not be used for any other purpose. It is extremely toxic to both humans and animals. It should be applied only by trained pest-control operators and should either never or rarely be used. The remains of the poison and the dead rodents must be burned in an open field or buried so that they cannot be dug up. Extreme caution must be utilized if 1081 is used in rodent control.

Norbormide, better known as raticate, is a dicarboximide. It is highly toxic for rats; slightly toxic for mice. It is extremely stable in all baits and environments.

Thallium sulfate is used as a slow-acting rat and mouse poison. It produces a variety of neurological, circulatory, and gastrointestinal symptoms. Because of the danger in the use of thallium sulfate, it has been banned since 1972 by the EPA and therefore should not be used.

Arsenic trioxide, sodium arsenate, and sodium arsenite are odorless powders used for mouse and rat control. They are fast acting, but tolerance to the chemical can develop. The arsenic compounds are extremely toxic to humans. They should only be used where absolutely necessary, away from children and animals, and should be used only by trained pest-control operators. Arsenicals are now banned.

Strychnine and strychnine sulfate are odorless compounds used in mouse baits. They may also be toxic to rats and are highly toxic to humans. It is extremely important that strychnine compounds not be used unless absolutely necessary and then away from humans, pets, and other animals, and used only by professionally trained pest-control operators. These chemicals have now been banned.

Zinc phosphide is effective against rats and mice. Since it is extremely toxic to all animals, including humans, it should only be used where absolutely essential and should be kept away from all animals and humans. It should be only applied by trained pest-control operators.

As can be seen, many of the rodenticides are extremely dangerous to humans or other animals. It is essential that rodenticides be selected with great care and that they be used by trained professional people. It is preferable, whenever possible, to utilize either anticoagulants, red squill, or raticate, rather than the other types of rodenticides, since the chance of harming nontarget animals and humans is reduced.

Herbicides

There are over 100 different chemicals that act effectively as herbicides. They affect plants either through contact, as a systemic poison, or as a soil sterilant. Herbicides are important in public health work because they are used for the control of weeds and therefore will reduce insect and rodent harborage, as well as decrease the amount of pollen present in the air. Contact herbicides kill plants through direct contact. They may be selective or nonselective and kill all plants. Systemic herbicides are also either selective or nonselective, and therefore pose a problem to plants other than those that one would want to destroy. Soil sterilants unfortunately may remain in the environment for long periods of time, and therefore pose a problem to the environment. Inorganic herbicides are derived from the inorganic acids in which hydrogen is replaced by a metal. They produce a burning effect when coming into contact with the plants. Examples are calcium arsenate, sodium chlorate, and sodium borate. The metal organic compounds include those that have a metal ion complex combined with an organic portion of the molecule. The herbicides are usually used to control large areas of weeds, such as on railroads and highway right-of-ways. An example is disodium methane arsenate.

A third group of herbicides are the carboxyl aromatic herbicides. This group has a carboxyl group and an aromatic group. They work as contact, systemic, and soil sterilants. They are categorized by five basic types. (1) Phenoxy herbicides, which are systemic in nature and usually last 30–60 days. They are only slightly toxic to humans and animals. Examples are 2,4-D (cautionary statement now added for grazing animals) and sesone. (2) Phenolactic acid is used for aquatic weed control and weed control in right-of-ways. (3) Benzoic acid compounds have a longer soil resistance and low toxicity to mammals. Examples are benzac and trysben. (4) Phthalic acid compounds act to prevent weed germination. They are persistent for about 30 days in the soil and are relatively nontoxic to mammals. Examples are dacthal and endothall. (5) Phthalamic acid compounds also prevent weed germination. They are relatively safe to humans and other warm-blooded animals. An example is alanap.

Other herbicides include aliphatic acid herbicides, which contain a carboxyl group and are temporary soil sterilants. Examples are dowpon and trichloracetic acid. Substituted phenol herbicides are used for contact killing and are applied by sprays in such areas as railroads and highway right-of-ways. These include dinoseb and pentachlorophenol. Their toxicity to mammals varies from moderate to very toxic. The nitrile herbicides are used for killing the seeds of broadleaf weeds. They are usually used in agricultural weed control.

The herbicides include chlorophenoxy compounds such as *2,4-dichlorophenoxyacetic acid (2,4,-D)* and *2,4,5-trichlorphenoxyacetic acid (2,4,5-T)*. These chemicals are used for the control of weeds and unwanted plants. They affect the kidney, liver, central nervous system, and skin. They cause chloracne, weakness or numbness of the arms and legs, and they result in long-term nerve damage. *Dioxin* or *2,3,7,8-tetrachlorodibenzo-p-dioxin (2,3,7,8-TCDD)* is an inadvertently produced trace contaminant in 2,4-D and 2,4,5-T and poses the most serious health risk of these chemicals. Dioxin is a colorless solid with no distinguishable odor. It does not occur naturally, nor is it produced intentionally by any industry, except as a reference standard. It is inadvertently produced during the incineration of municipal and industrial waste containing chlorinated organic compounds. It is formed when there are accidental transformer-capacitor fires involving chlorinated benzenes and biphenyls. Dioxin may enter the body through contact with the skin and contaminated soils or materials. It may also enter through the ingestion of food that is contaminated. This food includes fish, cow's milk, and other foodstuffs. Inhalation of contaminated ambient air may contribute a small amount of total body intake, however, the inhalation of particulates such as fly ash may constitute a major source of exposure. In humans, 2,3,7,8-TCDD causes chloracne, which is a severe skin lesion that usually occurs on the head and upper body. It is disfiguring and may last for years after the initial exposure. There is suggestive evidence that 2,3,6,7-TCDD causes liver damage in humans as indicated by an increase in levels of certain enzymes in the blood. There is also suggestive evidence that the chemical causes loss of appetite, weight loss, and digestive disorders in humans. In rodents, the chemical, if administered during pregnancy, reults in malformations of the offspring. 2,3,7,8-TCDD has been demonstrated to be a carcinogen in animals.

Biological Controls

A series of techniques is being tried on an experimental basis to sterilize male insects and set them loose in areas where females are present. The aim is to reduce the fertility of the insects and therefore utilize a biological control. *Bacillus thruingiensis* is a naturally occuring soil bacterium that produces toxins which cause disease in insects. This microbial insecticide is ideal for pest management because of its specificity to pests and its safety for humans and natural enemies of pests. To be effective, larvae must consume the bacteria during feeding. Predators have also been utilized to destroy insects. The only problem is that the predator may become a pest itself. In rodent control, biological control occurs when various environmental forces are utilized in the destruction of the rats by eliminating food supplies and harborage. The rat population density decreases. If a rat population is not tampered with by humans, it usually expands beyond the environment's ability to support the number of rats. The overcrowding results in disease and competition, and therefore an increased mortality rate and decreased natality rate. Eventually, the population drops to a size that the environment can support.

Other Types of Pesticides

In addition to the pesticides mentioned, fungicides and bactericides are used to prevent plant disease caused by fungi and bacteria. The nematicides are used to control nematodes, which attack plants. Molluscicides are used to control molluscs, which affect fishing areas or plant areas. Piscicides (fish killers) are used to treat public waters. Their objective is to remove rough or trash fish from restocking lakes or game fish lakes. Avicides (bird killers) are used to control birds and pigeons in areas where they are troublesome and damage crops.

PROBLEMS CREATED BY PESTICIDES

Pesticides in the Environment

Pesticides may enter the environment by means of the air route, the food route, and the soil route. The environment is contaminated through the indiscriminate, uncontrolled, unmonitored, and excessive use of pesticides by all types of people, including the owners of households.

The Air Route

Pesticides enter the air by means of aerial spraying, the use of mists and fog machines, and the application of pesticides by individuals using pressure containers. The pesticides, depending on the size of the particles and the volume discharged, the velocity of the air current, the temperature of the air, and other factors, may stay within a given area or may contaminate areas other than those intended. It is essential that great care be taken in the application of pesticides by the air route. The drift and weather conditions must be considered carefully. The human hazards are caused by inhalation, skin absorption, and ingestion of the pesticide. Pesticides may be transported over long distances if they attach to dust particles in the air. Further, they may be mixed in with other chemicals produced by a variety of air-polluting situations, causing secondary chemicals to form, which are in themselves very hazardous.

The Water Route

Pesticides enter surface waters by being washed from the surface of the soil or from plants, houses, and agricultural areas. Some pesticides percolate down into underground water supplies through a water flow, providing access to pesticides injected into the soil purposely for the control of insects, or through rain or snow, which washes the pesticides into the soil and slowly helps them percolate into the underground water. The use of pesticides must be carefully controlled and bodies of water should be regulated. Lakes and other bodies of water should be studied carefully before pesticides are used for either mosquito control or water-weed control, since the pesticides may end up causing more harm than good. It should be recognized that pesticides and fertilizers are used extensively by homeowners and farmers. All of these chemicals have a tendency to be washed into bodies of water through surface drainage and through storm sewer pipes.

The Food Route

Pesticides, from time to time, have caused disastrous consequences when stored in the same vehicles in which food was being transported. It is essential that this be forbidden, since any breakage or leakage would cause chemical food poisoning. Food treated with pesticides must contain the minimum quantity for effectiveness on insects and must not cause harm to humans.

Soil

The persistence of pesticides in the soil creates a situation in which not only is the soil contaminated, but additionally, the air may become contaminated by soil particles, or water may become contaminated by runoff. The chemicals used should not be hazardous and should degrade rapidly within the soil. Those chemicals that are taken up by plants in the soil and are hazardous to humans should not be utilized.

The persistence of a pesticide in the soil depends partially on how it is transferred to the soil. Is it done through leaching, erosion, evaporation, or through an uptake of plants. The persistence also depends on how the pesticide is degraded. Erosion is still another factor. The algae, fungi, and bacteria found in the soil may use the organic chemicals present as a source of energy, and therefore may reduce some of the amounts of pesticides found in the soil. Chemical reactions may destroy some of the activity of pesticides, while enhancing the activity of other pesticides. Diazinon is broken down in acid conditions, however, the opposite is true for malathion.

Household Use

Homeowners are often the ones most apt to contaminate their immediate environments or to provide opportunities for accidental poisoning of their children, themselves, and others around them. Many homeowners have no concept of the proper use of pesticides. They fail to read and understand the labels, use the pesticides under hazardous conditions, and generally store pesticides in areas where children play or can gain access, or in which fires may occur. Pesticides are misused in gardens, in homes, and in the care of household plants.

Effects of Pesticides on Humans

It is difficult to fully evaluate the human risks of chronic pesticide problems, since very few studies have been made in this area and because of the complex nature of the problems involved. Further, variables such as age, sex, race, socio-economic status, diet, state of health, length and state of exposure, and pesticide concentration level all profoundly affect the human response to pesticides. Although there are cases of acute pesticide poisoning, one should not extrapolate these results to chronic low-level exposures. The individual is not only exposed to pesticides in the environment, but also to dusts and a variety of environmental conditions that can alter the human response to any specific pesticide.

It is known that the organochlorine compounds, such as aldrin and dieldrin, can increase the excitability of the nervous system and may damage the liver. However, it is difficult to establish a correct diagnosis, since the symptoms vary. Certain compounds can penetrate the unbroken skin. Lindane is believed to cause hematological disorders.

The organophosphate insecticides inhibit cholinesterase enzymes. There is a great variation in acute toxicity from one compound to another. Organophosphates penetrate the skin easily. Carbamate pesticides also inhibit cholinesterase, but since the enzyme deactivates rapidly within the human body, it is difficult to measure the exposure based on this deactivation.

DDT is an organochlorine compound that becomes stored in the fat tissues of animals and humans. It also has considerable effects on fish and wildlife.

The difficulties involved in trying to gather adequate information on the human effects of pesticides contribute to the considerable uncertainty in determining which pesticides are safe and which are harmful. It is impossible to do a general epidemiological study, since all individuals have had some pesticide exposure; therefore, a control group cannot be selected. More often studies are made of either acute poisoning due to pesticides among occupational groups, children, or other individuals who have accidentally poisoned themselves. From these data, we can generally determine the kinds of effects that humans have in the event of acute poisoning. However, even in acute poisoning, inadequate amounts of data are available, and it is difficult to make careful judgements about the absolute danger of the given pesticide, unless the pesticide is taken into the body in large amounts. Chronic poisoning becomes an even more complex and confusing issue.

The routes of entry of the pesticide include absorption through the intestines due to ingestion, absorption through the lungs due to inhalation of airborne pesticides, and penetration through the intact skin or absorption directly into the bloodstream through broken skin. The route of entry depends very much on the group of individuals studied and the use of the pesticide. Absorption through the intestine occurs when the residues that remain on food are ingested. This is probably the major route through which pesticides enter the body. In addition, accidental poisoning of children generally occurs through ingestion. Inhalation occurs when bug bombs or aerosol sprays are used to control roaches and other pests in the homes or when individuals inhale particles from fogs and mists that are used to control mosquitoes in exterior areas. Most skin contamination occurs in occupational environments. Pesticides may also be hazardous to humans because of the possibility of fires or explosions. This subject has been previously discussed.

Considerable study is being conducted on the possibility of pesticides causing cancer. In the future, many of our existing pesticides may be banned because they are carcinogenic in animals. These studies have to be conducted with great care to ensure that false conclusions are not drawn. It will be important not to ban pesticides that are valuable to humans and, at the same time, to protect humans from a new burden of additional carcinogenic agents.

Resistance to Pesticides

Pests, particularly insects, develop resistance to pesticides. Some insects are less susceptible to certain insecticides, and some are affected but not killed by insecticides. Over a period of time, resistant insects survive exposure to insecticides and reproduce new generations of increasingly resistant insects. Generally, two forms of resistance develop in insects: physiological and behavioral. In physiological resistance, the insect develops an immunity to the poison. The exoskeleton becomes less permeable to the insecticide, the insecticide is detoxified into less toxic chemicals, or the insecticide may be stored harmlessly in the body tissues or be excreted. Behavioral resistance is the ability of the insect to avoid lethal contact with the insecticide because it has developed protective habits or behavioral patterns. This includes such things as mosquitoes changing resting places and the avoidance of baits by flies. Resistance of insects to insecticides is increasing. In fact, cross-resistance has been known to occur. Studies have shown that insects resistant to certain types of chlorinated hydrocarbons may also develop resistance to the organophosphates. Examples of these insects would be the housefly, and the mosquitoes *Culex tarsalis* and *Aedes nigromaculis*. It is necessary from time to time to conduct surveys to determine if insects are developing resistance to a given insecticide, and if so, the insecticide should be changed.

Economics

The major question in environmental economics as it relates to pesticides is whether the damage costs resulting from blighted crops, poor health, and higher death rates are greater or less than the potential benefits from increased crop yield, reduction of disease, and so forth. It is difficult to obtain a true picture of the cost-benefit ratio in the use of pesticides generally. Each pesticide needs to be studied and judged on an individual basis.

POTENTIAL FOR INTERVENTION

Intervention strategies include the use of the techniques of isolation, substitution, shielding, treatment, and prevention. The potential for intervention varies from poor to excellent, based on the understanding of the short-range and long-range problems associated with the particular pesticide; the preparation, storage, use, and disposal of the pesticide; and the training and ability of the pesticide applicator. Proper storage of pesticides to avoid fire, explosions, and contamination of individuals and their food are techniques of isolation. Substitution is the use of a less hazardous pesticide. Shielding is the use of safety glasses and protective clothing. Treatment is the technique used on an individual who becomes contaminated with a given pesticide. Prevention is the overall process of keeping pesticides out of the eyes, off the skin, away from the lungs, and away from clothing. Good housekeeping is part of the overall prevention technique.

RESOURCES

Scientific, technical, and industry resources include the Entomological Society of America, the Pest Control Operators Association, the American Public Health Association, the National Environmental Health Association, various land grant colleges and universities, and chemical manufacturers associations.

Civic resources include the National Audubon Society, the Wildlife Society, and the Environmental Defense Fund.

Governmental resources include state and local health departments, the Department of Agriculture, the National Bureau of Standards, Department of Defense, the Environmental Protection Agency, Department of Health and Human Services, the National Science Foundation, and so forth.

The area of pesticides continues to change so rapidly that what is used today may be banned tomorrow, and what does not exist today may exist tomorrow. Public health workers or environmental health workers should obtain a document from the U.S. EPA entitled, "Suspended, Cancelled and Restricted Pesticides." A second source is entitled *Handbook of Pest Control* by Arnold Mallis. The third set of references would be the current rules, regulations, and laws of the Environmental Protection Agency. A fourth resource is the Library of Congress, and a fifth resource is the U.S. Senate Agricultural Committee.

The National Pesticide Telecommunications Network may be reached by phone by calling 800-858-7378, which is a toll-free call. The Pesticide Producers' Association, 1200 17th Street, N.W., Washington, DC 20036, is another resource. The Synthetic Organic Chemical Manufacturers' Association, 1330 Connecticut Avenue N.W., Washington, DC 20036, is also another useful resource.

STANDARDS, PRACTICES, AND TECHNIQUES

Application of Pesticides

Pesticides are applied to standing water by the use of spray equipment. Oil solutions or emulsions may be utilized as well. The adult mosquito is controlled by the use of aerosol bombs, which are successful in the destruction of flying insects such as flies and mosquitoes. Fogging and misting techniques are used in large community programs as the major method of controlling the adult insect, especially the mosquito. Fog and mist are produced through specialized equipment mounted on driven vehicles and blown out into the open air from the equipment. Dusting is used largely in agricultural areas. The dust is usually applied by airplanes. Residual spraying is carried out by applying the spray to surfaces upon which the insects will alight or rest, usually from hand held equipment. Fumigation is the technique of releasing large quantities of aerosols quickly into an area. In catch basins, vapona strips are hung. This insecticide is a fumigant. It is slowly released from the solid material.

The type of equipment utilized varies from the hand sprayer, which may be 1–3 gal in capacity, to the aerosol bomb; to the compressed air sprayer, which is usually 3–4 gal in capacity; to a variety of power equipment; to aircraft application.

Pesticide Labels and Names

All pesticides must be registered for use. The registration can be revoked by the EPA if the pesticide is determined to be a hazard to the community. Pesticide labels and names are very important. The label on a pesticide container specifies the name of the manufacturer; the name of the product; the active chemical ingredients and percentages of concentration; the type of chemical, whether it is an insecticide, rodenticide, and so forth; recommendations for specific uses; directions for use; precautions in storage; and precautions during use by personnel.

Since many chemicals can be sold under the same brand or trade name, it is essential to identify the actual, active chemical ingredients. The generic or common name of the ingredients must appear on the label. The label must state specifically what pests are controlled and what rate of application of insecticide is needed to control the pests affected by the particular pesticide. Since the application of the pesticide is extremely important, instructions should be listed clearly and carefully regarding correct and safe methods of application. All labels should specify the necessary precautions to be taken by personnel. Insecticide labels must say, *Keep away from children.* Where highly toxic products are used, skull and crossbones are required on the label; plus the word *POISON* in red letters. The label should also contain the word *Warning*, instructions for handling, the antidote, and the statement, *Call a physician immediately*. Where the pesticide is not as toxic as stated above, the words *warning* or *caution* should be placed on the label, and it should be stated that a pesticide is a hazard and should be kept out of the hands of children. Common names of the pesticides should be used when possible.

Pesticide Formulations

To use pesticides properly, it is necessary to have some understanding of the form in which they come and also dilution specifications to achieve the proper formulation effective for a given pesticide in a given situation. Solvents dissolve insecticides, dispersing them evenly through the solution. The solvent is a carrier as well as a dilutant, depending on the amount used. Solvents may include volatile liquids, such as xylene, which evaporates after use and leaves a residual deposit, or nonvolatile or semivolatile liquids, such as petroleum distillates, which leave the surface coated with the solution of the toxicant. The emulsifiers are surface-active agents that allow liquids to be mixed within liquids. The emulsifier forms a thin film around each minute droplet of oil, thereby keeping the oil from coalescing or separating into oil and water. Wetting agents or spreading agents allow the pesticide to penetrate a surface. Adhesives are used to improve the quality of the deposits of the pesticide. An example of such a material would be gelatin or gum. Perfumes and masking agents are utilized to cover the unpleasant odors of certain pesticides. An example would be oil of wintergreen. Synergists are materials that help a pesticide work more effectively. Examples of synergists are piperonyl butoxide, sesamex, sulfoxide, and propylisome. Carriers and dilutants for dust are used to help deliver the insecticide more readily and in an inexpensive manner. Examples of these carriers would be aptapulvite, bentonite, calcite, diatonmite, and talc.

An insecticide is used in various forms. The technical grade insecticide is the insecticide in concentrated form plus inert material. It is basically the purest chemical form. The insecticidal dusts are prepared by blending toxic ingredients, such as diazinon, into an inert carrier, such as talc. Wettable powders are the toxic ingredients plus the inert dust and a wetting agent such as sodium lauryl sulfate. Water is added and the material is agitated to form a suspension. The material must be continually agitated so that the chemical will not settle out. Wettable powders are used frequently in out-buildings. Insecticidal solutions may be purchased or prepared either as a concentrate or as a finished spray. The concentrates are diluted with oil to make the usable solution, which becomes a finished spray. The solution is not soluble in water. Emulsifiable concentrates are concentrated solutions to which an emulsifier has been added. When the concentrate is mixed with water and the emulsifying agent, the insecticide spreads throughout the water. A finished emulsion is made by diluting a concentrate with sufficient water to form the concentration needed for use. Aerosol bombs typically contain 0.1–0.6% pyrethrums; 1–2% of malathion, or baygon; 10–12% petroleum distillates, and 85% propellant, such as petroleum distillates.

It is essential to have the correct dosage of pesticides for the proper type of pest control. Excess pesticides may cause unsightly residues, ruination of edible crops, toxicity to crops and animals, excess costs, and a specific hazard to animals and humans. When preparing pesticides for use, it is necessary to determine the safe time intervals for application; the amount of residue acceptable by the Food and Drug Administration; whether a combination of pesticides are safe or compatible; if products for animals are treated, the toxicity of the pesticide to be used, the manner in which the equipment was cleaned prior to pesticide use, the rate of application of the pesticide, the size of the area, and the safety precautions to be taken by the operators. Always follow the instructions on pesticide containers to determine the dilution of the pesticide to the proper concentration.

Pesticide Application and Equipment

Pesticides may be applied as solids, liquids, or gases. Solids are usually applied as dusts by means of hand dusting, shaker cans, bellows, vault dusters, or power dusters. Solids may be applied as granules, pellets, gelatinous capsules, or poisonous baits. Liquids may be applied by pouring, painting, spraying, ejecting, or through aerosol bombs. The method of application is based on which one effectively kills the pest, safety of the operation, the expense, durability of the pesticide, and potential problems. Hand sprayers range from something as old fashioned as a flit gun, used for bedbug control, to various types of compressed air or other sprayers. The sprayers are regulated for the type of spray desired, the quantity desired, and the most desirable pressure for delivery of the spray. It is essential that all parts of the sprayer, including the hose and all fittings, be checked carefully to make sure that they are thoroughly clean and do not leak. A wand is a slender metal tube that is placed on the sprayer and connected to a nozzle that extends the length of the range of the sprayer. The nozzle is the most important part of the sprayer. It has to be fitted so that it will effectively deliver the kind of spray and the concentration desired. Ultra-low volume (ULV) larviciding is very effective over land or water. It is necessary to have an air inversion with a 4–6°F difference between the hot upper air and the cooler lower air to have ULV work effectively. Late afternoon and early evening are the best times to larvicide with this technique.

Dusters are utilized in areas where rodent ectoparasites are found, in areas where a fire hazard results from the use of oils, and in areas where oil and water are not used advisably.

Powered sprayers or foggers are generally mounted on some type of mechanical equipment. The power unit distributes a large amount of spray at a high pressure. It is essential that this type of spraying be carefully controlled to avoid annoying residents of a home and also to avoid creating an accident hazard. Children have a habit of traveling in and out of fogging machine units, thereby creating accident hazards to the spraying crews, to automobiles, and to themselves. Wind direction and wind velocity must be taken into account whenever power fogging or spraying is utilized. The equipment should be operated to spray perpendicular to the wind and should be directed as close to the ground as possible. Mists and fogs are generally utilized to quickly knock down high concentrations of adult insects. They are economical and can be utilized to cover large residential areas in short periods of time. The mechanical fog generator breaks the insecticide into fine particles and then blows it into the air at high speed. Other types of fog generators are available that operate on either a pulsation principle, thermal principle, or steam principle.

Insecticides may also be injected into the soil to control a variety of pests, especially termites. Fumigation equipment is used to kill insects and rodents in large areas where people are not present. These would include railroad boxcars, the hulls of ships, certain aircraft arriving from overseas, in areas where arthopod-borne diseases are prevalent, in food storage warehouses, and in food-processing plants. Great care must be taken in fumigation.

Insecticides may also be applied by airplanes or helicopters (Figure 6.1). The insecticide must not drift into residential areas and affect animals or humans. Weather conditions, time of day, and so forth must be taken into account when aerial dispersion of insecticides is used.

Storage of Pesticides

Pesticides must be stored with great care, since they may be highly flammable, explosive, or toxic. Most insecticides, fungicides, and rodenticides can be stored in the same room. However, herbicides are quite volatile, and special precautions should be taken with their storage in order that they do not contaminate the storage area or escape to the outside, thereby damaging plants.

All storage areas should maintain the following precautions: Storage areas should be locked and located away from food, animal feed, plant seed, or water; pesticides should be stored in a dry, well-ventilated place, as directed on the labels; storage areas should be clearly marked as a pesticide storage area; a sign should be posted listing the types of pesticides stored and the hazards therein; pesticides should be kept in their original containers; containers should be checked periodically for leaks, tears, and spills; an inventory list should be kept in order that outdated materials are eliminated and shortages are accounted for.

In the event of a fire, it is extremely important that the firefighters and the general public be protected from the fumes, residues, or washings due to the fire. Fires may originate because petroleum distillates are present; aerosol containers

Figure 6.1.   Introduction of pesticides into the environment by aerial spraying.

become overheated and explode; other flammable or explosive solvents are present; finely divided dust or powders explode; chlorates, which are flammable or explosive, are present; ammonium nitrate fertilizers are stored; and/or calcium hypochlorites are present, and may cause spontaneous ignition and explosion if contaminated by organic substances.

The hazards that exist are due to the presence of organophosphates, carbamates, and chlorinated hydrocarbons, which are highly toxic. Further, fumes from solvents, the presence of gases, or any combination of substances may be toxic. Care must be taken that runoff water from firefighting, which may contain highly toxic pesticides in quantity, does not enter any of the environmental pathways.

In the event of a pesticide fire, a qualified physician should be available, hospitals should be alerted as to the type of potential hazard, and the firefighters should be protected from poisoning by use of proper protective clothing and special self-contained breathing equipment. Great care must be given to the cleanup after the fire to make sure the firemen have removed all traces of the pesticide from their clothing and boots. All of the areas where the pesticides are present must be cleaned with the utmost of care, and the resulting contaminated water must be trapped and treated.

Transportation of Pesticides

Transporting toxic chemicals can be very hazardous, with the possibility of accidents or leakage. It is essential that all regulations of the Interstate Commerce Commission and the Department of Transportation concerning proper identification of vehicles carrying hazardous chemicals be followed. The individuals operating these vehicles must be fully aware of procedures to follow in the event of an accident or emergency. Volatile pesticides should never be within the section of the vehicle holding passengers or the driver.

The transporting vehicles should be properly built in order that powders within paper bags will be protected from rain and will not be punctured or torn. A vehicle should be able to be easily cleaned. Where pesticides are in liquid form, they must be in tightly closed original containers. Glass containers are not recommended. If they must be utilized, they should be packed and transported in such a way as to avoid breakage. Since pesticides are affected by high and low temperatures, they should be removed from trucks as soon as possible after delivery and stored in safe, locked facilities.

In the event of an accident involving the vehicle transporting pesticides, the drivers should immediately avail themselves of protective clothing and respirators, and should inform the fire department, police, department, and health department of the accident. It is essential that fire, police, and health personnel understand the nature of the toxic material and the hazards therein. It is also extremely important that the public be kept as far away from the accident site as possible.

Disposal of Pesticides

Since pesticides in concentrated form or even in diluted form may constitute extremely serious hazards, unused pesticides, as well as the empty containers, must be disposed of in a safe manner. In all cases it is necessary to follow the laws, rules, and regulations stipulated by states, the federal government, and local legal bodies. Some types of disposal include ground disposal and incineration. Ground disposal may be very dangerous if the pesticide will contaminate either surface or groundwater supplies. Where this does not occur, it is possible to bury the pesticide in containers that will not deteriorate. This must be a deep burial in order that the pesticide will not be dug up inadvertently. Burial should not occur in normal landfills (see the chapter, Solid and Hazardous Waste).

Many pesticides are destroyed during incineration. Incineration is acceptable if the resultant fumes or resultant waste escaping into the air is scrubbed out of the gases, concentrated, and either reincinerated or finally buried in containers that will not deteriorate. Pesticide containers should be handled separately and should be buried only in areas in which the water percolating through the ground will not carry the pesticide into water supplies.

Ethylene dibromide (EDB), which has now been banned, is a perfect example of a chemical that is now considered to be very harmful. Leaking drums of EDB in Missouri showed how bad the disposal process had become.

MODES OF SURVEILLANCE AND EVALUATION

Surveillance and evaluation consists of the epidemiological approach, the major incident approach, analysis, and monitoring. The epidemiological approach is used to study the general population and its exposure to various pesticides over long time periods. It is used on specific populations, such as workers exposed to pesticides during production, use, and storage, or individuals in communities exposed to large amounts of pesticides, especially individuals living in small farm communities. The epidemiological approach is a technique evaluating mortality and morbidity records of these individuals.

The major incident approach is used in situations such as the contamination of the city of Hopewell, Virginia and its surroundings by the pesticide kepone. This particular incident caused widespread kepone poisoning. At least 70 victims were identified and 20 of these victims were hospitalized with untreatable ailments, including apparent brain and liver damage, sterility, slurred speech, loss of memory, and eye twitching. The National Cancer Institute reports that kepone causes cancer in test animals. Kepone was found in shellfish 60 mi down the James River from Hopewell, Virginia. This type of incident is dangerous to a given community. It does provide, however, large quantities of data for further analysis and evaluation. It is hoped that these data can be utilized in preventing other types of major incidents.

Analysis and monitoring provide information on the methods by which chemicals escape into the environment, the levels at which they are harmful, the types of controls currently utilized, and effective control techniques. Analysis consists of identification of the chemical, substance, and chemical entities, and the quantitative measurement of amounts present. It also provides toxicological evaluation of the product, its isomers, byproducts, secondary products, and unreacted intermediate products. During analysis, instruments capable of detecting concentrations in the range of 0.01 to 100 ppm should be used. For some chemicals it is necessary to detect even smaller quantities. Monitoring tracks specific chemicals through the environment. To monitor properly, it is necessary to consider a wide range of concentrations and potential toxicities and to understand the limit of detection of a given analytical method and the behavior of the chemical in the environment. How, when, and what to monitor, and when to stop monitoring, are complex decisions that must be made by competent professionals.

Chemical structure, reactivity, basic physical and chemical properties, proper analytical methods and monitoring strategies, gathering of reliable analytical data, storage of the data, and retrieval and use of the data are essential to the proper techniques of surveillance and evaluation.

CONTROL OF PESTICIDES

Safety

Pesticides must be used with great care, since they are hazardous to humans. Most victims of pesticide poisoning are either workers in the occupational preparation or occupational use area, individuals who inadvertently have been affected by improper use of pesticides, or children who have eaten the pesticide. Individuals may become chronically ill by being exposed for long periods of time or acutely ill by being exposed to large quantities over a short period. Pesticides follow the respiratory route, are absorbed through the skin, or are ingested with food.

The solvents used for dilution may also be toxic. Poisoning of children under 5 years of age by pesticides is a serious public health hazard. It is essential that these materials be stored out of the reach of children. Pesticides should not be stored in pantries, under sinks, or in garages where children can reach them.

Safety rules include proper reading and understanding of labels; proper preparing and applying of the insecticide; proper storage of the insecticide in the original containers with the original labels; mixing of pesticides in well-ventilated areas; mixing and applying flammable pesticides in such a way that they are not near fires, defective wiring, smoking, or hot areas; avoiding eating, drinking, or smoking where pesticides are used; wearing of appropriate clothing and headgear; avoiding contamination of food and water of humans and animals; avoiding inhalation of sprays and dusts; keeping equipment in good operation condition; avoiding the storage of partially used pesticides; proper disposal of pesticide containers; proper transportation of pesticides; proper storage of pesticides; and understanding of first aid measures where needed.

When accidental poisoning occurs, speed is the most essential concern. Proper treatment must be given at once and the individual taken to a hospital immediately. Poison information centers are available and usually can be reached by the telephone operator or local hospital. The poison information center or physician will supply information on immediate first aid, depending on the poison taken. It is essential that the label be read to determine the chemicals present and the antidotes and techniques used to counteract the poison. In no case should an individual be made to vomit if he or she is unconscious, in a coma, or convulsing, or after consuming petroleum products and corrosive poisons. If eyes have been affected, they should be washed immediately with cold water for at least 5 min. Any delay may result in permanent injury. If the individual is in a poisonous atmosphere, remove him or her as soon as possible to fresh air. It is best to reiterate that speed is absolutely essential in any type of pesticide poisoning. It is also essential to determine as soon as possible what chemicals are used and to get proper help from medical authorities.

Laws and Regulatory Agencies

The Federal Insecticide, Fungicide and Rodenticide Act of 1947 was superseded by the Federal Environmental Pesticide Control Act of 1972. The law forbids anyone, including the government, from using pesticides in a manner contrary to the label instructions and gave the EPA the authority to restrict use of certain pesticides to trained personnel in approved programs. It also extended control to all pesticides sold intrastate and interstate. In June 1975, regulations were published establishing a screening procedure. If a new pesticide was chemically suspected of causing cancer, heart defects, or mutations, it had to undergo testing before being declared safe. Any suspected pesticides that were already registered could retain their registration until a pending test proved that they were dangerous. These regulations also required that data be developed to determine when farm workers may reenter fields. On October 11, 1974, the EPA suspended the registration of aldrin and dieldrin because new evidence showed that they were an imminent hazard. Animal experiments indicated that they were carcinogenic. The courts held that the burden of proof of the safety of a pesticide rested on the registrant and not with the government. The court also accepted animal test results to indicate the cancer risk to humans. In July, the EPA issued a notice of intent to suspend the use of heptachlor and chlordane based on animal experiments indicating that these pesticides may cause cancer. Pesticides containing myrex and phenyl mercury compounds had also been cancelled. Although the EPA has the right to lift a ban in the event of a public health or other national emergency, generally they did not do so. There has been only one major exception involving the use of DDT that occurred in 1974 in the forests of the northwest. DDT has been banned since 1972.

Under the 1972 act, all pesticides distributed, sold, offered for sale, held for sale, shipped, or delivered were required to be registered with the EPA. Registration consisted of filing a statement with the EPA Administrator giving the name and address of the applicant or any other name appearing on the label; the name of the pesticide; a complete copy of the labeling of the pesticide, including directions for use; upon request, a full description of tests made of the complete formula; and a request for the type of classification, whether it be for general use, restricted use, or otherwise. The EPA Administrator then approves the registration of the pesticide. Another important part of the law stated that the pesticide must be used only by certified applicators. Therefore, pest-control operators have to obtain certification. Federal certification is handled by the EPA; state certification was submitted by the governor of each state to the EPA for approval. All operators after certification have to earn continuing certification hours credit to retain their license.

Registration of any pesticide is cancelled if the registrant does not reapply within a 5-year period. In addition the EPA can cancel the registration if the pesticide is found to be hazardous in any way. The EPA has the right to stop the sale and use, and to order removal and seizure of any pesticide that it deemed hazardous or being utilized in a hazardous manner.

The 1972 law (Section 171.3, categorization of commercial applicators of pesticides category number 8) stipulated that all state, federal, or other governmental employees using or supervising the use of restricted pesticides in public health programs have to be certified. Therefore, under the 1972 law, all pesticide operators and all public health workers involved in the control of pests of public health importance had to be certified by October 21, 1976, by the EPA and also by the state in which they were operating. The law was again updated in 1978 and in 1988.

The 1972 law was updated in 1975 to require impact statements and to require the EPA Administrator to submit actions concerning health matters related to pesticides to a scientific advisory panel. The comments of the advisory committee and the administrator have been published in the *Federal Registry* (see Current Issues for Update on 1988 FIFRA Act Amendments in this chapter).

SUMMARY

Pesticides move through the various ecosystems in the environment in a variety of ways. When pesticides are ingested or otherwise carried by the target species, they will stay in the environment. They may be recycled rapidly or further concentrated through bioconcentration as the pesticides move through the food chain. Most of the large volume of pesticides utilized do not reach their intended areas and therefore become contaminants. Pesticides are introduced into the environment by spraying or by surface application. The storage in body fat of a pesticide is based on its chemical nature, physical state, means of application, and atmospheric conditions. The persistence of pesticides in the air is influenced by both gravitational fallout and the washout caused by rain. They build up in the soil to concentrations that affect the various ecosystems. They may contaminate soil water, air, and various organisms. Pesticides can cause direct problems to humans through airborne contamination, drinking water, or contamination of food. The pesticide problem is increasing, despite the fact that the Departments of Health and Human Services, and the Environmental Protection Agency, are working diligently to remove dangerous pesticides from the market. However, without pesticides our society would be in great trouble, because insects and rodents would consume valuable food supplies and cause a variety of serious diseases.

It is obvious from the preceding material and concerns under discussion in this chapter that the problem of various pests is unresolved and will probably continue as long as we exist. It is also clear that pesticides are hazardous and must be used with the greatest of care. Public health officials and pest-control operators must be constantly on the alert to ensure that pesticide usage is proper. With new research, effective pesticides may be eliminated from the market because of their potential hazards as toxins, fire and explosion hazards, and because they are potential carcinogens.

RESEARCH NEEDS

Research is needed on all common pesticides used for the control of pests of public health significance. It is necessary to determine if the various chemicals can cause a resistance in the pest that might reduce or eliminate the effectiveness of large groups of pesticides. Chlorinated hydrocarbons must be studied to determine their reactions with food constituents and their potential harmful effect on humans. Recent epidemiological studies suggest that there might be a link between organochlorine compounds and breast cancer. Several pesticides have been shown to be altered by sunlight. It is important to study photochemical alterations to determine if the efficiency of the pesticide is reduced in the natural environment. What are the issues relating pesticides binding to macromolecular components of plant and animal tissues used as food? Little is known about the chemistry of these bound residues. Organophosphate pesticides may be affecting commercial sprayers, since they have a significantly higher anxiety score on standardized tests than the control groups. This area also needs further study. Laboratory animals have difficulty in learning and memory because of exposure to pesticides. Chemical agents, such as the anticholinesterase pesticides in high doses, acutely affect the nervous system. Studies should be made over long periods of time at low dose levels to determine if they cause irreparable damage.

Under The Federal Food, Drug and Cosmetic Act, the EPA, in cooperation with the U.S. Food and Drug Administration, sets allowable limits for pesticide residues in food. These limits, which are called *tolerance levels,* are supposed to protect human health while allowing for the production of an adequate, wholesome, economical food supply. Tolerance limits need to be reevaluated for many existing chemicals

Techniques of genetic engineering, as they relate to the production of genetically engineered microorganisms under the FIFRA Act, need to be studied further. Although many pesticides have been approved in the past, they need to be further studied to determine if more stringent standards should be applied to their use. Studies should be conducted on dioxins, which are byproducts of the production of pesticides. Studies should be conducted of the gas stream to determine the chemicals present and their potentials for causing serious health effects when pesticides are destroyed through incineration. Additional research is necessary in reducing the health risk of pesticides and also in reducing the risk to the environment. Evaluations of indoor air should be conducted to determine the level of pesticides present in the indoor air of the home and also the potential hazards. All new and previously registered pesticides should be screened for their potential to contaminate groundwater. Continuing studies are needed to determine the potential health effects of some 55 inert ingredients found in pesticide formulations. The EPA has invalid or fraudulent health data on 36 pesticides including 35 used on food. Additional studies in this area are necessary.

Chapter 7

THE INDOOR ENVIRONMENT

BACKGROUND AND STATUS

The indoor environment in its broadest sense includes housing, the immediate external environment of the structure, and other facilities that we may work in.

Each of us understands to some extent the importance of our home environment. However, the physical environment for millions of less fortunate families is conducive to disease, behavioral problems, accidents, and social problems. Many times during the course of history, attempts have been made to create more attractive housing and to eliminate health-related environmental hazards. This chapter is concerned with the origins of housing problems, the types of health problems related to housing, neighborhoods and their effects on housing, the types of housing, the environment within dwellings, the structural soundness of dwellings, various techniques for improving housing conditions, indoor air pollution, and injury control.

In 1626, the Plymouth colony passed a law stating that new houses must be roofed with board. In 1648, wooden or plastered chimneys were prohibited on new houses in New Amsterdam, and chimneys on existing houses had to be inspected regularly. In Charlestown in 1740, it was declared that all buildings must be made of brooker stone because of a disastrous fire that had occurred. The reason that these communities began passing housing laws was the frequent fires that destroyed many lives, houses, possessions, and food. During the 1600s, some Pennsylvanians were living in caves along the water banks. This was declared illegal in 1687. Most of the early housing codes were primarily concerned with the fireproof nature or size of the structure and not with the basic indoor environment, which was very poor. Outdoor privies were used. In 1657, New Amsterdam passed a law that rubbish and filth could not be thrown into the streets or canals.

In the 1800s during the Industrial Revolution, millions of immigrants arrived in the U.S. unable to support themselves and had to seek housing accommodations with relatives or had to move into housing in poor condition. The cities, particularly New York, grew extremely overcrowded, and housing became the source of numerous outbreaks of disease. Only the fire department had any authority, until 1867, when the Tenement Housing Act of New York City was passed. This was the first comprehensive rule of its kind in the country. It stated that each room used for sleeping or occupied by an individual had to have adequate ventilation by means of a transom window, if windows to the outside were not available. The roof, stairs, and banisters had to be properly maintained. Al least one water closet or privy had to be provided for every 20 persons. All houses had be cleaned to the satisfaction of the board of health, and all cases of infectious disease had to be reported. Authority was given to condemn houses unfit for human habitation. Additional laws were passed over the years to modify and strengthen the initial act.

In 1901, a new, improved Tenement House Act was passed, stipulating the amount of required space per person, the fireproofing of the structure, and the required number of bathrooms, room sizes, and so forth. Housing legislation was also enacted in Philadelphia, Chicago, and other cities. In 1892, the federal government passed a resolution that authorized an investigation of slum conditions in cities of more than 200,000 people. Very little money was allocated for the study, which limited the actual investigation. Although housing laws continued to be modified, enforcement of the codes was poor until the 1920s, when the state of housing in the U.S. grew extremely unsatisfactory, both within the cities and in the farm areas. By the 1930s, President Roosevelt reported to the country that one third of the nation was ill fed, ill housed, and ill clothed. The first federal housing law was passed in 1934, establishing a better system for residential mortgages through government insurance. The Federal Housing Administration (FHA) was created to carry out the objectives of the law. Many other governmental agencies became involved in the next few years in mortgage lending and in an attempt to clear slums and to increase urban renewal. However, it was not until after World War II that substantial housing acts were passed. The first one of these was the Federal Housing Act of 1949.

Industrialization, coupled with a substantial move from farms to cities, helped to create the severe neighborhood and housing problems within the last 100 years. Industries established themselves wherever they were most profitable. If they contaminated the land, water, or air, it really was not of much importance. Obviously, since people had no access to the modern automobile, they lived close to industry or within convenient public transportation distance. As new groups moved into areas to fill job needs created by industrial expansion, neighborhoods and houses were filled by a highly transient population. The nature of the population, the overcrowding that occurred, the nature of the industrial pollutants, and the cheaply constructed housing used as dwellings by workers were the factors contributing to the severe housing problems of the 20th century.

The 1940 census indicated for the first time the type of housing conditions found in cities. New Haven, Connecticut decided to assign penalty scores to four items of the census. The resultant maps showed the concentration of substandard and presumably slum conditions. On the basis of these findings, the city made a sample environmental survey of the dwelling units. This survey indicated the condition of the houses and the apparent condition of the neighborhoods. Points were given to such things as toilets, baths, means of exit, lighting, heat, rooms without windows, deterioration, number of people per room, and the sleeping area in square feet per person. This survey indicated where slum clearance and housing improvement was needed. It was evident that any rebuilding program or slum clearance program had to be carried out in relation to the overall city plan. Revision and enforcement of housing regulations and control of blight in nonslum areas were necessary parts of the new thoughts on housing. Problems common to most areas included overcrowding due to large families, poor quality of dwellings available for minorities, and substandard properties in relation to the amount of rent paid. Furthermore, the impact of decay during the years of World War II, the inability to build or improve housing from lack of materials, and the movement into the cities because of war production caused additional housing problems. At the end of the war, the birth rate expanded, marriages increased sharply, and the demand for new houses rose at an unusually rapid pace. Veterans moved to the suburbs with their large families, creating a new type of housing pressure in areas where people had previously lived on farms. The inadequacy of many of these new dwellings built immediately after the war, the inability of the soil to handle the enormous amount of on-site sewage, and the lack of proper water supplies contributed to a new wave of serious housing and environmental problems.

Poverty and poor housing are mutually inclusive; one feeds on the other. The poorer the individual, the more crowded the living quarters. Poverty further reduces the chances of attaining a better education, occasional recreation, and occupational skills. These limitations create feelings of insecurity and helplessness and deprive the poor of alternatives. In many cases, the poor have great difficulty maintaining proper sanitary conditions in their dwellings. There is either a serious problem before they arrive or they are unable to physically, mentally, financially, or emotionally cope with the upkeep of their dwellings. The situation tends to grow worse instead of better. Housing and poverty are not simple matters. In addition to poverty, the education background, lifestyle, health practices, outlook on life, age at marriage, understanding of marital responsibilities, and the single-parent syndrome must be understood. The housing of the poor leads to situations that intensify the potentially serious hazards that exist within their neighborhoods. Housing problems have been severely complicated by the presence of gangs and drug houses and operations.

Urbanization and suburbanization have become the predominant population shifts in the U.S. Lower income groups left the rural areas of Appalachia and the South to crowd into already overcrowded cities. This, coupled with increasing population growth, caused a severe housing problem, since the supply of new homes in these areas was minimal and the older properties continued to deteriorate at a rapid rate. Many of the people in the urban areas moved to the suburbs after attaining a higher socioeconomic status in the hope of finding a better way of life for themselves and their children. This has resulted in a huge suburban sprawl that has created innumerable environmental problems.

The inner city has, in many ways, become a jungle. Not only do people live with rodents and roaches and under the most severe conditions, but a new hazard has arisen. The arsonist is creating a hazard with a pricetag of 10–15 billion dollars annually. Residents are suffering injury and death because individuals either out of hate or for profit are setting fire to old structures. In Boston, 22 people were arrested including 6 attorneys, 11 real estate brokers, 4 public insurance adjustors, 1 police officer, and 1 retired fire chief for arson. These individuals were arraigned in Suffolk County Superior Court on charges varying from fraud to bribery and murder. These arrests were just an indication of the kinds of fire problems occurring in ghetto areas. Unfortunately, in some areas the only way owners could get any money out of their property was to have the property destroyed and to cash in on the insurance. New York City increased its force of investigators from 77 to 152 to try to cope with the arson problem. The destruction of property was increasing the serious housing problem that already existed.

Inflation is also placing a strain on housing. The American dream of securing a good education and a good home is very difficult to realize. Homes that 10 years ago sold for $40,000 are now selling for $75,000. In some areas homes that were selling for $40,000 5 years ago are now selling for $65,000. In other areas homes that sold for $38,000 14 years ago are now selling for $120,000. Due to this rise in cost, many individuals are unable to move to better living quarters and are forced to remain in overcrowded run-down dwellings. In existing homes, owners are having difficulty maintaining their property. Many homeowners have had to obtain second mortgages to simply repair or maintain their property. Although current mortgages rates are relatively low, they may increase in the next year and may still go higher. Many individuals become burdened with large debts and in many cases have to move to less expensive housing to survive. In the coming months and years, the situation is not expected to improve but rather to deteriorate. The federal government continues to put billions of dollars into housing, but still has not found a way to utilize this money effectively.

Current land-use patterns show a tendency toward further urbanization, which is consuming a significant amount of valuable land for agriculture and wildlife. Some communities have begun to renovate the older well-built structures found abandoned or under-utilized in the inner cities; however, most communities are still ignoring this vital resource. Improper land use, improper use of existing structures, and improper growth are serious environmental problems in our society today.

Indoor Air Pollution

During the last 25 years there has been a sharp increase and concern by the public and various levels of government about the quality of the air we breathe inside homes, schools, and where we work. In the 1970s energy conservation measures were widely supported by the federal government through income tax breaks and also by state and local government and the public in general. The cost of energy had become extremely high. Energy conservation meant the development and use of new insulation materials and "tighter buildings." Formaldehyde became part of the insulation material in many buildings. In the mid-1960s there was less understanding of potential cancer risks due to asbestos. Asbestos has been found in a variety of buildings, especially in schools. Radon gas was observed in the 1970s. Particularly high levels of radon were found in the Reading Prong geological formation that ran through New Jersey, Pennsylvania, and New York. Energy conservation also brought on the current problem of "sick building syndrome" or "tight building syndrome."

Indoor air pollutants were found in homes, offices, schools, hotels, restaurants, various buildings, buses, trains, and planes. In each of these environments, people were exposed to a wide variety of pollutants that came from smoking, building materials, rugs, furniture, appliances, pesticides, cleaning and deodorizing agents, etc. People were exposed to increased levels of carbon monoxide, nitrogen dioxide, metals, inorganic compounds, radioactive pollutants, and volatile organic compounds (VOCs). The best known inorganic compounds are mercury, chlorine, and sulfur. The particulates include the inorganic fiber asbestos and carbonaceous tobacco smoke. The volatile organic compounds include formaldehyde, benzene, and carbon tetrachloride.

Good indoor air quality includes: introduction and distribution of adequate ventilation air; control of airborne contaminants; and maintenance of acceptable temperature and relative humidity.

SCIENTIFIC, TECHNOLOGICAL, AND GENERAL INFORMATION

Neighborhoods and Their Effects on Housing

The neighborhood is that area comprising all the public facilities and conditions required by the average family for their comfort and existence. Residents of a neighborhood share services, recreational facilities, and generally an elementary school and shopping area. Some neighborhoods were artificially established in older cities, by a particular ethnic, religious, or racial group. Neighborhood community facilities include educational, social, cultural, recreational, and shopping centers; utilities and services include water, light, fuel, sewage, waste disposal, fire and police protection, and road maintenance.

The size of the neighborhood is difficult to assess, as it varies with the types of building structures, the density of the population, the identification of the population as a specific group, and the various services available. To establish limits for the area or population seems impractical. Small developments on the outskirts of a particular neighborhood should be included within the neighborhood if they utilize its services. Logical neighborhood boundaries also include such things as rivers, topographical barriers, interstate highways, parkways, railroads, industrial areas, commercial districts, and parks.

Site Selection and the Neighborhood Environment

When developing a new neighborhood, it is essential to consider not only the aforementioned services, but also the land-use trends for the area; the presumed availability of transportation, public utilities, and schools; and the legal controls placed upon the area by the local governing bodies. A new neighborhood should contain between 2000 and 8000 persons, with the most desirable size at about 5000 persons; it should be within a maximum walking distance of 0.5 mi from the local elementary school; it should not cover substantially more than 500 acres.

Site selection for the neighborhood is extremely important. The existing neighborhood may have to change, be razed, or be redeveloped to meet proper neighborhood and housing standards. The new neighborhood should avoid the problems of existing neighborhoods by locating on proper sites. In site selection, competent professional engineers should determine the geology of the land, the type of soil, the type of weather conditions prevalent in the area, whether disturbing conditions such as superhighways, heavy industry, or hog farms are nearby, and if the site is in a flood plain. Soil and subsoil conditions must be suitable for excavation, site preparation, location of utilities, weight bearing capacity, and so forth. Test borings should be made to determine the types of soils and subsoil conditions. Soil with a high clay content will shrink when extremely dry, causing cracks in foundations and within houses. The type of groundwater and drainage conditions should also be evaluated. Drainage conditions after construction should be forecast. Areas in which basement flooding may occur, swamps or marshes are present, or the groundwater table is so high that it will periodically flood the site should be avoided. It is also essential to plan a site that will not be flooded by surface waters from streams, lakes, tidal waters, or higher land. The land must not be too steep for proper grading and usage. Buildings should not be situated at an elevation above normal water pressure. The slope of the ground must also be evaluated. Adequate automobile and pedestrian traffic must be accessible to the community site. Sufficient land should be reserved for private yards, gardens, playgrounds, and neighborhood parks. The area should not be located in the vicinity of hazardous bluffs, precipices, open pits, or strip mines.

Water supply and sewage disposal are important keys to proper site selection. New sites in the community should not be approved unless public water and public or semipublic sewage are available. Today, due to poor planning or technical ignorance, millions of homes have overflowing on-site sewage disposal systems.

It is also necessary to determine the method of removal and disposal of solid waste. Since reasonably priced electricity is essential, the availability of electricity for the planned community must be determined before the site selection is approved. In addition, if gas is available, it should be included in these determinations. Telephone service must also be planned for the new site. The accessibility of firefighting crews and equipment, and the kind and amount of police protection, has to be considered before the site can be definitely approved.

Major accident hazards have to be eliminated from the site and adjacent areas. Accident hazards include aircraft, cars, trains, and so forth. The site should not be located near the use of fire and explosion hazards, such as oil or gasoline storage, or firearms. Dumps, rubbish piles, and sanitary landfills should not be adjacent to the site. The site should also be a safe distance from unprotected bodies of water, such as strip mines and quarries. It is essential that railroad crossings are well marked and automatic signals are used. Excessive noise, odors, smoke, and dust should be avoided. A site should be so selected that pollution from sewage plants, sewage outfalls, or farms will not contaminate any bodies of water running through the community. Adequate public transportation, and pedestrian and bicycle paths should be provided for proper recreation and access to community facilities out of the area of the new neighborhood. Other types of services, such as junior and senior high schools; large shopping centers; areas of potential employment; urban areas; outdoor recreation areas; public and private health services, including physicians, laboratories, and a hospital, must be accessible to residents. During the actual development of the neighborhood, proper grading techniques should be employed to ensure adequate surface drainage so that water does not stand and cause mosquito problems or flow in such a way to cause erosion.

Property Transfer Assessments

A property transfer assessment is a determination of those areas of potential legal and economic liability that a buyer may assume by purchasing a business or industry that has actual or potential environmental problems. There is a nation-wide trend in the federal and state legislatures to hold individuals responsible for cleaning up environmental contamination on pieces of property that they own. This becomes a serious problem for the purchaser of any piece of property, since the environmental health hazards may be hidden or buried. These hazards may consist of spilt chemicals that are polluting the groundwater, asbestos insulation, or buried waste. These are three major phases in determining the problem. They are as follows:

1. Examine available records to determine the prior land ownership and use. Contact appropriate agencies and review any hazardous materials handling practices that may have occurred at the site.
2. If necessary, conduct a more detailed investigation to determine how far local contaminants have already spread through the environment.
3. Develop and evaluate alternative site cleanup plans and costs. Only after the audit has been completed and the determination made concerning hazardous materials, should the buyer consider purchasing the property.

These properties are very much a part of the community and could be located near housing developments. If there are hazards present, they may constitute an unacceptable risk for the community. At the time of sale of the property, the community has an excellent means of bringing about change in the buildings if needed.

The Internal Housing Environment

The internal housing environment, which consists of heat, light, ventilation, plumbing, and so forth, affects the physical, emotional, and mental states of the occupants. An individual needs to be protected against the elements of heat, cold, disease, insects, and harmful chemicals. People need to know that when they arrive home they can leave the pressures of society behind them and can relax in safety and comfort in order to face the challenges of society again.

Heat

The ideal thermal environment for a given individual varies with age, sex, conditions of health, and so forth. Most housing codes and air-conditioning engineers recommend a minimum indoor temperature of 68°F for winter and 75°F for summer, however, for some this temperature is too cold or too warm. The recommended temperature must be evaluated in the light of the types of individuals, the amount of air flow, and the amount of energy available. With increasing energy problems and rising energy costs, the entire concept of air-conditioning may well change in the future.

The most important aspect of the thermal environment, regardless of the selected temperature, is that the individual should not suffer from undue heat loss and that the heat within the building is not so excessive as to be oppressive. The temperatures within rooms will vary tremendously from floor to ceiling depending on the window location, the type of heating devices, and the way in which the heating system is planned. Weatherproofing and insulating of buildings reduces heat loss and retains the cool air provided by air-conditioning during hot weather.

Light

The sources of lighting in a dwelling are both natural and artificial. Few dwellings have been built to maximize natural lighting from the sun's rays, which serve not only to illuminate, but can control temperature as well. Light is essential for cleanliness and avoidance of accidents, and it contributes to healthy mental attitudes.

Ventilation

Proper ventilation is necessary to maintain a thermal environment that allows adequate heat loss from the body, removes unnecessary chemicals, and permits proper aesthetic sensibility within the home environment. Heat loss is controlled by air temperature, relative humidity, air movement, and the mean radiant temperature of the surrounding surfaces. Cool moving air is valuable for promoting restful sleep. Odors, which are not removed in a poorly ventilated area, affect the well-being of the individual.

Electrical Facilities

The electrical code, the housing code, and, if these are not available at a local level, the National Electrical Code should be referred to during installation of electrical wiring in the home. Electricity is the conversion of mechanical energy into electrical energy, often by means of a generator. Electrical voltage increases by several hundred thousand to more than a million volts when it passes through a transformer. The high voltage is necessary to increase the efficiency of the transmission of power over long distance. This high voltage is then reduced to the normal 115- or 230-volt household current by a transformer located near the home Electricity is transmitted to the home by a series of wires called a service drop. In order for an electrical current to flow, it must travel from a higher to lower potential voltage. The current flows between hot wires, which are colored black or red, at a higher potential to the ground or neutral wire, which is colored white or green, at a lower potential. The voltage measures the forces at which the electricity is delivered. Electrical current is measured in amperes, which is the quantity of flow of electricity. Wattage equals the number of volts times amperes.

The Earth is an effective conductor. Near the house the usual ground connection is a water pipe of the city water system. Care must be taken that this ground is properly installed so that an individual will not be shocked by touching both sides of the pipe if the water meter is removed.

Plumbing

Plumbing is defined as the practice, materials, and fixtures used in installing, maintaining, and altering all pipes, fixtures, appliances, and appurtenances that connect with sanitary or storm drainage, a venting system, and the public or individual water supply system. (An in-depth discussion of plumbing can be found in Volume II.)

Structural Soundness

The structural soundness of a dwelling is related to the prevention of disease and injury, and the promotion of good health. The site selected should follow the requirements set forth under neighborhood site selection. The site must not be approved if a public sewage system is unavailable or the land is unable to absorb the fluid from on-site sewage systems.

The building must be placed on the site in such a way as to provide sufficient outdoor space, adequate air circulation, quiet, and safety. The location of the building may reflect certain conditions that cannot be corrected, such as the trajectory of the sun, the direction and velocity of winds, the general climate, and the nature of the water runoff from other properties or the potential for water problems created by heavy rains or snows. The ground slope is important, since this may well control the dryness and the safety of the structure. The site selection and the type of housing construction are dictated by local zoning, housing ordinances, and the amount of money available.

Daylight, direct sunlight, and the heat of the sun play an important role in the placement of the structure on the site. The building should be located on the site to maximize year-round exposure to sunlight in each of the rooms. This means at least 1 hr of sunlight on a clear day in each room. Daylight should be available to all habitable rooms during the daytime hours to reduce the additional energy requirements and also to provide a more natural setting.

In site selection and housing construction, it is essential to consider the free circulation of air around and through the structure. This helps to improve the indoor thermal environment and to reduce the cost of air-conditioning. During the winter, proper insulation will keep the cold air out of the house. In summer thermal comfort is a combination of air temperature, air movement, and relative humidity. The air movement within the building is a combination of the amount of air that penetrates into the building and the wind velocity. It is generally most comfortable if the effective temperature within the structure in the evening is not more than 72–75°F.

Site selection must also be based on the avoidance of excessive noise due to exterior problems such as factories, railroads, and highways. By orienting bedrooms away from recreational areas within the house and noisy outdoor areas, individuals are able to sleep uninterrupted. The noise level should not exceed 50 dB within the dwelling unit and 30 dB within study or sleeping areas.

In residential areas the site must be so oriented that the structure is placed an adequate distance from the road and other buildings.

Physical Structure

The basic physical structure of the house consists of the foundation, the framing, the roof, the exterior walls and trim, and roof covering. Since this chapter is not intended as a guidebook on building construction, many of the details that would ordinarily be covered are excluded.

Foundation refers to the construction below grade, such as the footings, cellar, and basement walls. It also refers to the composition of the Earth on which the building rests and other supports, such as pilings and piers. The foundation bed may be composed of solid rock, sand, gravel, or clay. Sand and clay are the least desirable, for they may shift, in the case of sand, or swell and shrink, in the case of clay, leading to sliding and settling of the building. The footings should distribute the weight of the building over a large enough area of ground to ensure that the foundation walls will stand properly. Footings usually are composed of concrete.

Foundation wall cracks occur for a variety of reasons, including the previously mentioned unfavorable subsoil, as well as improper construction or earthquake tremors. The foundation walls support the weight of the structure and transfer weight to the footings. These walls may be composed of stone, brick, or concrete. They should be moisture-proof, which involves the use of plastic sheeting joined with tar or asphalt and footing drains around the exterior of the walls to take moisture away from the property. The basement or cellar floor should have at least 6 in. of gravel beneath it and be composed of concrete. This protects the basement from rodents and flooding. The gravel distributes the groundwater moving under the concrete floor and reduces the potential of water penetrating the floor. Again a plastic shield should be laid before the concrete is poured. All basement doors and windows should be water-tight and rodent-proof.

HOUSING PROBLEMS

Health and the Housing Environment

There are approximately 12 million crowded substandard dwellings in the U.S. This comprises about 15% of all available housing. Four million are in such poor condition that they cannot be refurbished without major repairs.

Twenty-five million home accidents occur in the U.S. each year because of faulty appliances, electrical connections, poor lighting, broken furniture and equipment, and stairs, floors, and walls in need of repair. There are an estimated 60,000 cases of rat bite reported each year. Since many incidents of rat bites are unreported, the problem is far greater than the actual statistics indicate. Six thousand cases of rat-transmitted diseases occur each year. Accidental poisoning due to inadequate storage facilities in the home and improper storage of hazardous materials and chemicals causes 3000 deaths and one million injuries. There are over 400,000 children with unacceptably high levels of lead in their blood. Lead comes from paint chips, plaster chips, and windowsills and other woodwork in old houses painted with lead-based paint. Improperly constructed, installed, or maintained home-heating devices cause at least 1000 deaths and 5000 injuries from carbon monoxide poisoning annually. In 1986, there were 20,500 home accident related deaths, and 3,100,000 disabling injuries, at an estimated cost of 16 billion dollars.

In many areas, houses are abandoned because it is cheaper to leave them than to repair them or pay the property taxes. As a result, houses become dilapidated, vandalized, and eventually become places where young children, teenagers, and adults can get into serious trouble. Not only are such buildings aesthetically unpleasant, but they also become dumping places for garbage, trash, and other junk; breeding places for roaches, mice, and rats; and a meeting place for alcoholics, drug addicts, and criminals.

Disease Problems

Infectious diseases, chronic diseases, and environmentally created diseases are more prevalent in poor housing than in better housing. While congestion within the house can lead to increased upper respiratory diseases because of the close contact with contaminated individuals, and the spread of diseases occurs often in kitchens where food is prepared along with other activities taking place, such as baby diapers being washed, it is difficult to attribute any given disease to the housing problem. A number of additional variables should be considered. These include the basic personal hygiene practices and habit patterns of individuals, their susceptibility to certain diseases, and a vast variety of additional stresses. We know that within a defective housing structure there are various stresses, such as noise, improper lighting, inadequate space, improper ventilation, the presence of myriad insects and rodents, and a variety of solid and hazardous waste. These wastes include aerosols, oven cleaner, drain opener, nail polish, detergents, medicines, waxes and polishes, paints, pesticides, fertilizers, charcoal lighter fluid, gasoline, motor oil, and antifreeze. Surveys indicate that individuals living in substandard housing with the aforementioned problems have higher infant mortality rates, a greater level of disease, poorer health, more nutritional and dental problems, and a variety of other health defects. Even though a specific disease may not be traced to a specific type of housing problem, with the exception of an outbreak of typhoid fever being traced to a typhoid organism found in a defective plumbing system, it is still recognized that disease rates are higher among substandard housing dwellers.

Environmental stress not only reduces our ability to fight off infectious disease, but also causes rather specific environmentally related diseases. For the chronically ill, environmental stress creates further problems, contributing to deterioration of their conditions. Noise is a specific environmental problem. Today humans live in shelters that have increasingly high levels of noise. The noise may be due to a lack of soundproofing in poorly or cheaply constructed buildings, and neighborhoods in or near commercially zoned areas and traffic routes. Noise causes nuisances, irritability, and loss of sleep. Eventually the noise may lead to actual, temporary, or permanent hearing loss. Some of the greatest sufferers from noise pollution are those who live within the vicinity of noisy industries or airports. The level of sound well exceeds that recommended for a home situation. Thirty-five decibels of background noise is acceptable at night, with occasional sounds reaching 45 dB. Forty-five dB is acceptable during the day, with occasional sounds reaching 55 dB. Annoyance occurs in residential areas when the average continually exceeds 50 dB of noise in the background. Some data suggest that 60 dB may cause autonomic changes. Surveys indicate that individuals may get headaches or become nervous when the sound levels exceed 50 dB on a 24-hr average.

Ventilation is an important part of stress or the stress reduction factor within shelters. A minimum of 5 ft^3/min/person of uncontaminated air should be provided within the residential environment. The use of mechanical ventilation today may be good or bad, and if fresh air is added rather than inside air recirculated, then home ventilation may be good. In recirculated air, as many as 100 identifiable contaminants have been found. These include pesticides, cleaners, bacterial contaminants, and viral contaminants.

The thermal environment helps reduce or increase stress, depending on the combinations of temperature and relative humidity present and the velocity of the air. A proper temperature within the home should range from 68–75°F with relative humidity ranging from 20 to 60%. Temperatures or relative humidities above these levels tend to cause discomfort, and in extreme cases may even lead to death if an individual is already subjected to serious diseases.

Illumination levels are essential in providing a good pattern of daily activity. When lights are too low or glaring, they cause fatigue and may lead to a variety of accidents or additional stress upon an individual.

There are numerous cases where the external environment has contributed sharply to the problems within the home. In Texas, where a substantial Mexican-American population lives, there are many areas where silt and dirt filter through the cracks in the buildings into the kitchens. This contributes to improper handling of food and may lead to disease. In other areas, migrant workers have been periodically sprayed with pesticides used for the control of pests in the fields. These conditions are not unique. They may be found in every major or small city in the country.

Indoor Air Quality Problems

The serious concern over pollutants in the indoor air is due to the fact that indoor pollutants are not easily dispersed or diluted. The indoor air pollutant concentration is often many times higher than the outdoor concentration of these pollutants. At times indoor pollutants have exceeded the upper levels by 200–500%. Further, people spend as much as 90% of their time indoors, which subjects them to greater levels of risk. The degree of risk depends on how well the buildings are ventilated and the type, mixture, and amount of pollutants present in the structure. Improperly designed and operated ventilation systems add to these problems. Individuals complain of eye, nose, and throat irritations, fatigue, shortness of breath, sinus congestion, sneezing, skin irritation, dizziness, lethargy, headaches, nausea, irritability, and forgetfulness. Long-range health effects can include impairment of the nervous system as well as cancer. The harmful indoor air pollutants, in addition to the inorganic and organic compounds previously mentioned, include mold, viruses, bacteria, and fungi. In addition, there is a decrease in productivity, increase in deterioration of furnishings and equipment, and a decrease in good morale and human relations.

Once a substance enters the indoor air environment, it is difficult to remove. These substances are in constant motion, because of the air currents and people-type activities occurring within the building. The movement allows the substance or microbe to come in contact with other substances and either interact or become attached to other substances. Because there has been a reduction in ventilation rates from approximately 10 ft³/min to 5 ft³/min in these buildings, the amount of air turnover or potential removal of these substances has been decreased sharply.

The various pollutants or microorganisms may produce either acute effects if the contamination concentrations are high enough, or chronic effects if there is a prolonged exposure to the contaminants at low concentrations over an extended period of time. *Sick building syndrome* is a general term applied to a group of illnesses where an acute exposure has occurred. This syndrome was rare before 1977, but by 1985 it was responsible for 13% of all of the complaints received by NIOSH. There are five different forms of this illness. The first form is an allergic response called humidifier fever. The symptoms are breathlessness, flu-like symptoms, and a dry cough. The cause of the illness is the growth of microorganisms in the water that is used to cool the air conditioning units, or microorganisms in the water of humidifiers. The second form includes common allergies. The symptoms are runny nose, sneezing, and irritated eyes. The causes are fungi, mites, pollens, and dust that have entered the indoor environment and are present in the air. The third type is infectious diseases. These diseases are caused by microorganisms that are found in cooling towers, ice machines, shower heads, and water condensers. The symptoms include coughing, fever, and pneumonia. Legionnaires' disease is an example of this. The fourth type is due to allergic reactions caused by synthetic material. The symptoms are rashes, irritation, itching, and eye problems. The causes are probably the release of chemicals within the buildings. The fifth type, the largest one of all, is due to unknown causes. The symptoms are mucus, irritation, sneezing, headache, sore eyes, and sore throat. The symptoms appear when a person enters the building and disappears shortly after he or she leaves the building. Three possible explanations for this fifth type include the release in very low concentrations of VOCs, a psychological reaction by the employee because of a dislike of his or her work, or a psychological reaction to a traumatic situation. Building-related illness refers to illness brought on by a specific environmental agent such as the organism causing Legionnaire's disease.

Little information is available on the chronic effects of indoor air pollution. The problems of lung cancer caused by radon gas and the potential of cancer due to asbestos or health problems related to lead have been based more on what is occurring in industrial settings than that which is occurring in the pure indoor air environment of homes and businesses.

Age is a factor in determining how a person will be affected by indoor air pollutants. The groups most affected are the very young from birth to 10 years of age, and the elderly 65 and up. The reasons for this include the amount of time they spend indoors, the condition of their immune system, their physical condition, medications that they may be taking, and their current health problems. In small children there is a relationship between the weight of the contaminant, especially if it is a chemical or inorganic material, and the weight of the child. In the elderly, many of the individuals already have some form of upper respiratory distress, and therefore, are more affected by indoor air pollutants. Some individuals are highly sensitive and are allergic to the particular contaminant. When an antigen enters the body, there is an immune response, therefore, the greater the dose an individual receives, the greater the response, which in itself creates a harmful effect within the body.

The sources of indoor air contaminants include:

1. Outside sources
    a. pollen, dust, spores, industrial pollutants
    b. general vehicle exhaust
    c. odors from dumpsters
    d. reintrained (drawn back) air into building
    e. radon from soil, contruction materials, or water

      f. leakage from underground storage tanks

      g. chemicals from land disposal site

      h. pesticides

2. Heating, ventilation, and air conditioning equipment

      a. dust or dirt in ductwork

      b. microbial growth in drip pans, humidifiers, coils, and ductwork

      c. improper use, storage, disposal of chemicals

      d. leakage of refrigerants

      e. emission from equipment such as volatile organic compounds, ozone, solvents, toners, ammonia

3. Human activites

      a. smoking, cooking, body odors, cosmetics

      b. cleaning materials, deodorizers, fragrances, trash, dust from sweeping

      c. maintainence, airborne dust, voatile organic compounds, pesticides, microorganisms in cooling towers, animal dander, insect parts

4. Building components and furnishings

      a. carpeting, textiles, deteriorated materials, asbestos, new furnishings

      b. water damage, odors, molds, microorganisms

5. Accidental events

      a. spills, fire damage, soot, odors

6. Special use areas

      a. smoking rooms, laboratories, print shops, food preparation areas

The major route of entry for the indoor air pollutants is the respiratory system. The nasal passages are the major portals through which the air enters the body. Two filtering mechanisms are supposed to remove the particulate matter. The first is the nose, which includes its hairs. The second are mucous membranes that line the interior of the nasal passages. If these mechanisms do not work effectively, the particles may cause further harm. The oral route is the path through which the contaminants may go down into the gastrointestinal system and could eventually cause disease (such as cancer), depending upon the type of chemical ingested. The eye is also a portal of entry for indoor air pollutants. Various chemicals can be dissolved in tears and therefore move into the bloodstream.

Residential Wood Combustion

The problems associated with wood combustion include incomplete combustion, the type of wood, the type of combustion chamber, inadequate ventilation, and the burning of lumber that has been treated with chemicals. Carbon monoxide is probably the major pollutant emitted during the burning of wood. Nitrogen oxides are produced. They may be irritating to the eye and throat, and may cause coughing. Polycyclic Aromatic hydrocarbons (PAHs) may be produced and may be carcinogenic. Further, a large portion of the particulates from wood combustion are in the respirable particle range, which is less than 3 μm. These particles can be deposited in the deep alveolar region of the lungs and therefore help contribute to pulmonary disease.

Environmental Tobacco Smoke

Passive smoking is an involuntary second-hand inhalation of tobacco smoke by nonsmokers. The nonsmokers may receive the exposure either through the workplace or through the home. Two groups of individuals are within the category of nonsmokers. The first one is the exsmoker who has quit direct smoking but continues to be a chronic passive smoker, since he or she is still inhaling smoke from others. The second group are the true nonsmokers who are inhaling smoke from others. The two types of smoke entering the indoor environment are the "mainstream" smoke, which is inhaled directly through the butt of the mouthpiece during the pumping process and then exhaled into the environment. Second is the "sidestream" smoke that is produced while the tobacco is smoldering. The four places in which nonsmokers are subjected to passive smoking problems are in private homes, public building, public transportation, and office buildings. Passive smoking can result in at least seven different types of potential health problems. They are

1. Lung cancer — the risk of developing lung cancer is enhanced through passive as well as active smoking.
2. Chronic air flow limitation — passive smokers experience chronic air flow limitation when compared to a control group of individuals who have not been exposed to passive smoke.
3. Small-airway dysfunction — passive smoking increases the degree of small airway dysfunction.
4. Asthma symptoms — maternal smoking is a contributor to asthma development in the newborn child.
5. Acute cardiovascular response — nicotine is the primary constituent of tobacco smoke that affects the cardiovascular system. Acute cardiovascular alterations can result from the inhalation of smoke.
6. Pneumonia and bronchitis in infants — maternal smoking can increase the risk of infants developing pneumonia and bronchitis.
7. Carboxyhemoglobin affects on health — carbon monoxide is a gaseous constituent of tobacco smoke and it can combine with the hemoglobin in the blood to form carboxyhemoglobin.

Over 500 compounds are present in tobacco smoke. Proven animal carcinogens such as PAHs and N-nitrosamines are released from tobacco smoke. Carbon monoxide has been linked to arteriosclerosis and cardiovascular disease in people. Nitrogen oxides enhance the development of emphysema.

The potential dangers from the sidestream cigarette smoking to the nonsmoker are compounded by the amount of time of exposure to tobacco smoke, sex, age, ambient pollution exposure, size of family, occupation, drug consumption, alcohol consumption, and general health and immunological response of the individual.

Carbon Monoxide

In addition to cigarette smoke, which has been just discussed, carbon monoxide is also produced in the residential setting by improper combustion of fuel. On the average, the carbon monoxide levels appear to be higher in mobile homes than in regular housing. Improper ventilation in the tightly sealed mobile home and improperly functioning appliances are the two main reasons for higher carbon monoxide levels. One of the most common sources of emission is from the faulty furnace flues. The furnace is typically located very close to the living quarters, and therefore, any escape of carbon monoxide may enter the breathing space of the individuals very quickly. In addition, gas stoves and space heaters used in mobile homes may be hazardous. In the conventional house, carbon monoxide levels are increased by improper combustion and poor ventilation from gas stoves, gas water heaters, and gas furnaces. The carbon monoxide is produced by the incomplete burning of the gas, and once produced may be distributed rapidly through the forced hot-air system. If there is an attached garage and heating or cooling vents are present in the garage, carbon monoxide produced from the automobile may seep through the vents into the remainder of the house. One other exposure in the house would be the improper burning of fuel in a fireplace.

Radon

Radon is a naturally occurring gas that results from the radioactive decay of radium. The primary sources of radon gas in the home are from the soil and rocks upon which the house is built (Figure 7.1). Radon gas may also be present in well water. Occasionally, radon gas may come from building materials. The radon enters the buildings through cracks in the foundation. When inhaled, radon can adhere to particles and then lodge deep in the lungs. This increases the potential risk of cancer.

The EPA estimates that radon may be responsible for 5000–20,000 lung cancer deaths per year. Radon gas is a particularly bad risk for smokers who already have a health risk of cancer 10 times greater than nonsmokers. Radon-222 is a radioactive gas, with a half-life of 3.8 days. It is colorless, odorless, and chemically inert, but is fairly soluble in water. Radon gas is nine times heavier than air, and therefore, remains close to the ground, diffusing into the stagnant air of wells or crawl spaces under houses. Radon goes through a series of transformations, emitting alpha, beta, and gamma radiation. The radioactive elements that result from the decay of radon are called radon progeny or radon daughters. The radon

granite   water

cracks in concrete slab

cracks between poured concrete (slab) and Blocks

pores and cracks in concrete blocks

loose-fitting pipes

mortar joints

weeping tile

slab footing joints

exposed soil, as in sump

Fig 7.1.   Radon access routes into house.

daughters formed during the decay process include polonium-218, lead-214, bismuth-214, polonium-214, lead 210, bismuth-210, polonium -210, and finally the stable isotope lead-206. The amount of radon gas flowing into a home is primarily dependent on the radium content of the ground. Certain building materials, including concrete, brick, cement, granite, gypsum wallboard, and concrete blocks, have been found to contain radon. Radon daughters, unlike radon gas, are chemically active metals. These elements will stick to anything that they come in contact with, such as particulates in the air, on furniture, and in the room. When radon daughters are attached to particulates and come in contact with the air passageways in the lungs, they may adhere to the surfaces. The biological effects of radon can be attributed mainly to its alpha-emitting products.

Radon concentrations in indoor air are influenced by four main groups of factors. They include the properties of the building material in the ground, building construction, meteorological phenomena, and human activities. Construction materials, such as concrete, brick, and stone, can emit radon. Since radon originates from the bedrock, the levels of radon at the Earth's surface is affected by the thickness of the unconsolidated materials, the type of unconsolidated materials, and the level of uranium concentration in the bedrock. The thicker the unconsolidated material and the further away the bedrock is from the surface of the ground, the lower the radon level. Course grain sizes of gravel and sand are highly permeable and therefore allow more radon to escape. Mud and wet clays have low permeability and, therefore, allow little or no radon to escape. If there is poor water percolation, then there is a low radon availability, unless the ground is the site of a high uranium yield area. The meteorological phenomena relates to the amount of air that flows through the area and therefore is available for ventilating the buildings. The human activities include the rate of ventilation in the building, as well as the rate of smoking and burning of wood within the building.

Radon enters the building through dirt floors, cracks in concrete floors and walls, floor drains, sumps, joints, and tiny cracks or pores in hollow-block walls. Radon may also enter the house in water from private wells and may then be released into the home when the water is used in showers, in the kitchen, or in bathroom sinks.

Formaldehyde

Formaldehyde is a probable human carcinogen. It is responsible for a variety of acute health problems, such as eye and nose irritation and respiratory diseases. People who have lung diseases or an impaired immune system, children, and the elderly may be affected by formaldehyde. Formaldehyde is used in furniture, foam insulation, and pressed wood products, such as some plywood, particle board, and fiberboard.

Formaldehyde is a colorless gas with a characteristic pungent odor. Eight billion pounds of formaldehyde are produced in the U.S. each year. Formaldehyde is also found in cosmetics, deodorants, solvents, disinfectants, and fumigants. Urea-formaldehyde (UF) is a resin that is produced from a mixture of urea, formaldehyde, and water. It is most commonly used as an adhesive in plywood, particle board, and chipboard. The emissions from the UF boards and other materials are due to unreacted formaldehyde that remains in the product after its manufacture. It also occurs from the breakdown of the resin by its reaction with moisture and heat. Urea-formaldehyde foam has been used extensively as a thermal insulation in the walls of existing residential buildings. It is injected into the wall cavities through small holes that are sealed afterwards. The insulation is accomplished by mixing the UF resin with a foaming agent and compressed air. The material then looks like shaving cream and can be pumped through a hose and forced into the wall cavity, where it will cure and harden. Particle board is made by saturating small wood shavings with UF resin and pressing the mixture at a high temperature into its final form. Particle board can emit formaldehyde on a continuous basis for several months to several years. In buildings in which these products are used, either for partition walls or furniture, the formaldehyde can reach concentrations that will cause eye and upper respiratory irritation. Where the air-exchange rates are low, the concentration can reach 1 ppm or more. In contrast, exposure to formaldehyde at a concentration as low as 0.05 ppm in sensitive people, under certain conditions of temperature and humidity, may cause burning of the eyes and irritation of the upper respiratory system. Concentrations higher than a few parts per million, may produce a cough, constriction in the chest, and wheezing. Menstrual disorders are one of the most frequent problems caused by exposure to formaldehyde.

The environmental factors mentioned earlier relating to temperature and humidity are important in determining the overall affect of formaldehyde on people. If the average indoor temperature is raised by 10°F, the levels of formaldehyde will double. If the average indoor temperature is decreased by 10°F, there will be an approximate 50% decrease. An increase of relative humidity from 30–70% will cause a 40% increase in the formaldehyde present. During very cold and windy days, if the house is not well insulated, the level of formaldehyde will drop because of the infiltration of air and the mixing and dilution that will occur. Generally, the greater the temperature difference, the more ventilated air is available for formaldehyde dilution. Formaldehyde levels indoors increases during the winter months and decreases during the summer months because of the amount of air that is available.

Other Volatile Organic Compounds

Other VOCs commonly found in the indoor environment include benzene from tobacco smoke and perchlorethylene emitted from dry-cleaned clothing, paints, and stored chemicals. VOCs can also be emitted from drinking water. Twenty percent of water-supply systems have detectable amounts of VOCs, although only 1% exceed the 1990 Safe Drinking Water Act standards for VOCs.

Pesticides may be used indoors or outdoors for termite proofing. Preservatives may be used for the treatment of wood. Even when pesticides and preservatives are used properly, they may release the VOCs. Wood preservatives, which may cause cancer, include creosote, coal tar, inorganic arsenicals, and penta-chloro-phenol. Household cleaners, solvents, and polishes of all types may release VOCs.

Asbestos

Asbestos fibers have been shown to cause lung cancer and other respiratory diseases. Asbestos has been used in the past in a variety of building materials, including many types of insulation and fireproofing materials, wallboards, ceiling tiles, and floor tiles. When buildings containing asbestos materials are demolished, tiny asbestos fibers are freed, and these then can be inhaled by the individuals. Even during the normal aging process of materials, the deterioration may be released as asbestos fibers, which can be inhaled and accumulated in the lungs.

Buildings constructed between 1945 and 1978 may contain asbestos material. These materials were applied by spraying or trowling onto ceiling or walls. Asbestos has been used as a thermal insulator to protect walls, floors, and people from intense heat produced by stoves and furnaces. Some door gaskets and furnaces, ovens, and wood and coal stoves may contain asbestos. Pipe insulation manufactured from 1920 to 1972 contained asbestos to prevent heat loss from the pipes and provided protection from being burnt by hot pipes.

The reason why asbestos was used was because of its durability, flexibility, strength, and resistance to wear. It was also used because of its great insulating properties.

Asbestos fibers may remain airborne for periods of time depending on the fiber diameter, and to a lesser extent, the fiber length. An asbestos fiber will settle in stale air in a 9-ft high room in approximately 4 hr if it is 5 μm long and 1 μm in diameter. The same length of fiber with a 0.1 μm diameter will remain airborne up to 20 hr. If the air becomes turbulent, the fiber may stay airborne for long periods of time.

Asbestos has been found in office buildings where the practice used to be to fireproof these high-rise buildings with materials that contain 10–30% asbestos. It is still a major source of indoor-generated asbestos in these buildings, because the asbestos fibers are recirculated by the air-conditioning system. A study of carpet vacuuming showed that overall airborne concentrations after vacuuming was 4 times as great as before vacuuming. In schools the problem of asbestos has become acute. The EPA estimates that approximately 15 million students and 1.4 million teachers and other school employees use buildings that contain asbestos. School corporations are now in the process of trying to correct this problem.

Asbestos may cause four major types of health effects. Lung cancer, which may take as long as 20 years to develop, is known to cause 20% of all deaths among asbestos workers. The total amount of lung cancer present in these workers depends on their length of exposure, age, and how long they worked with asbestos. Smoking increases the risk of dying from lung cancer by 92 times as much as nonsmoking. The level of lung cancer among other individuals exposed at far lower concentrations needs to be explored. Asbestosis is a disease that involves a permanent scarring of the lungs caused by the inhalation of asbestos fibers. Asbestosis lung scarring may continue after the exposure to the asbestos ends. The symptoms include coughing, shortness of breath, and a broadening of the fingertips. Eventually heart failure may occur because of the lung obstruction.

Cancer can also develop in the digestive tract because of the consumption of asbestos in food, in beverages, and in the saliva that has been contaminated with cigarette smoke. Mesothelioma is a diffuse cancerous tumor that spreads very rapidly over the surface of the lung (pleura) and abdominal organs (peritoneum). This also may be due to exposure to asbestos.

Behavioral Problems

The density of individuals or the crowding factor within a dwelling can produce nervous irritation that is detrimental to mental and physical health. Crowding makes people more irritable, and causes interruptions and personal clashes that may end in deep-seated repressed bitterness or hatred. Crowding deprives individuals of personal space, which is a basic need of all animals. Privacy, another essential component for a healthy existence, is virtually impossible to obtain in crowded quarters when there is no place where an individual can go off and be alone. Sometimes it is necessary for people to "cool down" to avoid turning a misunderstanding into a disastrous argument. Even loving couples need time to be alone to assimilate what is occurring and to be able to redirect their thoughts and their desires.

Overcrowding increases environmental stresses by increasing noise levels, decreasing ventilation, and increasing odors, poor housekeeping, and solid waste materials. Behavioral problems resulting from inadequate control in crowded dwellings may increase infectious diseases through lack of personal hygiene. This obviously amplifies the major oral-fecal, airborne, skin, and respiratory routes of transmission of disease.

Alcoholism is often a problem in overcrowded housing, not as a direct result of population density, but rather because the alcoholic, unemployed, or chronically ill person is frustrated by his or her inability to provide adequate facilities necessary for a healthy, happy family life. Where population density is high, children lack the privacy for proper study, which creates additional problems in attempting to raise their educational level. Generally, the cycle that is formed by poor nutrition leading to disease, and disease leading to poor nutrition, may also be applied to education and lack of ability to find adequate places to study or concentrate. Poor education may lead to poor housing and poor housing, in turn, may lead to poor education.

Types of Housing

There are numerous types of dwellings. Those under discussion here include the single-family home, multiple-family dwellings, trailers, transient housing, dormitories, and emergency housing.

Single-Family Dwellings

The single-family dwelling is defined as one unattached dwelling unit inhabited by an adult person plus one or more related persons. Single-family homes vary from small cabins to mansions. The significance of the single-family home is that it contains only family members or close friends, thereby reducing the potential for conflict and neglect of property. Single-family homes are subject to housing inspections, depending on the housing code, in like manner to other dwellings. They vary in upkeep from poor to excellent, and in environmental problems from great to none. Some single-family homes lack water, adequate sewage, proper waste disposal, adequate lighting, and plumbing. Others have excellent physical facilities, including sufficient space to accommodate family members or close friends.

Multiple-Family Dwellings

A multiple-family dwelling is any shelter containing two or more dwelling units, rooming units, or both. The multiple family dwelling varies from the duplex, which is a single structure with two complete and separate units, to apartment houses, which contain many separate dwelling units. Apartment houses vary from condominiums, where the occupants own their separate units, to reconverted houses, which are single-family dwellings converted to multiple-family dwellings. The latter group is of greatest concern to housing authorities. A subdivided single-family home once inhabited by four to seven people may be converted to a dwelling inhabited by six to eight different families with as many as 50–60 family members.

Further difficulty arises from the enormous number of people crammed into a structure originally built for smaller numbers. These structures tend to deteriorate rapidly. The windows break, the wallpaper and paint peels, the stairs decay, and solid waste may be found everywhere — in the hallways, in the basements, and in the backyards. Unfortunately, the tenants of such a structure are generally poor and lack the capabilities to move to a better housing situation. The types of housing problems discussed earlier in this chapter are all found in these overcrowded converted structures.

Apartment houses, unless well protected by guard systems, may be very dangerous for residents. Burglars, muggers, and attackers frequent poorly supervised apartment houses. Many high-rise buildings, built for the urban poor in the inner city on redeveloped land, have become places for crime, dirt, and destruction. The concept of multistory inner-city housing is a disastrous failure in reality. In St. Louis, a development that cost many millions of dollars had to be destroyed because the occupancy rate had fallen so sharply. It is impossible for the inner-city poor to live under these conditions in a satisfactory manner.

Travel Trailers and Mobile Homes

The major environmental health problems are the same as in other forms of housing. In addition, mobile homes are being driven to out-of-the-way places and are being used as permanent dwellings. Specific problems relate to access to adequate water supplies and on-site sewage disposal. Travel trailers are being used as permanent residences in some areas. These units are neither large enough nor properly equipped to fill the needs and requirements of a permanent dwelling. In addition, dangerous storms frequently cause serious damage to this type of housing.

Transient Housing

Motels, hotels, and lodging houses are transient dwellings where occupants reside for a day, a week, and possibly longer. The problems emanating from this type of housing are similar to those found in hospitals, nursing homes, convalescent homes, and other short-term medical facilities. The major difference is the individuals within transient housing units are supposedly healthy individuals and are under no form of supervision. Many individuals fall ill when traveling, or may be carriers of a variety of diseases and spread the microorganisms throughout the transient housing environment.

The motel or hotel has the same basic kinds of problems as the medical facilities involving housekeeping, laundry, physical plant, food, solid waste disposal, insects and rodents, and so forth, but in a lesser degree.

Dormitories

College dormitories are basically similar to hotels, with the principal difference being that individuals occupy dormitories for several months rather than several days. The one problem that frequently occurs in dormitories is the spread of contagious or infectious diseases. In the event of illness, a dormitory resident should be removed to the student health center, placed in isolation, and properly treated.

Rehabilitation vs Redevelopment

In the past, redevelopment was the technique most widely used for the elimination of slum areas and for the reuse of land. Run-down and neglected structures were repossessed and razed by the local government using the right of "eminent domain." The land was then utilized for a variety of purposes, including the building of roads, parks, university buildings, and high-rise and low-rise housing for the poor. Several problems have developed as a result of redevelopment. The cost was extremely high. The families living in the area prior to redevelopment had to be relocated to neighboring areas, which in turn realized a devaluation in property and eventually were transformed into the same kinds of slums.

The high-rise multiple-family dwellings, commonly referred to as *projects*, quickly deteriorated, becoming sources of crime and dirt. Unfortunately, the occupants of these high rises, many of whom were relocated due to redevelopment, were not trained in the proper use of these new buildings and facilities. Some of the better properties, which were vital resources and which could, in fact, have been rehabilitated, were destroyed.

ACCIDENTS AND INJURIES

Accidents are the fourth leading cause of death among the total population and the leading cause of death for the age group of 1–34. Accidents cost more than 118 billion dollars a year in cash, lost time, medical treatment, and insurance. The home is the most frequent place of injury. One third of all nonfatal accidental injuries and one fourth of all fatal injuries occur in the home and its surrounding environment. Over 15 million people injured annually in home accidents need medical attention. The two most susceptible or highest risk groups are the very old and the very young. Accidental deaths and injuries occur in all socioeconomic groups and in all areas. The home environment has been altered tremendously over the last few years because of many new kinds of labor-saving and recreational devices. People have not kept up with the safety precautions necessary to use these products effectively. There are 3,100,000 disabling home injuries annually and 80,000 of these are permanently disabled. There are 20,500 deaths each year.

An accident may be defined as an event or incident resulting from the interaction of people with things and the environment. A home accident occurs within the home or its immediate environment. Injury is some hurt or harm done to the body as a result of the accident. A hazard is any source of danger.

Accidents are due to glass, gas, tools, pesticides, toys and other consumer products, and so forth. The Consumer Product Safety Commission estimates that 28,000 people die and 300,000 are injured each year from consumer products. In 1994 the EPA estimated that 81,000 children were involved in pesticide poisonings or exposures. Accidents do not occur spontaneously; people are involved in one way or another. People must understand the potential hazards of the home environment and the precautions necessary to avoid accidents. They must understand the problems related to alcohol and drug use and their relationship to accidents. Small children must be supervised properly.

The two most dangerous areas in the entire house are the kitchen and bathroom. In the kitchen, fires may be caused because of thoughtless situations or inoperative equipment. People leave dish towels, paper, and other combustibles on or near stoves. Frying foods are left without proper supervision, and if the temperature of the oils becomes too high, they may ignite to cause a grease fire. People wear clothing with long, loose sleeves, resulting in food being overturned or garments catching fire. Pot handles are turned outward and in reach of toddlers, who may easily upset hot liquids and suffer severe burns. Unfortunately, the kitchen becomes a storehouse for most aerosol-type cans. Although many of the labels clearly indicate that the cans should be stored away from direct sunlight and temperatures over 120°F, individuals do not read the labels, creating serious potential hazards. Since the aerosol can is under pressure, it will explode like a hand grenade at high temperatures and fragments, including flammable materials, may shoot out in all directions.

Houses using LP or any other gas must be checked carefully to ensure that there is no leakage. Gas leaks in stoves and ovens may be due to faulty or improperly cleaned equipment. Carbon monoxide poisoning is another danger that will occur when there is inadequate combustion of the gas in the stove. If the flame is irregular, slow, and yellow, the burner is not working properly and carbon monoxide may be produced. Other incidents of carbon monoxide poisoning occur when heaters are improperly serviced or chimneys or flues are clogged. All equipment used for heating other than that serviced by the central heating system should be checked carefully, since the equipment may not be vented properly to the outside. As houses become better insulated, the incidence of carbon monoxide poisoning will increase.

Poisoning is a particularly serious hazard within the kitchen, bathroom, and garage. In these areas, children often have access to furniture polish, waxes, all types of household cleaners, bleach, medications, gasoline, and a variety of other materials stored for household use. Within the kitchen, falls occur from spilled grease, water, or other liquids. Individuals also create hazards by standing on rickety boxes or chairs to try to reach objects on cabinet shelves. The bathroom is a notorious area for accidents. Very serious falls occur because of wet and slippery floors, groggy individuals entering the bathroom at night, and narrow spaces in many bathrooms. Poisons are mistakenly swallowed by half-awake individuals when there is little or no illumination.

In new homes there is special concern about fireplaces. The fireplace is a source of danger, especially to children. Individuals making fires must be certain that sparks do not ignite a room and that the children are kept at a safe distance. Handrails, stairs, and carpeting within halls and on stairwells should be checked for good condition to ensure that they are not torn, worn, or made slick. Adequate lighting is essential for these areas. Particular attention should be paid to the presence of childrens' toys on staircases.

Garage and other storage areas are extremely hazardous. Unfortunately, flammable liquids, including gasoline, paint, paint thinners, turpentine, and other such things, are kept in glass containers, improper storage areas, and within reach of children. Other garage and storage area hazards include tool boxes and tools, pesticides, pesticide sprayers, garden equipment, lawnmowers, glue, poison, and other material that an individual saves, stores, or utilizes.

Electrical Hazards

There are several common electrical violations found during housing inspections. (1) The power supply is not grounded properly and does not have adequate capacity for major and minor appliances and lighting. (2) The panel box covers or doors are not sealed to prevent exposure to live wires. (3) The switch outlets and junction boxes are not covered to protect against electrical shock. (4) Frayed or bare wires are present as a result of drying out and cracking of the insulation around the wires, or constant friction and rough handling of the wires. (5) Electrical cords run under rugs or other floor coverings creating potential fire hazards. (6) Bathroom lights are not permanently installed in ceiling or wall fixtures and controlled by wall switches. (7) Inadequate light is used in stairwells and hallways and within habitable rooms. (8) Octopus outlets or several different appliances or lights are hooked into one outlet. (9) Excessive or faulty fuses are used. (10) Hanging cords or wires are used. (11) Long extension cords are used. (12) Temporary lighting is used in areas where the lighting comes from other types of electrical fixtures.

Heating Equipment

In the use of any of the fossil fuels it is important that carbon monoxide is not produced by faulty combustion. Coal also produces many excess volatile hydrocarbons and sulfur dioxide. Gas, which is colorless, may be detected by odors inherent or added to the gas. Gases form explosive mixtures when there is a leak, causing additional hazards. Oil is as hazardous as the other two fuels, and in addition where oil leakage occurs, it may be a fire hazard. Inadequate amounts of air added to any of the fuels creates hazardous byproducts.

Hot-water heaters may be especially hazardous, as gas can escape if the flame in the gas hot-water heater extinguishes, and if the pressure relief valve is not functioning properly.

Housekeeping

Areas where people live should be kept uncluttered and clean to promote good general emotional health and to prevent infestations of insects and rodents. A badly cluttered and dirty residence not only implies potential or actual infestation of roaches, mice, and rats, but also gives some indication as to the lifestyle and emotional state of the individuals living within such premises. Solid waste and clutter, although in themselves environmental problems, are also symptomatic of other types of problems within our society, within the family, and with the individual. Since this book is not intended as a housekeeping manual, it is only necessary to state that keeping property free of solid waste, improperly stored foods, and clean, using adequate detergents and friction, will satisfy general housekeeping requirements.

Animal Bites

Animal bites, especially among children, come from dogs, cats, rodents, and occasionally wild animals, especially raccoons. Animal bites, which are generally preventable, cause trauma, wound infection, psychological problems, a potential for the hantavirus from the deer mouse, and a potential for rabies. The trauma involves lacerations, and/or puncture wounds typically in the extremities. However, in small children, the head, eyes, ears, and lips may be involved. Bite wounds may become infected with a variety of microorganisms. People at special risk include those who are immuno-compromised or have a history of problems related to the spleen, diabetes, blood circulation, or other chronic conditions. Serious bites can cause profound psychological problems, especially in children. Care must always be taken in wild animal bites, as well as domestic animal bites, that rabies is not present. All bites should be reported to the local health department.

Rabies is a viral encephalomyelitis with a headache, fever, malaise, sense of apprehension, paralysis, muscle spasm, delirium, and convulsions after the bite of an infected animal. The incubation time is 2–8 weeks or as short as 10 days. It is caused by the rabies virus which is a rhabdovirus which is found worldwide. Reservoirs of infections include wild and domestic dogs, foxes, coyotes, wolves, cats, raccoons, and other biting mammals; also found in bats. It is transmitted in the saliva from the bite of a rabid animal. It is communicable in dogs and cats for 3–5 days before the onset of clinical signs and during the course of the disease there is a general susceptibility. It is controlled by vaccinations of all pets for rabies, ten-day detention for observation of dogs and cats who have bitten people, immediate evaluation of heads of animals for virus isolation in the brain when suspected of being rabid, and immediate immunization of an individual bitten by a rabid animal.

Epidemiology of Injuries

Epidemiology, which is the fundamental science for studying the occurrence, causes, and prevention of disease, can also be used on a theoretical basis to study the problems related to injuries. In order to do this it would be necessary to determine the time, place, person injured, age, sex, nature of the injuries, factors involved in injury causation, other environmental factors, and other human factors.

Injuries like diseases do not occur at random. The elderly and persons age 15–24 have the highest fatality rates. The risk of fatal injury is $2^1/_2$ times greater for males as it is for females. Males are also at greater risk for nonfatal injuries. Fracture rates are highest among older women who have osteoporosis or bone decalcification. Death rates, with the exception of homicides and suicides, are highest in rural areas, possibly because or differences in socioeconomic status, types of occupational and other exposures, and the lower availability of rapid emergency care. Socioeconomic factors influence the incidence of homicide, assaultive injury, pedestrian fatalities, and fatal housefires. These occur at the highest rate among the very poor. Among the wealthy, the greatest number of injuries are caused by falls and home swimming pools. Other factors causing injuries include high-risk jobs, poor housing, older cars, space heaters, and housing related problems. These are especially true among the poor. The use and abuse of alcoholic beverages have a huge influence on all types of injuries, especially among young teenagers. About half of the fatalities related to injured drivers or pedestrians are alcohol related. Alcohol is frequently detected in the blood of individuals who have been involved in a variety of different types of injuries. Usually the more severe the injury, the higher the level of alcohol. Other types of behavioral factors are very difficult to determine, since they may be very transient.

The environment may contribute to the level of injuries that occur. Poor weather, as well as improperly maintained equipment, may contribute to injuries. The speed of emergency medical care is essential in controlling the level of the injury that has occurred. This emergency medical care system helps reduce the levels of fatalities among individuals.

OTHER ENVIRONMENTAL PROBLEMS

Noise

Noise is unwanted sound. Within the home environment, sleep or comfort is disturbed by loud television sets, grinding truck gears, fan motors, children, radios, dishwashers, garbage disposals, vacuum cleaners, washing machines, knocking pipes, and so forth. There are a myriad of sources of noise within the home. Noise is transmitted directly through the walls or indirectly through walls, ceilings, and floors. Each person has a unique feeling about a comfortable sound level and the definition of noise. For this reason, special consideration should be made of neighbors and family members when creating loud sounds at varying times during the day and night.

Pests

Roaches, mice, flies, lice, ticks, rats, and mosquitoes are in the interiors and exteriors of housing units. The infestation may be brought to the unit by grocery bags, clothing, and equipment, or may be caused by migration of insects and rodents from their normal habitats.

Sewage

Sewage is overflowing in many housing developments in various areas of the country. Currently there are over 10 million homes with septic tank systems. Considerable research is needed to determine the proper techniques for sewage disposal for these properties. An in-depth discussion on individual sewage disposal can be found in Volume II.

Drainage

Pools of stagnant water accumulate in poorly drained properties. These are excellent areas for mosquito breeding and may create health hazards. Electrical hazards must also be considered when standing water is encountered. Excessive dampness or water causes wood to rot and structures to deteriorate.

Solid Waste

The storage and disposal of solid waste in a proper manner is the responsibility of the occupant of the property. In many areas the local governmental unit provides adequate solid waste removal and disposal. In some areas the occupant must make private arrangements. In any case, where solid waste is not handled properly, insects and rodents, odor, and air pollution problems due to the burning of solid waste, may occur. (Further discussion concerning this subject will be found in the appropriate chapter on solid and hazardous waste in Volume II.)

Environmental Lead Hazards

Lead poisoning is one of the most common preventable public health problems related to children today. Lead paint, the major high-dose source for children, is present on an estimated 30–40 million houses in the U.S. It will continue to pose a hazard to children for many years to come. In addition to lead due to paint, lead may be found in house dust, garden or backyard soil, flaking paint of automobiles, and industrial emissions. Children are constantly being exposed to lead hazards. Lead may also be found in water because of old plumbing systems where soldered lead joints were used. Lead can leach into the water from the joints or from lead pipes. Acidic or corrosive water will cause this leaching. Lead may be present in food such as canned fruits, vegetables, and juices. The lead may be taken up by crops, from the soil, or may be deposited on the crops from lead air contamination. Lead may also be available in pottery, fish sinkers, or lead weights, jewelry making, antique ceramic doll making, other toys, metal sculpture soldering, special powders, and Chinese Herbal Medicines.

Workers may bring lead home to their children on their clothes. Children are more at risk for lead exposure than adults because they have more hand-to-mouth activity than adults, they absorb more lead than adults, and they are smaller than adults. Lead is a poison that effects virtually every system in the body. Whereas before the mid-1960s, a level above 60 μg/dL of blood was considered to be toxic, today above 15 μg/dL of blood is considered to be toxic. In fact, there may be adverse health effects above 10 μg/dL of blood. In children, very severe exposure, blood levels of 80 μg/dL, can cause coma, convulsions, and death. Blood lead levels as low as 10 μg/dL, which do not cause distinctive symptoms, are associated with decreased intelligence and impaired neurobehavioral development. Other symptoms of lead poisoning include colic, abdominal pain, constipation, cramps, nausea, vomiting, anorexia, weight loss, hepatic effects, renal effects, interference with growth, and erratic behavior.

The External Housing Environment

The exterior premises in many ways are as important as the interior premises. Improper solid-waste storage, poor pest and weed control, and accident hazards create housing problems. Exterior premises with piles of old lumber or other materials, dilapidated shacks or garages, abandoned automobiles, or other kinds of junk or waste are both hazardous and unsanitary. The exterior solid waste problem is often an indication of the interior housing environment.

Fires may be caused by improper storage or disposal of flammable liquids and by obstruction of electric power lines by trees, birds nests, and so forth. Other safety hazards on the exterior grounds include stepoffs, holes in the ground, rocks, broken glass, power equipment, and storage of LP gas.

Role of Government

The many policy inconsistencies that exist as a result of changes in federal programs and tax laws have adversely affected housing. For example, federal deductions are permitted for building depreciation. It might be wiser to allow a tax deduction for building maintenance to encourage investment in deteriorating buildings. The federal highway administration provides funds for the construction of radial highways that often cut through valuable urban neighborhoods. These plans might well be altered to avoid interference with neighborhoods, thereby reducing the potential for deterioration. In the past, federal housing policies have emphasized new building rather than the revitalization of existing buildings and neighborhoods.

Housing Code Enforcement

There are many problems related to housing code enforcement. These include absentee landlords, properties belonging to estates, effective court procedures, the court and tenant, political support, fiscal support, proper personnel, code enforcement, and relocation of families. Unfortunately, a high percentage of the difficult housing code violations involve absentee landlords or estates. The landlord may not correct the violations, either because of the cost involved or because previous corrections were destroyed by the tenants or the property simply serves as a tax write-off and it is simpler to take a tax cut than to spend capital on expensive repairs.

Tenants are also taken to court for violations of the housing code, since tenants may create as many problems as landlords. Where the tenant refuses or will not take care of the housing environment, the court needs to exercise its powers in the same way that it would exercise them over property owners.

It is one thing to notify the property owner of violations and quite another to secure compliance. Although many individuals will correct violations because they are naturally law-abiding citizens, there are those who will attempt to avoid doing this. Unfortunately, politics may enter the situation. Political leaders generally are very careful about supporting vigorous code enforcement, since the code enforcement may in effect cause endless complaints from citizens and eventually result in the loss of votes.

Code enforcement at times cannot be carried out because the individuals in violation lack the money to make the necessary repairs. In the past, federal agencies, and in some cases state and local agencies, were able to provide funds for some corrections. These funds, however, are very indefinite and tend to vary from year to year based on the current political situation in Washington or in the various localities.

As a result of the uncertain political attitude toward code enforcement, inadequate funding is provided for the health department or housing department to carry out necessary code enforcement programs. This tends to deprive departments of trained competent individuals capable of performing necessary tasks and also deprives the department of having adequate numbers of individuals to make a program workable.

One of the difficulties of code enforcement is that when a property is found to be unfit for human habitation, a family must be relocated. Necessary assistance must be provided to help these families relocate in a shelter that is acceptable for proper living. In some areas, the properties involved are so numerous that it becomes a very difficult task to try to remove the families from substandard housing and to relocate them in other areas. Either the new housing is unavailable or the family is unable to live its normal life in a new situation because of transportation and shopping difficulties, and difficulties in reaching friends or relatives.

Impact on Other Problems

Health statistics demonstrate a relationship between morbidity or mortality rates and physical surroundings. People living in economically depressed neighborhoods with poor facilities are less healthy than those living in the higher socioeconomic areas. Obviously inadequate nutrition, medical care, and stressful social conditions contribute to these health problems. However, there still appears to be a relationship of physical facilities to infectious disease and infant mortality. As the facilities and levels of cleanliness improve, the rate of disease occurrence seems to drop. The role of housing and neighborhood conditions on juvenile delinquency, psychiatric illness, mental retardation, and learning disabilities is seriously questioned. There certainly are concerns about overcrowding and the effect that it may have on individuals. Substandard housing also contributes to air, water, and land pollution and to solid-waste problems.

Economics

The economic costs of housing are related to high taxes for services, congestion, streams loaded with silt, air pollution, the destruction of open space, and the provision for a vast variety of services. The social costs are certainly difficult to evaluate but must be included. Other costs include the provision of schools, necessary streets and roads, and utilities. Noise, energy consumption, and erosion of land are added factors. Personal aspects must be considered in any cost analysis, including travel time to and from work, traffic accidents, crime, and the cost involved in resolving emotional or psychiatric problems.

POTENTIAL FOR INTERVENTION

The potential for intervention is very difficult to assess in the area of housing. It is obvious that the technique of isolation can be utilized through the tearing down of unsafe structures. Substitution can be utilized by providing low-rise structures for high-rise ones in lower socioeconomic areas to eliminate problems that occur in the high-rise structures. Shielding is not an effective technique. Treatment includes renovation or improvement of the structures and the neighborhoods. Prevention is the most effective approach, since land planning coupled with adequate housing inspections and code enforcement, as well as good zoning, can help prevent problems from occurring. As has been pointed out repeatedly, the federal government has spent an enormous amount of money yet still some 12 million crowded, substandard housing structures are standing. Prevention and control of indoor air pollutants and injuries can be accomplished through isolation, substitution, shielding, and treatment.

RESOURCES

Scientific and technical resources include the National Architectural Accrediting Boards Incorporated, the Association of Collegiate Schools of Agriculture, various graduate professional schools of architecture, the American Public Health Association, the National Environmental Health Association, construction and general contractors, and so forth.

Civic associations consist of a vast variety of homeowners associations, land-owners associations, block councils, and community councils found in many cities or county areas.

The governmental resources include local and state health departments, local and state housing departments, the National Bureau of Standards, the Department of Defense, the Department of Energy, the General Services Administration, the Department of Housing and Urban Development, the Law Enforcement Assistance Administration, the National Aeronautics and Space Administration, and the National Science Foundation.

An excellent book that may be used for a study of housing codes and enforcement procedures is *Basic Housing Inspection*, published by the U.S. Department of Health, Education, and Welfare, Public Health Service, Publication #(CDC) 80-1979. A resource on safety is the Office of the Secretary, Consumer Product Safety Commission, Washington, DC 20207.

The American Public Health Association—Centers for Disease Control recommended minimum housing standards, as updated in 1986, may be obtained from the American Public Health Association at 1015 15th St., N.W., Washington, DC 20005.

The Marion County Health Department has an excellent housing program. Their ordinance is entitled *Housing and Environmental Standards Ordinance*. Information may be obtained from the Marion County Health Department, 3838 Rural Street, Indianapolis, IN 46205.

The U.S. Department of Housing and Urban Development is responsible for large numbers of programs related to community planning and development and housing. The material from them may be obtained by writing to the U.S. Department of Housing and Urban Development, Washington, DC 20410-4000.

Philadelphia has been at the forefront of injury prevention in homes. The Philadelphia Injury Control Program material may be obtained from the Philadelphia Department of Public Health, Division of Environmental Health, 500 South Broad Street, Philadelphia, PA.

Injury prevention is a serious concern that has been addressed by the Childhood Injury Prevention Resource Center, Harvard School of Public Health, Department of Maternal and Child Health, 677 Huntington Ave., Boston, MA 02115. "Injury in America," a major report of the Board of the National Research Council of the National Academy of Sciences, the National Academy of Engineering, and the Institute of Medicine, has been published by the National Academy Press, 101 Constitution Ave., N.W., Washington, DC 20418. The Centers for Disease Control, U.S. Department of Health and Human Services, Public Health Service, Atlanta, GA 30333 has programs and have published reports on injury control. The Consumer Product Safety Commission's Product Safety Hotline is 1-800-638-2772.

Indoor air pollution, radon gas reduction, lead contamination, and other associated programs and problems are under the control of the U.S. Environmental

Protection Agency. Information may be obtained from them by contacting the U.S. EPA, 401 M Street, S.W. Washington, DC 20460; the National Institute for Occupational Safety and Health, 1-800-356-4674.

Additional resources for injury control include: the Surgeon General's 1990 Injury Prevention Objectives for the Nation and Model Standards; A Guide for Community Preventive Health Services; Developing Childhood Injury Preventions; and Administrative Guide for State, Maternal and Child Health Programs; Strategies for Injury Prevention; Injury Prevention in New England; and the Surgeon General's Workshop on Violence and Public Health Report. (Contact the Centers for Disease Control, Atlanta, GA.)

STANDARDS, PRACTICES, AND TECHNIQUES

Rehabilitation

It is true that in some areas redevelopment has been necessary, but in other areas a better concept, known as rehabilitation, has been attempted. In rehabilitation, existing residential properties that are deteriorating are renovated with the aims of (1) ensuring improved housing that is livable, safe, and physically sound; (2) ensuring low-cost housing; (3) providing an acceptable minimum level for housing; (4) encouraging innovation and improved technology to reduce construction costs; and (5) utilizing instead of destroying resources that need upgrading. Obviously rehabilitative construction standards are not the same as new construction standards, since the work is being done on existing buildings. However, all safety precautions do have to be taken into consideration and followed carefully. All rehabilitated properties must have access to the property from the road by emergency vehicles, living units that exist independently of other living units each with its own entrance and exit to the building, and practical living units. Under rehabilitation, all highly dilapidated properties that cannot be rebuilt efficiently are torn down and the ground space is given to the community for use as playgrounds, parks, or parking lots, as decided by residents of the community. Adequate night lighting and open space are site criteria that must be provided; also, the land must be made free of flooding, sewage problems, and solid-waste problems. The materials and products used for replacements or additions have to be of good quality, conforming to generally accepted practice. A building must be structurally sound. The exterior walls must be able to support the weight of the roof and also must be moisture free.

It is important in rehabilitation that individuals have sources for adequate food, hot water, baths, and proper sewage and solid waste systems. One must recognize that rehabilitated property may not look as good as new property. However, low-income families have the opportunity to live in good shelters that provide all of the essentials for prevention of disease, elimination of behavioral problems and accidents, and promotion of health.

Standards for Housing Utilities and Construction

Heat

The APHA-Public Health Service (CDC) housing code states that the temperature should be at least 68°F at a distance of 18 in. above the floor level. If the person needs a higher temperature because of age or physical condition, 70°F is required. Others feel that this temperature should be maintained at different points above floor level, such as at 5 ft. If carbonaceous fuel is used, the heating devise or hot water heater must be vented to the outside and receive adequate air for proper combustion of fuel. In any case, the local housing code requirement will be the one followed by health department personnel.

Light

In most communities, every habitable room must have at least one window or skylight facing outdoors, and the window area must be at least 8% of the total floor area of the room. Daylight should be used wherever possible in such a way that the amount, distribution, and quality of the light aids in, rather than detracts from, visual tasks.

Artificial light varies tremendously from area to area, and from room to room. Artificial light must always be provided in public halls and stairways, and should be adequate for all use within the house. The house that is properly illuminated enables the family to function more efficiently. Switches should be in convenient places, and the type of lighting should serve the various needs of the people during the day and night. Obviously, such areas as kitchens and bathrooms should be well lit to prevent accidents. Such places as furnace rooms, laundry rooms, and other areas where appliances are kept should have adequate lighting to make the task less demanding and to prevent accidents. Proper lighting is also very essential in preventing fatigue.

All electrical units, wiring, and appurtenances with electrical connections must be installed by electricians following electrical codes to avoid the occurrence of fires.

Plumbing

In the house, the piping for water service should be as short as possible, and elbows and bends should be reduced to maintain water pressure. The water line to the house should be at least 4 ft below the soil to prevent freezing. Valves are usually located outside the building so that the building supply may be turned off when it is necessary to service the building. Hot water heaters should be thoroughly evaluated and properly installed and ventilated. Usually a .75 in. pipe is the minimum size required for water that must rise from the basement level to other parts of the house. The drainage system should have a main house drain a minimum of 6 in. in diameter. The house drain has to be sloped toward the sewer, usually at a level of .25 in. fall per 1-ft length. Determination of the amount and size of drainage lines depends on the types of rooms within the house. Traps are used on sinks and toilets to prevent sewer gases from entering the property. All plumbing systems must be adequately ventilated directly through the roof. The most essential part of the plumbing is the proper size of the lines, adequate installation, proper ventilation, and proper drainage to avoid cross-connections or other situations in which the potable water supply becomes contaminated or the sewage backs up through the toilets and various drains within the house.

Ventilation

The usual requirements for ventilation is that 45% of the minimum window area be easily opened or an approved mechanical means of ventilation be installed. Every bathroom and toilet room must comply with the various housing codes and rules on adequate ventilation. Ventilation must be to the outside. In many housing codes, two to six changes of air per hour are required.

Space Requirement

One hundred and fifty square feet of total, habitable, room area per person is required by the APHA-CDC code. Where second-person occupancy occurs, the requirement is reduced to 100 ft^2. The code also requires that no more than two people are permitted in a habitable room. With the exception of rooms that have sloping ceilings, the habitable room should be a minimum of 7 ft in height. Where a room is less than 5 ft in height, it should not be used for living purposes. Individuals should not live in rooms, except in emergencies, that are totally below the ground surface. There should be some exposure to the outside. The exception to this rule is where the construction is waterproof, there is adequate window area, and the other rules and regulations relating to light and ventilation are followed. Pipes, ducts, or other obstructions must not be less than 6 ft 8 in. from the floor. Where two people share a room, at least 70 ft^2 of floor space is required for the first person and 50 ft^2 of floor space for the second. Living quarters must have easy access to a bathroom, and an individual must not have to pass through another sleeping room to get to a bathroom.

Water

All housing codes require that adequate quantities of running water be provided for a dwelling. This generally means 1 gal/min of hot and cold running water per each fixture in the house. Generally, 120°F is accepted as the maximum hot-water temperature. However, the temperature may need to be lowered to prevent burns.

Kitchen Facilities

All kitchen facilities must contain sinks, cabinets or shelves, stoves, and refrigerators. The sink must be large enough for kitchen use and not a small hand-washing sink. As the kitchen sink is used in the preparation of food and the cleaning of dishes, utensils, and equipment, it must have hot and cold running water under pressure and must be so built as to be utilized in an effective manner. The stoves must be adequately built, installed, and maintained to avoid electrical, fire, or carbon monoxide accidents. A refrigerator is needed to keep food under 41°F at all times. Freezers are highly desirable.

Physical Structure

Houses are framed in many different ways. However, all of them contain foundation walls and provide support for the outside walls of the building. The flooring system is made up of girders, joists, subflooring, and finished flooring, which may be composed of concrete, steel, wood, composition material, and so forth. Studs, which are usually 2 × 4 in. or 2 × 6 in., are used to provide a wall thick enough to allow the passage of waste pipes. Usually studs and joists are spaced 16 in. on center. All openings, such as windows or doors, must be framed with studs to provide support for that portion of the wall. The interior wall finish varies from plaster to wallboard, paneling, wood, and so forth. Interior stairways must be greater than 44 in. in width with a maximum rise of 8.25 in. and a minimum tread of 9 in. The handrail and all other parts of the stair enclosure should be no lower than 80 in.

The four basic types of windows are double-hung sash windows, which move up or down; casement windows, which are hinged at the side; awning windows, which are in panes on a horizontal axis; and sliding windows, which slide past each other. There are many types of doors and door finishes.

The roof framing consist of the rafters, which support the roof and also create place for roofing material. The exterior walls and trim may be made of a variety of materials. Their function is to enclose and weatherproof the building. Exterior walls also serve as weight-bearing walls. They are composed of wood, brick, stone, and so forth.

The roof covering must keep the house dry and intact. Roof covering composition includes asphalt shingles, asphalt built-up roofs, tar, slate, tile, and wood. Flashing is the metal that joins two portions of the roof where it forms a valley. This keeps water and air from getting into the building. Gutters are used to take water off the roof and away from the house.

Insulation materials is extremely important to keep the house cooler in the summer and warmer in the winter. Good insulation prevents the loss of vital energy.

Fire Safety and Personal Security

All dwellings should have at least two means of egress leading to a safe and open space at ground level. Individuals should not have to exit through someone else's dwelling unit. All entrance doors into a dwelling or dwelling unit should be equipped with a deadbolt or locking device. The entrance doors in a multiple dwelling should be equipped with a device that allows the occupants of the unit to see a person at the door without fully opening the door. All exterior windows and other means of egress should be equipped with locking hardware. Every dwelling unit should have at least one functioning smoke detector located on or near the ceiling in an area immediately adjacent to a sleeping area. These smoke devices should be tamper-proof.

Travel Trailers and Mobile Homes

The travel trailer is a portable structure used as a temporary dwelling for travel, recreation, and vacation. The body width must not exceed 8 ft and the trailer must not exceed 4500 lb in weight of 9 ft in length. Travel trailers vary considerably in size, types of sleeping accommodations, and actual living facilities. The trailer may simply be built for sleeping or may have self-contained hot and cold running water, full kitchen, baths, gas tanks, air-conditioning, and lighting systems. The travel trailer has to be fitted for hooking into sanitary stations for removal of sewage. The trailer has toilets aboard and must also be fitted for taking on water. The facilities used for travel trailer parking areas have to be carefully selected for good drainage, gentle sloping, and no obstructions. The roads around the parking areas must be curved to reduce the speed of the drivers. A separation of 15 ft between trailers is an absolute minimum. In trailer areas there must also be provisions for service buildings where individuals can take showers, do laundry, and also use toilet facilities. Solid waste has to be stored in tight-fitting containers that are waterproof, fly-proof, and rodent-proof. These containers must be removed daily to prevent pest infestation. When a travel trailer area plan is ready to be approved, it must indicate the area in site dimensions; the number, location, and dimension of the trailer spaces; the location and width of roads and walks; the location of the service building, sanitary station, and other facilities; the location of water and sewer lines; and the location of storm drains and catch basins.

The number of mobile homes has increased enormously within the last 15 years. Mobile homes may be on wheels or modular in nature. The mobile home is being used in many cases as low-cost housing. Mobile homes may be placed on individual sites or in mobile home parks. The home placed on an individual site must follow all rules and regulations of the health department for proper water, sewage, site locations, and so forth. The site has to be zoned for this particular use. It should be well drained and free from topographical hindrances. It should not be located near swamps, marshes, heavy industries, in flood plains, or on steep slopes. It should also be placed, if possible, in such a way that in the event of tornados, it will be protected as much as possible. The owner of the home has to either hook into an existing sewage system and obtain city water plus other utilities, or he or she must dig a well and provide an adequate on-site sewage disposal system. (Well-drilling and on-site sewage disposal will be discussed in further detail in Volume II.)

Where a mobile home court is developed, there has to be a community development on the same order as any other type of community. All the roads have to be convenient and of adequate width for the parking and traffic loads. Streets should be approximately at right angles. Street intersections should be at least 150 ft apart, and the intersection of more than two streets should be avoided. The grades of the streets should not exceed 8% whenever possible, although for a short run as much as 12% is acceptable. The streets must be of all-weather construction and of hard dense easily drained material. Off-street parking is necessary. Walkways at least 3.5 ft wide must be provided within the area.

A mobile home should be on a minimum size lot of 2800 ft^2. A double-wide unit should be on a lot which is a minimum of 4500 ft^2. The mobile home should not occupy more than one third of a total lot area. Small lots contribute to overcrowding and create a poor aesthetic image of the area. If an area is developed for five or more mobile homes, a recreational area that is safe and free of traffic hazards should be made available for use by children and adults. A recreational area can be located next to recreation or service buildings. The minimum space should be .66 acre. Where swimming pools are constructed, they must comply with all of the regulations of the various state and local health codes.

All buildings in a mobile home park must conform to the housing codes of the local and state areas. They must be kept in a sanitary way, be screened, properly ventilated, and properly lighted. Separate bathroom facilities have to be provided for men and women. If a laundry facility is present, it has to be of adequate size, properly installed, and the wastewater must go into adequate wastewater disposal units.

Before a mobile home development can be approved, the plans have to be submitted to the health department, zoning department, and other responsible authorities. Detailed construction plans, and the site and the location of all roads, buildings, and so forth have to be evaluated. It is essential, wherever possible, that the public water supply and public sewage disposal be used. If these are not available, a package treatment plant should be built. Access to firefighting equipment, police protection, and schools have to be determined before the site is approved.

In the mobile home park, a mobile home stand has to be provided for adequate foundation and anchoring of the mobile home. The stand must be properly graded and of such material that it can support the weight load, regardless of weather conditions.

The anchors must be cast in place in concrete in order that they will not be ripped out in the event of storms. Particular attention should be paid to this in those areas where tornados frequently occur. Since tornados will knock over mobile homes or cause damage to them, all water and sewage hook-ups and electrical and gas hook-ups have to be installed in such a way as to avoid hazards and to provide adequate service. Telephone service is also needed. Further discussion concerning water and sewage will be found in other chapters in Volume II.

Storm-water drainage systems have to be installed in such a way as to avoid flooding of the mobile home site. The drainage system must remove the heavy rain waters from the area. Solid-waste containers, storage containers, and collection techniques have to meet the requirements of local and state codes. Further discussion concerning this subject can be found in the chapter on solid waste in Volume II.

Insect and rodent control must also be considered in a mobile home park. Of particular importance is mosquito control, since many mobile home parks are in outlying areas where weeds or other breeding places, such as depressions in the ground, may be present. In addition, the areas must be free of cans, jars, buckets, old tires, and all other junk that serve as breeding places for mosquitoes. Fly- and rodent-control procedures are to be followed. It is essential that all garbage be properly stored and then disposed of. Insect and rodent control are discussed in further detail in other chapters of this book.

Electrical wiring, equipment, and appurtenances should be installed and maintained in accordance with local and state codes. Where these codes do not exist, it may be necessary to consult the National Electrical Code. If possible, distribution lines should be installed in underground conduits, which are placed at least 18 in. below the ground surface and at least 1 ft away from water, sewer, gas, or phone lines. If overhead power lines are installed, it must be done in such a way as to prevent accidents and should be available for service in emergencies. Grounding is necessary for all exposed, non-current-carrying metal parts of mobile homes and equipment. Adequate lighting has to be provided on exterior walkways, streets, and buildings.

The handling or storage of fuels, such as natural gas, liquified petroleum gas, fuel oil, or other flammable liquids or gases, have to meet federal or state ordinances or the National Fire Protection Association Standards. It is best to have central storage and underground distribution of fuel. All fuel oil containers or other fuel containers on the lot where the mobile home stands are to be securely fastened to prevent accidents or fires.

It is essential that adequate fire protection be provided in mobile home parks. Applicable state and local codes and laws should be followed. If they do not exist, standards established by the National Fire Protection Association may apply. A water supply of adequate quantity must be available for fire hydrants within 500 ft of all mobile homes, service buildings, or other structures. A minimum of two 1.5-in. hoses must be available for hooking up to the water supply. At least 75 gal of water per minute per nozzle must be available at a pressure of 30 lb/in.^{2}. Where hydrants are not available, the types of equipment that can be utilized are limited to those that meet the National Fire Protection Associations' standards for adequate fire extinguishers. Class A extinguishers have to be provided for mobile homes.

Since carbon monoxide is a constant danger within the mobile home, it is essential that heating and cooking equipment be properly vented. If there is any question concerning this, the standards of the local or state authorities should be followed or the health department or fire department should make an evaluation of the equipment and the adequacy of the venting system.

MODES OF SURVEILLANCE AND EVALUATION

Housing Inspections and Neighborhood Surveys

Housing inspections and neighborhood surveys are used to determine the quality of housing and neighborhoods within a community. The American Public Health Association has done considerable work in this area. They have developed tools of evaluation, including comprehensive forms, and have established point systems for determining into which category a neighborhood falls. (For more detailed information on housing, the reader is referred to the latest methods, techniques, and publications of the American Public Health Association.)

The now defunct Bureau of Community Environmental Management of the U.S. Public Health Service developed a system of forms used by federal and local governments to evaluate neighborhoods and housing structures. The forms were entitled, "Neighborhood: An Environmental Evaluation and Decision System Interior Interview," form number 85-R-0134. Two additional forms, entitled "Neighborhood: An Environmental Decision System Exterior Premises Analysis," form number 85-R-0044, and "Neighborhood: An Environmental Evaluation and Decision Block Analysis," form number 85-R0044, can be utilized for an in-depth evaluation of neighborhoods, blocks, and houses. Since housing inspection is a long and complex procedure, the forms developed for this purpose could be utilized, or at the local level, new forms should be developed based on the federal forms to meet the individual needs of the community.

Indoor Air Pollution Surveys

Indoor Air Quality Profile

When an indoor air pollution problem occurs or if you want to determine the current status of air quality in the building, follow this proceedure:

1. collect and review existing records, including heating, ventilation, and air conditioning records, complaint records, design, construction and operating documents
2. conduct a walkthrough inspection of the building
3. collect detailed information on pollutant pathways, pollutant sources, occupants
4. make an inventory of the heating, ventilation, and air conditioning system repairs, adjustments, or replacements needed
5. obtain a set of Material Safety Data sheets (see Table 7.1) and an inventory of significant pollutant sources and their location

The walkthrough inspection should include schedules and procedures on facility operation and maintanence, housekeeping, and pest control. Pressure relationships between special use areas and other rooms should be identified and any special circumstances related to areas where complaints have been received. Look for odors, dirty or unsantitary conditions, visible fungal growth, mold odors, drain pans, cooling towers, filters; staining and discoloration, smoke damage, presence of hazardous substances, improper storage of chemicals and cleaning substances; inadequate maintanence, occupant discomfort, overcrowding, blocked airflow, crowed equipment areas; new furniture, rugs, equipment, renovations, and painting.

Residential Wood Burning

Air samplers can be used to determine the level of particulate matter present in the air. Grab samples can be used for obtaining a small representative sample of the gases that may be present in the air from wood burning. The polycyclic aromatic hydrocarbons (PAHs), such as chrysene and benzo(d)fluoranthene, are found in residential wood smoke. These PAHs are adsorbed predominantly on particulate matter, which is collected on a filter during the air sampling process. The PAHs can be extracted from the filter. They are then analyzed by the use of column gas chromatography (GC) and high-performance liquid chromatography (HPLC). Further biological monitoring techniques are useful for detecting whether an environmental or occupational exposure has occurred to the PAHs. PAHs are present in blood, mammary, adipose, liver, and brain tissue from rats. PAHs may also be found in the urine of exposed humans.

Passive Smoking

Passive smoking can be determined by observing the smoking habits of individuals within the buildings and whether or not there is adequate ventilation or shielding of the nonsmokers from the smokers.

MATERIAL SAFETY DATA SHEETS

Under OSHA regulations, responsible parties are required to document information on potentially hazardous products. These Material Safety Data Sheets (MSDSs) may be of limited help in identifying some products that may pose IAQ concerns. However, professional judgment and collection of additional information may be necessary in order to make full use of the MSDSs. The following table summarizes some of the issues to keep in mind when deciding whether information from MSDSs is applicable to emission sources and exposures of concern in a building.

Item	Possible Uses	Comments
Substances Covered	• MSDSs may identify significant airborne contaminants	• MSDSs may not be available onsite for many products
		• some components are listed as proprietary and are not disclosed
		• MSDSs do not always highlight products most likely to be airborne
		• contaminant byproducts inadvertently formed during manufacture won't always be listed
Personal Protection/ First Aid	• may suggest precautions for conducting source inspection	• usually relates only to high-level, worst-case exposures in general industry
Health Effects	• generally presents types of health effects that may be expected primarily at high level (e.g., industrial) exposures	• symptoms listed may not occur at low-level concentrations found in indoor air
		• MSDSs may not include more subtle IAQ aspects such as nuisance factors and sensitivity to mixtures
Physical Data	• odor description may help identify sources	• reference material on how to use physical data information to predict IAQ impacts may be scarce
	• volatility may suggest which products are likely to be airborne	
	• contaminants to expect in event of a fire or decomposition may be listed	
	• reactivity data may suggest potential problems with storage or use	
Control Measures	• identifies proper storage and packaging procedures	• many office chemicals are kept in much smaller amounts than found in industrial settings
	• identifies steps for cleanup of gross spills	• spill cleanup may not eliminate airborne contamination
		• does not specify routine emission controls

A reasonable effort should be made to collect available MSDSs during IAQ profile development. Care should be taken to consider information that is relevant to IAQ concerns. Other important indicators of how a particular product may affect IAQ are available from direct odor and dust observations, a review of work practices, and interviews with operators and occupants. The manufacturer is a good source of follow-up information on a given product (phone number should be included on each MSDS).

Table 7.1.   Material Safety Data Sheet. Source: *Building Air Quality–A Guide for Building Owners and Facility Managers* (Washington, D.C. 20460; U.S. Environmental Protection Agency, Indoor Air Division and U.S. Department of Health and Human Services, National Institute for Occupational Safety and Health, 1991), p. 39

Carbon Monoxide

Carbon monoxide is measured by infrared spectroscopy. The test is sensitive in ranges from 0.1 to 50 ppm. Water vapor and temperature must be controlled in order to ensure accurate readings. There are three types of personal monitors that are currently being used. Two types have color indicators, while the third has an alarm system that indicates when the threshold limit value has been reached. Grab sampling can also be used to determine what is occurring at any given time. The sample should be taken near the floor, since carbon monoxide will sink because of its density.

Radon Gas

Radon gas sampling can be accomplished by using charcoal canisters or alpha track detectors. A trained technician may also test for radon gas levels within the home. Charcoal canisters and alpha track detectors require exposure to the air in the homes for a specified period of time and then are sent to the laboratory for analysis. For charcoal canisters, the test period is 3–7 days. For alpha track detectors the test period is 2–4 weeks. Technicians can do a direct radon measurement at the time of analysis. These include real-time radon measurement, continuous radon readers, and integrating readers. The biggest problem with measuring radon is that the levels of radon fluctuate considerably during different time periods.

The appropriate technique for using radon detectors is to first do a screening measurement. This gives you an idea of the highest radon level in the home. It also advises you if it is necessary to continue with radon testing. The screening measurement should be made in the least livable area of the home, probably the basement if one exists. All the windows and doors should be closed for at least 12 hr prior to the start of the test and kept closed as much as possible throughout the testing period. This will help keep the radon level relatively constant throughout the testing period. It is recommended that the short-term radon measurements be done during the cool months of the year. Next, determine the need for further measurements. In most cases the screening measurement is not a reliable measure of the average radon level to which the family is exposed. The screening measurement only indicates the potential for a radon problem, since radon levels vary from season to season and from room to room, based on the level of ventilation that is present. If in the screening measurement the following results occur, then these results act as a guide to further measurements.

1. If the screening measurement result is greater than about 1.0 WL (working levels) or greater than about 200 pCi/l (picocuries per liter), it is necessary to perform follow-up measurements as soon as possible. Expose the detectors for no more than 1 week. Consider taking immediate action to reduce radon levels in the home.

2. If the screening measurement result is about 0.1 WL to about 1.0 WL or about 20 pCi/l to about 200 pCi/l perform follow-up measurements. Expose the detectors for no more than 3 months.
3. If the screening measurement result is about 0.02 WL to about 0.1 WL or about 4 pCi/l to about 20 pCi/l, perform follow-up measurements. Expose detectors for about 1 year or make measurements of no more than 1 week in each of the four seasons.
4. If the screening measurement result is less than about 0.02 WL or less than about 4 pCi/l, follow-up measurements are probably not required.

Formaldehyde

Instruments used to detect formaldehyde include Infrared Spectrometer photoionization detectors and electrochemical sensors. A source emission test that is conducted is the extraction test and the static chamber test. Extraction tests measure the total releasable formaldehyde content under typically harsh environmental conditions. The static chamber test determines the steady-state formaldehyde concentration.

Other Volatile Organic Compounds

Sampling techniques used for other volatile organic compounds include the evacuated bomb and the sampling bag, which are two means of grab sampling. In both cases, the air is taken back to the laboratory for analysis. For a better study, a quantitative time-weighted average test should be conducted. This type of test would include charcoal adsorption, silica gel, and bubblers.

Asbestos

Air sampling is used to determine the amount and size of asbestos fibers present in the air.

Environmental Lead Survey

The environmental lead survey is called for when a child's blood level exceeds 15 μg/dL. The survey is two part: an epidemiological study consisting of a field environmental investigation and an on-site analysis of paint samples. The two major routes of exposure for children to lead is through ingestion and inhalation. The environmental health specialist determines if the child:

1. Ingests, chews, or puts into his/her mouth painted articles, paint chips, printed material, matches, tobacco items, etc.
2. Consumes folk remedies that may be contaminated with lead.

3. Plays or lives in, or has access to, any areas where cans of lead-based paints, pesticides, or other substances are used or stored.
4. Eats from containers or utensils which are glazed or decorated with paint.
5. Lives and/or sleeps in areas where there is crumbling paint or plaster.
6. Lives near heavily traveled highways or lead-based industries.
7. Lives near areas of old building renovation.
8. Drinks water supplied by old plumbing.

A rapid on-site method for measuring the amount of lead in paint is called XRF, or portable x-ray flourescence ananlyzer, and is used to determine the presence and levels of lead in paint. The device works on the principle of x-ray flourescence. It contains a small radio isotopic source. When the unit is placed against a surface, pressing the trigger opens a shutter that allows radiation to fall on the paint. If the paint contains lead, then the radiation stimulates the lead atoms to re-emit characteristic x-rays. These are sensed by the detector in the unit. Solid-state electronic circuits convert the signals into the final reading.

CONTROL OF HOUSING

Zoning and Land Use

Zoning is a means of ensuring that community land will be used in the best possible way for the health and general safety of the community. To prevent disorderly patterns of growth, a community must plan. It would be very unfortunate if heavy industry and superhighways were placed in the center of residential areas. When a house is built where a zoning ordinance is in effect, the builders must comply with the zoning regulations, and they in turn are assured that the other houses in the area comply with the same regulations. A single-family dwelling, for example, cannot be converted into multiple-family units unless zoning laws permits this type of structure. It is important that public health officials, housing inspectors, developers, and planners have a good understanding of the plans of a community and the applicable zoning laws.

Zoning regulations have been in effect for several hundred years. They were originally passed to keep gunpowder mills and storehouses away from heavily populated areas of town. Later zoning was used to form fire districts and to prohibit certain types of highly flammable structures. Zoning has been used and misused in a variety of places. Proper usage would mean that zoning was tied to an overall community plan. Improper usage is to change the zoning by using variances, so that special interest groups or political groups can utilize structures or land as they see fit. The objectives of zoning are to regulate the height, bulk, and area of structures; to avoid undue levels of noise, vibration, and air pollution; to lessen street congestion through off-street parking and off-street loading requirements; to facilitate adequate provision of water, sewage, schools, parks, and playgrounds; to use areas that are not subject to flooding; and to conserve property values. Zoning cannot correct existing overcrowding of substandard housing or change materials and methods of construction, since these are controlled by building codes. Zoning laws do not affect the cost of construction, since this is based on the economy. Zoning cannot be used to develop special regulations or covenants, nor can it design and layout subdivisions, since this is controlled under other rulings. Zoning is concerned with lot size, usage, depth, and width; the amount of open space; and the types of buildings and alterations to buildings placed on the lot.

Role of Government

There are four major problem areas in the role of government in land use and housing. These include the proper use of federal government program and federal policy, state government in action, the local taking of land by right of "eminent domain," and adequate municipal financing. There is a need for coordination and consistency between the federal, state, and local governments in land use programs and policies. State governments should have greater input in land use matters affecting large local areas. The regional metropolitan units of government should be concerned with land use and its effects on outlying, as well as immediate, areas. It is important that a sharp distinction be made between public interest and the rights of private ownership.

Since local governments' major source of revenue is generally the property tax, it is important that land-use decisions be made in accordance with good, sound scientific and socioeconomic political decisions. Not only must the metropolitan area be concerned with proper land use, but also with misuse. Many unutilized buildings, if properly redesigned, could once again be used for shopping centers, office buildings, and so forth. Recycling of structures such as warehouses, garages, factories, and schools is not frequently considered. They are therefore abandoned or torn down. Recycling of these structures could possibly be done at a lesser expense than redevelopment, thereby aiding the rebuilding of the nuclear city, which is so necessary to the metropolitan area.

Since the Tenth Amendment to the Constitution grants the states the right to plan land use over nonfederal land, the states should devise uniform zoning laws and delegate their authority to proper local governments. They should also provide adequate supervision and evaluation. If this were done, instead of having an enormous number of local decisions made on zoning that affects land use, decisions might be made on a more rational basis and the decisionmakers would be responsible to some higher authority. Zoning would cease to be a political game and serve its proper function of providing adequate land use for the greatest number of people. Considerable legislative and judicial actions concerning the right of private property vs the public interest have been taking place. It is essential that this question be viewed in detail and that firm long-range decisions be made to provide the greatest good for the greatest number of people and still protect private interests.

It is also wise to consider whether property-tax revenue is an adequate base from which to obtain taxes to run the government. Should other major sources be utilized? If property does continue to be the principal source of taxation, then decisions must be made to provide proper land use and adequate taxing for a given area. Assessment practices vary from community to community and industries often receive tax breaks to attract them to a given area to provide jobs. All of these things must be considered in the evaluation of the proper use of land and the role of government in this use. The federal government would do well to integrate all of the various programs from housing through road construction to environmental controls in such a way that the states would have a better understanding of how land-use patterns could be properly determined. The federal government should coordinate land-use patterns with the states by the establishment of adequate interstate highway construction, pollution-abatement facility construction, home-mortgage guarantees, proper federal income-tax deductions, and the planning of areas surrounding federal lands.

The state governments should not only carefully plan and develop land-use and zoning laws, but should also establish standards for the use of flood plains, coastal areas, wetlands, agricultural lands, highways, and airports. They should further set up a governing body to coordinate the innumerable local government agencies involved in land-use planning. It is essential that land use becomes not a single problem of one governmental unit but a coordinated approach by all agencies at all levels of government.

Housing Codes and Housing Laws

The state may regulate the use and enjoyment of individual property for the good of its citizens. This does not usually conflict with the constitutional protections given to the owner or occupant of a property. The power that the state has in establishing housing codes is based on the state's police power, which has never been precisely defined. However, the general authority comes from the Constitution, where it is stated that the state has the power to protect the order, safety, health, morals, and general welfare of the society in which its citizens live. The courts have stated that municipalities may impose restrictions on individual property owners if these restrictions protect the public health, safety, or welfare. In some court cases, it has been clearly held to be constitutional that housing codes may require windows and ventilation, screens, hot water, and other necessities of life. Local housing codes have been generally upheld by the courts.

In order for someone to conduct a housing survey, it has been generally agreed by the Supreme Court that an individual search warrant is not needed in a systematic area inspection unless the individual refuses entry to the environmentalist. It is not necessary for the health department to establish due cause, providing that there is a reasonable need to conduct periodic areawide inspections based on the passage of time, the nature of the buildings, or the condition of the entire area. However, under the Fourth Amendment to the Constitution, warrantless nonemergency inspections of residential and commercial premises cannot be carried out without the occupant's consent, unless a complaint has been lodged because of conditions supposedly existing within a given dwelling. In all cases where entry is refused to the environmental health practitioner, the county or city attorney should be consulted on the proper procedure used to gain entry to conduct the housing inspection.

It is difficult to determine what is a typical housing code. Although the American Public Health Association and the U.S. Public Health Service has issued, from time to time, a book entitled "APHA-CDC Recommended Housing Maintenance and Occupancy Ordinance" (the 1986 edition is called "Housing and Health, Recommended Minimum Housing Standards"). Local government can and does determine what it feels is a proper housing code. Therefore, codes differ from one area to another. As the housing codes are amended and changed by the local authorities, the tendency is toward decreasing stringent requirements to meet the needs of the governmental units. Where codes have been made stricter, it is because of federal requirements for obtaining urban renewal or other housing funds. It should be emphasized that housing requirements are minimum standards and that they affect the owners or landlords more than the tenants, although the tenants are minimally involved.

The entire establishment of the housing code and its subsequent use and enforcement cannot be based on any given political area, since it is necessary to look at the geographic rather than political boundaries. However, in reality, political boundaries still control the types of housing codes that are in effect. Although houses outside the city boundary are on a street adjacent to the city boundary, the housing code may be completely different. Most housing codes are municipally centered and therefore provide for the control of properties within the municipality and not outside it. Unfortunately, the contiguous areas become blighted because of lack of proper control. Some states have attempted to impose statewide housing codes, including California, Maryland, Arkansas, and Georgia. Generally, these housing codes grant counties the authority to establish codes for their own area. A small number of counties have already done this. Unfortunately, the political nature of most counties and their varying subdivisions are such that conflicts arise and housing code enforcement becomes very difficult. An exception to this is Marion County, Indiana, with Indianapolis as it center city. The Marion County Health Department exercises the authority to regulate housing throughout the entire county. As a result of this, Marion County exercises greater control over housing code violations than many other areas of the country.

Honolulu, Hawaii; Denver, Colorado; and Dade County, Florida have city-county housing programs. Urban renewal, through the use of federal funds, has helped establish housing codes in a variety of jurisdictions. As a result of this, many urban areas have had an opportunity to raze large blighted areas and to reuse the land for other purposes; examples include Kansas City, Philadelphia, Baltimore, and St. Louis.

Effective court procedures may be needed to ensure compliance with housing codes. The procedures should be brief and court hearings should be held immediately following code violations. Conclusive evidence should be presented to enhance the procedure. Difficulties occur from long delays in hearing cases and because certain courts or judges are not particularly concerned with housing code violations.

A specific housing court would be able to exercise careful and quick judgement when needed. The court's objective would be to correct the violation rather than to punish the individual. A special court could also provide specific time blocks that could be utilized for better total enforcement. The court could use the punitive measures as a last resort and court appearances as a final attempt at an educational approach to correct the situation.

Personnel enforcing housing codes should be well-trained environmental health practitioners, who are assisted by environmental technicians. The practitioners should be responsible for the overall area in housing surveys. The technicians should be responsible for the house-to-house surveys. It is necessary for the technicians as well as the practitioners to understand the role of the housing code inspection in the overall problem of the environment. They need to understand how to work effectively with people to avoid the kinds of problems that lead to securing a warrant to enter a property. Good community and public relations on a one-to-one and one-to-community basis are absolutely essential to obtain good results. The function of the environmental health technician is to attempt to improve the housing environment, not to be a punitive police officer, or to gather evidence to cause problems for a family. He or she is there to evaluate the environmental health problems and to attempt to instruct the individual and family on corrections needed to improve their housing environment. Good, sound personal communications coupled with scientific educational approaches presented in a simple manner derive far better results than any series of court procedures. Once a court hearing occurs, partial defeat is suffered. Time that should have been used for helping people has been lost in courtrooms, and the individual forced to act against his or her will is not prone to maintain his or her property properly unless repeatedly penalized.

Community Programs in Housing

Housing programs work best when the community is involved and when agencies other than the health or housing department are also involved. The other agencies include planning, zoning, welfare, low-rent public housing, building inspection, and urban renewal.

Community groups have grown in numbers and importance in the last 50 years. They were formed in the cities for the purpose of neighborhood or community maintenance. Many of these groups are either ethnically or racially homogeneous. Individuals are often property owners and the groups are called *Improvement Associations*. With the change in neighborhoods, particularly in the inner city areas, community organizations have assumed different roles. Some of them are comprised of individuals with a civic consciousness who want to help others. These individuals may or may not live within the specific neighborhood and may include business associations, clergy, educators, or other professionals. They may also include individuals who want to improve a neighborhood. Other neighborhood organizations form for specific interests, either because they are not getting something they want in the way of services or because they are seeking some means of providing a better place to live or a better social atmosphere. Unfortunately, neighborhood associations are often difficult to maintain. They are funded by contributions and often where such organizations are most needed, residents are unable to contribute. There also appears to be a general situation of public apathy. If 5% of any given block or area actively participates in a neighborhood association, a considerable amount can be accomplished. The active group will carry out the necessary implementation and planning, and with good public relations techniques, other residents will become involved in special projects for the neighborhood. It is difficult to establish a metropolitan or regional organization that is of value to a given neighborhood, since these organizations are made up of individuals who are the presidents, chairpersons, or executive directors of large agencies. They have a solid understanding of their own agencies, but generally lack adequate understanding of given communities.

In attempting to maintain or build a community group, it is necessary to have trained public health educators working with the leaders or potential leaders in the planning process and throughout the entire process of the projects undertaken within the community. It is important that these individuals maintain a low profile. They should be facilitators rather than leaders, consultants rather than decision-makers. They should help and guide the organization when guidance is needed or requested. In any community organization, it is important before any type of program is started, such as a housing program or rodent control program, that meetings of the community group be called to identify problems. Although the concern of the official agencies may be rodents or housing, the immediate concerns of the community may be police relations, recreational facilities, street repair, or better lighting. If the citizens receive an opportunity to voice their concerns and realize results, then they are more apt to work within their group to achieve the kinds of goals of interest to official agencies. The health educator should be involved in fact finding for and with the citizen groups, dissemination of information to the public, discussion of information among small groups, decision making on the basis of recommendations, testing decisions, establishing objectives, and helping to establish the program and the necessary follow-up and evaluation.

The community must identify its natural leadership and utilize this leadership in accomplishing ends established by the community and the official agencies. The community can identify those individuals who are most influential or powerful. These include church and business leaders, and professional, financial, governmental, educational, or labor leaders. These individuals can be utilized by the community in top coordinating positions. Other types of individuals, who actually get the work done, are the individuals to whom the community members turn when specific needs or problems arise. If these individuals exist in a community, they will be readily identified by other community members. When several individuals identify one person, a natural community leader has been found. This individual is probably most effective in helping get a program across, because he or she lives, works, and plays within the community, and is respected by its members. This individual can help determine the best approach to be used in establishing and implementing a program. It is wise to recognize and utilize such a person whenever and wherever possible.

INDOOR AIR POLLUTION CONTROLS

General Controls

To prevent or control indoor air pollution problems, the following strategies may be used: source control, ventilation, air cleaning, exposure control. Source control would prohibit or limit smoking indoors, relocate contaminant-producing equipment, select products or raw materials to be used with lesser contamination potential, modify occupant activities, seal or cover the source, modify the environment, and remove the source, such as a chemical spill. Ventilation can be used to dilute the contaminants with outdoor air, and isolate or control contaminants by modifying air pressure relationships within areas. Air cleaning can be accomplished by using particulate filtration, electrostatic precipitaiton, negative ion generators, and gas sorption. Exposure control occurs when contaminant-producing activities are scheduled for off-hours and susceptible people are relocated.

Specific symptom patterns and suggested means of dealing with these problems can be found in Table 7. 2. there are three general types of air cleaners used to remove particles from the air. They are:

1. mechanical filters installed in ducts to trap dust
2. electronic air cleaners which use an electrical field to trap charged particles
3. ion generators which use static charges to remove particles from the air

Table 7.2.   Symptom Patterns for Indoor Air Pollution Problems

SYMPTOM PATTERNS	SUGGESTIONS
THERMAL DISCOMFORT	• Check HVAC condition and operation. • Measure indoor and outdoor temperature and humidity. See if extreme conditions exceed design capacity of HVAC equipment. • Check for drafts and stagnant areas • Check for excessive radiant heat gain or loss.
COMMON SYMPTOM GROUPS **Headache, lethargy, nausea, drowsiness, dizziness**	If onset was acute (sudden and/or severe), arrange for medical evaluation, as the problem may be *carbon monoxide poisoning.* • Check combustion sources for uncontrolled emissions or spillage. Check outdoor air intakes for nearby sources of combustion fumes. • Consider evacuation/medical evaluation if problem isn't corrected quickly. • Consider other pollutant sources. • Check overall ventilation; see if areas of poor ventilation coincide with complaints.
Congestion; swelling, itching or irritation of eyes, nose, or throat; dry throat; may be accompanied by non-specific symptoms (e.g., headache, fatigue, nausea)	May be *allergic,* if only small number affected; more likely to be *irritational* response if large number are affected. • Urge medical attention for allergies. • Check for dust or gross microbial contamination due to sanitation problems, water damage, or contaminated ventilation system. • Check outdoor allergen levels (e.g., pollen counts). • Check closely for sources of irritating chemicals such as formaldehyde or those found in some solvents.
Cough; shortness of breath; fever, chills and/or fatigue after return to the building	May by *hypersensitivity pneumonitis* or *humidifier fever.* A medical evaluation can help identify possible causes. • Check for gross microbial contamination due to sanitation problems, water damage, or contaminated HVAC system.
Diagnosed infection	May be *Legionnaire's disease* or *histoplasmosis,* related to bacteria or fungi found in the environment. • Contact your local or State Health Department for guidance.
Suspected cluster of rare or serious health problems such as cancer, miscarriages	• Contact your local or State Health Department for guidance.
OTHER STRESSORS **Discomfort and/or health complaints that cannot be readily ascribed to air contaminants or thermal conditions**	• Check for problems with environmental, ergonomic, and job-related psychosocial stressors.

Source: *Building Air Quality–A Guide for Building Owners and Facility Managers* (Washington, D.C. 20460; U.S. Environmental Protection Agency, Indoor Air Division and U.S. Department of Health and Human Services, National Institute for Occupational Safety and Health, 1991), p. 56

Specific Controls

Residential Wood Burning

The impurities or pollutants of residential woodburning can be reduced by modifying the woodburning stoves or fireplaces for greater efficiency; by providing adequate air for proper burning; by using dry wood; by using smaller pieces of wood; by avoiding burning of any rubbish, garbage, plastic, or colored newspapers; and by cleaning the furnace, chimney, and fireplace regularly.

Passive Smoking

Passive smoking can be prevented or controlled by adequate sources of natural or forced ventilation. Further control will occur if there is a physical separation of nonsmokers from smokers. Many public laws today require that smoking take place in only certain very limited designated areas.

Carbon Monoxide

Carbon monoxide poisoning can be prevented in three specific ways:

1. Through proper ventilation, which will provide adequate air for burning, if burning is taking place, or will dilute the carbon monoxide that is present in the air.
2. Using proper equipment and methods to exhaust flue gases from the environment.
3. Physically separating all heating units and combustion appliances from people. Oil and gas heating units may be placed outside the house or vented to the outside.

Radon Gas

Radon gas can be controlled through natural and forced ventilation. This is done by ventilating the lowest level of the house. Either fans are used to force outdoor air into the home or ventilation grills are put into areas to allow outside air to flow in by itself. A major disadvantage of this is the increased cost of heating and cooling. Another technique used to control radon gas is to seal the cracks and openings through which the gas passes from the soil into the house. This would mean to seal all openings around utility pipes; the tops of concrete block walls, chimneys, joints; and any other cracks or openings. It would be impossible to find every crack or dent. Another technique is to heat recovery ventilation. In this case, replace the radon-laden air with outdoor air. The indoor air is warmed or cooled by the radon-laden air exhaust. This procedure is similar to convection where heat energy flows from one gas to another gas. There is still a loss of indoor heating and cooling energy. Another technique is through drain-tile suction. A drain title is installed completely around the house and a fan is used to draw the radon from the drain tile. A subslab suction system may be used to reduce the radon from the slab. There are also techniques utilized in block-wall ventilation and the prevention of house depressurization. In house depressurization, it would be necessary to consider discontinuing the use of wood stoves and fireplaces, since they lower the air pressure in a house by consuming air and exhausting it to the outside. One other technique would to be have more pressure in the house by actually pressurizing the house and using fans blowing upstairs air into the basement.

Formaldehyde

There are seven different approaches that may be used in trying to alleviate the formaldehyde problem. They are as follows:

1. Use of interim measures, such as alternative housing or temporary housing, until the formaldehyde problem can be resolved. This is especially important for families with infants or young children, who are very sensitive to the irritating symptoms caused by formaldehyde. If the formaldehyde levels reach above 0.01 ppm, this measure should be considered.
2. The control of the indoor temperature and relative humidity can help significantly reduce formaldehyde levels. The recommended temperature levels for maximum formaldehyde reduction include lowering of the household temperature to 68°F during the heating season in northern states and the use of an air conditioner to maintain a constant indoor temperature of 70–72°F in the southern and southwestern states. The air conditioner also helps to dehumidify the air.
3. Source removal is the most effective control measure for formaldehyde. This would mean the removal of particle board, subflooring, and urea-formaldehyde foam insulation. This is very difficult and very costly. Other sources of formaldehyde emission include paneling, furniture, and cabinets. The removal of these sources are less costly and less difficult and can help reduce the amount of formaldehyde in the air.
4. Source treatment is an alternative to source removal but is generally less effective. This method consists of applying a coating material to the sources that emit formaldehyde. The coating material acts as a barrier and prevents formaldehyde release and also prevents moisture from entering the material, which helps promote the release of formaldehyde. Source treatment can be applied to unfinished wood products, particle board subflooring, plywood paneling, particle board shelving, cabinets and countertop undersurfaces. The two primary coating materials used are polyurethane and a nitrocellulose-based varnish.
5. Ammonia fumigation, if done properly, can cause a reduction of 50–60% of the formaldehyde. With this approach the whole interior of the house is enclosed and exposed to ammonia gas over a 12 hr period at an indoor temperature of 80°F. It is very essential that trained individuals do this, since ammonia is very toxic. At present this is the only method that is practical to reduce formaldehyde in environ-

ments that contain high concentrations. All surfaces must be vacuumed very carefully, since the ammonia combines with the formaldehyde to form hexamethylene tetramine, which is a fine dust.

6. Pure air purification systems remove formaldehyde from the air by absorption or adsorption. The effectiveness of the system is based on the chemical, the system design, the capacity to move air, and the levels of formaldehyde present. For adsorption, activated charcoal is relatively inefficient in reducing formaldehyde levels.

7. Forced ventilation systems will push fresh air into the contaminated building and therefore reduce or dilute the level of formaldehyde present. The amount of reduction is based on the amount of air flow.

Other Volatile Organic Compounds

Other volatile organic compounds used in a house, such as solvents, paints, polishes, household cleaners, etc., should be only utilized in areas where there is plenty of ventilation. The amount of material used should be strictly limited to that which is recommended on the container. Volatile organic compounds are hazardous. They should be treated as such.

Asbestos

Two techniques necessary for control of asbestos include

1. suppression
2. removal

Suppression is the best way to avoid dust, since the asbestos dust is not created. Removal is a complicated task that must be carried out by individuals who are highly trained. The EPA and OSHA have specific guidelines for the removal of asbestos from buildings. In any case, the asbestos materials are prepared for removal. They are wetted with a water and surfactant mixture sprayed in a fine mist, allowing time between sprayings for complete penetration of the material. Once the material is totally wetted, it is then removed from the building and placed in thick plastic bags and sealed in such a way as to make them leaktight. They are then doubled bagged and placed in plastic-lined cardboard containers or plastic lined metal containers and removed to the disposal site.

COMMUNITY PROGRAM AND INJURY CONTROL

Injury prevention is based on three general approaches. They are

1. Persuade the individuals at risk or injury to alter their behavior. An example of this would be the installation of smoke detectors.

2. Require that there be an individual behavioral change by law or administrative rule. An example of this would be a law that smoke detectors must be installed in houses or apartments.
3. Provide automatic protection by changing the product or by some form of design correction. An example of this would be the installation of built-in sprinkler systems.

The Philadelphia Injury Prevention Program is called the *Safe Block Intervention Program*. It is directed at reducing the occurrence, severity, and consequences of injuries in an economically disadvantaged community by creating safe blocks. Its goals are to conduct home modification programs, inspections of all homes in the target area, and education of the residents about injury prevention methods. The safe block concept includes the use of a public health team who inspects residences for problems and then has the environmental "modifiers" to help implement the prevention program. After the initial visit, a second visit will is made about 6 months later to determine whether the modifications remain intact, whether recommendations have been followed, and what the results are in injury control at this point. Obviously, data has to be gathered initially on the level of injuries, types of injuries, etc., or prior to the intervention program.

An intervention program consists of modifications of the home in the following ways:

1. Place stickers for regional poison control centers and 911 on telephone.
2. Provide fire extinguishers for the kitchen.
3. Provide smoke detectors for homes without the detectors. It may also be necessary to inspect existing smoke detectors and replace batteries.
4. Provide bathtub nonslip strips.
5. Provide ipecac and instructions for safe use if a poisoning occurs.
6. Reduce the temperature of the hot-water heater.
7. Provide light bulbs of higher intensity if there is inadequate lighting on the stairs.
8. Rat-proof the homes.
9. Install cabinet locks if young children are present.
10. Store poisons, medicines, and cleaning supplies out of the reach of children.
11. Install staircase gates at the top of the stairs and the bottom if children are present.
12. Inspect staircase hazards.
13. Inspect wiring or other hazards within the home.
14. Check space heaters and whether or not they are operational.
15. Inspect storage areas for newspapers, gasoline, and gasoline storage.
16. Provide education for the individuals in the home on how to avoid injuries from a variety of environmental sources.

This program is considered to be operating at a good level.

Role of the Federal Government in Housing

Congress stated in the Housing Act of 1949 that the general welfare and security of the nation and the health and living standards of its people require housing production and related community development to correct the serious housing shortage, to eliminate substandard and other inadequate housing through the clearing of slums, and to provide decent homes and suitable living environments for all American families. To obtain this national housing objective, Congress provided for private enterprise to carry out the major portion of the program; government assistance to private enterprise to carry out the major portion of the program; government assistance to private enterprise to be given where needed; establishment of positive local programs; government assistance to eliminate substandard and other inadequate housing through clearance of slums and other related areas; government assistance for decent, safe, and sanitary farm dwellings. Federal government departments were given the responsibility of facilitating the steps leading to this national housing objective. They were to do so by encouraging and assisting in the production of housing of sound standards of design, construction, livability, and size; reducing the cost of housing without sacrificing standards; using new designs, materials, techniques, and methods of residential construction; developing well-planned integrated residential neighborhoods; and stabilizing the housing industry.

The 1965 Department of Housing and Urban Development Act stated that, as a matter of the general welfare and security of the nation and the health and living standards of the people, the national purpose was to have a sound development of the nation's communities and metropolitan areas. To carry out this goal and in recognition of the importance of housing and urban development in the national life, Congress established an executive department to handle the problems of housing and urban development.

The 1968 Housing and Urban Development Act reaffirmed that national goal set forth in 1949 of a decent home and suitable living environment for every American family. Congress recognized that the goal had not been fully realized and that there were many suffering low-income families. They declared that special provisions be made to assist families with incomes so low that they could not decently house themselves. In August of 1968, under Title Sixteen, Housing Goals and Annual Housing Report, Congress declared that the supply of housing was not moving ahead rapidly enough to meet the national housing goal of 1949. It stated that there was a need for construction or rehabilitation of 26 million housing units, six million of which were for low- and moderate-income families. It stipulated that the President had to set forth a plan to be carried out for a period of 10 years from June 30, 1968 to June 30, 1978 to eliminate all substandard housing by the latter date. The Department of Housing and Urban Development Act, Public Law 89-174, granted the Department of Housing and Urban Development all of the functions, powers, and duties of the Housing and Home Finance Agency, the Federal Housing Administration, and the Public Housing Administration. In addition, the National Mortgage Association and other departments relating to housing were placed under the control of the Department of Housing and Urban Development. This department was to have vast powers to improve housing in the U.S.

The Housing and Community Development Act of 1974 stipulated that there should be an allocation of housing funds to provide for local housing assistance plans. The Department of Housing and Urban Development, which received an extremely large budget from Congress, was responsible for the following areas: (1) the Council for Urban Affairs; (2) environmental and consumer protection; (3) rent supplements; (4) disaster assistance; (5) establishment of advisory committees; (6) civil defense, as it related to vulnerability to attacks; (7) handicapped individuals and accessibility to buildings; (8) assignment of emergency preparedness functions; (9) the Federal Council for Science and Technology; (10) coordination of federal urban programs; (11) the National Institute of Building Sciences; (12) housing renovation and modernization under Title 1 of the revised July 31, 1975 basic laws and authorities on housing and community development; (13) mortgage insurance, Title 2; (14) war housing insurance, Title 6; (15) insurance for investment in rental housing Title 7; (16) armed services housing, Title 8; (17) national defense housing, Title 9; (18) mortgage insurance for land development, Title 10; (19) mortgage insurance for group practice facilities, Title 11; (20) FHA and VA interest rates; (21) the Rehabilitation Act of 1973; (22) closing of military bases and mortgage defaults; (23) special assistance programs such as emergency homeowner relief, housing for the elderly, college housing rehabilitation loans, rent supplements, urban homesteading, training and technical assistance; (24) community development assistance programs such as block grants, urban renewal, public works planning, public facilities loans, public facilities grants, model cities, historic preservation, Lead-Based Paint Poisoning Prevention Act; and (25) rural and other nonhousing and urban development community development programs, such as the Rural Development Act, Headstart, regional action planning commissions, and Appalachian regional development.

In 1974 the Emergency Home Purchase Assistant Act was passed. The Emergency Housing Act of 1975 included Emergency Home Owner's Mortgage Relief Title I. In 1976, the Housing Authorization Act was passed by Congress.

In 1977, the Housing and Community Development Act was passed by Congress. It included special titles on Community Development; Housing Assistance and Related Programs; Federal Housing Administration; Mortgage Insurance and Related Programs; Lending Powers of Federal Savings and Loan Associations and Secondary Market Authorities; World Housing; National Urban Policy; Community Reinvestment; and miscellaneous provisions.

In 1978, 1979, 1980, 1981, 1983, 1984, and 1987, additional housing laws were passed. However, in the 1980s, in reality, many housing programs were reduced in scope.

Programs of Housing and Urban Development 1988–1989

The Housing and Urban Development Department (HUD) is responsible for a huge multitude of housing programs to help the citizens of the U.S. There are some 94 specialized programs handled in the Housing and Urban Development Department. They range from housing for the poor and the homeless, to the elderly, to nursing homes and hospitals, to public housing developments, to Indian housing. There are policy, development, research, and mortgage guarantee programs. The programs are extensive and extremely complex.

As can be seen, the federal government has made an enormous commitment to better housing for the American public. The ultimate goal has not yet been achieved. Considerable additional effort and money will be needed. But beyond this, considerable research is required in the area of attitude change and human behavioral patterns toward the housing environment. The best housing may be destroyed readily unless there is a thorough understanding of the effect of environment on human behavior. It is strongly urged that environmental health practitioners, other public health officials, citizens bodies, and legislators work toward obtaining proper funds for behavioral analysis and training. It will then be necessary to institute the training that will maintain the new environment in a proper manner for the individuals who will be living within it.

Emergency Housing

Emergency housing consists of the use of trailers provided by the Department of Housing and Urban Development, gymnasiums, churches, or any other available buildings for use when normal housing has been destroyed, made unsafe, or has been inundated by floods. Emergency housing should be closely supervised by environmental health and other public health authorities to prevent the outbreak of contagious or infectious diseases. Adequate water, food, clothing, and blankets must be provided. Generally, the American Red Cross is most helpful in this type of situation.

ROLE OF THE FEDERAL GOVERNMENT IN INDOOR AIR POLLUTION

The EPA has been very concerned with the problems of indoor air pollution. They are currently operating under a variety of environmental laws, including the Toxic Substances Control Act; the Federal Insecticide, Fungicide, and Rodenticide Act; the Safe Drinking Water Act; the Resource Conservation and Recovery Act; the Asbestos in School, Hazard Abatement Act of 1986; and the Uranium Mill Tailings Radiation Control Act. The EPA was given specific directions in establishing an indoor air quality program in the 1986 Superfund Amendments and Reauthorization Act.

Since 1982, the EPA has conducted research programs on indoor air quality. The research has been directed at increasing the understanding of personal exposure, emissions, health effects, and mitigation techniques. The Radon Gas and

Indoor Air Quality Act of 1986 was directed at having EPA conduct research, implement a public information and technical assistance program, and coordinate federal activities on indoor air quality.

The EPA has taken regulatory action on asbestos, VOCs in drinking water, and certain pesticides. They require schools to inspect for asbestos, to prepare management plans, and to take action when the schools find friable (easily crumbled) asbestos. The EPA has issued maximum contamination levels for the VOCs in water supplies that serve more than 25 people. They have also acted to control indoor exposure to pesticides, as well as requiring childproof packaging for certain pesticides. They prohibit indoor applications of the wood preservatives pentachlorophenol and creosote.

Environmental Lead Hazard Abatement

Laws

The 1971 Lead-Based Paint Poisoning Prevention Act coordinated the efforts of health and housing people in dealing with the problem of lead-based paints in houses. It provided for the environmental management and control, as well as prevention, of lead-based paint illness. It also provided for screening and medical management of children who had ingested lead. In 1988 the Lead Contamination Control Act required the following:

1. The identification of water coolers that are not leadfree.
2. The repair or removal of water coolers with lead-lined tanks.
3. A ban on the manufacture and sale of water coolers that are not leadfree.
4. The identification and resolution of lead problems in schools' drinking water.
5. The authorization of additional funds for lead screening programs for children.

The reason for the enactment of this law is because of the very dangerous nature of the ingestion of lead from water for children. The harmful health effects include serious damage of the brain and central nervous system, kidney, and liver. Obviously inhalation of lead dust is also a serious problem.

Abatement Procedures

Abatement procedures consist of emergency repair and permanent repairs. Emergency repairs consist of the following:

1. scraping of all peeling, chipping, or flaking lead-based lead surfaces
2. sanding with a rough grade of sandpaper on all lead-based paint surfaces
3. applying inexpensive contact paper to lead-based paint surfaced with wallboard, plywood, etc.

Permanent repair methods include the removal of all lead in paint, scraping and sanding, the use of liquid paint removers, the use of heat to remove leaded paint, and covering all areas where lead may be on the walls. Lead in dust may be removed by the use of proper dustless cleaning techniques. There are hazards associated with paint removal. An open-flame torch or heating device may volitalize lead. Machine sanding produces lead dust, as does sand blasting. Chemical strippers containing methylene chloride are extremely toxic. Lead in water may be removed by altering the plumbing system and removing the lead.

A Childhood Lead-Prevention Program

The City of Philadelphia has developed a childhood lead prevention program that includes screening of a large number of children in high-risk populations; a referral system that ensures a comprehensive diagnostic evaluation of every child with a positive screening test; a program that ensures identification and elimination of the child's lead exposure source and a system that monitors the adequacy of the treatment of the child. Children from 6 months to 5 years of age are screened for undue lead absorption and or lead toxicity. These children come from high-risk areas. Medical management includes evaluating the medical aspects of the program; evaluation of the timeliness and appropriateness of referrals; evaluation of case records; consultation with physicians; and conduction of educational programs. All individuals in high-risk areas are followed up after the initial determination has been made concerning lead poisoning.

Environmental management is carried out through:

1. Investigations to determine if lead-based paint is present or other sources of lead are present.
2. Abatement through the removal of lead hazards, especially paint. Abatement is difficult because of the total number of houses that have to be taken care of and the cost of doing this. Cost estimates for removing lead-based paint for an average three bedroom house range from $8000 to $15,000 per house.

An education program is carried on by health professionals to teach parents the problems of lead, to make the general community aware of lead problems, and to provide an outreach program to assist individuals, hospitals, and physicians, as well as day-care facilities, in understanding the risks of exposure to lead. The program that has been identified is operating in a successful manner. Other communities have also developed childhood lead programs.

SUMMARY

The indoor environment is a complex human environment. It encompasses facets of all other environmental problems. The indoor environment may include anything from an eight-family, three-story dwelling to a huge home with a swimming pool. The environment has an impact on the shelter, and the shelter has an impact on the environment. In addition, to fully understand the indoor environment, one must appreciate the problems of poverty, joblessness, welfare, socioeconomic status, racial and ethnic origins, and the feeling in America that an individual's home is his or her castle. A house in good condition provides a satisfactory environment for psychological, physical, and mental well-being. The house and neighborhood in poor order may furnish the impetus for crime, poor education, poor health, and fires, and creates a society that is helpless in its dealings with the rest of the world. Further, indoor air pollution has caused additional stress and potential of actual disease. Injuries are an overwhelming concern.

There are many new theories on the relationship between housing and health based on nonspecific responses, in which a given cause may create a variety of health patterns and a given syndrome of physical or emotional maladjustment may stem from a variety of causes. It is necessary to consider the causes and to carry out an analysis incorporating all of them. New housing development, including new technology, materials, construction, and a variety of consumer products, have added complexities to the housing environment. In some areas this has provided for the greater good of the individual. In other situations, this has created additional problems, since the new product, for example, when broken must be serviced, and the service cost in many cases exceeds the dollars that an individual has to take care of the problem.

Once again, it must be emphasized strongly that despite everything that the federal, state, and local governments, and various community groups, may do to improve the indoor environment, the basic changes that occur will be brought about by the resident. It is recommended that the resident obtain a copy of "The Effect of the Man-Made Environment on Health and Behavior," by L. E. Hinkle, Jr. and W. C. Loring, Center for Disease Control, Public Health Service, U.S. Department of Health, Education, and Welfare, Atlanta, GA, 30333 (Department of HEW, publication number (CDC) 77-8318, 1976). This publication may be obtained from the Government Printing Office, Washington, DC 20402, stock number 017-03-00110-8. In the future, conferences must be held between the specialists that examine the people-made environment and members of this environment for a fruitful exchange of ideas that we may move forward to the goal proclaimed by Congress in the Housing Act of 1949 — the ultimate goal of a decent home and suitable living environment for every American family.

RESEARCH NEEDS

It is necessary to thoroughly evaluate the existing types of planning used in community development and redevelopment. From this evaluation, measures should be formulated to determine the best approaches to housing development and redevelopment. Research techniques are needed to determine the manner in which improved housing can reduce social problems. Behavioral studies of individuals living in poor housing should be conducted to determine the underlying sociobehavioral disturbances that occur in these environments and also to determine why certain individuals will move into good housing and lower the physical structure to the level of their previous housing. Behavioral studies are needed to determine the best means to educate individuals to improve their housing environment and also to learn to live within an improved housing environment. Studies are needed to determine the critical variables in establishing new communities.

Indoor air has become a serious concern within the home as well as within institutions. The indoor air may contain a variety of pollutants created within the property or outdoor pollutants that have entered the property through cracks and openings, especially in older structures. Carbon monoxide, particles, sulfur dioxide, and oxidants enter buildings. A variety of odors created by external pollution sources or indoor pollution sources contribute to the air problem. The indoor pollutants consist of particulate matter from activities such as cooking, tobacco smoke, and a vast variety of cleaning materials, synthetic floor coatings, polishes, cosmetics, pesticides, paints, vapors, glues, and other contaminants that come from the workshop or other parts of the property.

It is necessary to conduct research and demonstration projects on how to evaluate the techniques that are used to reduce radon levels in new and existing homes. It is also necessary to develop new and more effective techniques for determining if radon gas is present within the home. Additional research should be conducted on effective and inexpensive techniques for the removal of existing indoor air pollutants.

Surveys are needed to determine the patterns of polluting equipment and materials used in homes. Research priorities should then be developed to determine which areas need to be studied. Maximum exposure levels need to be determined and this information should be correlated with the individual's occupation to determine if he or she is receiving additional stress from the home environment. Studies are needed of the best techniques to control indoor concentration of air contaminants by improving air-cleaning devices and developing inexpensive techniques for the removal of air pollutants at low air pressures.

Injury control in the indoor environment is essential. Further research is necessary on the causes of accidents and injuries in the indoor environment, their controls, and how best to limit the degree of the injury. Research emphasis must be placed on an economical approach to altering the environmental conditions leading to accidents and injuries, as well as the best techniques for product modification to enhance safety. New public health education tools need to be developed and evaluated to determine if they are in fact reducing the levels of injuries in the indoor environment.

Chapter 8

INSTITUTIONAL ENVIRONMENT

BACKGROUND AND STATUS

The successful design and practice of an environmental health and safety program in an institutional setting is a comprehensive undertaking. It involves the coordination and integration of a broad range of people, both professional and nonprofessional, and their diverse and complimentary resources and skills. The institution is in effect a small community and therefore contains the numerous environmental, community health, social, emotional, and psychological problems attributed to or found in any small community. In addition, there are a considerable number of people congregating in a limited amount of space. The typical institution continues to grow in size and complexity. The budget of the institution and means of financing have become problems, since costs rise as new techniques and equipment are developed and as the inflationary economy continues to spiral. Without any technical changes, the dollar in 1967 now must be replaced by about $4.23 in 1995 to purchase the same goods and services. The actual cost of health care is much higher than inflation.

People congregate in institutions. This congregation comprises a shifting population that may use the facilities for parts of a day, the entire day, or for an entire 24-hr period. As in any other situation where groups of people gather, the potential for the spread of disease is increased. Institutions contribute to air and water pollution, and solid-waste problems. Further, each institution has its own unique set of occupational hazards and other types of safety hazards. Some problems in institutions are limited to the basic environmental health problems, including water, air, solid waste, food, and shelter. Others have many shops, or even the equivalent of comprehensive industries, and therefore face all the occupational

377

hazards found in these settings, including chemical, microbiological, and radio-logical hazards. A further complication is the fact that many institutions receive federal and state financing.

Institutions include primarily hospitals, nursing homes, convalescent homes, old age homes, schools, and prisons. (For a detailed presentation of the institutional setting, the reader is referred to "Environmental Health and Safety," by this author, Pergamon Press, 1974, now out of print.)

Nursing and Convalescent Homes

Nursing homes exemplify in many ways a failure of public policy, public concern, and private enterprise. Although a good proportion of the patients are aged and have a variety of infirmities, including senility, there are still many patients with active minds who should be receiving the type of care to which they are entitled. It is much easier to medicate individuals to conform their behavior to the standards set by the nursing home than to allow them to maintain their dignity as human beings. Some nursing homes today are as bad as the pesthouses of the 18th century. Overcrowding, unsanitary conditions, and potential firetraps exist. Each year a serious fire occurs in which unfortunate older people die or are injured. The basic causes behind this dilemma are the social, health, and economic trends of the past 85 years.

The U.S. became a rapidly growing industrial nation during the early part of the century. People moved to the cities to find jobs and often to smaller houses with fewer rooms. They were unable to keep with them the grandparents or other family members who in the past had a secure place to live in their old age. Industrialization created more jobs that were filled by women and younger people. Older persons were frequently left out of the job market. If they became disabled, their opportunity to get a job was almost totally lacking. This situation with the aged survives today and grows worse, since long-term inflation has sharply reduced or totally eliminated the savings of a group of the aged, thereby creating additional problems.

Unfortunately, in the past, nursing homes did not evolve in an orderly manner but rather erupted, almost in the way an earthquake causes disruption within a community. In order to provide homes for the aged or sickly, many ill-equipped reconverted private houses were pressed into operation. Unfortunately, this contributed to the severity of the nursing-home problem. Nursing homes were set up in run-down residential areas, in poor and inadequate facilities, and in difficult areas for family members to reach, unless they had access to private transportation. The nursing home movement received further impetus from the Social Security Act of 1935. This act provided for the elimination of public poorhouses and public assistance for residents of these institutions. The residents went to private institutions, since these were available.

Today there are an estimated 20,000 nursing homes in the U.S. of which some 77% are proprietary or operating for profit. About 7% are government sponsored and about 16% are run by various nonprofit organizations, including religious groups. In some of these private nursing homes, individuals do not necessarily receive care commensurate with the amount of money that they pay. There are many fine proprietary nursing homes in the U.S. However, there are others interested primarily in making profit and not in providing adequate care for the patients. Congressional hearings in New York City demonstrated that, despite federal regulations for skilled nursing-home facilities that were required by Medicaid and Medicare, innumerable abuses were found. Obviously, because the average age of the people in institutions is approximately 80 years old, they seldom challenge their treatment and care.

Nursing homes, in effect, are medical facilities that should have the same type of personnel and programs as hospitals, but to a lesser degree. Obviously such things as operating rooms and emergency rooms do not exist. However, nursing homes should provide good patient care, medical and nursing supervision, physical therapy, occupational therapy, dental services, eye services, necessary drugs, and effective humane care. Nursing homes should be able to handle emergencies to save the lives of patients. The facilities should meet the requirements of skilled nursing home facilities under Title 19 of the Social Security Act. These requirements are updated from time to time by the Department of Health and Human Services. For current requirements, the Department of Health and Human Services should be contacted.

Old Age Homes

Old age homes encompass every type of shelter from a simple house where a few elderly people live to extensive beautiful facilities where the individuals not only have shelter and food, but also adequate recreational facilities, medical care, and occupational and physical therapy. The old age home is an outgrowth of the same kinds of conditions that helped to create the nursing home.

Schools, Colleges, and Universities

Healthy school living requires considerable effort to provide good physical, emotional, and social well-being. The school or university administrator, along with the Board of Trustees and the various other administrative support and professional staffs, have as a goal the proper education of students at all levels. It is necessary to provide an environment that is wholesome and supportive of learning. This environment includes all of the problems, programs, and services found in other institutional environments, but with some exceptions. In grade schools, the young age of the students increases the likelihood of accidents. In high schools, the students not only face potential accidents from various play areas, but also from shops and laboratories. In colleges everything that applies to the high school continues to be a problem but is multiplied manyfold by the sheer numbers of students, by increasingly sophisticated and complex laboratory experiments and chemicals used, and by the fact that many of the students reside on the university grounds.

Prisons

Each correctional institution should provide a healthy place for a person to live. Since a healthy environment includes all of the aspects of a good institutional environment, and as custody means more than possession, the prisons of our country have a definite obligation to fulfill in taking care of prisoners. In the past there have been considerable problems in correctional institutions because of overcrowding, numerous existing health hazards, and a poor environment.

The conditions in all types of institutions continue to deteriorate. Specific types of concerns are discussed under the section on problems. To avoid repetition in this chapter, most of the material discussed will relate to the most complex type of institution, which is the hospital. In other institutions where problems, techniques, and so forth vary from the hospital setting, they too will be discussed.

SCIENTIFIC, TECHNOLOGICAL, AND GENERAL INFORMATION

Hospitals

The hospital today is more than a place where people come who are terminally ill. The use of intensive care units, cobalt therapy, artificial kidneys, organ transplants, inhalation therapy, and so forth has made the hospital significantly different from its predecessor. It is a continually and rapidly changing environment in which diagnostic and treatment facilities are improving, staffs are better trained, and costs are soaring. The functions of the hospital are to give care to the sick and injured, provide necessary health education, conduct realistic research, and promote community health.

Hospitals, generally, are either governmental, proprietary (run for profit), or voluntary nonprofit. The voluntary nonprofit hospital accounts for about 48% of the nation's hospital beds. These hospitals are usually brought into operation by community groups, churches, or other organizations. Governmental hospitals, which have about 40% of the hospital beds, are operated under federal, state, or local budgets. Proprietary hospitals, which contain about 12% of the hospital beds, may either be individually owned or owned by a partnership or corporation. Hospitals may also be listed as general, special, short term, or long term.

Organization

The Board of Trustees or Board of Directors is the highest authority within the hospital. This board determines the direction that the institution will take and establishes policies that are consistent with its goals. It usually represents a cross-section of businesspeople, educators, health leaders, and other distinguished individuals. They dedicate a significant part of the working time to the institution, usually without pay. Boards vary in size, although an adequate-size group probably would be a maximum of seven to nine individuals. The board forms committees and so forth. The board evaluates and approves the hospital budget and also gives guidance to the chief executive officer. It is responsible for providing a bridge between the administration, the medical staff, and the public.

The chief executive officer, who may have a variety of titles from director to administrator, is the day-to-day chief administrator of the hospital. His/Her job involves managing, coordinating, and directing the activities performed by the employees. He/She presents the hospital's problems to the board, implements the board's decisions, and interprets the board's actions to the hospital community. It is essential that the chief executive officer be a fine administrator, keep in close touch with all parts of the institution, and have a good working relationship with the many chiefs and supervisors of the various parts of the institution.

The medical staff directs patient care. It is responsible for giving professional care to the sick and injured, and for making careful and proper diagnosis. It is also responsible for maintaining its own efficiency, governing itself, and choosing new members. It participates in educational programs and gives advice to the chief executive officer and the Board of Trustees. The medical staff is divided into a series of committees that oversee the various operations of the hospital, including medical records, the surgical department, pediatrics, radiology, and so forth.

For the hospital to operate properly, it is essential that the three major groupings — the Board of Trustees, medical staff, and chief executive officer — cooperate and work together efficiently. The other professionals, such as the nursing staff, environmental practitioner, and safety officer, generally work through the chief administrative officer and therefore are extensions of his administrative and supervisory control.

Electrical Facilities

Electrical energy is the major source of power on which almost every part of the hospital depends. Electrical supply must be adequate and always dependable. It is essential to have electrical supplies that switch on automatically when there is a loss of normal service. Electricity must be available on a 24-hr basis.

Heating, Air Conditioning, and Laminar Flow

A thermal environment should be maintained within the hospital to permit adequate heat loss but prevent excessive heat loss from the human body. At the same time, this environment must maintain an atmosphere that is reasonably free of chemical and bacterial impurities and odors, and one that is conducive to the best physiological and psychological comfort of the individual.

The function of the ventilation system, apart from providing adequate temperature and humidity control, is the control of the number of microorganisms in a given area. These microorganisms come from other areas into the target area, come from equipment and supplies, from outside air, from individuals within the immediate environment, from infectious hazardous waste, other solid waste, blankets, or dirty linens. The function of mechanical ventilation in removing these organisms is carried out by literally blowing the organisms away from the areas that are to be kept free of contaminants. An example of this would be the wound or operating site on a patient. There have been several methods proposed to remove the air that may contain microorganisms from these critical areas. The first method directs a stream of air either horizontally or obliquely toward the operating table to create turbulent air flow directly over the patient. The second method introduces air at the ceiling in such a way that only mild turbulence is created by the supply air mixing with the room air, and then it is exhausted. This creates a constant dilution of the air. The third method is the same as the second, except that it produces minimal turbulence when mixing with the room air and is then removed by downward displacement of the air to ports located on the walls near the floor. The most effective technique for removing microorganisms appears to be the third system. The turbulent system, although it removes a considerable amount of contaminated air, also brings more contaminated air across the wound site. Along with the air system, is the air filtration system which is used to remove particulate matter from the air.

Laminar flow of air provides the best total control of the air environment, which includes temperature, humidity, cleanliness of the air, and direction of flow. Laminar flow is simply the introduction of large volumes of clean air through a very large diffuser or perforated panel, which reduces the velocity of the incoming air, thereby preventing agitation and reintroduction of settled contaminants. It also decreases the contamination blown off of personnel and equipment. At the opposite side of the room, the air is removed through a perforated area of the same size as the inlet diffuser. This is a single-pass technique in which the air either comes from the wall or the ceiling, passes through ultra-high efficiency particulate air filters, moves uniformly across the room and through the outlet to return ducts, which then go through the final filter and back through the room again. The amount of air flowing depends on the size and shape of the room and the height of the ceiling.

Heat and proper air control of temperature is important. Heat balance in the human body is affected by the rate of metabolic heat produced in the body, the rate of storage or change in body heat, the rate of heat loss or heat gain by convection, the rate of heat loss or heat gain by radiation, and the rate of heat loss by evaporation of sweat. It is essential that the ventilation, rate of flow of air, and the actual heating and cooling of the areas be done in such a way that the individual will be comfortable. Comfort is difficult to define, since it is not only based on the previously mentioned items, but also on the sex, age, and physical condition of the patient, and the sex, age, and type of activity of the worker. Generally speaking, within the hospital a temperature of 75°F would be considered a comfortable temperature for most patients. The room temperature, however, may have to be adjusted to the individual.

Housekeeping

The function of the housekeeping service is to provide and maintain a clean, safe, and orderly environment that helps prevent disease and accidents, and promotes health. The individual, whether a hospital employee, visitor, or patient, is affected by aesthetic surroundings. It is important that all areas appear clean and in fact are clean to avoid hazards. Industry practices proper cleaning by using adequate procedures and techniques, proper task-oriented detergents, disinfectants, and proper equipment. It is important for institutions to follow this example.

Maintenance

The maintenance department is responsible for the maintenance and operation of the physical plant of the hospital. The size of the department and its functions vary with the size and complexity of the hospital. Maintenance usually includes the operation of the power plant, electrical service, plumbing service, heating, and maintenance; maintenance personnel include refrigeration mechanics, welders, carpenters, brick masons, roofers, gardeners, locksmiths, plasterers, painters, and air-conditioning specialists. The maintenance department is also responsible for all keys, especially the special keys that fit the narcotics cabinets, medicine cabinets, psychiatric areas, and prison areas. The proper operation of the maintenance department and the proper maintenance and repair of all equipment and parts of the physical structure is an essential part of a good environmental health program.

Plumbing

Hospitals probably have more piping systems than any other type of building. The common piping systems include cold potable water, chilled and recirculated drinking water, distilled water, fluid suction systems, vacuum cleaning systems, oxygen, fire sprinkling and stand pipes, lawn irrigation, air-conditioning, refrigeration systems, recirculated cooling water, drainage systems, soil and waste systems, vent systems, stormwater systems, and building sewers.

Radioactive Materials

The use of radioactive isotopes in hospitals and other institutions has greatly increased in the last 40 years. Radioactive materials are found in dental clinics, nuclear clinics, pharmacies, diagnostic and therapeutic radiology laboratories, and patient rooms where radionuclides are used.

Water Supply

The water supply for hospitals and other institutions must be of high quality, physically, chemically, biologically, and radiologically. The potable water must meet the drinking water standards of the "National Primary Drinking Water Regulations" of 1989, as updated through 1991. Water must be provided in adequate quantities under proper pressure. It may then be further purified within the institution, if it is used in various processes. The water may be distilled, it may be turned into steam, or it may be sterilized. It is essential that the plumbing utilized in the distribution of water within the system be set up in such a way as to avoid any cross-connections or submerged inlets.

Emergency Medical Services

The initial emergency medical service program began in 1966 when the Highway Safety Act provided federal grants for this purpose. Its primary objective was to establish self-supporting systems at the state and local levels. Each state provided a commission to receive the federal highway safety funds and to utilize them by establishing standards for ambulance services, their equipment and personnel, and for institutions engaged in the training of ambulance personnel. The current laws, which were initially funded in 1975 and are currently being funded, also provide funds for the certification and training of emergency medical technicians. Any person who serves in an ambulance and who administers emergency care to patients must be certified as an emergency medical technician by the commission. (See Chapter 8, Volume II, for more information.) The individual must obtain specific continuing education credits annually and be recertified on a yearly basis.

Fire Safety Programs

Fire prevention, protection, and suppression are an integral part of an institution's health and safety program. An effective fire prevention program includes early detection, prompt extinguishment, and evacuation of people. To accomplish this, the institution must define its needs in conjunction with skilled fire prevention specialists. It also must assign responsibilities for each part of the program to appropriate staff members; establish practical regulations; establish uniform procedures for fire prevention, fire protection, and fire suppression; give emergency instructions; provide appropriate training courses and educational programs; have adequate fire design control; make necessary inspections and corrections; and carry out fire drills.

Hand Washing and Hospital Environmental Control

Nosocomial infections (hospital-acquired infections) occur in approximately 6% of all patients admitted to acute-care hospitals in the U.S. The cost of these infections run into the billions of dollars. The CDC put out guidelines concerning hand washing and hospital environmental control in 1982 and then updated them in 1985. The guidelines contain six sections. They include hand washing; cleaning, disinfecting, and sterilizing patient care equipment; microbiological sampling; infectious waste; housekeeping; and laundry.

Direct contact is considered to be the primary means of transmission of nosocomial infections. Hand washing is considered to be the most important means of preventing these infections. The hand washing materials are extremely important, since they aid in the cleaning of the hands and the removal of not only soil, but also microorganisms. The mechanical friction that is used in hand washing is also of considerable importance. The use of antimicrobial agents are of particular significance when personnel are dealing with newborns, between caring for patients in intensive care units, before caring for patients whose immune system has been depressed, after caring for patients who have nosocomial infection, before labor and delivery, before operation, and before any type of procedure where the body will be invaded in any way. See Chapter 3, Food Protection, in this volume for specific handwashing proceedures for food handlers.

The cleaning, disinfecting, and sterilizing of patient care equipment varies with the three categories of equipment being used. These categories include equipment for critical patients, semi-critical patients, and noncritical patients. Disinfectants used in the health-care setting include alcohol (to be used on patients only), chlorine and chlorine compounds, formaldehyde, glutaraldehyde, hydrogen peroxide, iodophors, phenolics, and quaternary ammonium compounds. These disinfectants are not interchangeable, and therefore the user must be aware of how the disinfectant works in different settings. Sterilization, which is the complete elimination or destruction of all forms of microbial life, is accomplished either by means of steam under pressure or dry sterilization. Ethylene oxide, which has been used as a cold sterilization gas, is now listed under CERCLA as a potential carcinogen. However, it still may be used if proper control techniques are put in place. The major control is ventilated through a dedicated exhaust system which serves the sterilizer area only and routes the ethylene oxide directly to the outside air, away from people and building intake air systems.

Under the current CDC guidelines, the only microbiological sampling that is recommended is the placing of spore strips in sterilized materials to determine if sterilization has occurred. In addition, the water used to prepare the dialysis fluids and the dialysate needs to be checked on a monthly basis. However, microbiological sampling could still be an effective technique if used when there is a massive outbreak of nosocomial infections in an institution and there is no known cause for the outbreak. It might be worthwhile to take air samples in the area where patients are becoming infected.

The area of infectious waste will be discussed later. Typical nosocomial pathogens include *Pseudomonas aeruginosa, Enterobacter* sp., *Klebsiella* sp., *Group B streptococci,* and *Staphylococcus* sp. Other organisims include *Listeria, monocytogenes, Corynebacterium, Propionibacterium,* and *Bacillus* sp.

Housekeeping is an essential area for the reduction of nosocomial infections. The cleaning of hospital floors and other horizontal surfaces helps remove microbial contamination on these surfaces. Appropriate cleaning is another measure used to prevent the microorganisms that have landed on the surfaces from being reintroduced into the air environment.

Hospital laundering will reduce significantly numbers of microbial pathogens. The laundry then may be sterilized if used as drapes or other necessary items in specialized situations, such as the emergency room, operating rooms, delivery rooms, and other areas where sterile sheeting is required.

Human Immunodeficiency Virus and Hepatitis B Virus

The Health Omnibus Programs Extension Act of 1988 requires that guidelines be developed for the prevention of transmission of human immunodeficiency virus (AIDS) and hepatitis B virus for health-care and public-safety workers. These guidelines are essential to reduce the risk in the workplace of becoming infected with the etiological agent for acquired immune deficiency syndrome and for determining the circumstances under which exposure to the etiological agent may occur. The mode of transmission for hepatitis B virus (HBV) are similar to those of human immunodeficiency virus (HIV). The potential for HBV transmission in the occupational setting is greater than that for HIV. There is a larger body of experience relating to the control of the transmission of HBV in the workplace than the transmission of HIV. The general practices used to prevent the transmission of HBV will also minimize the risk of transmission of HIV. Other blood-borne transmissions of disease will be interrupted by adhering to the precautions used for protecting the workers against HBV and HIV.

The workers involved not only include hospital personnel, but also fire-service personnel, emergency medical technicians, paramedics, and law-enforcement and correctional-facility personnel.

HBV and HIV transmission in occupational settings occur only by percutaneous inoculation or contact with an open wound, nonintact (chapped, abraded, weeping, or dermatitic) skin, or mucous membranes, blood-contaminated body fluids, or the concentrated virus. The probability of infection by the health-care worker is related to the potential for the organism being present in the patient's blood, etc. The risk of infection changes rapidly with the type of patient who is being dealt with. The normal risk of HBV infection from the general population varies from 1 to 3 per 1000, whereas in the drug or homosexual population, the risk ranges from 60 to 300 per 1000 population.

From the 1990–1992 viral hepatitis surveillance program, the CDC reported the number of HBV infections in the U.S. to be 21,000 for 1990, 18,000 for 1991, and 16,000 for 1992, with approximately 25% of infected people developing acute hepatitis. Approximately 6–10% will become HBV carriers and will be at risk of developing chronic active hepatitis and cirrhosis, and spread the infections to others. From the viral hepatitis surveillance program of 1990–1992, the CDC reported that the infection rate in health care workers was 3.7/1000 in 1990, 3.5/1000 in 1991, and 3.9/1000 in 1992. These workers' jobs entailed exposure to blood. The degree of risk not only to the health-care workers, but also to the emergency medical and public-safety workers varies with the amount of exposure to blood that these individuals come in contact with.

Because of the serious concern from the federal government about the spread of HBV in the general population as well as health care workers, several prevention programs were funded by Congress. The programs were developed in 1989 and the funding continues to the present. Areas of prevention include screening all pregnant women, infants, high-risk children and adolescents, and using vaccines and hepatitis B immunoglobulin where appropriate. Special precautions for health care workers were instituted by the CDC.

Resistant Organisms

A resistant organism is one which is resistant to two or more unrelated antibiotics to which the organism is usually considered susceptible or to certain key drugs, such as *Staph aureus* resistant to methicillin. Virulence of the microorganism is based on its ability to:

1. attach to mucosal surfaces
2. survive and/or penetrate the skin or mucous membrane
3. multiply within the body
4. inhibit or avoid host defenses
5. produce disease or damage the host

It then colonizes the area of the body site by growth and multiplication, which in fact may cause a carrier state. Colonization increases the reservoir of resistant organisms.

Resistance to antimicrobial agents has been going on for at least 30–35 years. Although it started in large teaching hospitals, it has spread readily to all hospitals, other health care facilities, and the community at large. Resistance can develop because of:

1. Antibiotic pressure due to prolonged or excessive exposure to certain antimicrobial agents.
2. Disruption of the normal flora and colonization with more resistant organisms.
3. Severity of illness and immuno-suppression.

Bacteria can produce antibiotic-inactivating enzymes, change cell wall permeability, alter target sites, change their susceptible metabolic pathways, or change cell wall binding sites to prevent antibiotic attachment. Although resistant organisims are not more virulent than non-resistant organisms, they cause more disease because they cannot be treated properly.

Resistant organisms are primarily spread by contact with hands of personnel, through droplets from coughing, sneezing, suctioning, ventilation equipment, and contaminated fomites.

PROBLEMS IN INSTITUTIONS

Hospitals

Since the institution is in effect a small community, all environmental health problems and necessary resulting services that would be found in any small community are found within the institution. These include air, food, housekeeping, insect and rodent control, laundry, lighting, noise, solid waste, water, and liquid waste. In addition, because of the uniqueness of the institution, other areas may develop specific environmental problems. These areas include the laboratories, the surgical suites, and emergency rooms. Unsafe and/or careless practice of environmental protection in the institution not only endangers individuals within the institution, but may also endanger the community at large.

Air

A significant difference between the normal population and the patient population is the increased susceptibility of the patient to infection and also the increased risks of spread of infection from the infected patient to others. This susceptibility to infection varies tremendously with the patient, the disease process, the chemical and radiological therapies, and the other techniques of treatment used. Pathogenic microorganisms, chemicals, and other contaminants are transmitted through the air. The transmission may be from humans or operating techniques to susceptible individuals. The organisms are found on clothing, bedding, and floors, and may readily become airborne. The rate at which the particles settle out is determined by their size. It has been observed that particles released from basements of buildings travel quickly trough the ventilation ducts to the upper stories of the buildings. This would indicate that microbiological or chemical contamination, which may occur anywhere, could easily be distributed to any or possibly all parts of the institution. The types of air contaminants include dust, particles that range in size from 0.1 to about 100 µg; fumes, solid particles formed by condensation of vapors from metals ranging in size from 0.001 to 2 µg; smoke, solid or liquid particulates produced by the incomplete combustion of fuel ranging in size from 0.1 to 30 µg; mists, small liquid droplets produced by atomizing, boiling, sneezing, and so forth, ranging in size from 0.5 to about 10 µg; vapors, gaseous phase of liquids such as gasoline; gases, produced by chemicals found in the institution, ranging in size from 0.0001 to 0.001 µg; odors, caused by a variety of processes or illnesses, ranging from 0.00001 µg to a much larger concentration; and microbial contaminants, ranging in size from 0.005 to 40 µg.

All of these contaminants are produced by some form of personnel activity or operation of a variety of pieces of equipment. Further, these contaminants may readily be caused by carriers of specific organisms or shedders or dispersers of organisms. Some examples of patient equipment or other hospital equipment as the source of air contamination include microbial contaminants found in humidifier water, microbes found on cooling coils and other parts of the air-conditioning and ventilation system (*Pseudomonas aeruginosa*), nebulizing equipment, and inhalation equipment. Other air contaminants include skin scales, textile fibers, and dust particles.

The use of lasers may cause a potential hazard within the institutional environment. The laser vaporizes or fragments harmful viable or nonviable substances and may cause an airborne health hazard or a direct hazard to tissues. The principle indirect hazards and sources of air contamination are solvents used in the cleaning of equipment and gases, and fluids associated with the use of the equipment. Some of the toxic substances used in the laser facilities include benzene, carbon disulfide, carbon tetrachloride, cyclohexane, nitrobenzene, pyridine, toluene, xylene, mercury, and chlorine. The direct sources of air contamination vary with the type of laser. The carbon dioxide laser has a high potential for generating air contamination. The beam can readily melt, vaporize, and burn a wide variety of materials, depending on the power density and the ability of the material to absorb power. The hazards are very similar to those produced in welding. Some of these hazards include the production of nitrogen oxide and ozone.

Some additional potential environmental health air hazards involve the degree of filtration of exhaust air from vacuum cleaners; aerosols produced from solid-waste compactors; aerosols produced from bedpan flusher units; aerosols from centrifuges, blenders, and other laboratory devices; aerosols from hydrotherapy tanks; air flow around doorways of isolation and reverse isolation rooms; airborne contamination from housekeeping materials, including polishes, solvents, detergents, disinfectants, and pesticides; and airborne concentration from rehabilitation and occupational therapy areas, including glues, solvents, and materials such as fiberglass, dust from grinders, shavers, and sanders.

Equipment Design and Construction

Equipment can readily be contaminated by patients, professional staff, or hospital employees. The equipment may then become a source of infection for other individuals. Additional problems occur because of improper design and/or construction, making the equipment difficult to clean, disinfect, and sterilize. Faulty design of equipment may also contribute to occupational hazards or cause hazards to the patients or visitors. Because of the great variety of equipment found in an institutional setting, it is impossible to discuss all of the potential problems. Therefore, only a few pieces of equipment that may contribute to the spread of infection or cause other hazardous conditions will be discussed.

Hemodialysis machines have been associated with the spread of hepatitis C to patients and the potential spread of hepatitis to staff and employees. Worldwide, the average prevalence of antibody to hepatitis C virus among hemodialysis patients is 20% with a range of 1–47%. Bronchoscopes, cystoscopes, and endoscopes are pieces of equipment that are not easily cleaned or decontaminated. Many types of electronic equipment also present a cleaning and decontamination problem.

Other problems related to electronic monitoring devices include the effects of the devices on patients and staff when they have been exposed for long periods of time. Plastic furniture and drapes are rapidly being introduced into the hospital environment. There is a poor understanding of what might occur if the drapes and furniture would start to smolder. Fumes from smoldering equipment mixed with gases utilized within the patients' rooms could present a toxic hazard to patients, staff, employees, and visitors. Hydrotherapy tanks are a constant source of potential infection problems. The tanks are difficult to clean because of the design and construction of the turbine pump and the agitator. The so-called bedpan sterilizer is actually a bedpan flusher unit that is inefficient, tends to become inoperative, is hard to clean, and becomes a serious potential source of infection. Bedpans, after they go through the flusher unit, may be more contaminated than before the cleaning operation.

Solid-waste compactors push enormous quantities of contaminated waste together. They are difficult to clean, often leak, and become a source of insect and rodent contamination, and air contamination. Fiber optic instruments are built in such a way that they are not readily sterilized. Inhalation therapy equipment and nebulizing, anesthesia, and infant resuscitation equipment have been known to cause pulmonary necrotizing infections.

Food

The dietary department is similar to a large commercial food operation. In addition, it has the problems of preparing diets for sick and debilitated individuals, and for preventing massive outbreaks of food-borne disease. (An in-depth discussion on food appears in the chapter on food protection.) This section is concerned with some of the problems that can be found in food operations within hospitals and other institutions. They are as follows: (1) inadequate supply of light, contributing to worker fatigue, inadequate removal of soil and wastes, and increases in insects and rodents; (2) poor housekeeping resulting from inadequate management of the large food service area that must operate on a least a 12-hr minimum basis and in some cases up to 24-hr a day; (3) accumulation of food particles, dirt, dust, and other soil of food equipment, such as cookers, stoves, ovens, deep-fat fryers, slicers, can openers, conveyors, sinks, cabinets, shelves, storage areas, pots, and pans; (4) coating of surfaces with greasy residues; (5) dirty and corroded refrigerators with worn gaskets on the doors; (6) poor ventilation in warewashing rooms; (7) corroded and clogged warewashers and spray arms within the warewashers; (8) inadequate steam pressure for washing, rinsing, and cleaning; (9) potential chemical food poisoning due to inadequate and improper storage of chemicals, insecticides, cleaning materials, and other poisonous materials; (10) inadequate and improperly placed hand-washing facilities; (11) food handlers who are working although they have or have recently had diarrhea, vomiting, severe upper respiratory illness, or skin infections; (12) carrying over of leftovers from one meal to another; (13) improper thawing of turkeys, meats, and other foods outside of refrigeration; (14) improper cleaning of equipment, such as grinder heads and tenderizer blades; (15) reusing disposable water carafes and glasses; and (16) improper care and cleanliness of food carts that have refrigerator and heating units on them.

Other problems revolve around the preparation, storage, and use of infant formulas and tube feeding. Great care should be given to adequate cleaning, evaluation, and supervision of all food areas to prevent an outbreak of food-borne disease. Unfortunately, too many times in the past, thousands of patients have become ill from avoidable outbreaks of food-borne disease.

Hazardous Chemicals

There are many thousands of raw materials used in industry today. The toxicity of these substances, the ways in which they are used, the chemicals to which they may be bound, the degree of susceptibility of individuals, and the methods of control are all intimately related to the possibility of toxic responses. The National Institute of Occupational Safety and Health lists over 8000 substances as toxic. It has been noted that over 630 different chemicals are used in hospitals. Of this number, about 300 are of unknown toxicity. About 30 are safe and about 300 have hazardous properties. These chemicals can either be toxic or carcinogenic, flammable or explosive. They include acids, alkalis, ammonia, assorted oxidizers, organic and inorganic peroxides, oxides, permanganates, nitrates, nitrites, toluene, ether, lithium, potassium, sodium, and more.

Areas of the hospital that have the greatest usage of chemicals and the greatest potential for hazardous situations include the departments of anatomical pathology, clinical pathology, renal dialysis, and various storage ares. The laboratories contain many solvents, fixants, and volatile hydrocarbons. In the patient rooms chemicals are used for cleaning, degreasing, and decontamination. These many affect the patients and the staff and personnel. The housekeeping department uses pesticides, polishes, solvents, and detergent disinfectants. The physical plant stores gasoline, paints, paint thinners, and a variety of materials used in the repair process.

The potential hazards of chemicals are enormous. It is necessary to determine the kinds of chemical mixtures that are in the air. Will the mixtures be explosive, flammable, toxic, or carcinogenic? There is, for example, a very serious potential problem related to anesthetic gases. Anesthesiologists have an unusually high incidence of headaches, fatigue, and irritability. Further, the incidence of spontaneous abortion and a high incidence of abnormal pregnancies have occurred among females in the operating room. It has also been postulated that an exposed male can transmit the defect to his unexposed wife. There has been an increase in the incidence of embryotoxicity, mutagenesis, carcinogenesis, and liver disease among females working in the operating room. There is a lack of knowledge as to whether these anesthetic gases will cause problems in patients receiving repeated doses of anesthesia who are receiving immunosuppression therapy or coronary care. More than 424,000 dentists, dental assistants, and dental hygienists are potentially exposed to nitrous oxide used as an anesthetic. Occupational exposure to N_2O causes adverse effects, such as reduced fertility, spontaneous abortions, and neurological, renal, and liver diseases.

Housekeeping

Housekeeping procedures are related to the direct or indirect transfer of pathogenic organisms from person to person. Many special problems arise within the institution. The housekeeping activities must be adapted and scheduled in such a way as to avoid the activities of medical staff, nurses, and other auxiliary personnel. Some rooms are overcrowded because of the use of essential equipment. Other rooms contain food spillage, dripping urine, broken glass, silver nitrate, chewing gum, infected or contaminated bandages, linens, and other materials. The housekeeping attendant, his/her equipment, and any other material that he/she may utilize can become a vehicle for the transmission of disease.

Since the institution is a vast conglomeration of buildings, floors, corridors, windows, and assorted rooms used for a variety of functions. the housekeeping task becomes even more difficult. The equipment and techniques utilized in bathrooms certainly must be modified for use in kitchens. X-ray areas, store-houses, laundries, garages, outside grounds, intensive-care units, cardiac-care units, and surgical units are all independent and have unique problems.

Insect and Rodent Control

Insects and rodents cannot be tolerated in an institutional setting. Not only are conditions favorable for their entrance and multiplication, but also their normal propensity for transmitting disease is sharply increased by a substantial quantity of virulent organisms present in the environment. Insect and rodent control is an important part of the overall program of preventive medicine within the institution. These pests, which include German and American roaches, bedbugs, ants, flies, rats, mice, and fleas, may be found in food service areas, housekeeping areas, janitorial closets, nurses stations, clinical areas, patient areas, animal labo-ratories, conveying equipment and shafts, morgues and autopsy rooms, and the laundries. A complete description of the life cycles, problems, and controls of these insects and rodents appears in other chapters of this book.

Structural problems are of concern, since wherever the maintenance depart-ment is involved in correcting plumbing or other types of repairs, holes may be left open in walls, floors, and ceilings. These openings provide a natural pathway for the invasion of rats, mice, or insects.

Food areas are a primary harborage for insects and rodents. These areas include preparation, serving, storage, dishwashing, and equipment. Food residues, crumbs, and other food materials may accumulate under sinks, around walls, on the bottoms of shelves, under tables, under equipment, and behind and under all other types of standing equipment within the preparation and serving area. In storage areas, there are frequently accumulations of spilled flour, sugar, beans, and other foods. Storage areas provide an excellent source of food for insects and rodents, since these areas are generally warm and have sufficiently enclosed spaces for harborage. Insects and rodents breed in the residues from the evening meal, visiting-time snacking, encrusted step cans, uncovered wastebaskets, and baby formulas or foods dumped into sinks. Visitors frequently bring food into the institution and at times may even bring roaches in with their clothes or in food. The nurses stations, clinical areas, informal staff rooms, and other places where people congregate tend to be collection stations for food and become primary sources of insect and rodent problems. Animal laboratories are of special concern, since animal food is always available for the feeding of experimental animals. Roaches and rodents have a constant source of food, water, and harborage in these areas. Conveyors, tray carts, laundry carts, elevator shafts, plumbing, and electrical shafts are also sources of harborage and paths of travel for insects and rodents. Insects and rodents find powerful attractants in the laundry, since soiled and blood-encrusted linens are a food supply for insects and rodents. The solid-waste disposal areas are frequently nothing more than open dumps and become sources of insect and rodent problems.

Laboratories

In addition to the fire hazards occurring in laboratories there are also biohazards. Biohazards are more prevalent due to the considerable amount of experimental work conducted today. Two key points to recognize are that it is necessary to identify and classify all biohazards, because any accident occurring in the presence of biohazard materials may result in infection, and many biological agents have unknown or incompletely understood etiologies. The environment may impinge upon laboratory animals or workers, on biohazardous agents or insects, on physical or chemical substances, and vice versa. Reported laboratory acquired infections only represent a fraction of those that have actually occurred. Many times unusual viruses are sent to laboratories in the U.S. for analysis. In 1994, at the Yale School of Medicine, a researcher was exposed to a rare and potentially lethal disease caused by the sabia virus when a laboratory accident occurred. Other recent viruses are marburg and ebola viruses in Africa and the Junin, Machupo, and previously mentioned sabia from South America. Virus "X" emerged from the rain forest in Southern Sudan, killed thousands, and disappeared.

Laundry

The laundry is a large service unit within the total hospital whose function is the collection and processing of soiled linens and the distribution of clean linens. Laundry and linens flowing to and from the laundry room can readily become a source of infection or a source of insect and rodent infestations. A well-run, reliable laundry service helps create an aesthetically clean environment. A poorly run service may become dangerous. There are various ways in which linens are processed. They may be cleaned within the institution or in commercial laundries. The institutional cleaning problems include the actual facility, which may be dusty and dirty. This facility may also have large quantities of lint on equipment, walls, and ceilings. The areas for sorting of soiled and blood-encrusted linens, if improperly kept, become a powerful attractant for insects and rodents. In many of these laundries, adequate screening is lacking. The physical flow of linens is very important. Where laundry chutes are used, the exhaust air driven back into the nursing units by the piston effect of a falling laundry bag can contain staphylococci and other organisms. Trash chutes create the same type of problem. Where the linens are taken by cart to the laundry, the carts tend to become dirty and highly contaminated. In special areas where the linens are coming from rooms where the individuals may have infections, the linens need to be kept separate from the rest of the laundry, transported separately, and then washed in the bags before they are sorted and rewashed. Studies conducted in various hospital areas indicated that the blankets, mattresses, pillows, bed, and linens are highly contaminated by varying patient discharges including urine, blood, feces, and saliva. Bed-making and linen-handling procedures, when carried out in a vigorous manner, tend to circulate these organisms, creating a potential airborne hazard.

Commercial laundries may cause a link of infections from hospital to community to hospital to community and to other types of institutions. The commercial laundries have to have separate delivery and receiving trucks for dirty and clean linens to prevent contamination. The hospital linens must be washed separately from other commercially washed linens; otherwise there is an opportunity of infections.

Lighting

Illumination is a very complex subject of exceeding importance to any environment. It is well known that poor lighting results in fatigue, an increase in the number of accidents and injuries, poor housekeeping, and increased risk of infection. Much of our available light is not necessarily lost because of poor design, but rather because of poor maintenance of fixtures and lack of replacement of light bulbs. The color and shade of paint on walls, ceilings, and floors affects illumination. Dark colors absorb and decrease light; light colors reflect and utilize light properly.

Noise

Noise is unwanted sound. Environmental noise may cause temporary or permanent hearing loss, physical and psychological disorders, interference with voice communications, and disruption in job performance, rest, relaxation, and sleep. The increased use of therapeutic equipment and poor maintenance of electrical and mechanical equipment, coupled with lighter weight construction materials, tend to increase noise problems. Normal human activities increase the noise level. Sources of noise in hospitals or other institutions include autoclaves, communications systems, compactor units, dishwashing machines, electrical equipment, heaters, ventilators, air conditioners, and transportation systems. In addition, noise is created by comatose patients who are suffering from specific pains or who may be upset by the strange environment. On the exterior of the hospital, noises come from automobile traffic, compactor trucks, equipment, and emergency vehicles.

Patient Accidents

Many patients have accidents within institutions. The most frequent of these accidents are falls occurring within the immediate area of the patient's bed.

Plumbing

Although various plumbing systems are installed properly during hospital construction, changes are made frequently as hospital space is either utilized for different purposes or as problems occur within the institution. The vast number of activities undertaken within the hospital and the formidable kinds of microbiological and chemical contamination present makes the hospital a potentially hazardous area. Cross-connections, submerged inlets, and back-siphonage problems occur. Plumbing fixtures become corroded, chipped, cracked, and blistered. As new hospital equipment is placed into operation, hospital plumbing surveys must be conducted to determine whether the equipment will put such demands on the plumbing system that it will create improper pressure differentials and excess loads of waste that cannot be handled by the existing plumbing system. The potentially hazardous fixtures and equipment include bathtubs, laundry tubs, sinks in all areas — including the scrub rooms and operating rooms, hydrotherapy tanks, soup kettles, drinking fountains, laboratory sinks, urinals, laundry machines, development tanks for X-ray and other films, and all tanks used for dilution. (An in-depth discussion of plumbing appears in the chapter on plumbing in Volume II.)

Radioactive Material

There are numerous concerns about the use, storage, and disposal of radioactive materials found in body wastes. In the event of fires, special care must be taken to avoid contaminating firefighters and the general public. Other problems include improper shielding of medical X-ray and fluoroscopic machines, improper size of the X-ray beam, and improper beam filtration and alignment. Most of the newer dental X-ray machines have excellent built-in protection mechanisms. However, in the older machines there is still considerable stray radiation. Other sources of radiation are lasers, ultraviolet tubes, microwaves, and ultrasound.

Solid and Hazardous Wastes

Hospital wastes are discarded materials and byproducts of normal hospital activities. The principal types of solid wastes are garbage, paper, trash, other dry combustibles, treatment-room wastes, surgical-room wastes, autopsy wastes, noncombustibles, such as cans and bottles, and a variety of biological wastes, including wound dressings, sputums, placentas, organs, and amputated limbs. In the last 25 years the number of disposable items have increased from a few to hundreds. They range from paper bedding to cardboard bedpans. This has caused a sharp increase in the volume and weight of solid wastes that must be disposed of on a daily basis. The average patient contributes an estimated 8.5 lb or 0.7 ft^3 of solid waste daily. Containers and equipment into which this waste goes are unclean and are not easily cleaned. The waste is usually dumped into plastic bags and then removed to either trash rooms or outside to large metal containers. The containers tend to be dirty, filled with organic material, difficult to clean, and frequently left improperly cleaned. Unfortunately, hospital personnel leave open the doors to the metal units, creating open dumps, instead of a desirable means of temporary storage of solid waste. Numerous additional problems are created by the presence of syringes, needles, examination gloves, catheters, enema basins, and petri dishes. This material is frequently contaminated or contains sharp items, causing safety and infection hazards. The solid waste of individuals with infectious or contagious diseases become considerably more dangerous to the other patients, the community, and the infected individual. Transportation and disposal of hospital solid waste and research animals to community incinerators and landfills deserves investigation. Many of the communities in the U.S. have unsatisfactory incineration systems or landfills.

Rubbish or trash chutes and trash rooms present additional hazards, since solid waste falling down a chute creates a piston effect, which forces air back into the patient areas. Further, as plastic bags hit the floor of the trash room they burst open and the organisms spew into the air, readily capable of recirculating through the institution.

Hazardous waste that exhibit ignitability, corrosivity, reactivity, and EP toxicity may be found in the institution. (See chapter on solid and hazardous waste in Volume II.)

Surgical Suites

The surgical suite in the modern hospital is potentially more hazardous today than in former years, because of the number of new surgical techniques and the increased risk of infection due to modern drug therapy. Today many hospitals perform various organ transplants, such as kidneys, hearts, and parts of the arteries. It is critical that infection is prevented, patients and personnel are safeguarded from technological hazards, shock and trauma to the patients are minimized, physical comfort and emotional support is provided, and the staff delivers effective patient care in the surgical suite.

Some of the specific problems regarding the patients and the prevention of infection are elimination of cross-traffic or back-traffic of staff, patients, or soiled materials; separation of clean from dirty cases; preparation of operating rooms for handling general surgical procedures; establishment and observation of rules concerning aseptic techniques; immediate change of operating room garb by surgical staff; elimination of the possibility of contamination through air systems; provision of restricted flow zones; and proper cleaning and decontamination of operating rooms.

Hospital-Acquired Infections

There are three basic kinds of infections related to the hospital. The first is the infection of the patient present upon admission to the institution. This infection can spread from patient to patient. The second type of infection appears during hospitalization and most frequently is hospital acquired. This may also spread from patient to patient. The third type of infection appears after the patient is discharged from the hospital. The incubation period for the infection occurs during the hospital stay. The individual, once released, is in a position to spread the infection to members of his/her family, the community, and of course from the community back to the hospital. An example of the third type of infection would be some types of staph infections where the postpartum woman does not appear to have a breast abscess until she arrives home.

The hospital environment presents a special risk of infection to employees, patients, visitors, and the community. The emergency room, admission areas, outpatient clinics, surgical, medical, and obstetrical units are potential foci of infection. Patients, visitors, employees, supplies, equipment, and the physical structure of the institution may either aid in harboring pathogenic organisms or introducing them to the environment. The organisms are transmitted by improper patient care, mishandling of contaminated materials and equipment, poor housekeeping, and inadequate physical facilities.

Hospital-acquired infections differ depending on hospital size, type, and patient usage; usage of antibiotics; employee supervision; time of year; and the quantity of virulent organisms that are introduced into the hospital environment. Some general types of infections found in institutions include gram-negative bacteria from urinary catheter-associated infections and infections following prostatectomies, burn infections, and infections of the newborn; *Staphylococcus aureus* infections from nurseries, surgical suites, surgical recovery areas, and other areas of the institution; infant diarrhea; infections caused by salmonella and shigella or other enteric organisms; *Klebsiella* infections due to a gram-negative bacteria found in thoracic surgery, intensive care units, and on floors and other surfaces; *Serratia marcescens* infections due to organisms found in hexachloraphene dispensers; aerosols from ultrasonic cleaning devices; *Pseudomonas aeruginosa* found in the air and on surfaces; and *Bacillus anitratum* found in frozen plasma, water baths, respiratory equipment, and in hospital food-service, solid-waste, and laundry operations.

Emerging Infection Problems

Gram-positive bacilli may cause serious infections. *Corynebacterium* sp. and *Bacillus antbrachis* can cause a significant disease in an immunocompromised person. *Listeria monocytogenes* can cause septicemia and meningitis. *Clostridium perfringens* can cause gas gangrene or necrotizing fascitis in all types of patients.

Antibiotic-resistant organisms have created a new challenge in infection control. Methicillin-resistant *staphylococcus aureus* is generally difficult to eradicate once it invades a hospital facility. The organism can remain viable in the environment for a long period of time. The organisms frequently cause bloodstream infections, pneumonia, osteomyelitis, and bacterial endocarditis. The infections typically occur in larger hospitals where there is overcrowding. Methods of control include isolation, bathing with antimicrobial agents such as chlorhexidine, and aggressive antibiotic treatment.

Enterobacter cloacae, Enterobacter agglomerans, and *Enterobacter aerogenes* can cause a wide variety of infections including bacteremia, meningitis, brain abscess, osteomyelitis, pneumonitis, and urinary tract infection. They become resistant during B-lactam antibiotic therapy. The increasing prevalence of nosocomial *Enterobacter* species infections may be related to the wide use of broad-spectrum antibiotics. These organisms are typically transmitted from patient to patient by hospital personnel.

Enterococcus faecium has become vancomycin resistant. It is also resistant to ampicillin. The major problem appears to be the large number of patients receiving broad spectrum antibiotics. There have been several outbreaks among immunocompromised patients. Sepsis resulted in death. A resistant strain was involved in an outbreak spread by electronic thermometers.

There have been worldwide new epidemics of cholera and dysentery. In 1994 on cruise ships there have been outbreaks of intestinal disorders. Virulent strains of streptococcus-A-infections have caused many thousands of severe illnesses in the U.S. and Europe. There has been a sizable increase in resistant tuberculosis. In Cincinnati an epidemic of whooping cough occurred in 1993.

Modern travel and global commerce quickly spread known and unknown organisms. This coupled with antibiotic resistance creates serious new hazards for hospital employees and patients and may lead to sizeable increases in hospital-acquired infections.

The extent of these infections and the susceptibility of the patients, whether they are burn patients, under special chemotherapy, newborn, elderly, or immunocompromised, makes it necessary to understand the principles of epidemiology of the spread of these organisms, the means of determining potential hazards, and the institution of environmental, personnel, and patient controls.

Hospital-Wide Problems

1. Lack of training and inadequate supervision are often principal factors in the spread of infection.
2. Movement of personnel and supplies contribute to the dissemination of infectious organisms.
3. Failure to appropriately mark the chart of each infectious case can lead to contamination of unaware personnel.

4. Improper transportation and storage of sterile supplies can cause contamination.
5. Improper transportation and storage of infectious hazardous waste can cause contamination.
6. Failure to properly segregate clean and dirty linen, and failure to properly clean and disinfect bedding may increase the risk of infection.
7. Careless handling of any waste may introduce pathogenic organisms into the environment.
8. Since hospital food is often ground, creamed, and prepared in large quantities, the opportunities for contamination through improper handling and storage is increased.
9. Bedpans and bed urinals improperly removed, handled, stored, washed, and disinfected may cause a cross-infection.
10. Supplies, instruments, and equipment may directly or indirectly contaminate patients.
11. Failure in any detail of isolation technique creates a risk of infection from employees, visitors, and other patients.
12. Failure in any detail of proper catheterization technique or follow-up care can cause urinary tract infections.
13. Other procedures, such as inhalation therapy and intravenous cut-downs, may be contributors to infection.
14. The flow of traffic in special areas is of unique importance, since the mere movement of traffic can lead to outbreaks of disease through the distribution of microorganisms.
15. Antibiotic-resistant organisms may colonize the skin and therefore improper or ineffective treatment may quickly spread infection to others.

Routes of Transmission

Many factors influence the spread of infections within the institution. Therefore, it is frequently impossible to determine the actual causes surrounding the spread of a given disease. However, epidemiological studies should be conducted in all cases, since vital information can be gathered that is of value in efforts to reduce the numbers of organisms in the environment and in the patients. The routes of transmission of disease include droplets or droplet nuclei from person to person, enteric or cutaneous contact, fomites, food, water, insects, and rodents. The disease organisms may come from draining wounds, lesions, secretions, as well as healthy skin. They may come from patients, staff, visitors, or indirectly through the environment. It is known, for instance, that strep and staph infections come from nasal discharges. Staph infections also come from draining wounds, boils, infected decubitii ulcers, sores, hangnails, and infected pimples. Enteric organisms, which cause salmonellosis or shigellosis, are spread by fecal material. Most disease organisms within the institution are probably spread through improper or poor medical-nursing techniques. A smaller, but still a significant proportion, are spread through the environment.

Occupational Health and Safety

Hospitals have a real potential for occupational health and safety hazards, patient hazards, and hazards to visitors. The hospital contains not only all of the problems and services of a small community, but also the additional hazards brought to the institution by the need for special types of equipment, gases, chemicals, and operating procedures. Some specific problems include excessive heat from laundries, kitchens, boiler rooms, and furnace rooms; inadequate illumination in numerous areas of the institution; variety of chemicals, many of which are toxic, explosive, carcinogenic, or possible fire hazards; considerable noise from operational procedures in the physical environment; radioactive materials found in a variety of treatment and diagnostic areas; and research areas where animals contribute to disease and injuries. An in-depth discussion of particular safety hazards would take many pages. It is best to understand that the hazards exist and that they are found not only in the settings discussed, but also in any setting within the institution (see Chapter 10 in this volume).

Fires may be caused by problems related to housekeeping, smoking, electrical appliances and their installation, flammable liquids, heating units, explosive materials, painting units, compressed gases, and so forth. Fires are caused by poor housekeeping and storage of solid waste, including combustible debris, oil- or paint-soaked rags, and wood scraps. Smoking is hazardous, Another serious hazard within the institution is the use and storage of flammable and explosive liquids and gases.

Nursing Homes

Special Environmental Problems

Sanitary conditions within nursing homes may change very rapidly as a result of inadequate budgets, insufficient numbers of employees, poor administration, or inadequate building structures and facilities. Personnel are often lacking proper in-service training. There is a need to improve supervision of personnel, establish standard operating procedures, and develop a better understanding of cleaning techniques and environmental controls. In some instances, the quality of food is poor, not only because of inadequate preparation but, more importantly, because cheap food is purchased and may come from areas where there had been salvage operations. Salmonellosis, staph food poisoning, and other types of food-borne diseases appear regularly within the nursing home. Contributing to the environmental health problem is the nature of the patient and his or her mental and physical condition. More than half of the patients in nursing homes are disoriented at least part of the time. About 20% are in a state of confusion most of the time. The confused patient not only contributes to the accident problem, but also contributes to the spread of disease and infection. An estimated one third of the nursing home patients suffer from urine or feces incontinence or both. Obviously these wastes are excellent sources of contamination within the institutions. When incontinency combines with mental confusion and lack of adequate staff supervision, the chance of spreading environmentally related diseases increases sharply.

Further, in many of the older homes, there is a problem with insect and rodent infestation. Even in the newer homes, infestations occur and can be spread because of lack of personnel and personnel training, and hoarding of food, providing a constant source of insects and rodents.

Special Safety Problems

The average age of the nursing home resident is 80 years. Residents generally have debilitating or crippling diseases, making them more prone to accidents. About 40% of the patients have cardiovascular conditions, and about 75% have some heart or circulatory difficulties. Sixteen percent have the residuals of paralytic strokes, and about 10% are receiving care for fractures, primarily hip fractures. All of these conditions may be coupled with arthritis, mental disorders, and diabetes. The physical conditions of the patients, plus their frequent intake of medications, make them prime subjects for accidents within the nursing home.

Other Special Problems

There is a perceived shortage of nursing personnel either because of lack of trained people or financial cutbacks. This causes unskilled staff to assume duties requiring training. The general quality of patient care is affected by this shortage. Unskilled labor is used within the nursing home to take care of many of the tasks of running the facility and for patient care. The individuals who fill these positions receive low pay, are overworked, improperly trained, and unfortunately are often improperly screened to determine their temperament and emotional stability.

A lack of communication exists between hospitals and nursing homes. Records are not transferred quickly enough or in enough detail to the nursing home for the nursing home or the hospital emergency personnel to be able to properly care for the patient. Physicians, because of large workloads, make infrequent visits. When emergencies do occur, it is difficult to reach the necessary physicians to provide adequate patient care.

Problems arise with the patients, since they are prone to chronic illness and multiple impairments and disabilities. Many feel rejected by friends and family. Patients may feel they have come to the nursing home to die and will lose the will to care for themselves and to recover and utilize their physical resources in the best way possible.

Enforcement is difficult because of small numbers of trained public health personnel, lengthy court procedures, frequent changes in federal regulations, budgetary problems, and general confusion regarding proper operation of the nursing homes. The high cost of nursing-home care is alarming. The number of state and federal dollars supplied is substantial.

Old Age Homes

The types of problems found in the old age home vary with the number of people and the condition of the facility. They may be simple problems of inadequately prepared food, insects, and rodents, up to the many complicated problems found within the nursing-home situation. It is often difficult to determine when an individual simply needs assistance with everyday life and when an individual needs nursing care. Since the population of the U.S. will increasingly fall into the 65 and over category, proper evaluation of old age homes will be essential to determine if they are performing their necessary function in our society.

Prisons

Environmental health practitioners have recognized the problems in prisons for many years. Prison riots, such as those in Attica, New York, have further emphasized the problems and needs. The U.S. District Court for Alabama in the case *James vs Wallace* held that the conditions of confinement within the Alabama penal system violated the judicial definition of "cruel and unusual punishment," and therefore improvements had to be made within the prisons. The improvements included reducing overcrowding, improving general living conditions, improving food service, education, recreation, and rehabilitation programs. The court not only defined the problems, but also stipulated that the prisons comply with the recommendations of the court or the court would order the prisons closed.

Several groups, including the APHA-NEHA Jails and Prisons task force, the Federal Bureau of Prisons, the federal government, and the state of Kentucky, have conducted studies and made recommendations to improve penal institutions. A model code proposed include specific details on food, water, air, liquid waste, solid waste, vectors, recreational environment, housing, accidents and safety, industrial hygiene, public facilities, and emergency plans. It was obvious from the discussions and studies that an environmental health specialist was needed in penal and correctional systems to assist in changing the environment of the institution.

Bedding

Frequently, bedding is unclean or in bad repair. Many institutions do not provide sheets or pillows. Linens and blankets, when available, are often dirty.

Food

Food problems are intimately related to the food-service facilities and to the prisoners, who are the food handlers. Food is often contaminated from improper purchasing and storage, as well as inadequate training in food preparation, food hazards, and personal hygiene. The floors, walls, ceilings, and ventilation hoods are in poor repair in many institutions and contribute to the overall lack of cleanliness and the potential spread of food-borne disease.

Heat and Ventilation

Many buildings, because of their age and lack of repairs, provide inadequate heat and improper ventilation. Windows behind the bars have been smashed or are nonexistent. Proper heating and ventilation of buildings is necessary in providing a reasonable environment.

Insect and Rodent Control

Insects and rodents are found in about 25–35% of the institutions. Again this is due to the poor maintenance of buildings, to the age of the buildings, and to the nature of the institutions. In many cases, inadequate control procedures are used because of lack of proper understanding of insect and rodent control.

Laundry

Laundry service is not available in certain types of institutions unless friends or relatives bring clean clothing to the prisoners. Where laundries are present, they tend to be extremely hot and poorly ventilated.

Lighting

About one half of all the institutions surveyed in the state of Kentucky had inadequate illumination. About 45% of the light fixtures were in poor repair. Poor lighting was typically found in many institutions.

Plumbing

Plumbing problems abound within the penal institutions because of their age, overuse, and overcrowding. There is a serious lack of toilets and sinks. In many cases, the toilets, sinks, and shower facilities are in bad repair, dirty, or almost totally inoperative.

Safety

Many institutions lack adequate fire extinguishers or the proper techniques available to take care of prisoners in the event of fires or accidents.

Occupational Health

Occupational health practices need improvement to conform with OSHA regulations.

Schools

Environmental Health Problems

Many schools ranging from elementary through college are substandard because of the existence of environmental health problems.

Food

The basic food protection problems found in the hospital applies also to the school. The additional concern that exists for children in substandard urban areas is that the school lunch may be the only nutritious meal received within the course of a day. For this reason, it is even more important that food-service personnel provide a wholesome, safe meal in an attractive manner.

Within schools, mass feeding leads to leftovers, and leftovers and poor food-handling practices lead to serious outbreaks of food-borne disease. Bagged lunches become a particular hazard, since they are stored in school lockers until lunchtime. If food is perishable, the student may contract a food-borne disease. Often dust, dirt, roaches, and mice may get on the food bag or into the food. In universities, students have refrigerators in their rooms or take food back to the room for later snacking. This also leads to problems of potential food-borne disease and insect and rodent infestations.

Gymnasiums

Many accidents occur as a result of poor supervision, poor equipment, improperly trained individuals, poor lighting, horseplay, and so forth. Locker rooms present hazards, particularly through slipping, and a source for the spread of fungal and bacterial infections.

Laboratories

The problems of school laboratories are similar to those of hospitals, except that in junior and senior high schools fewer dangerous substances are used. The hazards are not only related to the types of substances used and the by-products of these substances, but more importantly to the inexperience of the students.

Plumbing

The special problems encountered in schools include the students stuffing the toilets with toilet paper, paper towels, and other objects and wanton destruction of the facilities.

Roughly half of the schools lack adequate hand-washing facilities, soap, and single-service towels. About 10% lack adequate toilet paper. Since hands are probably the most common means of transmission of infectious diseases, it is important that these items be provided in the bathrooms.

Shops

Depending on the type of school, there are a variety of shops being used. These shops constitute a hazard for the students.

Ventilation, Heating, and Air Conditioning

Air heating, air cooling, humidity control, and air distribution is essential for the removal of body odors and for the comfort of the individual. Where there is inadequate air circulation, temperature of the room is raised by body heat, humidity is raised by occupants' breath and perspiration, organic matter is released into the room causing odors, oxygen content is reduced, and carbon dioxide is increased. All of this contributes to stuffiness and odors, and to an uncomfortable environment.

Water

In the school a specific problem occurs when water fountains are misused. All water fountains should be of an angular rather than bubble type. A student should not have to put his or her mouth directly on the water source since this becomes a vehicle of contamination for other students.

Accidents

Accidents are the leading cause of death in all student age groups. About 16,000 students of school age lose their lives each year. Motor vehicle-related deaths account for 58% of these deaths; drownings, 15%, and fires and burns, 7%. Of these accidental deaths, 43% are associated with social life. About half of these occur in school buildings. The most hazardous areas are the gymnasium, where one third of the school-related deaths occur; the halls and stairs, where one fifth of the school-related deaths occur; shops and laboratories, where one fifth of the deaths occur; and other kinds of classrooms, where one seventh of the deaths occur. About 40% of the deaths occur on school grounds. Football is the most hazardous of sports.

In a study of 1,900,000 students, it was determined that 24% of the injuries occurred within school buildings, 28% on school grounds, 5% going to and from school, 20% at home, and 23% in other types of situations.

The college is a complete, highly active community. It contains all of the community safety problems, all types of industrial safety problems found in laboratories and industrial shop areas, and various types of home and recreational safety problems. In a study conducted in 22 colleges and universities serving 207,000 students, roughly 15,000 injuries were reported to the student health services. Of these injuries, 75% were incurred by men and 25% by women. Fifty-two percent of the campus injuries were either in the athletic or recreational facilities, 20% were in the residence halls, 15% in the academic buildings, 11% on the campus grounds, and 2% in motor vehicles. It is obvious from these studies that schools, regardless of the level, can be dangerous places.

Bus Safety

Each year school buses are involved in accidents and in fatal injuries. This occurs because of the numbers of school buses traveling on the roads, poor roads and road conditions, bad weather, and in some cases inexperienced bus drivers. There are many school buses that are old, improperly maintained, and therefore, hazardous. Constant checks should be made of the equipment by utilizing systematic maintenance programs. All buses should be checked for levels of carbon monoxide at the back of the bus; carbon monoxide may seep in through the emergency door.

Hazardous Materials and Equipment

The storage and disposal of this material should be controlled by knowledge-able individuals. Laboratories present a special area of concern because of the lack of training of the student and the presence of potentially flammable liquids, explosive reactions, compressed gases, electrical units, and toxic substances.

POTENTIAL FOR INTERVENTION

The potential for intervention is based on isolation, substitution, shielding, treatment, and prevention. All these techniques can be utilized to a greater or lesser extent in the institutional environment. The potential for intervention in hospitals is very good. In nursing homes and old-age homes, the potential varies with the type of facility, the kind of staff, and the amount of administrative control. In schools the potential for intervention should be excellent. In prisons, it is poor.

Within all these areas isolation is a technique used to separate the contaminated or infected individual from the rest of the population. Isolation also refers to the separation of hazardous substances and gases, such as radiological material, hazardous chemicals, and explosive gases. Substitution is the use of a less hazardous chemical or technique for a more hazardous chemical or technique. Shielding is the utilization of special cabinets with special hood systems where highly dangerous or infectious materials are being used. Shielding also refers to personal protective devices, such as gloves, gowns, masks, caps, and safety glasses. Treatment is utilized in all institutional settings for exposure to disease processes and chemical, physical, and radiological hazards. Prevention is the principal technique that should be used within the institutional setting.

RESOURCES

Scientific and technical resources include the national, local, and state chapters of the American Medical Association; the American Public Health Association; the American College of Hospital Administrators; the American Hospital Association; the Catholic Hospital Association; the Joint Commission of Accreditation of Hospitals; medical schools within all of the states; and a variety of environmental health practitioners.

Civic resources consist of the various auxiliaries to the medical associations and to the hospitals, and a variety of foundations that provide funds for hospitals, such as the Ford Foundation and the Rockefeller Foundation.

Governmental resources include state health departments, the National Bureau of Standards, the Department of Defense, the National Institutes of Health, the Food and Drug Administration, the National Science Foundation, the Veterans Administration, the Department of Health and Human Services, and so forth.

Numerous codes and standards should be utilized in the actual construction of an institutional facility or in the modernization of existing facilities. These codes and standards may be obtained from the following sources:

1. The Superintendent of Documents, U.S. Government Printing Office, Washington DC 20402 (for all government documents)
2. American National Standards Institute, 1340 Broadway, New York, NY 10018
3. American Society for Testing and Materials, 1916 Race Street, Philadelphia, PA 19103
4. American Society of Heating, Refrigerating, and Air Conditioning Engineers, United Engineering Center, 345 East 47th Street, New York, NY 10017
5. Compressed Gas Association, 500 Fifth Avenue, New York, NY 10036
6. International Conference of Building Officials, 50 South Los Robles, Pasadena, CA 91101
7. National Association of Plumbing-Heating-Cooling Contractors, 1016 Twentieth Street N.W., Washington, DC 20036
8. National Council on Radiation Protection, 4210 Connecticut Avenue N.W., Washington, DC 20008
9. National Fire Protection Association, 60 Batterymarch Street, Boston, MA 02110
10. Underwriters Laboratories, Inc., 207 East Ohio Street, Chicago, IL 60611

11. National Bureau of Standards, U.S. Department of Commerce, Washington, DC 20234

12. American Hospital Association, 840 North Lakeshore Drive, Chicago, IL 60611

For an in-depth discussion of biohazard control, it is recommended that the reader obtain the book entitled National Institutes of Health Biohazards Safety Guide, U.S. Department of Health, Education and Welfare, Public Health Service, National Institutes of Health, Superintendent of Documents, U.S. Government Printing Office, Washington, DC 20402.

Very little has been written in the area of penal institution environmental health. It is recommended that the reader explore the publication "Corrections" by the National Advisory Commission on Criminal Justice, Standards, and Goals, Washington, DC, published January 23, 1973, as a base of study. The most knowledgeable source on prisons is the Federal Bureau of Prisons, 320 1st Street, N.W., Washington, DC 20534.

STANDARDS, PRACTICES, AND TECHNIQUES

Hospitals

Physical Plant and Site Selection

The site of the medical facility must be well drained and free of obstructions or other natural hazards (see site selection for communities in chapter on the indoor environment). Its location should be accessible for fire apparatus, ambulances, and other types of service vehicles. Public transportation systems should be available to the patients and staff. Competent medical and surgical consultation from private practitioners or from medical schools should also be readily available. Paved roads need to be provided from the main entrance to the buildings and the emergency section. Paved walkways are necessary for pedestrian traffic. Off-street parking should be provided and easily accessible to patients, staff, and visitors. Special curbing facilities and walkways are needed for wheelchair patients and the physically disabled. Parking spots located close to the entrance should be reserved for the handicapped. All public facilities, including bathroom facilities, should be designed for use by the handicapped. The medical facility should be accessible to major highways and conveniently located within reasonable distance of interstate highways to expedite the flow of patients or emergency vehicles to the institution. Heavily trafficked areas should be avoided. In planning the location, it is necessary to consider the path and proximity of existing air-polluting industries, future construction of industries, and the possibility of flooding and natural disasters due to fault lines, underground mines, and other potential hazards.

According to the National Environmental Policy Act — Public Law 91-190, the site and the project of the hospital must be approved before construction can begin. An impact statement should be prepared, and any possible adverse environmental problems should be shown and techniques of control explained.

Interior Structural Requirements

The following items are requirements for the various parts of the typical hospital. The general nursing unit should include:

1. Patient rooms should have a maximum capacity of four patients.
2. Minimum floor area of 100 ft² for a single bedroom plus 80 ft² for each additional bed.
3. Each room should have a window that opens and is above the floor or grade level. The openings should be limited or screened to prevent accidents.
4. A nurses' call system should be provided in each room and in the bathrooms.
5. Each patient room should have its own lavatory or be attached to a bathroom so that the patient does not have to go into the central corridor to utilize the toilets or sinks.
6. Room should be provided for wardrobes, lockers, or closets.
7. Provision should be made for visual privacy of the patients.
8. A series of service areas should be provided for the nursing unit to function properly, including a nurses' station; nurses' office; storage and supply room; convenient hand-washing facilities; charting facility; staff lounge and toilet facilities; staff lockers; conference rooms; patient examining rooms, with a minimum floor space of 120 ft²; clean workrooms and holding rooms; dirty workrooms and holding rooms; drug distribution stations; clean-linen storage closets; nourishment stations; equipment storage; parking areas for stretchers and wheelchairs away from the normal traffic flow; patient bathing facilities; and emergency equipment storage facility.

The nursing facility also needs isolation rooms that are completely and totally separated from all other patients and patient activity rooms. Isolation rooms must have private toilet facilities, sinks, and a dressing area for change of clothing and decontamination of soiled isolation gowns, masks, and so forth. Each hospital needs rooms for emotionally or mentally disturbed patients.

The intensive care unit has critical space requirements, since it is used for seriously ill medical, surgical, or coronary patients. These patients are acutely aware of their surroundings and may be easily affected by them. As a result, it is necessary to eliminate all unnecessary noise, to provide individual privacy in keeping with the medical needs of the patient, and to provide adequate observation by the staff. Patients should have some outside view through a window. The window should not be higher that 5 ft above floor level.

The intensive care unit should include the following:

1. Coronary patients should be housed in single bedrooms.
2. Medical and surgical patients should be housed in either single or multi-bedrooms.
3. All beds must be under direct visual observation by the nursing staff.

4. The clearance between beds in a multi-bedroom must be a minimum of 7 ft.
5. In a single bedroom there must be a minimum clear floor area of 120 ft² and at least 10 ft as a minimum width or length.
6. A single bedroom has to be provided for each patient for medical or surgical isolation.
7. Viewing panels, which can be covered by curtains for privacy, must be installed in the doors and walls to provide adequate visual coverage by the nursing staff.
8. Intravenous solution supports must be provided for each patient so that they do not hang over the patient.
9. A hand-washing sink is required in each patient room.
10. A nurses' calling system is needed and should be required.
11. Each coronary patient must have direct access to a toilet facility.
12. The location of the nurses' station should permit direct visual observation of the patient.
13. Hand-washing facilities should be conveniently located at the nurses' station.
14. Charting facilities must be separated from the monitoring service.
15. The staff restrooms should contain toilets and adequate hand-washing facilities.
16. Storage areas must be close to nursing personnel
17. A clean workroom and system of storage and distribution of clean and sterile supplies is mandatory.
18. A soiled or dirty workroom for proper storage of soiled equipment and supplies is necessary.
19. Bedpan flushing and sanitizing units are required.
20. A convenient 24-hr distribution system and station for medications is needed within the nursing area.
21. A clean-linen storage closet is needed.
22. A nourishment station should be provided.
23. A storage room with easy access is needed for emergency equipment, such as inhalators and "crash carts."
24. Patient lockers must be furnished for patients' personal effects.
25. A separate waiting room, including toilet accommodations, seating accommodations, and telephones, is needed for visitors.

Newborn nurseries can be a serious hazard to newborn infants if the design is poor and special precautions are not taken to protect the infants. The following criteria are essential in a newborn nursery:

1. One nursery should not open directly into another.
2. Nursery should be conveniently located to the postpartum nursing unit and obstetrical facilities.
3. Nurseries should contain hand-washing sinks at a ratio of one to eight bassinets; emergency nurses' calling system; observation windows to permit viewing of the infants in public areas; charting facilities; full-term nurseries containing a maximum of eight bassinets, or as many as 16 if the bassinets are isolets; the minimum floor area for a bassinet should be 24 ft²; special-care nursery for high-risk infants and infants in distress; a minimum of 40 ft² per bassinet in the special-care nursery; and workroom space.

4. Examination and treatment rooms or space for infants containing workcounters, storage, and hand-washing sinks.
5. Infant formula rooms should be provided if on-site formula preparation is carried out. However, it is recommended that, for safety and also for economic reasons, commercially prepared formulas should be utilized.
6. A housekeeper's closet for storage of all housekeeping supplies.

The pediatric and adolescent unit is separated from the adult units to provide a quieter setting for the adults and to prevent the spread of childhood disease to the adult population and conversely. The pediatric nursery follows the same requirements as the other nurseries. The pediatric rooms conform to the same requirements as the other rooms within the general nursing facilities, with the additional need for an area used for dining, educational, or play purposes. Also, special storage closets and cabinets for storage of toys (that can be sanitized), and educational, or recreational equipment should be installed.

The psychiatric unit should be constructed to provide a safe residence for patients and one in which patients are unable to hide, escape, injure themselves, or commit suicide. It must be flexible to care for the ambulatory inpatients and also meet the needs of various types of psychiatric therapy. The unit should convey a noninstitutional atmosphere to facilitate recovery. The following requirements for construction are as follows:

1. Maximum of four patients to a room.
2. Minimum room area of clear floor space for single rooms should be 100 ft^2; for multiple-patient rooms, 80 ft^2 per bed.
3. Each room should have a window with limited opening space and security screens.
4. Sinks must be provided in each patient's room.
5. Bathrooms must be provided in such a way that the patient does not have to go into the corridor to go to the bathroom. The bathroom door should unlock from the outside.
6. All other usual facilities found within a normal nursing unit, with the exception of all items that could be hazardous. These should be locked away and out of reach of the patients.
7. A special space for dining, recreation, and occupational therapy, with a minimum floor space of 40 ft^2 per patient.

The number and types of operating rooms are based on the anticipated surgical load for the hospital. The surgical-suite construction should be designed to prevent the possibility of contamination. The surgical suite must be located in a separate traffic area to exclude individuals unassociated with surgery. Surgical suites should meet the following requirements:

1. A minimum clear floor area of 360 ft^2, exclusive of all cabinets and shelves, per room, with the minimum dimension of 18 ft.
2. An emergency communication system.
3. Two X-ray film illuminators for each room.

4. Special storage space for splints and traction equipment in orthopedic surgery rooms.
5. A minimum clear floor area of 250 ft² for cystoscopic and other surgical endoscopic procedures.
6. Recovery rooms arranged for visual control of patients, drug distribution, handwashing, and charting.
7. Additional recovery space for outpatient surgical patients.
8. Special service areas containing control stations, for visual surveillance of all traffic entering and leaving the operating suite; supervisors office and station; sterilizing facilities with high-speed autoclaves, conveniently located near the operating rooms; drug distribution station; two scrub stations adjacent to and contiguous with the entrance to each operating room; dirty workrooms for all soiled equipment and materials; fluid-waste disposal facilities convenient to the operating room; clean workrooms and supply rooms for clean and sterile supplies; anesthesia storage facilities, where flammable gases can be stored without causing a hazard; anesthesia workrooms for cleaning, testing, and storage of anesthesia equipment; medical gas storage areas; special equipment storage areas; emergency equipment storage areas; staff locker rooms, with adequate space for cleaning, changing, and resting; outpatient surgery change areas; holding areas for patients prior to surgery that are under visual control of the nurses; stretcher storage areas out of the direct line of traffic; staff lounges; housekeeping closets for storage of housekeeping supplies.

Maximum capacity needs must be anticipated when constructing delivery and labor rooms to provide adequate facilities. Construction requirements for obstetrical facilities are as follow:

1. Delivery rooms should have a minimum clear floor area of 300 ft² with minimum dimension of 16 ft.
2. Rooms should have emergency communication systems.
3. Resuscitation facilities for newborn infants, including oxygen, suction, and compressed air.
4. Labor rooms with a minimum floor area of 80 ft² per bed.
5. A minimum of two labor beds per delivery room.
6. Hand-washing facilities within the labor room.
7. Adequate visual observation of the patients.
8. Recovery rooms with a minimum of two beds and all ancillary facilities.
9. The usual service areas, as stipulated in the surgical suites.

The emergency suite is an extremely critical part of the entire hospital. When an emergency occurs, this is the place where a life is saved or lost. The emergency suite is a combination outpatient clinic, operating room for minor injuries, admissions area, and surgical suite. It is essential not only that the emergency room staff is properly trained and efficient in dealing with patients, but also that the facilities are designed to facilitate emergency care. The emergency room facility should contain the following:

1. A well-marked entrance at grade level that is sheltered from the weather and easily accessible to ambulances and pedestrians.

2. A reception and control area conveniently located near the entrance.
3. Public waiting areas with bathroom facilities, telephones, and drinking fountains.
4. Stretcher and wheelchair storage areas.
5. Treatment rooms equipped with hand-washing facilities, containers, cabinets, and work counters.
6. Treatment rooms containing medical suction, storage of emergency equipment such as emergency trays, defibrillators, cardiac monitors, and resuscitators.
7. Staff work and charting areas.
8. Clean supply storage areas.
9. Sterile supply areas.
10. Daily workroom areas for supplies and equipment.
11. Patient restrooms.
12. Communications equipment and an area where adequate communications between physicians at the hospital and ambulances can be provided.
13. Adequate emergency surgical care and treatment in the event of a serious immediate surgical need when the individual cannot be taken to the operating room.

The communications system is of vital importance, since the physician can provide guidance to the emergency medical technicians (EMTs) or paramedics on the care of a given patient. It is also useful in alerting the hospital to the type of arriving patient and the specific needs, in order that the physician and nurses may be prepared to immediately start treating the patient.

The hospital also provides outpatient areas, clinics and other services, including diagnostic clinics, laboratories, physical therapy suites, occupational therapy suites, mortuary and autopsy rooms, pharmacies, administrative areas, public areas, medical records units, central supply, linen services or laundry, housekeeping, food areas, and solid-waste disposal areas.

Electrical Equipment

All electrical equipment and facilities must adhere to the standards of the Illuminating Engineering Society, The Institute of Electrical and Electronic Engineering, Inc., and the codes and standards of the National Fire Protection Association. The hospital should have at least two separate sources of electricity, so that in the event that one fails the second one can immediately go into effect. The emergency system may be either operated as a separate service from the electric company or operate on portable generators or storage batteries. It is important that the emergency system be tested regularly to ensure working order should the regular system become inoperative. Since motors operate at a specific voltage, safety precautions should be provided in the event that voltage drops or increases sharply. The telephone and electric power conductors are not permitted in the same raceways, boxes, cabinets, and so forth. Any switchboards or panel boards should be protected to prevent shock. All hospital wiring must be placed in conduits to facilitate alterations and repairs. The conduits must be large enough to allow for expansion when needed. Receptacles should be installed in all required areas and at convenient heights. All receptacles should be grounded and explosion proof in locations where anesthetic gases are used. Patient rooms should have at least three double receptacles for a single bedroom with two outlets near the head of the bed. Multi-bedrooms should also have two double receptacles at the head of each bed. A variety of clocks are used in institutions; two types are wired and electronic. In the wired system individual clocks are controlled by a master clock; in the electronic system no control wiring connection is required. In locations where anesthetic or other hazardous gases are used, conductive flooring is required. The electrical resistance of these floors should be only moderately conductive to provide a conductive path for neutralizing static charges and to act as a resistor to limit current to the floor from the electrical system in the case of an electrical short. The possibility of electrostatic sparks causing ignition of the gases must be eliminated. As a further precaution, the electrical distribution system should be ungrounded in the operating and delivery rooms, with the exception of fixed nonadjustable lighting fixtures eight or more feet above the floor and permanently installed X-ray tubes. All operating room floors and delivery room floors have to be monitored periodically for the buildup of static electricity. Generally, improper cleaning or removal of detergents and detergent residues contribute to an increase in electrostatic electricity.

Lighting fixtures must be installed to meet the comfort and work requirements of the particular workspace. Since these requirements vary, it is important to obtain from the Illuminating Engineering Society the current lists of lighting levels recommended for hospitals or other institutions. The voltage supplied to the X-ray unit should be nearly constant so that images and pictures will be uniform and consistently reproducible. It is recommended that an independent feeder with capacity sufficient to prevent a voltage drop greater than 2% be used. Elevators should be manufactured, inspected, and maintained in strict excellent working condition. Skilled elevator mechanics should be utilized for repairs. In the event of an emergency, the alternate source of electrical service in the elevator should automatically be shifted to elevator usage. Only one elevator should be so arranged.

The communications system within the hospital may include loudspeakers, chimes, coded bells, radio communications, telephones, buzzer systems, light signals, closed circuit television, and independently carried receivers. Since the communications system is so essential to the operation of the hospital, it is necessary that emergency electricity be provided for its operation in the event that the normal system goes out of order.

In addition to the ones mentioned, modern hospitals have remote dictation services. Fire alarm systems are necessary in every hospital. A manually operated system should be available and utilized and the system tied in with the local fire department. Special electrical installations include the facilities of the electro-encephalographic areas and medical electronic equipment areas.

Heating and Air Conditioning

Heating and air-conditioning systems should be designed to provide the following temperatures and humidities: operating room, 65–76°F at 45–60% relative humidity; delivery room, same as operating room; recovery room, 75–80°F at 50–60% relative humidity; nursery (observation for full term), 75–80°F at 50% relative humidity; nursery (premature), 75–80°F at 50–60% relative humidity; intensive care, 70–80°F at 30–60% relative humidity; and all other occupied areas at 75°F minimum temperature.

The ventilation system must be mechanically operated, with a minimum separation of 25 ft from air intake to air exhaust. The outdoor air intake should be a minimum of 8 ft above ground level, preferably on the roof. If on the roof, it should be 3 ft above roof level. The ventilation system must be designed to provide the general pressure relationships required in various areas of the hospital. For example, the air pressure should be positive within the operating room and negative within the corridors; and negative within the isolation room and positive in the corridors. In the operating room, the fresh air supply should be delivered from the ceiling and the air exhaust at floor level. Corridors should not be used to supply or exhaust air, with the exception of bathrooms or janitors' closets, which open directly onto the corridors. The ventilation systems in the operating and delivery rooms should have final filters located downstream from the main coil systems and a 99.7% efficiency rating in removing micron-size particles. All other areas can have an efficiency rating of 80–90%. A manometer used in the central air systems will measure the flow of air through the system. All isolation rooms, laboratories, and other areas where potential environmental hazards exist must have ducts to the outside. Air should not be recirculated from these rooms to any other part of the hospital. In food preparation areas, the minimum exhaust rate is $100 \text{ ft}^3/\text{min/ft}^2$ of hood space area. The ventilation system in anesthesia and other rooms where hazardous gas is used must meet requirements of the National Fire Protection Association. Boiler rooms require sufficient outdoor air to maintain adequate combustion in the equipment.

Flammable and Explosive Liquids and Gases

In storing flammable and explosive materials, the following precautions should be used:

1. Gasoline, other flammable liquids, and gases should be handled and stored in specially designed containers and should be kept in well-ventilated areas. All flammable substances should be appropriately labeled.
2. Cleaning fluids should have high flash points.
3. Containers with flammable liquids should always be sealed when not in use, and caps should be replaced promptly after use.

4.   All flammable liquid tanks should have vent pipes and accessible shut-off valves.
5.   Flammable liquids and gases should not be kept in areas where individuals could trigger an explosion or fire.
6.   All spills of flammable liquids should be cleaned immediately.
7.   Approved fire extinguishers should be kept in areas where flammable liquids or gases are utilized.

Structural Features

The structural features of the hospital building must comply with established building, fire, and electrical codes. All footings must lie below the frostline, and the foundations must rest on solid earth or on pilings or piers that have been predetermined to be able to withstand the weight of the structure to avoid detrimental settling. The building materials must in all cases be fireproof. The minimum corridor widths must be 8 ft, 10 in., and the minimum door widths should be 3 ft, 10 in. All doors leading to patient rooms and bathrooms should be easily opened from the outside in the event of an emergency. Thresholds and expansion joint covers must be flush with the floor. Drinking fountains, telephone booths, vending machines, and other objects should not block corridors or exits. All parts of the building below the grade level must be waterproof and built to prevent the entrance of surface water. Storage rooms, patient bathrooms, and so forth must have a minimum width of 7 ft, 6 in. All patient rooms should have at least one window leading to the outside.

Boiler rooms, food-preparation areas, mechanical equipment rooms, and laundries need to be insulated and ventilated to prevent the floor surfaces from exceeding normal floor temperatures. It is important that the boiler rooms not be located underneath patient or patient service areas. In the event of a fire or explosion, this could cause serious loss of life or injury.

Adequate cooling systems are needed for all areas of the hospital. Indirect cooling is most desirable in surgery suites, delivery areas, nurseries, emergency, and intensive care areas. The exhausting of air from various critical areas should be done in such a way that it does not interfere with the air movement patterns of the rest of the institution.

Where solid-waste or linen chutes are used, they must comply with the applicable National Fire Protection Standards, which include approved-glass service openings; chute openings located in rooms with a minimum of 1-hr fire-resistant construction; the minimum diameter of the chute must be 2 ft; chutes must discharge directly into the refuse room, or in the case of linen into the linen room; the chutes must have sprinklers and provisions for washing the chutes and appropriate areas.

Housekeeping Services

One of the most effective techniques for organizing housekeeping services is to establish good work plans. The first step in planning work is to identify the type of surfaces to be cleaned and the amount of congestion, plus the potential hazards within the area. Once these have been identified and specified, time allotments are made for the amount of work to be produced by the average housekeeping person. The actual work scheduling is set up based on the competency of the individual, the kind of equipment available, and the type of problems encountered during cleaning. A daily cleaning schedule should be established for floor washing, dusting with treated cloths, and cleaning of bathrooms and other specialized rooms. This schedule should be followed on a daily basis. In the daily cleaning, a general-purpose, good detergent disinfectant should be utilized. Equipment used in a contaminated area should not be brought into other areas, as this is a direct means of spreading organisms from one area to another. Housekeeping personnel entering contaminated areas must take the same precautions in the use of caps, gowns, gloves, and masks as other personnel, to prevent the spread of infection.

Periodic cleaning or heavy duty cleaning should be performed in areas where patients can be moved or in areas where patients are not present at all times. This includes cleaning of windows, walls, and stripping of wax and sealers from floors by the use of mechanical equipment. Electrical equipment should be utilized for this work, and the individuals should be specifically trained in the cleaning procedures and proper use of the equipment.

Special cleaning procedures should be carried out in isolation rooms or when terminal cleaning of the room is called for, after the patient with an infection or contagious disease has been transferred, discharged, or has died. Special cleaning techniques are also used in operating rooms, delivery rooms, nurseries, and in rooms where reverse isolation has taken place. In special cleaning, it is essential that all equipment and materials be fresh, clean, and free of contaminating organisms. The cleaning equipment must not be transferred from special cleaning areas to other areas of the hospital. All disposable materials and equipment should be marked, and placed in special plastic bags to be disposed of separately. Equipment should be washed, cleaned, and sanitized before reuse when possible. All personnel should perform a 2-min medical scrub before entering and after leaving these areas.

Lighting

The desired levels of lighting vary tremendously from the areas where social activities and gatherings may take place with a minimum of 10–15 foot candles, to the cystopscopic tables, where a minimum of 2500 foot candles of light is required. It is suggested that the environmentalist obtain lighting standards from the various Illumination Engineering Society reports that are current for a given time period. Since there are innumerable places within the institution with different lighting requirements, it is difficult to keep an updated list in any text.

Plumbing and Water

All sinks used for patient care or in-service areas must have a water spout a minimum of 5 in. above the rim of the fixture. In surgical areas and in patient isolation areas foot controls are necessary for the staff to hand cleanse before and after patient care or treatment. It is particularly important that water storage, such as distilling tanks, is tested frequently, since the water could become contaminated. Beside the normal water supply, institutions need a minimum pressure on the upper floors of 20 lb/in.^{2} during maximum demand periods. Additional water supply and water sources are needed for firefighting. These water supplies should be kept separate from the potable drinking water. (Further information concerning water, plumbing, sewage, cross-connections, and submerged inlets is found in Volume II.)

Emergency Medical Services

Typical ambulance specifications include front-disk power brakes, eight-cylinder engine, a speed of at least 70 mi/hr when fully loaded, three forward transmission gears, and heavy-duty double-action shock absorbers. Another requirement is ambulance rescue equipment, including a number 10 ABC dry chemical fire extinguisher, primer wrecking bar, ropes, bolt cutters, and so forth. A third requirement is emergency care equipment, such as portable suction apparatus, bag mask ventilation unit, and oral phyrayngeal airways for adults, children, and infants.

Training requirements include emergency medical technician courses successfully completed and a passing grade on a written and practical examination, 24 hr of in-service instruction each year, and U.S. Department of Transportation emergency medical technician refresher training courses.

Fire Safety

The fire program should come under the control of a fire marshall or fire director trained by the fire department. The fire marshall should be a member of the overall safety committee and should be responsible for evaluating all fire hazards within the institution. The fire marshall, in conjunction with appropriate administrators, should determine the institutional needs and provide the following: supervision of fire prevention, protection, and suppression programs for the staff; administrative supervision and support for all subareas within the institution; coordination of all activities of all subareas; final authority on all matters related to fire prevention, protection, and suppression; ensure that periodic inspections are made of all parts of the building and that all equipment, facilities, procedures, and regulations are in keeping with good fire prevention policy; and act as administrative head should fires occur.

The subunit fire staffs, comprised of individuals from various operating departments of the hospital, have responsibility for working directly in all program activities or emergencies under the fire director; conducting training of fire protection and suppression staffs; understanding the effects of extinguishing agents, such as cooling, smothering, anticatalytic, or inhibiting; scheduling and conducting fire-prevention inspections; reviewing and monitoring all activities that constitute fire hazards; holding regular conferences with all individuals involved when fire incidents occur; reviewing and approving plans and drawings for construction and alteration of buildings to ensure adequate fire protection; reviewing and approving storage plans for flammable materials; approving requests for open flame operations, such as welding; making technical investigations concerning fire incidents; preparing and publishing seasonal fire protection and prevention material; enforcing fire standards and rules in various work areas; helping develop, coordinate, and present plans for emergency evacuation of personnel, fire control, and salvage of property; conducting fire and evacuation drills; evaluating workers' knowledge of fire extinguishers, fire practice, and how to report fires; developing and posting operational instructions for extinguishers and fire hoses; posting appropriate posters and signs concerning fire problems; and working closely with the fire department.

In the event of a fire at an institution, a specific code should be used to alert the institutional personnel concerning the whereabouts and severity of the fire. At the same time, the fire department should automatically be notified of the existence of a fire and its location. Immediate action should be taken to remove individuals from the fire area and to extinguish the fire if possible. Generally, however, in the event of serious fires, fire department personnel should be the ones responsible for extinguishing the fire. Evacuation plans should be fully understood and personnel well trained in necessary evacuation procedures.

Fire equipment on the premises should be checked periodically to ensure that it is in proper operating condition. This should be done by institutional personnel and by the fire marshall's office.

Typical fire prevention rules in an institution include keeping stairwell doors closed at all times; keeping room and corridor doors closed; discarding all solid waste promptly; placing waste chemicals in special safety cans and removing them through specialized hazardous waste collection companies; providing adequate ashtrays; and keeping corridors free of obstructions.

Patient Safety

Patient safety rules include keeping the adjustable height beds in the low position, except during actual treatment; the use of safety or siderails on the bed; the use of safety vests that restrict but allow movement; proper indoctrination of the patients to the hospital environment; a good communications system, including signals within the bathrooms; proper positioning of bedside tables and other furniture;

proper use and instruction in use of patients aids such as walkers, wheelchairs, and canes; installation of shower bars; stools within showers; adequate illumination during the day and night; special attention to patients most frequently prone to accidents, such as those on medication, the elderly, and those with diseases causing dizziness; proper medication at proper times to avoid medication errors.

Radiation Safety

X-ray equipment should be monitored by competent individuals for stray radiation and improper functioning. Factors in radiation safety include the X-ray beam size, X-ray machine filtration, shielding, scatter measurements, tube-housing leakage, and fluoroscopic exposure rates. Surveys are conducted to determine the level of leaking radiation and to suggest recommendations for necessary corrections. The handling of radioactive material is of grave concern. The materials must be handled in such a way that the operator and the patient are not contaminated. Several specific controls are involved, including restricted areas for use of radioactive material; elimination of pipetting by mouth of radioisotopes; use of impervious gloves; proper cleaning techniques; storage of radioisotopes in containers that limit exposure to 100 mrem/hr maximum at a distance of 1 ft from the container. Hot radioactive materials must be transported in shatterproof containers and shielded, depending on the type of radiation. Radioactive wastes are stored in metal cans and removed to restricted areas, where they are disposed of using special techniques. Animals containing radioactive materials are tagged, and all waste material, including their excreta, are collected and disposed of as all other radioactive material would be. The carcasses are wrapped in absorbent material and disposed of in a safe manner. Human sources of radioactive material include patients containing radionuclides that are used for treatment or diagnostic purposes. The internal beta-gamma emitters must be kept at the lowest practical level. Patients should not receive a total dose exceeding 10 gray in any 12-month period. External beta-gamma emitters must not exceed the levels of the internal beta-gamma emitters. Since internal alpha emitters are particularly hazardous, special consideration must be given by a committee on radionuclides before they are used on humans. Nursing care for patients receiving radioactive isotopes must be carried out with the greatest caution, since the individual or the nurse may have contaminated skin, be contaminated by inhalation or ingestion of the radioactive materials, or may have their body irradiated by these materials. These problems are largely eliminated by good housekeeping, proper hand-washing techniques, and clean work habits. In addition, equipment must be handled with great care.

Infectious Waste Management

An Infectious Waste Management program provides protection to human health and environment from infectious waste hazards. An infectious waste management system should include the following:

1. Designation of infectious waste.
2. Handling of infectious waste, which includes segregation, packaging, storage, transport and handling, treatment techniques, disposal of treated waste.
3. Contingency planning and staff training. The program needs to be designed for the particular situation. Such things as the number of patients, the nature of the infectious waste, the quantity of infectious waste generated, the availability of equipment for treatment on site and off site, the regulatory constraints, and the cost are important factors in planning. The method of treatment will vary with the waste type. The waste must be evaluated for its potential to cause disease. Such characteristics as the chemical content, density, water content, and microbiological content must be evaluated. Steam sterilization may be used for laboratory cultures, whereas incineration would be used for pathological waste. When possible, it is best to handle all infectious waste in the same way in order to cut down on cost. In any planning program, it is necessary to determine the local standards for incinerators, regulations for water quality, the regulations and standards relating to chemical pollutants, thermal discharges, biological oxygen demand, total suspended solids, regulation of sharps, and any other environmental regulations at the local or state level.
4. A responsible person at the facility should prepare a comprehensive report that outlines the policies and procedures for the management of all infectious and hazardous waste in the facility. Reports should contain detailed procedures for all phases of the waste management program at the facility, including research, clinical laboratories, autopsy rooms, and other types of infectious waste from a variety of areas of the institution.

It is necessary to determine which of the waste at the institution will be considered to be infectious waste, and specific policies and instructions concerning this waste need to be presented in written material. The infectious waste should then be separated from the general waste stream to make sure that these wastes will receive appropriate handling and treatment. The infectious waste should be segregated from the general waste stream at the point at which it is generated, either in the patient areas or the laboratories. Infectious waste should be discarded directly into containers or plastic bags and marked with the universal biological hazard symbol, which is three circles with a gap overlaying a smaller circle. The symbol is fluorescent orange or orange red in color.

The infectious waste should be packaged to protect the waste handlers and the public from possible disease and injury. The packaging must be preserved through handling, storage, transportation, and treatment. In selecting the packaging container, it is necessary to determine the type of waste, the handling and transport of the packaged waste before treatment, the treatment technique, special types of plastic bags, and the package identification. Liquid waste should be placed in capped or tightly stoppered bottles or flasks. Containment tanks can be used for large quantities of liquid waste. Solid or semisolid waste, such as pathological waste, animal carcasses, and laboratory waste, may be placed in plastic bags. Sharps should be placed directly into impervious, rigid, and puncture-resistant containers to eliminate injury. The clipping of needles is not recommended, unless the clipping device contains the needle parts. Otherwise, an airborne hazard may occur. Sharps should be placed in such containers as glass, metal, rigid plastic, wood, and heavy cardboard. The sharps' containers should be marked with the universal biohazard symbol. A single bag for containment of infectious waste is not adequate. Either a double plastic bag should be used or a single bag should be placed in a bucket, box, or carton that has a lid or a seal. Containers of sharps and liquids can be placed in other containers for transportation and storage. When other plastic bags of infectious waste are handled and transported, care must be taken to prevent tearing of the bags. Plastic bags containing infectious waste should not be placed in a trash chute or on a dumbwaiter. The proper practice for the handling of these plastic bags include loading by hand, transporting in such a way that minimal handling occurs, and putting the plastic bags in rigid or semirigid containers to prevent spills or breakage of the bags. Where recycled containers are used for transport in the treatment of bag waste, they should be disinfected between use. If the waste is to be incinerated, the containers must be combustible. If the waste is to be steam sterilized, the packaging materials must allow steam penetration and evacuation of air. Interference with the steam sterilization treatment may occur because of high-density plastics, which prevent effective treatment of the air and therefore will not allow the contents of the bag to reach the appropriate temperatures. Low-density plastics will enhance steam penetration and allow air evacuation from the waste.

There are several types of plastic bags available for waste disposal. The quality of the plastic bag and its suitability to contain infectious waste is based on the raw materials used during manufacture; and the product specifications suitability is based on the fitness and the durability of the plastic. It is most important that the plastic bag be tear resistant until the waste is treated. Tear resistance can be improved by not placing sharps, sharp items, or items with sharp corners in the bags or overloading the bag beyond its weight and volume capacity.

Although it is preferable to treat the infectious waste as soon as possible, it may not be practical to do so. Therefore, the waste will have to be stored in an appropriate manner. There are four important factors related to the storing of infectious waste. They are the packaging, storage temperature, duration of storage, and location and design of the storage area. The packaging must provide containment of the waste throughout the waste-management process and must deter rodents and insects from invading the packaging, and thereby potentially becoming vectors of disease. As the temperature increases, the rate of microbial growth and putrefaction increases. This will cause bad odors to be associated with the waste because of the retained organic matter present. Time is also a major consideration here. It is suggested that the waste be stored under refrigeration in special units at 34–45°F for a maximum of 3 days. The storage areas should be extremely secure, have limited access, be free of rodents and insects, and display the universal biohazard label.

The infectious waste when transported on or off site should be in appropriate packaging, which has already been mentioned, and in containers that are leak-proof, rigid, and appropriately marked.

The infectious waste is treated to reduce the hazard associated with the presence of infectious agents. This will be discussed further under Controls.

Once the infectious waste has been effectively treated, it is no longer biologically hazardous and may be mixed with and disposed of as ordinary waste, providing it does not pose any other hazards that are controlled by federal or state regulations. Treated liquid waste may be poured down the drain to the sewer system. Treated solid waste and incinerator ash may be disposed of in a sanitary landfill. Needles and syringes must be rendered nonusable before disposal. Treated sharps can be ground up, incinerated, or compacted. Body parts should be incinerated in special incinerators or buried, depending upon religious law.

The institution should set up a contingency plan, which would provide for the following situations:

1. Spills of liquid infectious waste and the cleanup procedures utilized by personnel. Also, procedures should be available for the protection of the personnel and disposal of the spilled residue.
2. Rupture of plastic bags or other loss of contaminants and the necessary cleanup procedures, protection of personnel, and the repackaging of the waste.
3. Equipment failure and the alternate arrangements for waste storage and treatment.

Staff training is essential in order that all technical, scientific, housekeeping, and maintenance personnel will be protected. The training programs should be used for all new employees whenever infectious waste methods and practices change and when a new program is put into operation. Continuing education, including refresher courses, should be given periodically to protect the employees of the institution.

The Medical Scrub

The medical scrub, which should take a minimum of 2 min should be practiced by all personnel when hand washing procedures are indicated. The medical scrub consists of the following steps:

1. Remove all jewelry, rings, watches, etc.
2. Turn on the water (running water is essential).
3. Wet hands.
4. Soap hands and wash them. Use considerable friction. Begin with hands, lead to wrists, then to elbows (be sure to clean well between fingers and around thumb).
5. Rinse (rinse hands, arms to elbow, letting water run off the elbow.)
6. Clean nails under running water. Use orange stick and discard.
7. Repeat steps 4 to 6.

8. Dry hands. Wipe water from hands to elbow with one paper towel. Use second paper towel to finish drying.
9. Turn off faucet. Use second paper towel if knee-action control is absent. Use paper towels to turn off the faucets.

Essential Practices Related to AIDS Virus and Hepatitis Virus

The essential practices related to the prevention of the transmission of the HIV and HBV viruses include the following:

1. classification of the work activity
2. development of standard operating procedures
3. provision of training and education
4. development of procedures to ensure and monitor compliance
5. workplace redesign

Further, the worker needs to have available a program of medical management. This program should reduce the risk of infection by HBV and HIV. All workers involved in any exposure to blood or other body fluids should be vaccinated with hepatitis B vaccine.

Once an exposure has occurred to blood or other infectious body fluids and the person has not been adequately protected, a blood sample should be taken from the individual (with the individual's permission) to determine if the surface antigen of hepatitis B or the antibody to HIV is present. If the individual has not previously been given the hepatitis B vaccine, a single dose of hepatitis B immune globulin is recommended within 7 days of exposure. The individual should also receive the vaccine series. If the person is possibly exposed to AIDS, it is recommended that the Centers for Disease Control in Atlanta, Georgia be contacted and the most recent techniques approved by the Centers for Disease Control be followed.

Employees need to take precautions to prevent injuries from needles, scalpel blades, and other sharp instruments or devices that they use during various procedures and that they may handle during cleaning, disposal, or removal of these objects.

Schools

Physical Plant and Site Selection

Site selection is based on present and projected student population; distance from areas served; distance from sources of noise, air pollution, and other hazards; type of surface drainage; type and nature of industrial or highway construction; amount of acreage needed for athletic fields, parking, landscaping, and other recreational areas; access to good highways; suitability for a given neighborhood.

The site is also determined by the type of soil, soil conditions, and the recommendations of school architects, engineers, public health officials, and community groups.

The physical structure or structures should be designed and built to facilitate learning and to reduce the potential for hazards. The building construction must take into account all necessary lighting and acoustics, heating and ventilation, water supply and waste disposal, laboratories and gymnasiums, workshops, and classrooms. Before the plans are approved, the school architect submits them to an advisory committee of school personnel, community groups, and finally to the state departments of education and health.

Housekeeping

A clean environment is conducive to a happy school day and therefore promotes good emotional health and the desire to learn. Many of the housekeeping tools and techniques described under the section on hospitals are appropriate for schools. (The book entitled "Environmental Health and Safety" by this author mentioned earlier contains detailed housekeeping procedures.) Proper organization of the housekeeping or custodial staff, adequate scheduling, proper training of personnel, and the use of proper cleaning agents and detergent disinfectants are essential to a good housekeeping program. The program should include, at the minimum, the following responsibilities for housekeeping personnel: maintenance of a clean, safe, and aesthetically satisfying environment; maintenance of grounds to prevent accidents and to promote aesthetic appeal; maintenance and orderliness of all apparatus and equipment; prevention of fires and promotion of fire safety; proper operation of heating, ventilation, and all other types of equipment; adequate storage of all materials utilized in construction; separate and proper storage of all hazardous equipment and materials, including chemicals and other potential fire hazards; and maintenance of proper records concerning housekeeping.

A variety of floors are used within the school as within other kinds of institutions. These floors include concrete, terrazzo, ceramic tile, rubber or vinyl tile, asphalt tile, magnesite, cork, and a variety of woods. Concrete floors are quite common in school buildings. They are easy to clean, yet they sometimes pit or crack. Adequate maintenance is important. If the concrete is painted, it should be thoroughly cleaned with solvents to remove all oils and greases. Terrazzo floors are a mixture of cement and marble, or cement and granite chips. They are strong and durable, but acids, abrasives, and strong alkaline cleaners cannot be used because they will deface the terrazzo. It is important to put a sealer on this type of floor to prevent penetration. Ceramic tile floors, rubber or vinyl tile floors, and asphalt tile floors are not resistant to grease, oils, turpentine, or petroleum-based waxes. These floors look very attractive and must be given special care. Magnesite floors are somewhat like concrete and should be treated similarly. Cork floors may be cleaned by sanding lightly with steel wool and applying light water-emulsion wax. Wood floors consist of a variety of woods, ranging from very hard to soft. Although they are very pretty if maintained properly, many wood floors have difficulty withstanding the constant pressure of footsteps.

A special problem within schools are chalkboards. Black slate boards are difficult to clean and pose a problem for use in classroom instruction. It is preferable to use green or blue boards or other kinds of special materials. Erasers are best cleaned by using a special electric vacuum eraser cleaner. This should be carried out one or twice per week.

The following lighting standards are recommended: 10 footcandles of light for open corridors and storerooms; 15 footcandles for auditoriums, cafeterias, locker rooms, and stairways; 20 footcandles for reception rooms, gymnasiums, and swimming areas; 75 footcandles for classrooms, study halls, lecture rooms, libraries, shops, and laboratories; and 150 footcandles for drafting rooms, typing rooms, sewing rooms, and special learning rooms.

Maintenance is an essential part of lighting. A program should be established to include periodic cleaning of lamps and fixtures, replacement of burnt-out bulbs, and repainting of room surfaces where needed. Student seating should be arranged to eliminate shadows and glare. Window shades, drapes, or other types of window coverings should be provided to give balanced lighting and to eliminate glare problems.

Noise

Noise prevention starts during construction of the school. Site location, classroom design, and the use of special noise-control materials are important. Noise must be suppressed, since it is a source of irritation, distraction, and emotional strain, and fosters inefficient performance in school children, teachers, and employees. Noise levels should not exceed the following: classrooms, 35–50 dB; cafeterias, 50–55 dB; outside noise, 70 dB; other specialized rooms, 40 dB.

Space Utilization

Each elementary school should have a minimum of 12 acres of land for the first 200 students and 1 acre of land for each 100 additional students. In the kindergarten, 1050 ft^2 of space is needed for a maximum of 30 students. In grades one through three, 875 ft^2 are required for 30 students, and in grades four through six, 750 ft^2 per 30 students. In the high schools, which include junior and senior high schools, 12 acres of ground are required for the first 300 students, and 1 acre for each 100 additional students. Also, in the high schools, 85% of the classrooms must have a minimum of 750 ft^2 for each group of 30 students. Other classroom sizes vary with the types of activities. All of the above requirements are minimum standards that can and should be exceeded by the school system for special types of study or where more space is desirable.

Plumbing

There should be one sink for every 30 students in elementary schools, and one sink for every 40 students in secondary schools. In the elementary schools, a toilet is required for every 40 male students and one for every 30 female students. In the secondary schools, one toilet is required for every 75 male students and one for every 45 female students. There should be at least one shower head for every five students and one drinking fountain for every 100 students.

The recommended classroom temperatures during the winter should be at 68°F with slightly lower temperatures in gymnasiums. The temperature should be approximately 70°F in elementary schools. Fresh air should be added to the school at a rate of not less than 15 ft^3 of air per minute per person per room. Where the activity level is higher, the amount of room air should be increased to 20 or 30 ft^3 per minute per person. The air must be distributed within the room to reduce a chilling effect and to prevent discomfort. The velocity of air should not exceed 500 ft/min at the duct outlets. A recommended relative humidity would be 50%. The type of heating system varies with the school, its location, and the type of fuels available.

MODES OF SURVEILLANCE AND EVALUATION

Infections

There are several techniques used to determine the extent of possible hospital-acquired infections. The first technique is to make in-depth intensive microbiological studies of all critical areas and other areas of the hospital environment. These samples are taken through the air or by means of swabs and Rodac™ contact plates used on surfaces. As a general rule, the microbiological sampling technique is not effective and should only be utilized where a specific organism is being sought and the cause of the disease outbreak cannot be determined. Microbiological sampling is also used on a research basis to determine the levels of microorganisms in given types of operations or where special study techniques have been used and the sources of the outbreak of disease cannot be traced.

A second technique involves the development of a surveillance and reporting program in which a nurse epidemiologist regularly carries out the following functions:

1.   reviews daily charts indicating that the patient has an infection or is suspected of having an infection

2. holds conferences on a daily basis with nursing supervisors and nursing units to determine potential infection problems
3. reviews all laboratory reports confirming infections and correlates the laboratory studies with the observations of the physicians and nurses
4. reviews special reports submitted by physicians or supervising nurses where suspected infection problems are occurring
5. lists and tries to determine the location and source of specific types of infections
6. evaluates necrospy reports to determine if infections not previously reported are found upon autopsy
7. maintains a close liaison with the employee health center and evaluates employee infections as they may relate to hospital acquired infections
8. prepares monthly statistical reports indicating levels of infection as determined by charts and laboratory tests and pinpoints where the higher levels of infection occur

The nurse epidemiologist may work for the hospital environmental health specialist, whose responsibility would encompass the overall environmental health, infection control, and occupational health and safety programs of the institution.

The third type of technique is a programmed approach for a team comprised of the supervising nurse, the physician in charge of the unit, and the environmental health specialist, using specifically established study forms to determine the potential sources of infection, within the institution. This team approach, systematically carried out, helps to establish the potential sources of infection, thereby assisting the individuals within their own unit and also in the hospital at large in determining the kinds of potential infections occurring within the institution. The programmed approach and the study form are found in Diagram 8.1 and Figure 8.1.

Accidents

The hospital building should be carefully evaluated and supervised to minimize accidents and injuries. It is important to evaluate all accidents and illnesses to determine if any hazards were involved and to make the necessary recommendations to eliminate the hazardous situation or unsafe behavioral pattern. When investigating the incident, all possible facts should be collected. The equipment operation should be fully understood; processes, special conditions, the fatigue factor, and other environmental factors related to the individual should also be known. Always determine the direct causes of the accident, such as faulty operating equipment or the use of improper techniques, and the indirect causes, such as poor eyesight and ignorance. There may be additional contributory causes, such as inadequate standards, inadequate training, lack of policy, lack of enforcement, faulty design, faulty maintenance, or failure of supervisors to perform their functions. Reporting and recording of information notifies the institutional management of the incident, the results of the incident, the contributory causes, and in this way helps them conduct proper evaluations and take necessary actions to

Diagram 1   Programmed Approach to Hospital Infection Control

APPRAISE	REVIEW	CONTRIBUTE
TEAM Nurse Doctor Env. Health Specialist	Review the Programmed Approach for the area to be visited.	Each team member contributes information previously gained about the area.
Critically appraise all factors in the patient care routine, support function, and physical plant.	Use of the tool. Question involved personnel about details of select procedures. If the team decides a problem exists, a recommended solution is discussed and recorded.	The environmental health specialist writes a draft of a report. Draft of the report is reviewed by other team members. Final consensus report written: 1. stating the problems, 2. recommending solutions, 3. indicating areas of further study related to personnel utilization, materials, and methods.
	Restudy made at appropriate intervals	Report sent to hospital director

prevent future injuries. Inspections, special studies, and investigations before accidents occur are of vital importance. A good inspection should determine specific unsafe or unhealthy acts and conditions; determine the need for specific safeguards for people, equipment, and materials; detect inefficient, inadequate, or totally lacking safety programs; determine the condition of work areas, structural components, and service facilities; determine the condition of equipment; determine job procedures; identify the various types of materials, chemicals, and other

_____ Hospital

PROGRAMMED APPROACH TO HOSPITAL INFECTION CONTROL

S=Satisfactory   I=Improvement   U=Unsatisfactory

Building Name and No._____ Floor_____ Date _____Time _____

Names of Study Team_____

ADMINISTRATIVE PROCEDURES

_____ 1.   Does any employee have a patient history of being a carrier of pyogenic infections?
_____ 2.   Do all employees report pyogenic infections?
_____ 3.   Is the employee who reports an infection immediately removed from patient care?
_____ 4.   Is there a restriction of traffic within the given patient area?
_____ 5.   Is there routine daily recording of infections on charts?
_____ 6.   Is indiscriminate antibiotic use discouraged?
_____ 7.   Are the interns given instruction in hospital infection control?
_____ 8.   Are the residents given instruction in hospital infection control?
_____ 9.   Are the nurses given instruction in hospital infection control?
_____10.   Are nonprofessional employees given instruction in hospital infection control?
_____11.   Are visitors restricted in time and number?
_____12.   Do visitors observe special isolation procedures when indicated?

STERILIZATION PROCEDURES

_____ 1.   Is each autoclave tested monthly for its ability to destroy bacterial spores?
_____ 2.   Are autoclave indicators or dates used?
_____ 3.   Are routine cultures made of utensils, trays, packs, instruments, catheters, gloves, sutures and parenteal solutions?
_____ 4.   Are records kept of all isolations of significant bacteria such as "staph"?
_____ 5.   Are all autoclaved items distinctly marked as sterile?
_____ 6.   Are autoclaved items stored separately from nonsterile items?
_____ 7.   Are autoclaved items protected from contamination while in storage?

LAUNDRY AND LINEN PROCEDURES

_____ 1.   Are linens washed in such a manner that all pathogenic bacteria are destroyed?
_____ 2.   Are bacteriological samples taken of the clean linen to insure that all pathogenic bacteria are destroyed?
_____ 3.   Are the clean, ironed linens exposed to the dirty linens in such a way that there may be a transfer of pathogenic bacteria?

Figure 8.1.   Programmed approach study forms. Source: _Environmental Health and Safety_ (Elmsford, NY: Pergamon Press, 1974), p. 140. With permission.

_____ 4. Are clean linens protected during transportation to ward areas?
_____ 5. Are clean linens transported on the same truck as dirty linens?
_____ 6. Are all linens, blankets, and other bed covers gently removed from beds to prevent air dispersion of bacteria?
_____ 7. Are all linens from infected patients placed in a specially marked laundry bag and closed at once?
_____ 8. Do all linens from infected patients receive special handling?
_____ 9. Are laundry bags from isolation placed in clean outer bags?
_____10. Are all blankets sterilized before reuse?
_____11. Are mattresses washed with a germicide between patients?
_____12. Are pillows washed with a germicide between patients?
_____13. Are beds washed with a germicide between patients?
_____14. Are nonporous covers used for mattresses and pillows?
_____15. Are the laundry trucks and carts cleaned daily?
_____16. When not in use, are the trucks stored in an area which is free of gross contamination?

TRASH AND SEPTIC WASTE REMOVAL PROCEDURES

_____ 1. Are infectious materials either autoclaved or placed in sealed plastic bags before being discarded?
_____ 2. Is food waste from infected patients placed in sealed plastic bags?
_____ 3. Are paper bags used in isolation rooms for discarded tissues, dressings, tongue blades, and other disposable items?
_____ 4. Are they immediately closed and placed in a sealed plastic bag?
_____ 5. Are bags containing infectious materials marked "Danger, Infectious Materials Within"?
_____ 6. Do employees who remove infectious materials discard their gowns at the completion of the waste disposal and then follow proper hand washing procedures?
_____ 7. Do employees who remove infectious materials wear gowns?

FOOD HANDLING PROCEDURES FOR INFECTED PATIENTS

_____ 1. Are all clean dishes, trays, and eating utensils protected against contamination during transportation or storage?
_____ 2. Are disposable single service paper units discarded after each use?
_____ 3. Are all remnants of meals scraped into a plastic bag in the isolation room?
_____ 4. Are nondisposable utensils thoroughly and immediately disinfected after each use by infected patients?
_____ 5. Are all dishes, trays, and eating utensils washed and sanitized before reuse?
_____ 6. Is ice making equipment and ice storage equipment kept clean and sanitary?
_____ 7. Are there individual water containers and glasses for each patient?
_____ 8. Are these containers and glasses thoroughly washed and sanitized before reuse?
_____ 9. Are utensils for medications thoroughly washed and sanitized before reuse?

Figure 8.1. continued.

BEDPAN HANDLING

_____ 1. Are all bedpans, urinals, enema containers and other equipment for collection of excreta sterilized before reuse by other patients?
_____ 2. Are all such contaminated utensils handled in an aseptic manner after use by patients?
_____ 3. Are bedpans stored off the floor in a cabinet?
_____ 4. Is toilet paper stored off the floor in a sanitary manner?

EQUIPMENT HANDLING AND SANITIZATION PROCEDURES

_____ 1. Is there an individual thermometer for each patient?
_____ 2. Are thermometers stored in an adequate concentration of germicide to kill all bacteria?
_____ 3. When equipment such as portable X-ray units, flashlights, otoscopes, opthalmoscopes, stethoscopes, ice bags, etc., are used for infectious cases are they cleansed thoroughly with a germicide solution before reuse?
_____ 4. Is there a Graduate Nurse assigned to the use of the dressing cart?

ISOLATION PROCEDURES

_____ 1. Are all patients with draining wounds isolated?
_____ 2. Is reverse isolation strictly enforced when indicated?
_____ 3. Are patients isolated in such a way and in such areas that proper procedure can be carried out?
_____ 4. Are isolated patients properly identified?
_____ 5. Is a suspected infected case reported the day it is noted?
_____ 6. Are clean gowns put on before entering isolation rooms?
_____ 7. Are the gowns discarded in containers in the isolation area?
_____ 8. Are hands washed with an acceptable technique before and after handling any contaminated material or patients?
_____ 9. Are all areas in which infectious cases have been treated thoroughly cleaned and sanitized before reuse?

CATHETERIZATION PROCEDURES

_____ 1. Are the genitalia thoroughly cleaned with an appropriate soap in preparation for the catheter?
_____ 2. Does the catheter come from a sterile package?
_____ 3. Does the syringe used for irrigation of the catheter come from a sterile package?
_____ 4. Are hands washed before donning sterile gloves?
_____ 5. Do the gloves used for sterilization come from a sterile package?
_____ 6. Does the urine drainage bottle overflow?
_____ 7. Is the urine bottle suspended above the floor?
_____ 8. Is the urine bottle autoclaved?
_____ 9. Is there a cotton plug in the air vent of the catheter?
_____10. Is an antiseptic present in the urine bottle?

DRESSING OF WOUNDS

_____ 1. Are dressings put on with gloved hands?

Figure 8.1. continued.

_____ 2. Are hands washed before and after dressing patients' wounds?
_____ 3. Are dressing trays and carts arranged to prevent contamination of sterile and clean equipment?
_____ 4. Are dirty drapes or dirty dressings kept on dressing carts?
_____ 5. Does cold sterilization take place on the dressing cart?
_____ 6. Are wounds dressed at specifically scheduled times?

SURGICAL FACILITIES

_____ 1. Are the operating suites separate from each other and other patient areas?
_____ 2. Is cross traffic through these units from other hospital areas eliminated?
_____ 3. Are the dressing rooms for physicians and nurses accessible directly from the rest of the hospital and do they open directly into the operating suites?
_____ 4. Is there adequate separation between utility areas used for clean and those used for contaminated equipment, supplies, or waste?
_____ 5. Are all surgical wastes placed in plastic bags for disposal?
_____ 6. Are sub-sterilizers (rapid instrument sterilizers) accessible to operating rooms?
_____ 7. Are facilities for scrubbing directly accessible to operating rooms?
_____ 8. Are these facilities kept clean at all times?
_____ 9. Are locker rooms kept clean at all times?
_____10. Is there adequate space in clean areas for storage of sterilized items?
_____11. Are there facilities in the recovery room for isolation of infected cases?
_____12. Are there closed containers for all contaminated and soiled dressings and other contaminated material?
_____13. Are there adequate closed containers for disposal of gowns, masks, and booties in the locker room?

ANESTHESIA AND INHALATION THERAPY

_____ 1. Inhalation
a. Is the equipment cleaned and sterilized after use by patients?
_____ 2. Intravenous
a. Do the needles and syringes come in a sterile package?
b. Does the plastic tubing come in a sterile package?
_____ 3. Spinal
a. Is detergent used in cleaning needles?
b. Is all equipment sterilized?
c. Are solutions sterilized?
d. Are contents of ampoules sterile?
_____ 4. Conduction Anesthesia (local)
a. Are gowns, masks, gloves, worn when anesthesiologist does conduction anesthesia?
_____ 5. Is the anesthesia table stored outside of the O.R. within a traffic area?
_____ 6. Are endotracheal tubes cleaned and then sanitized in 70% alcohol plus Formalin? (1 part Formalin to 250 parts alcohol.)
_____ 7. Is the endotracheal tube then placed in a boiling sterilizer to boil off the Formalin?

Figure 8.1. continued.

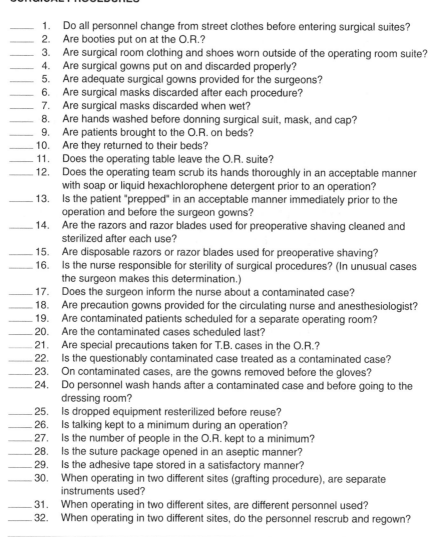

_____ 8. Is humidification maintained during inhalation therapy?
_____ 9. Is the equipment cleaned properly?
_____10. Are the nasal catheters sterilized?
_____11. Is the humidifying solution sterile?

SURGICAL PROCEDURES

_____ 1. Do all personnel change from street clothes before entering surgical suites?
_____ 2. Are booties put on at the O.R.?
_____ 3. Are surgical room clothing and shoes worn outside of the operating room suite?
_____ 4. Are surgical gowns put on and discarded properly?
_____ 5. Are adequate surgical gowns provided for the surgeons?
_____ 6. Are surgical masks discarded after each procedure?
_____ 7. Are surgical masks discarded when wet?
_____ 8. Are hands washed before donning surgical suit, mask, and cap?
_____ 9. Are patients brought to the O.R. on beds?
_____ 10. Are they returned to their beds?
_____ 11. Does the operating table leave the O.R. suite?
_____ 12. Does the operating team scrub its hands thoroughly in an acceptable manner with soap or liquid hexachlorophene detergent prior to an operation?
_____ 13. Is the patient "prepped" in an acceptable manner immediately prior to the operation and before the surgeon gowns?
_____ 14. Are the razors and razor blades used for preoperative shaving cleaned and sterilized after each use?
_____ 15. Are disposable razors or razor blades used for preoperative shaving?
_____ 16. Is the nurse responsible for sterility of surgical procedures? (In unusual cases the surgeon makes this determination.)
_____ 17. Does the surgeon inform the nurse about a contaminated case?
_____ 18. Are precaution gowns provided for the circulating nurse and anesthesiologist?
_____ 19. Are contaminated patients scheduled for a separate operating room?
_____ 20. Are the contaminated cases scheduled last?
_____ 21. Are special precautions taken for T.B. cases in the O.R.?
_____ 22. Is the questionably contaminated case treated as a contaminated case?
_____ 23. On contaminated cases, are the gowns removed before the gloves?
_____ 24. Do personnel wash hands after a contaminated case and before going to the dressing room?
_____ 25. Is dropped equipment resterilized before reuse?
_____ 26. Is talking kept to a minimum during an operation?
_____ 27. Is the number of people in the O.R. kept to a minimum?
_____ 28. Is the suture package opened in an aseptic manner?
_____ 29. Is the adhesive tape stored in a satisfactory manner?
_____ 30. When operating in two different sites (grafting procedure), are separate instruments used?
_____ 31. When operating in two different sites, are different personnel used?
_____ 32. When operating in two different sites, do the personnel rescrub and regown?

Figure 8.1. continued.

_____ 33. Are special precautions taken during an intestinal anastomosis? (Is a subset of instruments used?)

_____ 34. Are the proper types of drapes used?

_____ 35. Is the proper thickness of drapes used?

_____ 36. Are dry drapes used?

_____ 37. Do they get wet during the procedure?

_____ 38. Are instruments retrieved from below the patient level and used?

_____ 39. Is sheet wadding sterilized when used on open procedures?

_____ 40. Are gloves changed when needles puncture them?

_____ 41. Are personnel with significant respiratory illnesses or with overt lesions excluded from the operating and delivery suites?

_____ 42. Are Kelly pads being used on the O.R. table?

_____ 43. Is the O.R. table cleaned properly between operations?

_____ 44. Is the operating room equipment cleaned properly?

_____ 45. Is the recovery room clean?

_____ 46. Is the recovery room maintained with the same basic technique as the operating room?

_____ 47. Is the operating room thoroughly scrubbed and washed with a germicide solution and allowed to sit for at least 24 hours after an infected case?

_____ 48. Are the walls spot cleaned between all operations?

_____ 49. Are the floors and other horizontal surfaces cleaned between all operations?

_____ 50. Is a germicide solution used in the above mentioned procedures?

_____ 51. Is there routine bacteriological surveillance of aseptic techniques and sterilization in the O.R.?

_____ 52. Do surgeons wash their hands, put on masks, and don sterile gloves before performing intravenous cutdowns?

_____ 53. Are intravenous catheters inserted in an aseptic manner?

_____ 54. Are intravenous catheters removed promptly when not needed?

HOUSEKEEPING
See self-inspection sheet for Housekeeping Program in Housekeeping Section.

COMMENTS:

Figure 8.1. continued.

hazardous substances being used; identify the controls that should be employed to prevent accidents; determine if these controls are actually operable; determine all of the potential problems related to the environment, such as lighting, ventilation, heat, obstructions, slippery surfaces, and so forth; determine the kinds of existing hazards and the types of controls used for each type of operation.

Evaluation by the occupational health specialist includes routine surveys of chemical laboratories, storage, and handling areas; sampling and detection of airborne contaminants; evaluation of noise, lighting, temperature, ventilation, and so forth; regular testing of chemical fume hoods and biological safety cabinets to ensure proper functioning; evaluation and investigation of employee and student accidents; consultation services for special problems to make necessary recommendations for correction; routine surveys of shops, laboratories, and other specialized areas; special studies concerning special hazards.

The Food Service Self-Inspection Concept

Consistent unbiased evaluation of items requiring correction is essential to the improvement of a hospital food-service program. The best approach to be used by the food-service managers is the self-inspection program. The food-service manager, on a weekly basis, should systematically inspect his or her own facility; identify problem areas; determine whether improvements are being carried out each week; and record the facts, which can be used for program changes or additional budget requests for training and/or workforce. In order for the food-service managers to be able to carry out a good self-inspection of the facility, these individuals should be put through a training program, which would be conducted by the local health department, state health department, or an in-house environmental health practitioner. The recommended program would have 17 sessions of 2 hr each. Each session would be conducted as a seminar, with the trainer getting necessary preliminary information, citing case histories and solutions, and then bringing the entire group of managers into the discussion concerning actual problems at the institution and potential solutions for them. The recommended training program is as follows:

- Session 1 – introduction by hospital administrative staff as well as introduction of the program itself
- Session 2 – review of general hospital problems related to infections, injuries, and hospital practices
- Session 3 – review of general hospital problems related to infections, injuries, and hospital practices (continued)
- Session 4 – the self-inspection form and technique
- Session 5 – field experience in use of the form
- Session 6 – additional field experience in use of the form
- Session 7 – review of food-service department problems

- Session 8 – review of food-service department problems and start using forms on a weekly basis
- Session 9 – general housekeeping principles and practices and discussion of results of inspections
- Session 10 – general housekeeping principles and practices and discussion of results of inspections (continued)
- Session 11 – general food service equipment and warewashing techniques, and discussion of results of inspections
- Session 12 – bacteriology and communicable disease control and discussion of results of inspections
- Session 13 – food preparation, storage, serving, and transportation and discussion of results of inspections
- Session 14 – insect and rodent control, and discussion of results of inspections.
- Session 15 – supervisory techniques.
- Session 16 – supervisory problems.
- Session 17 – final exam.

At the completion of the course, individuals should be given a certificate indicating completion or completion with honors (see self-inspection form in Figure 8.2).

Ventilation

To evaluate a ventilation system, it is necessary to evaluate the air patterns, air velocities, room air ventilation rates, and types of air supply and exhaust systems for moving the contaminated particles throughout the room. In this evaluation, the most important factor is the rapidity with which widespread contamination can be removed. The removal rate is determined by taking airborne samples of particles and microbial particles from various parts of the room during various operations.

Housekeeping

It is the function of the housekeeping department to determine the special housekeeping problems and, in consultation with the environmentalist, to determine those matters most critical to the environment. In-depth studies should be made of existing housekeeping practices, work plans, work scheduling, activities of house-keeping personnel, employee training, and work and time evaluations. It is essential that there be understanding as to how organisms are spread, and the daily periodic and special cleaning techniques required to fight the spread of these organisms.

Supervisors, after gaining familiarity with the total cleaning program, will evaluate the work of the personnel and give them necessary additional in-service training when such training is called for. The function of the supervisor is one of inspection, evaluation, training, encouragement, and disciplinary action, in the event that such action is needed.

Various housekeeping evaluation techniques have been proposed. The most effective appears to be the use of housekeeping self-inspection forms, where supervisors evaluate their own areas and the information is then tabulated by the environmentalist. This is supplemented by the use of a white-glove technique on surfaces to determine if they are dirt free and the use of either contact or swab bacteriological samples or air samples. Since sampling is a very costly and an inaccurate means of determining the level of housekeeping in an institution, it is recommended that its use be limited to the evaluation of an actual outbreak of disease. Under ordinary circumstances, the use of self-inspection forms and a wet, white towel smeared across surfaces would be excellent indicators of the degree of cleanliness of a given area.

All hospital problems and problems of other institutions are evaluated by trained environmental health practitioners, who use the techniques and study forms discussed in other areas of this book.

The housekeeping self-inspection program is similar to the food-service self-inspection program. The housekeeping training program for supervisors and managers would be similar to the one discussed under food service, but the specific topics would change in food to housekeeping. (See Housekeeping self-inspection form, Figure 8.3.)

INSTITUTIONAL CONTROLS

The institutional environmental health specialist is the staff person most involved in carrying out comprehensive studies and presenting reports with combined recommendations for various committees and individuals. The environmental health professional has specific training in epidemiology; a broad knowledge of biological, physical, and chemical hazards; extensive public health and environmental health training and experience; and an intimate knowledge of the institutional environment, its risks, and problems. The function of this specialist is to promote better patient care and better care of all individuals in institutions by reducing the risk of infection and the opportunity for accidents, and by limiting the extent of injuries and improving the physical environment.

Environmental Control Measures

Environmental control measures include architectural considerations; proper installation of equipment; control of airborne contamination; control of surface contamination from linens, and solid and liquid wastes; and control of other environmental factors.

A good part of environmental control of infections in hospitals is based on initial architectural considerations. Basic problems of hospital design and construction always include traffic patterns; systems used for handling materials, equipment, and liquid and solid wastes; ventilation systems; special airflow

FOOD SERVICE INSPECTION REPORT

(Submit Weekly to Sanitation Officer)
Environmental Health Practitioner

_____ (name) _____ HOSPITAL
ENVIRONMENTAL SANITATION

KEY:
2–Satisfactory
1–Improvement Needed
0–Poor

BUILDING NAME AND NO.	FLOOR	DATE	TIME	TOTAL SCORE

POINTS	ITEM
	1. Perishable foods used within a 2-hour period of preparation or immediately refrigerated.
	2. Perishable foods are not to be reused at any time.
	3. All refrigerators kept below 41°F.
	4. Ice making machines clean and operating properly.
	5. Food carts washed and sanitized at ward kitchens.
	6. Food protected from contamination: Preparation Storage Transportation
	7. Hands thoroughly washed and cleaned after use of toilets and before preparation of food.
	8. Clean outer garments worn at all times.
	9. Soap, paper towels, and waste receptacles available at all times at handwashing sink.
	10. Lavatory facilities clean and in good repair.
	11. Dressing rooms, areas, and lockers kept clean.
	12. All food handlers with diarrhea removed from duty until negative stool culture.
	13. Food utensils, dishes, glasses, silverware, and equipment scraped, pre-flushed, and soaked.
	14. Silverware, dishes, and glasses clean to sight and touch.
	15. Silverware, dishes, and glasses sanitized.
	16. Dishwasher wash and rinse temperatures checked daily (insert temp. rinse) (insert temp. wash)
	17. Food utensils, such as pots, pans, knives, and cutting boards clean to sight and touch.
	18. Grills and similar cooking devices cleaned daily.
	19. Food-contact surfaces of equipment clean to sight and touch.
	20. Tables and counters clean to touch.
	21. Non-food-contact surfaces of equipment clean to sight and touch.
	22. Washing and sanitizing water clean.
	23. Single service food articles properly stored, dispensed, and handled.
	24. Containers of food stored off floor on clean surfaces.

25. No wet storage of packaged food.

26. Sugar in closed dispensers or individual packages.

27. Poisonous and toxic materials properly identified, colored, stored, and used; poisonous polishes not present.

28. Clean wiping cloths used and properly restricted.

29. All late trays served after 8 p.m. scraped, rinsed, and stacked in previously cleaned sink.

30. Garbage and rubbish stored in metal containers with lids; containers adequate in number.

31. All garbage rmoved from area at end of working day.

32. Containers cleaned when empty.

33. When not in continuous use, garbage and rubbish containers covered with tight-fitting lids.

34. Garbage and rubbish storage areas adequate in size and clean.

35. All outer openings screened.

36. Absence of rodents.

37. Absence of flies.

38. Absence of roaches.

39. Floors kept clean.

40. Floors in good repair.

41. Floors and wall junctures properly cleaned.

42. Walls, ceilings, and attached equipment clean.

43. Walls and ceilings properly constructed and in good repair; coverings properly attached.

44. Walls of light color; washable to level of splash to shoulder height.

45. Lighting adequate.

46. Light fixtures clean.

47. Areas clean; no litter.

48. Ample quantities of cleaning supplies available.

REMARKS:

Signature of Dietician or Nursing Supervisor

Figure 8.2. Food service inspection report. Source: *Environmental Health and Safety* (Elmsford, NY: Pergamon Press, 1974), p. 34-35. with permission

HOUSEKEEPING CHECKLIST
Prepare in Duplicate

This form based on a design by Herman Koren, R.S., M.P.H., Chief of Environmental Health and Safety at Philadelphia General Hospital. Philadelphia Dept. of Public Health. Printed and distributed by Economics Laboratory, Inc., 250 Park Ave., New York, NY 10017

RATINGS:	
U–Unsatisfactory	(0)
S–Satisfactory	(1)
N–Not Apply	

BUILDING NAME AND NO.	FLOOR & NAME	DATE	TIME

SUPERVISOR MAKING INSPECTION (Sign)

CUSTODIAN OR HOUSEKEEPING AIDE (Sign)

TOTAL SCORE

$$\frac{S's}{S's + U's} \times 100 = \quad \%$$

Fill in all parts of the form. If the item is "Unsatisfactory" show a "U" under the **RATING** and nothing under the **AREA CODE**. If an item is "Satisfactory" show an "S" under the **RATING** and the code under the **AREA CODE**. For example, if the floors are found dirty in the Nurses' Station, Laboratory, and Supply Room, put a single "U" under **RATING** on line #2 (Floors-dirty). Then list N.R., Lab #3, and S.R. Use "N" if the item is not applicable. Under **TOTAL SCORE**, count up all the U's plus the S's, then divide by the S's plus the U's to give you the percentage grade.

AREA CODES

W–Ward–Give#	U.T.–Utility Room
So–Solarium	T.R.–Treatment Room
P.R.–Patient Room–Give#	E.R.–Examination Room
N.R.–Nurses' Station–Give#	J.S.C.–Janitor Supply Closet
B.R.–Bathrooms	L.C.–Linen Closet
	S.S.C.–Sterile Supply Closet

S.A.–Storage Area	A.R.–Animal Rooms
W.R.–Wait. Rm. & Lobby	K–Kitchen
Of.–Offices–Give#	S.R.–Supply Room
Co.–Corridor	I.C.R.–Instr. Cleaning Rm.
Lab.–Laboratory–Give#	A.U.–Autoclave Room
L.R.–Locker Rooms	M.R.–Miscellaneous Rm.

L.D.R.–Labor & Del.–Give#	
O.R.–Operating Room–Give#	

FLOORS	RATING	AREA CODES
1. Dust		
2. Dirty		
3. Litter		
4. Spillage		
5. Stains		
6. Warning signs		
WALLS		
7. Dust		
8. Splattering		
9. Cobwebs		
10. Dirty		
WINDOW SILLS, VENTILATORS		
11. Dust		
12. Spots		
RADIATORS		
13. Dirty		

LAVATORIES	RATING	AREA CODES
32. Toilet Paper		
33. Wash Bowls		
34. Soap		
35. Paper Towels		
HOPPER ROOM		
36. Dirty		
JANITORIAL SUPPLY CLOSET		
37. Adequate supplies		
38. Clean		
39. Equip. stored properly		
PAILS		
40. Dirty		
MOP WATER		
41. Dirty		
MOP HEADS		
42. Dirty		

DOOR LEDGES & FRAMES, TOPS OF CABINETS				
14. Dirty				
WARD SINKS				
15. Wash bowl				
16. Wall				
17. Soap				
18. Towels				
WASTE RECEPTACLES				
19. Need emptying				
20. Dirty				
EQUIPMENT				
21. Dust				
FURNITURE				
22. Dust				
23. Spillage				
LAVATORIES				
24. Walls and windows				
25. Partitions				
26. Floors				
27. Tubs				
28. Showers				
29. Urinals				
30. Toilet Seats				
31. Toilet Bowls				

	43. Poor Condition			
REPAIRS				
44. Repair reported below				
INSECT AND RODENT CONTROL				
45. Mice				
46. Roaches				
47. Ants				
48. Flies				
LIGHTS				
49. Fixture dirty				
50. Bulbs need replacement				
WINDOWS AND SCREENS				
51. Dirty				
STAIRWAYS				
52. Litter				
53. Dust				
54. Spillage				
55. Needs repair				
56. Needs bulbs				
ELEVATORS (give number)				
57. Floor				
58. Walls				
59. Tracks				
60. Doors				

REPAIRS REPORTED	LOCATION
a. Wall needs repair	
b. Broken window	
c. Broken or missing window screens	
d. Ceiling needs repair	
e. Peeling paint	
f. Floor tile loose	
g. Floor tile broken	
h. Broken lighting fixtures or burnt out bulbs	

REPAIRS REPORTED	LOCATION
i. Stopped up floor drains	
j. Hopper broken or stopped up	
k. Wash basin faucets leaking	
l. Wash basin stopped up	

Figure 8.3. Housekeeping self-inspection form. Source: *Environmental Health and Safety* (Elmsford, NY: Pergamon Press, 1974), p. 52-53. With permission

control; and the use of easily cleanable materials for surfaces. Traffic patterns are established in such a way as to keep the majority of individuals away from crucial areas, such as the surgical suites, obstetrical areas, isolation and reverse isolation areas, and nurseries. Further, the emergency room and its traffic patterns must prohibit movement of individuals to other critical areas of the institution, unless the patients are to be admitted to the institution. Even at that point, the emergency room patients, visitors, and families should never enter the critical areas that have been identified.

Materials handling, which is of great concern within the institution, must be so organized to avoid the possibility of disease spread. Mechanical conveyors are designed for easy access for cleaning and inspecting to determine if insect and rodent problems may exist. Certainly where such materials as bedpans and their contents are moved, specific consideration must be given to avoid taking them into clean rooms and contaminating other areas and personnel. The ventilation system should be designed such that potentially pathogenic or even questionable organisms, dust, lint, respiratory droplets, and dirt will not be transmitted from dirty areas to clean areas. Laminar air flow is utilized in rooms where patients have or may have infections.

The materials used in construction should have good acoustical properties, durability, fire resistance, pleasant appearance, and be easily cleaned. Many other environmental controls are discussed in other sections of this book.

Control of infectious diseases for personnel and patients parallel each other. The reservoir of infection, that is, the individual who has the infection or the insect or rodent carrying the organism, the secondary reservoirs of infection, must be either treated, controlled, or removed from the presence of individuals who are susceptible to the infection. This is done using a surveillance program to determine the potential or actual carriers of disease; using medical aseptic precautions in the care of patients with communicable diseases, such as mumps or various staph infections; controlling secondary reservoirs of infection through good housekeeping, the removal of infected linens and solid waste, the destruction of rodents and various insects, the filtering and cleaning of ventilation systems, and the control of other fomites that may cause infections to spread throughout the institution.

Personnel are exposed to infectious disease, since they treat diseased patients or they are involved in either removing waste products or cleaning the environment. All personnel will be protected if adequate isolation techniques and procedures are utilized. Where masks are required, they must be clean, dry, and changed frequently. Masks are not an absolute barrier for droplets or aerosols. If gowns are donned and removed properly, and changed after leaving the patient's room, the opportunity for infection by organisms will be reduced. Hand washing is still probably one of the most effective techniques used for proper protection of personnel and patients. All staff members, including all housekeeping and ancillary personnel, must perform a medical scrub before and after they leave the room where a patient is isolated. Gloves should be used when indicated and should be disposed of properly to avoid contaminating personnel or patients. Equipment and supplies used in patient care should stay within the isolation room until time for disposal. At that time, all equipment that cannot be properly sterilized or adequately sanitized should be disposed of in special double-bagged sacks marked *isolation*. Linens coming from isolation areas should also be stored in double bags and clearly marked *isolation*. Employees can further protect themselves by reporting to the hospital infirmary as soon as they have a suspected infection of any type, including diarrhea, vomiting, skin disorders, and so forth. Personnel should receive vaccinations for specific diseases in those areas where they may be subjected to the diseases. They should also be tested frequently in highly critical areas such as tuberculosis areas.

Control of Infectious Waste

Infectious waste may be controlled by steam sterilization (autoclaving), incineration, thermal inactivation, gas-vapor sterilization, chemical disinfection, and sterilization by irradiation. Incineration and steam sterilization are the most frequently used techniques for the control of infectious waste. Steam sterilization utilizes saturated steam within a pressure vessel (known as a steam sterilizer) at temperatures that will kill the infectious agents present. The two types of steam sterilizers are: the gravity displacement type, in which the displaced air flows out the drain through a steam-activated exhaust valve; and the prevacuum type, in which a vacuum is pulled to remove the air before steam is introduced into the chamber. In both cases, the air is replaced with pressurized steam. The temperature of the treatment chamber continues to increase as the pressurized steam goes into the chamber. The treatment of the infectious waste by steam sterilization is based on time and temperature. The entire waste load has to be exposed to the proper temperature for a defined period of time. The decontamination of the waste occurs primarily because of the steam penetration. Heat conduction provides a secondary source of heat. The presence of residual air within the sterilizing chamber prevents effective sterilization, since the air will act as an insulator for the material. Three factors that can cause incomplete displacement of the air include use of heat-resistant plastic bags; use of deep containers; and improper loading of the chamber. The type of waste, the packaging and containers, and the volume of the waste load is important to know when establishing how best to treat the waste during steam sterilization. Infectious waste with low density, such as plastic, is better for steam sterilization than high-density waste, such as large body parts and large quantities of animal bedding and fluids. Steam sterilization can be used for plastic bags of low-density, metal pans, bottles, and flasks. The infectious waste also include other kinds of hazards, such as toxic material, radioactive material, or hazardous chemicals. Steam sterilization should not be utilized to treat these wastes. The individuals who operate the steam sterilization process should be highly skilled in proper techniques to minimize exposure of people to the hazards due to the waste. Protective equipment, minimization of aerosol formation, and prevention of spillage of waste during loading of the autoclave are important factors to be considered. A recording thermometer should be used to make sure that the proper temperature is maintained through an appropriate period of time during the cycle. All steam sterilizers need to be routinely inspected and serviced. Steam sterilizers used for infectious waste must be in an area away from the central sterile supply and should never be used for sterilization of equipment or instruments.

Incineration is a process that converts combustible material into noncombustible residue or ash. The product gases are vented to the atmosphere through the incinerator stack. The residue may be disposed of in a sanitary landfill under proper conditions. The advantage of incineration is that it reduces the mass of the waste by as much as 95%, and therefore substantially reduces transport and disposal costs from the site of the incinerator. Incineration is especially good when used for pathological waste and contaminated sharps. If the incinerators are properly designed, maintained, and operated, the microorganisms present are killed in an effective manner. If the incinerator is operating poorly, then microorganisms may be released into the environment. When incinerating waste, it is necessary to consider the variation in waste composition, the rate of feeding of the waste into the incinerator, and the combustion temperature. The amount of moisture contents and heating value of the waste affects its combustion. The rate at which the waste is fed into the incinerator affects the efficiency of the burn. The combustion temperature of 1600°F will allow for a complete destruction of the microorganisms in the infectious waste and will also allow for a considerable reduction of the material. The amount of air and fuel that is used has to be adjusted to maintain the combustion temperatures at the proper level, depending on the type of waste and volume. If the infectious waste has special hazards attached to it, such as the inclusion of antineoplastic drugs, then the waste should be incinerated only in special incinerators that provide high enough temperatures and a long enough time to completely destroy these compounds. The plastic content of the waste is also important, since many incinerators can be damaged by temperature surges caused by combustion of large quantities of plastic. Polyvinyl chloride and other chlorinated plastics are important, since one of the combustion products is hydrochloric acid, which is corrosive to the incinerator and may cause damage to the lining of the chamber and the stack.

Thermal inactivation includes treatment methods that utilize heat transfer to provide conditions that reduce the presence of infectious waste. The waste may be preheated by heat exchangers, or heat may be acquired by steam jacket. The amount of heat is predetermined, and the material is kept in a vessel for at least

24 hr. However, the temperature and holding time depends on the type of pathogens present in the waste. After the treatment cycle is complete, the contents of the tank are discharged.

Dry-heat treatment may be applied to solid infectious waste. The waste is heated in an oven that is operated by electricity. Since dry heat is less efficient than steam heat, it is necessary to have higher temperatures to accomplish the same type of kill. The typical cycle for dry-heat sterilization is treatment at 320–338°F for 2–4 hr.

Gas-vapor sterilization is an option that could be used for treating certain types of infectious waste. The two most commonly used chemicals are ethylene oxide and formaldehyde. However, there is considerable evidence that both of these chemicals are probable human carcinogens, and therefore, considerable caution must be taken if the materials are used as gas-vapor sterilizing agents. In fact, it is best not to use ethylene oxide gas because of its toxicity and the fact that other options are available for treating the infectious waste. If formaldehyde gas is used to sterilize certain disposables, such as HEPA filters from biological safety cabinets, then only trained persons should do the work.

Chemical treatment is appropriate for liquid waste, but also may be used in treating solid infectious waste. To use chemicals effectively, it is necessary to consider the following: type of microorganism; degree of contamination; amount of protein materials present; type of disinfectant; concentration and quantity of disinfectant; contact time; temperature; pH; mixing requirements; and the biology of the microorganism.

A new technology involving the use of ionizing radiation sterilization includes nominal electricity requirements; no steam; no residual heat in the treated waste; and the performance of the system. The disadvantages include the high capital cost; the requirement for highly trained operating and support personnel; a large space requirement; and the problem of the ultimate disposal of the decayed radiation source. The ionizing radiation must be properly monitored.

Biological indicators are used to monitor the treatment process when some form of heat is utilized. These indicators are used in steam sterilization, incineration, and thermal inactivation. Spores of a resistant strain may be used as an indicator. An instantaneous reading may be gained by using a chemically induced color change when a particular temperature has been reached. The chemical indicators are not good for the sterilization process, since the temperature is only part of the sterilization effort.

Control Measures for Transmission of HIV and HBV

Control measures for transmission of HIV and HBV for health care workers and public-safety workers include the general principles that have been discussed previously in the relationship to infectious body fluids and in the use of personal protective equipment. Precautions need to be taken with the individuals who are exposed to blood or body fluids through procedures related to CPR, IV insertion, trauma, and delivering babies.

Disposable gloves must be used by individuals in institutions, especially when coming in contact with patients where blood or body fluids are involved. This is particularly true in the emergency room. It is also true for fire and emergency medical service personnel. If public-safety workers are involved in any type of body searches, they ought to also wear disposable gloves. The disposable gloves should provide for dexterity, durability, and fit in the task being performed. Where large amounts of blood are likely to be encountered, the gloves must fit tightly at the wrist to prevent blood contamination of the hands around the cuff. Where multiple trauma victims are encountered, the gloves must be changed between patients. Contaminated gloves should be removed properly and placed in bags that are leakproof and can then be appropriately disposed of, as has been discussed earlier.

Masks, eye wear, and gowns should be utilized in institutions and in emergency vehicles as needed. The protective barriers will be necessary in order to protect the workers from diseases that are borne by blood and other body fluids. The masks and eyewear (safety glasses) should be worn where blood or body fluids will be splashed. Changes in gowns and/or clothing should be available if necessary.

Although HBV and HIV infection has not been shown to occur during mouth-to-mouth resuscitation, there still may be a risk of transmission of such diseases as *Herpes simplex* and *Neisseria meningitis* during mouth-to-mouth resuscitation. It is also a theoretical risk of HIV and HBV transmission during artificial ventilation of trauma victims. Disposable airway equipment for resuscitation bags should be used. Pocket mouth-to-mouth resuscitation masks designed to isolate emergency response personnel from contact with victims' blood and blood-contaminated saliva, as well as respiratory secretions and vomitus, should be utilized by emergency personnel.

Blood, saliva, and gingival fluid from all dental patients should be considered infectious during dental procedures. Gloves, masks, and protective eyewear should be worn during examination and treatment. Hand pieces should be sterilized after use with each patient. Blood and saliva should be cleaned from material used in the mouth and the material should be disposed of or sterilized. Dental equipment and surfaces that are difficult to disinfect, such as lights or x-ray unit heads should be wrapped in impervious-backed paper, aluminum foil, or clear plastic wrap.

Laboratories need to consider all blood and body fluids as infective and take the following precautions:

1. All samples must be put in a secure, non-leak container with a lid for transportation.
2. All technicians should wear gloves and masks as well as protective eyewear should mucous membrane contact occur.
3. Biological safety cabinets should be used if aerosols can be generated.

4. Mechanical pipeting should be used only.
5. Only use needles and syringes when needed.
6. Decontaminate laboratory work surfaces with appropriate chemical germicide.
7. Decontaminate or dispose of infective waste properly.
8. Decontaminate scientific equipment contaminated with blood.
9. Use medical scrub before eating and/or leaving lab.

Patient Control Measures

The techniques utilized in employee control measures plus proper environmental controls will be a major step toward proper patient infection control procedures. In addition, patients who are highly susceptible to infections should not be kept in the same areas with patients who have infections. If in doubt, remove the patient from the area. Patients should be taught good personal hygiene and body care to avoid recontamination by the discharged organisms. In hazardous situations, such as where blood-bank blood is used, special techniques must be employed to avoid contaminating the patient. Whenever special procedures are used, such as hemodialysis treatments, tracheotomies, intravenous cutdowns, or change of surgical dressings, careful aseptic techniques must be employed.

Managing Patients with a Resistant Organism

Not all patients with resistant organisms will show symptoms of such. Therefore all suspected patients should be evaluated to determine whether or not he/she is colonized or infected with a resistant organism. If so, the normal infection control procedures may not work. Controls should include:

1. Very strict handwashing (medical scrub with an antimicrobial agent).
2. Use of gloves when touching non-intact skin mucous membranes, moist body substances, or items contaminated by these items.
3. Proper glove removal, disposal, and handwashing between patients.
4. Wearing a mask when working within 3 feet of a patient who has a lower respiratory infection involving a resistant microorganism.
5. Wearing appropriate gowns and discarding same between patients.
6. Notifying a facility when a patient with a resistant organism(s) is being admitted.
7. Appropriate use of antimicrobial agents.
8. Giving additional infection control training to health care workers.

The Infection Control Committee

The infection control committee is a working committee of the institution. It is necessary that the committee consist of representatives of the medical staff, nursing staff, environmental health staff, infection control staff, and various members of the administrative staff. The function of the committee is to evaluate current and past levels of infection and to make necessary recommendations for changes in either medical-surgical techniques or other techniques used within the institution that may contribute to infections. They also act as the supervisors and recipients of all special studies related to infections that are conducted within the institution. Once they have evaluated the study, they forward it, along with their recommendations, to the executive director of the institution for implementation.

Occupational Health

The occupational health and safety program consists of program planning; establishing program objectives to reduce frequency and severity of occupational hazards, increase inspection coverage, charge supervisors with safety responsibilities, coordinate health and safety activities, plan and evaluate all safety activities, and plan and direct an adequate accident prevention program; safety and health training, education, and promotion; accident and illness investigation, reporting, and analysis; safety and health inspections; safety and health committees; safety and health engineering for hazard control; safety and health standards and regulations; fire prevention, protection, and suppression; motor vehicle safety; medical help and first aid; accident cost control; program design, application, and evaluation.

Health education is the process of transmitting knowledge and developing skills to produce safe practices. In order to accomplish this, the trainers must evaluate and the trainees must learn about hazardous conditions, safe and healthy practices, general rules of health and safety, special on-the-job problems, and job safety analysis.

The engineering approach is necessary in the initial layout and construction of the institution. Proper engineering will eliminate many of the hazards already mentioned and will provide not only a safe, but also an aesthetically comfortable environment.

All institutions should provide pre-employment and preplacement examinations. These examinations should not only determine the employee's medical history, but also determine if there are specific problems with specific types of environmental conditions. It is well to determine if the applicant can read and interpret written instructions and is able to react to posted instructions concerning safeguards. In addition, periodic medical examinations should be given to the employee to determine if he/she is currently suffering from any health problems and needs to be removed from his/her current environment and placed in a different type of job. An institutional occupational health program should include the diagnosis of occupational disease and injuries, the treatment of occupational disease and injuries, the treatment of nonoccupational illness and injuries, and the evaluation of the kinds of health problems, occurring over time, that may become serious health or safety problems. Special concern should be given to hearing, vision, the employment of pregnant women in special hazardous situations, the prevention and control of alcoholism, and the establishment and use of emergency control centers in the event of toxicity.

Biohazard Controls

Some of the measures employed to achieve maximum biohazard control include air-pressure differentials; filtration of supply air; filtration of exhaust air from laboratories, animal rooms, and so forth; air locks and pass-through autoclaves between clean and contaminated areas; change rooms; isolation and separation of utility systems between clean areas and biohazard areas.

Housekeeping

The selection of housekeeping personnel is important. Individuals should be selected for their desire to work, their intelligence, experience, age, physical makeup, character, and ability to read and understand housekeeping procedures.

Housekeeping training programs need specific formats. Individuals should be taught the various housekeeping techniques in the classrooms and in the institution. They should also be taught the basic fundamentals of the spread of infection and the seriousness of the use, storage, and handling of dangerous and hazardous materials and chemicals. There should also be instruction in isolation techniques from nursing personnel. By the completion of the course, new housekeeping employees should have a solid understanding of where they fit into the entire institutional process, what are their specific functions, how they should carry out the tasks, the materials and chemicals to be used, how they can prevent accidents, spread of infectious diseases, and promote better health and welfare for all. After completing a week of this intensive training, the individual should receive 6 to 9 months of probationary work under close supervision. He or she should take a refresher course of at least 1 day every 6 months.

Supervisors and administrators generally reach their positions because of the good work done at a technical level. It is important that the supervisors and administrators themselves receive training. They need to learn everything that their staff learns and, in addition, need a good understanding of how to best obtain an adequate quantity of quality work from the individuals working for them. It is one of the important functions of a supervisor and administrator to determine what the area looks like and how well the job is being done. This could be accomplished by using, on a weekly basis, the housekeeping self-inspection sheet found in Figure 8.3.

A good housekeeping program will integrate all housekeeping techniques with a self-inspection program, where the individual supervisor will have an opportunity to evaluate the actual work being carried out in a given area. The supervisors and administrators may also want to earn continuing education credits while learning about basic supervision and administration by taking corresponding courses written by the author. If so, contact Continuing Education Department, Extended Services, Alumni Center, Indiana State University, Terre Haute, IN 47809 or phone 1-800-234-1639, ext. 2522.

Financing and the Role of Government in Institutions

Nursing home care is paid for by a combination of funds from the individual, federal government, and private organizations. Nursing homes wishing to participate in federal government programs must adhere to the rules and regulations of the U.S. Department of Health and Human Services. Regulations exist in the areas of Medicaid and also Medicare. These encompass all types of nursing home activities, including environmental health activities. For example, standards have been established for the maintenance of sanitary conditions in the storage, preparation, and distribution of food. Unfortunately, out of a booklet of some 49 pages on specific requirements for nursing home facilities under the Medicaid program, only one quarter of one page was given to the environment, and this portion concentrated on food and wastes. All other areas of the environment seemed to be ignored.

SUMMARY

It is obvious that the previous discussions of topics within the institutional environment are, of necessity, limited in scope due to the nature of the material being presented. All these areas are evaluated by one or more public agencies, including the local and state health departments and the local and state welfare departments. In addition, various areas of the hospital environment are supervised and comply with the standards of the various professional societies that are specialists within the given area. Further, all hospitals must be accredited to operate. Accreditation comes from the Joint Commission on Accreditation of Hospitals, which evaluates physicians, facilities, techniques, and so forth. The Joint Commission, after duly studying a hospital facility, will make certain recommendations that must be carried out by the institution to maintain its accreditation. This accreditation is extremely important for the operation of the institution.

The institutional environment and its problems are so complex that it takes not only the environmental health specialist, but also a true team effort on the part of medical, nursing, administrative, and other personnel to properly operate the institution and to provide a better, safer and disease-free environment for all.

RESEARCH NEEDS

There is need for continued assessment of potential sources and transmission of institutionally acquired infections in individuals and staff. New techniques should be developed to rapidly assess the modes of the spread of infection. New accident-prevention techniques are needed. Studies to determine the impact of noise on relaxation and sleep are needed. Studies are also needed to determine the quantity and exposure to chemicals in institutions. These studies must include epidemiological, toxicological, monitoring, and laboratory analysis. The air levels of chemicals associated with housekeeping, laboratories, and other areas where solvents are used should be assessed.

Chapter 9

RECREATIONAL ENVIRONMENT

BACKGROUND AND STATUS

Recreation is a vital part of public health. Recreational activities provide necessary exercise for individuals to control weight problems, strengthen hearts, and provide an emotional release from the day-to-day cares faced in the home and at work. With increasing numbers of middle-class people, the development of the interstate highway system, and the overcrowding of many areas, individuals have sought to obtain their fun through recreational activities that take them into state parks, federal parks, various national forest reserves, and local park areas. Each of these areas, while providing the kind of necessary escape from tension and troubles, also provide an environment highly conducive to the spread of disease and injury. Individuals come from different areas, states, and even countries. They congregate in recreational areas and bring diseases from home. These diseases spread readily within the recreational areas because of existent environmental health problems. The public has increased sharply the number of visits it makes to various recreational areas. In 1950 approximately 110 million people visited state parks. In the 1980s over 500 million people visited state parks. The National Park Service in 1950 had approximately 25 million people visit their various park areas. In the 1980s over 125 million people visited national parks. The U.S. Forest Service had approximately 30 million people visit forest service areas in 1950. In the 1980s about 200 million people visited these areas. In the 1990s these numbers continued to grow even though there are now proposals to limit the number of people entering National Parks to keep from destroying the environment. Facilities are being torn down and areas are being made off-limits. These figures do not include the millions of individuals who visit local and county park and recreational areas.

Water-oriented recreation has also increased at an unusually rapid rate. In 1960, 19 million people bought 23 million state fishing licenses. In 1970, 31 million licenses were purchased. In 1970, it was estimated that there were nine million boats of various types in use. Since the end of World War II, there has been an enormous change in recreational activities in all categories. In 1970, it was estimated that 150,000 seasonal homes were built at a cost of 1.2 billion dollars. Since then, there has been an even sharper increase in the number of individuals who wish to build homes on any type of water body. It has been estimated by the Bureau of Outdoor Recreation that by the year 2000, the summertime participation in swimming will increase 207% above 1965 levels, 78% in fishing over 1965 levels, 215% in boating over 1965 levels, and 363% in water skiing over 1965 levels. There is also an anticipated increase in camping of 238%, in picnicking of 127%; and in sightseeing of 156%. It is obvious from this smattering of data that water recreational areas have become of enormous value to our society from both the aesthetic and financial points of view.

The American public make an estimated 7.8 billion 1-day visits each year to a variety of recreation areas, including national parks, parks of other federal, state, county, and local agencies, and private-sector facilities. This is an estimated increase of 40% over 1970. The tremendous number of visitors to these areas has intensified sharply the various pollution, land, and wildlife problems, and problems at historical sites. The recreational environment is becoming an increasing source of concern to public health officials, ecologists, and private groups.

SCIENTIFIC, TECHNOLOGICAL, AND GENERAL INFORMATION

The scientific and technological background varies with the specific type of recreational environmental problem being discussed. See the appropriate chapters within this textbook and Volume II for each of the problem areas.

PROBLEMS IN RECREATIONAL AREAS

Because of sheer numbers of people, lack of adequate facilities, and subclinical carriers of a variety of diseases readily spread through the environment, it is essential that there be a better understanding of recreational environmental health. An example of the kinds of hazards encountered include improper drainage and soil permeability, inadequate sewage facilities, inadequate kitchen equipment and improper cleaning practices, improper refrigeration of food, contaminated water supplies, unsuitable solid waste disposal, vector problems, swimming pool hazards, other accident hazards, and inadequate training and physical fitness of individuals visiting recreational areas.

Recreational areas have peculiar problems that generally are not found in the average city or community. They include seasonal operation, problems of public behavior, vector and animal problems, noxious plants and weeds, remote locations, and protection of wildlife from humans and humans from wildlife.

Seasonal operation is one of the most difficult problems faced in a recreational area. As a result of this, equipment is outdated or inadequate. This equipment includes chlorinators, warewashing machines, swimming pool filters, food facilities, sewage treatment, and water facilities. Further, it is difficult to get experienced personnel, since they are not anxious to work for a 3- to 4-month period during the year usually at minimum wage. As a result of this, many of the individuals are either college students with no specific training in the environmental areas or job hoppers who have difficulty keeping jobs and do not do an adequate job within the recreational area. The maintenance of equipment, even where good equipment exists, generally is poor, because the problems have to be discovered by the operators of the equipment and frequently are not, and because it takes time for maintenance personnel to come from some central area to make necessary corrections within the recreational area. Another difficulty caused by seasonal operation is that of inspectional procedures. Since the environmental health personnel have many other functions apart from the recreational areas, by the time the studies have been completed and corrections made, the season may be over. It is necessary then to start again the following year with new management and new personnel.

Public behavior is a difficulty because some visitors are highly irresponsible. Some people dump garbage outside of their cabins or trailers, leave solid waste scattered everywhere, and contribute to fly and mosquito problems. In addition, some visitors like to take souvenirs along to show where they spent their vacations. As a result of this, they cause damage that is expensive to repair. The money spent could have been used for improving environmental conditions.

In the recreational environment, the individual is exposed to animals, reptiles, and insects. Infection by rabid bats, ticks causing Rocky Mountain spotted fever, mosquitoes causing encephalitis, bees, and hornets, as well as poisonous snakes, become troublesome problems. In some areas, individuals have even been mauled by bears. The outdoor environment contains weeds causing hayfever and other allergic reactions, such as poison ivy and poison oak. Since in some areas electrical power is not available and roads are not accessible, it is difficult to obtain a proper water supply and to treat an injured person.

As has been mentioned, communicable disease is a very definite hazard, since the agent or reservoir of infection may be present or may have been brought into the environment. The vector or vehicle of disease is present in the form of insects, rodents, or inanimate objects. It is known that when large numbers of people gather together and associate closely, disease spreads rapidly. This, coupled with the lack of adequate environmental protection measures, contribute to the spread of communicable disease. When such a disease does occur, it should be immediately reported to the nearest health department and also to the park authorities.

Impact on Other Problems

The quality of the recreational experience is being eroded because of the vast number of individuals who are invading recreational areas. There are constant shortages of motels, campgrounds, stores, laundry facilities, gas stations, and space. The sewage systems and other utilities cannot service the increased numbers of people. Noise and physical damage has been caused by the use of light planes within the national parks. The alpine tundra close to access roads have been damaged by excessive use. Crime has increased sharply, and vandalism will eventually destroy our Indian cliff carvings and many of our natural wonders if it is not controlled. The national and state parks used to be free of pollution. This is no longer true. The automobile has brought air pollution to the park areas. Water pollution has increased sharply as a result of human interference with various watersheds, animal, and plant life. Congestion has become a part of the peak use periods of recreational sites. The increased numbers of people have created a transportation problem, since the existing systems cannot service them and there is a lack of adequate access to transportation. Problems have been created by haphazard private development of areas adjacent to recreation areas.

Economics

There is no real way currently of measuring the economic cost of misuse of the recreational environment. We do know that when conditions are poor, visitors will not return to the recreational areas, resulting in a tremendous loss of money to highly specific types of businesses. However, there is little understanding and information available concerning the impact of economics as a problem of the recreational area.

POTENTIAL FOR INTERVENTION

The potential for intervention in the variety of environmental problems varies with the given problem and with the number of individuals who literally invade a recreational area. Human impact on the recreational environment can only be reduced by either increasing facilities and constraints, which may have tremendous economic and/or political costs, or by reducing the number of visits that can be made to recreational areas. The full potential for intervention can only be understood when enough research is conducted to determine what can and cannot be done within this unique environment.

RESOURCES

The resources for each of the environmental health problem areas are identified by problem area in other parts of this book and in Volume II. See the appropriate problem area for the resources desired. In addition, many of the civic organizations, such as the Sierra Club, are concerned with recreational environmental problems. Governmental agencies include the Corps of Engineers, the National Park Service, the Department of the Interior, and the Tennessee Valley Authority.

STANDARDS, PRACTICES, AND TECHNIQUES

In planning for a recreational site, it is essential to gather and analyze all pertinent data concerning the land, the environmental systems present, the environmental controls needed, and the use of the land. This type of preplanning concerns not only the recreational agencies and the health agencies, but also the Department of Interior, the Department of Mines, and other agencies concerned with the health of individuals involved in the environment.

Although it is seldom possible to meet all desirable site criteria, a good many of these and, of course the most important ones, should be considered when developing a new campsite or altering a former recreational area. It is important that the area be well drained; gently sloping; free from topographical and other hindrances; accessible to sources of water supply, sewage disposal, and solid-waste disposal; away from heavy traffic and noises; conveniently located near major highways or helicopter landing strips to remove injured or seriously ill; away from swamps and marshes, where mosquitoes may breed and cause annoyance and disease; provided with adequate facilities for proper food storage and handling; and provided with adequate housing.

Water Quality

Water quality depends upon watershed management, water supply, water usage, sewage, and plumbing. Watershed management is concerned with the supervision, regulation, maintenance, and use of water in the drainage basin to produce the maximum amount of water of desirable quality and to avoid erosion, pollution, and floods. The condition of the soil, the use of the land, and the construction of such things as roads all affect watershed management. To protect water at its source and to ensure its proper usage, it is necessary to preserve undeveloped areas and to prevent abuse of the watershed. In other areas, the quality of the watershed can be restored by controlling these problems. Preservation of existing quality or improvement of existing quality in effect means control of adjacent areas involved in construction, logging, raising of livestock and wild game, mining, land development, and waste disposal.

The construction activities that affect the quality of water are roads, railroads, power transmission lines, mines, and dams. These activities increase the amount of silt or the quantity of chemicals, oil, gasoline, and solid waste. Soil erosion occurs when construction is not carefully preplanned and too many trees are cut down.

It is essential that logging operations be carried out in such a way that erosion will not occur. Products of logging activities, decayed vegetation, pesticides, oil, and gasoline must not enter the water sources. All areas where construction roads are built must be carefully controlled to avoid erosion and watershed contamination.

Erosion is also caused by overgrazing of recreational areas by livestock and game. Grazing should be supervised by the proper authorities and wildlife controlled by hunting-season programs.

Waste disposal of liquids and solids is a serious watershed problem. People who build cabins and do not install proper sewage systems cause sewage to seep into the watershed and watercourses. Garbage and solid waste are frequently carelessly discarded, becoming contaminants. Industrial operations must be controlled to avoid inadvertent contamination. If sanitary landfills are used, they should not leach into the watercourses.

Uncontrolled mining and processing of ores have destroyed many streams. The mines and the mine wastes not only spoil the land, but also drain into bodies of water. Abandoned mines must be protected to avoid seepage into nearby bodies of water. In addition, roads, strip-mining operations, and stockpiling of various chemicals used in processing contribute to the water problem. Mining may cause the streams to be clogged with silt, colored matter, minerals, acids, and chemicals. Abandoned strip mines should be leveled and covered with the original soil or new top soil, and then seeded with an approved vegetative cover.

Pesticides, as explained in a previous chapter, are toxic to humans and animals. Depending on the pesticide and its use, it may become a contaminant of the watershed. It is important that the persistence and concentration of the pesticide, as well as its lethal effects, be thoroughly understood before it is used. Usage of pesticides in recreational areas is necessary, but must be handled carefully to prevent contamination of the watershed areas. Where recreational areas exist in watersheds, it is important that careful environmental controls be followed to avoid human contamination.

Fires either caused by carelessness, lightning, cooking, or bonfires must be carefully controlled to avoid damage to the forested areas and the watersheds. The greater the quantity of trees, the better the chances of avoiding erosion and maintaining the watershed. Conservation activities are carried out by a variety of agencies, including various park services, fire services, corp of engineers, and bureaus of land management.

Water Supply

There must be an adequate supply of pressurized water for drinking and other household purposes. The water supply must meet the bacteriological, chemical, physical, and radiological requirements of the "National Primary Drinking Water

Regulations" of 1989 as updated through 1991. These standards stipulate that the water supply system must include all the collection, treatment, storage, and distribution equipment needed for a proper, safe water supply. For firefighting purposes, water may be utilized from streams or creeks. To meet the requirements of proper water supplies, water should be taken from wells, springs, or infiltration galleries that are not subject to contamination. The water should be examined regularly to determine the level of safety. When water comes from underground or surface areas and it has been determined that low degrees of contamination are present, the water must be chlorinated. Where water contains impurities that require coagulation, sedimentation, and filtration for removal, the water should receive complete treatment. In inaccessible areas, water must be hauled to the places of use in tank trucks. In this case, the water must come from a safe source and the following precautions taken to prevent contamination:

1. The tank interior should be devoid of all defects. It must be thoroughly cleaned and disinfected before use.
2. The tank lining must be nontoxic.
3. The tank must be used for no other purposes.
4. Chlorine should be added to the water in the tank to ensure that a residual of at least 0.4 ppm of free chlorine is available after the chlorine has been in contact with the water for at least 20 min.
5. Only one hose that is utilized strictly for filling and unloading the tank should be available. The hose should be stored in a covered container in the truck when not in use and then thoroughly flushed with clean water before each use.

Further discussion of water supply and water treatment is found in the chapter on water in Volume II.

In the event of emergencies, water suspected of being contaminated with bacteria may be used if any of the following procedures are followed:

1. Boil vigorously for 10 min.
2. Add five drops of 2% U.S.P. tincture of iodine to each quart of cold water and allow to stand for 30 min.
3. Add iodine tablets, one tablet for each quart of water and allow to stand for 30 min.
4. Add chlorine bleach at two drops to a quart of water and allow to stand for 30 min.
5. Add chlorine tablets at one tablet per quart of water.

It is essential that drinking fountains be constructed so that the individual's mouth does not come into direct contact with the water fountain itself.

Sewage

Safe disposal of human and domestic waste is essential in recreational areas. Improper sewage disposal contaminates the water, thereby ruining the water for its many recreational and drinking purposes. Sewage treatment systems must have adequate capacity and must meet all of the necessary requirements set forth in federal and state water pollution control laws. Water carriage systems must be kept in good repair to prevent leakage into the underground water supplies. In many areas, water carriage systems are available, but sewage treatment systems are not. Septic tanks and subsurface disposal systems are utilized. In these cases, it is necessary to keep the on-site sewage disposal system away from wells, streams, or other sources of water supply. It is important that they do not overflow into the creeks or other water areas draining the campsite. The septic tanks and systems must not be permitted to flow downhill into water sources. (Additional discussions on this subject will be found in the chapter entitled On-Site and Public Sewage Disposal in Volume II.) Nonwater carriage sewage disposal facilities are also necessary in recreational areas. These include such things as chemical and burn-out toilets and pit privies. Again it is essential that these units be so constructed that they do not contaminate the water supply nor become a source for insect breeding. It is important that privies be correctly built and be properly maintained. (Further information on this area will also be found in the chapter entitled On-Site and Public Sewage Disposal in Volume II.)

Plumbing

Plumbing includes all of the practices, materials, and fixtures used in installing, maintaining, extending, and altering any pipes, fixtures, appliances, and appurtenances. (Plumbing fixtures and materials, as well as an additional discussion on plumbing, will be found in the chapter on plumbing in Volume II.) It is known that back-siphonage and cross-connections cause contamination of potable water supplies. Plumbing systems should be installed carefully by registered plumbers who adhere to the minimum requirements of the National Plumbing Code. The systems should be checked carefully to ensure their proper functioning. Air gaps, non-pressure-type vacuum breakers, or back-flow preventors are utilized to prevent problems from occurring. Cross-connections are most prevalent in the following recreational situations: direct connection of the water supply to cooling or condensor systems in refrigerators or air-conditioning systems; fish ponds with submerged inlets; fire hydrants with drains connected to sewers; frostproof water closets; kitchen and laundry fixtures with common waste and supply lines; underground water sprinkling systems that do not have vacuum breakers; and ice-cube machine drainage lines connected to the sewer lines.

Comfort stations are mobile flush toilets used for the public. They are frequently used in recreational areas. It is necessary that comfort stations be constructed in such a way that any materials removed from them will not contaminate water supplies or ground areas.

Shelter

Shelter consists of the site locations, structures to be utilized, and various campgrounds, picnic areas, and facilities.

Site Selection

The site selected must be away from swamps and mosquitoes, and must be off the main thoroughfares to reduce potential accident hazards. The access roads should be constructed with turns and loops so that vehicles have to reduce their speed. If feasible, one-way traffic patterns should be utilized. The roadways should be of all-weather construction where possible. The sites should be removed from sources of noise, air pollution, odors, or other kinds of contaminants. Recreational areas should be located on gently rolling well-drained land that is not subject to flooding and natural hazards. If possible the camps should have an eastern exposure for the campers to receive the early morning sun and afternoon shade. Lightly wooded areas provide additional shade and help make the area more comfortable. If public utilities are available, they should be utilized.

Structures

The construction of buildings is important from a public-health viewpoint, since people spend many hours within them. All units should meet the minimum requirements of the proposed housing ordinance of the American Public Health Association. The following items are important for good and proper housing:

1. A habitable room should have at least one window or skylight facing outdoors.
2. Window or skylight should be readily opened.
3. Every bathroom or water closet should have adequate quantities of hot and cold running water.
4. Heating facilities should be available in those areas where temperatures are low at night. It is essential that the heating facility be well constructed, well engineered, and well ventilated.
5. All hallways and stairways should be properly lit.
6. Mosquito screening should be supplied in all housing units.
7. Foundations, floors, walls, ceilings, and roofs should be weathertight, watertight, and rodent-tight.
8. Plumbing fixtures and water and waste pipes should be properly installed and in good working order.
9. There should be at least 150 ft² of floor space for the first occupant and 100 ft² of floor space for every additional occupant.
10. Cellars should not be used as rooms.
11. The buildings should be provided with or have easy access to adequate water supplies and sewage disposal.
12. Fire protection devices should be provided.

Additional material on housing will be found in the chapter on the indoor environment.

Trailers

Trailer parking facilities must be so arranged that adequate water and sewage connections, and adequate shower and sanitary stations, are provided. The sites must be well drained and free of obstructions. Further information on trailers is provided in the chapter on the indoor environment.

Campgrounds and Picnic Areas

Campgrounds and picnic areas are part of the area of shelter. These need to be selected in such a way that they are well drained, free of heavy undergrowth, and on solid and gently sloping ground. The layout should be such that cars have to travel slowly. Individuals should be able to walk comfortably to all areas of recreation, water supply, and sewage disposal and solid-waste disposal. There should be adequate supervision within the park to maintain these areas in good working condition.

Food

The handling of food tends to be less stringent and the facilities less modern and well maintained than in other types of situation. In a recreational setting, food may either be prepared for one individual, a family, several families, or a very large camp. It is essential that all of the rules, regulations, codes and good practices associated with food service be followed in the camp situation. At times it is difficult to refrigerate in this setting, however, since proper refrigeration of food is so essential to avoid food-borne disease, this is an absolute must.

Food Service

To conduct an effective food service program in recreation areas, whether it be for a family or for a large group, it is necessary to start with good clean products, keep them well refrigerated, prepare them under proper conditions, and utilize them within adequate time spans. The whole area of food protection will be discussed in detail in the chapter "Food Protection". However, for the recreational area it is well to remember some of the most frequent food problems. Food must be prepared with clean utensils, pots and pans of nontoxic materials. It is essential to maintain proper cleanliness. Handwashing is a must for the people who prepare and serve the food. Food that is left open should be disposed of rather than saved for other meals. Food should be served either on single-service plates, which are then thrown away, or on dishes and plates that are thoroughly washed and then sanitized. Solid waste from food should be stored in tightly closed containers to prevent breeding of flies and rodent problems. Since vending machines are used extensively in recreational areas, particular care should be given to these units. Vending machines must comply with the requirements of the vending of food and beverages codes. If foods are refrigerated, the temperatures must be adequately low. The machines should be cleaned carefully on a daily basis.

Milk

Milk, milk products, and frozen desserts must come from certified sources. Milk must be stored below 41°F at all times. Once the milk has been poured into pitchers, it is essential that it be utilized completely. It should not leave the original container and then be kept in a refrigerator for reuse. All milk, milk products, and frozen desserts should be inspected by the various county or state health departments.

Solid Waste

Solid waste consists of materials and entrails from fish cleaning, stables, eating, and life-associated processes. Solid-waste handling is important in the prevention or reduction of flies, other insects, and rodents. People have a tendency, especially in recreation areas, to throw their solid waste on the ground or dispose of it in careless ways. It causes a hazard and a nuisance and may lead to fires, odors, and unsightliness.

Fish Handling

Fish cleaning should take place in screen-enclosed facilities with nonporous floors. The washings from the fish should go into trapped floor drains. The traps should be cleaned thoroughly and the floor drains flushed out. It is important that the fish cleaning is done on impervious, nonabsorbent sloping surfaces to ensure proper cleaning. Pressurized water is necessary for final cleanup. All the materials — entrails, fish heads, and so forth — should be put into garbage-grinding units or plastic bags that are fastened and then enclosed in metal cans with tight-fitting lids. The fish materials should be removed daily to an incinerator or a properly operated landfill. These materials should be buried as soon as they are deposited within the landfill. Fish-cleaning facilities must be scrubbed down very carefully to avoid attracting insects and rodents and becoming an odor nuisance.

Stables

Since horseback riding is popular in recreational areas, and since cattle may also be found in certain recreational areas, it is important that the stables are properly maintained. Manure must be collected on a daily basis and put into concrete manure storage bins. These bins must be tightly closed to prevent fly breeding and odors. A second technique is to have the stables located on well-drained, gently sloping sites where the floors are made of concrete. The floors in the horse stalls should be paved with wooden blocks and sealed with asphalt or another impervious material. The rooms should be hosed down and the drainage should either go into a special holding tank to be pumped out later and removed to sewage treatment systems or should be put into waste stabilization ponds away from the recreational areas. It is essential that all areas of the stables be kept extremely clean at all times and effective insect and rodent control be practiced. Stables should be screened and rodent-proofed.

Other Solid Wastes

The volume of garbage and other solid waste produced depends on the geographic location, the season, the number of people, the kinds of facilities, and so forth. It is essential that all solid waste be stored in watertight, rust-resistant, nonabsorbent, durable containers, which are covered with tight-fitting lids. The containers should be lined with plastic bags. The garbage, if not ground up in a garbage disposal unit, should be stored in the same way as other solid waste. It is important that the solid waste at campsites be removed on a daily basis to prevent rodents and other scavenger animals from knocking over the containers and scattering the solid waste. The material should be removed to either an incinerator or sanitary landfill. Further information concerning solid waste disposal can be found in the chapter on solid and hazardous wastes in Volume II.

Swimming and Boating

Swimming takes place in artificial lakes, natural lakes, and swimming pools. An in-depth discussion on all types of swimming pools can be found in a separate chapter on swimming pools in Volume II.

Swimming Pools

Swimming pools in recreational areas must be designed and constructed in such a way as to prevent accidents and the inflow of sewage or other types of contaminants. The water supply must be pure and feed into the pool through a system where an air gap exists. The sewer system must be separate from the swimming pool and in no way connected with it. Dressing rooms, toilets, and showers should be provided for the individuals using the pool. It is essential that all pools have lifeguards and that children or adults not be permitted to swim unless adequately trained lifeguards are present. It is also essential that the swimming pool water maintain a pH of 7.2–7.8 with a chlorine residual of 1 ppm. All pools should be clear at the deepest point and properly filtered. The types of records necessary, water quality, and so forth will be discussed in the chapter on swimming pools. Outdoor bathing places, such as lakes, streams, rivers, and tidal waters, must be carefully checked before they are used for swimming. It is quite possible that the coliform aerogens group of bacteria from human feces may be present in the water. It is also possible that many organisms are present from runoff that comes from animal feces and from agricultural fields. It is important in outside bathing areas that all potential sources of harmful pollution are identified and these sources blocked or the bathing areas closed. These sources include sewage from boats, outlets from dwellings and other establishments, public sewer systems, leakage from improper dumping of solid waste or landfills, and the runoff from land. In artificial areas, it is also essential to recognize that the bottom of the swimming area may be muddy, and therefore it may be difficult to view or it may have stepoffs. Proper safety is an integral part of swimming area programs.

Boating Areas

Boating has become one of the largest industries in the recreational field. There are more than 10 million pleasure boats used for recreation in the U.S. Many of these boats have galleys and toilet facilities. It is necessary to ensure that debris and garbage from the galley and the human fecal material from the various toilet facilities does not end up in the lake areas. These materials should be taken back to the marinas and emptied into holding tanks, where they are removed to sewage disposal plants for proper disposal. Other problems created by boats include waste oil, fuel, and any other refuse material that may be purposely or accidentally discharged into the waters. In addition to the environmental problems involved in disposal, additional environmental problems occur in the obtainment of adequate quantities of safe water. It is important that the tanks holding pure water are not cross-connected to sewage systems within the boat. These tanks must be properly washed, rinsed, and chlorinated. Water should be obtained from safe drinking sources. Food must be stored aboard the boats in refrigerated lockers until used. In addition to the problems created by motorboats or boats driven by gasoline, there are many sailboats and rowboats on the waters. All boat users must have a complete understanding of boating safety requirements and know what to do in the event of an emergency. It is essential that all water areas be clearly marked for hazardous situations.

MODES OF SURVEILLANCE AND EVALUATION

See the appropriate chapters for surveillance and evaluation by environmental health area.

CONTROLS

Recreational officials and agencies must work closely with public health agencies in an attempt to reduce potential and existing hazards. Individuals traveling into remote areas should have proper training in camping and hiking to prevent additional health and safety hazards. (See appropriate chapters for given environmental health areas in Volumes I and II.)

Safety

Recreational safety encompasses all forms of safety problems including problems related to site selection, types of buildings and equipment, fire and electrical wiring, heating systems, structural hazards, campgrounds, playgrounds, food service, swimming pools, bathing areas, and refuse disposal. In addition to these hazards found in all phases of life, it is essential to realize that the individuals who go out into the recreational environment are out there for fun. Many of them are not equipped to operate efficiently, effectively, and safely within this environment. They take risks, they do things out of lack of understanding, and they lack proper equipment. Fire safety is a special concern, since many buildings are made of wood and it is difficult to get proper firefighting equipment into these areas. Individuals hike and climb without understanding proper techniques and get into trouble, resulting in lost lives or injuries. Many facilities in recreational areas have the kinds of structural hazards and poor electrical wiring that would never be tolerated within the home. The individual is on vacation and is unconcerned about it. However, it is just these types of hazards that lead to many serious accidents. Further, within the recreational environment, people hunt and many hunters get killed each year. People drown because they do not understand water safety.

SUMMARY

To summarize the problems of the recreational environment and to properly introduce the kinds of programs that are necessary to resolve these problems, it is important that the individual read an entire book on the basic principles of environmental health and apply each section as it fits into a segment of the recreational environment. The difference between the recreational environment and other environments is that the former is more dangerous.

Humans have made increasing use of the environment for recreational purposes. These recreational uses by large groups of transient individuals cause destruction to the environment and intensify the potential for the spread of disease and injury. Limitations in the number of individuals using a given area, along with improvement of facilities and good educational approaches, are needed to enhance the recreational environment for humans and to protect it for succeeding generations.

RESEARCH NEEDS

See appropriate chapters by area of concern for research needs.

Chapter 10

OCCUPATIONAL ENVIRONMENT

BACKGROUND AND STATUS

The occupational environment can be simply defined as any place, indoors or outdoors, where people work in return for financial or other remuneration. The prevention of occupationally related illnesses and injuries among workers is an important issue. Accordingly, occupational health and safety principles and practices are emphasized to identify, evaluate, and control sources and causes of hazards in the workplace environment and, in turn, to decrease risk of illness and injury to workers.

The hazards include chemical, physical, biological, ergonomic, mechanical, electrical, and psychological agents or factors that can potentially cause illness or injury to workers. The chemical stresses include toxic, flammable, reactive, or corrosive solids, liquids, and gases. These materials may become airborne in the form of dusts, fumes, mists, fibers, smoke, gases, and vapors. The physical hazards include ionizing radiation, nonionizing radiation, noise, vibration, and extreme temperatures and pressures. Biological hazards include insects, animals, bacteria, viruses, protozoans, fungi, and helminths. Ergonomic hazards include unusual body positions, repetitive motion, and poor equipment and workplace design. Mechanical and electrical hazards include instrumentation and/or conditions that may cause accidents and related injuries among workers. Psychological factors focus on emotional stressors and pressure encountered by workers, fatigue, monotony, and boredom.

Historical Perspective

Historically to the present, health and safety hazards have been associated with certain occupations. Effective action to protect the workers from related illnesses and injuries has ranged from being either totally lacking or inadequate to extremely comprehensive in scope.

In the first century A.D., the hazard of working with sulfur and zinc was described and protective masks were used by workers. The Industrial Revolution in England and the resultant increased exposure to toxic materials and dangerous occupations were noted, but little was done to improve the industrial environment. Conditions continued to deteriorate as the push for industrial expansion increased sharply. In the U.S. prior to 1911 an employee had to sue his/her employer to collect damages for illness or injury due to his/her job. In 1911, the first state compensation laws were passed. More recently, workers have received increased protection through the following federal acts: Walsh-Healey Public Contracts Acts of 1936; Maritime Safety Amendments, Public Law 85–742 of 1958; Construction Safety Amendment, Public Law 91–54 of 1969; and other acts and administrative orders.

The Williams-Steiger Occupational Safety and Health (OSHA) Act, the most comprehensive and far-reaching federal legislation, was passed in 1970. The basic law was amended in 1974, 1978, 1982, and 1984. This legislation covers about 75% of the civilian labor force. Further special provisions of the act pertain to the 2.7 million federal government civilian employees.

The OSHA Act is a general law providing a broad spectrum or powers for use by the Secretary of Labor to reduce exposure to hazardous conditions in the occupational environment. The secretary has the authority to promulgate, modify, and revoke safety and health standards, make inspections, issue orders, and invoke penalties, require employers to keep adequate safety and health records, ask the courts to stop potentially dangerous situations, and approve or reject various state plans for OSHA programs under the act. The Secretary of Labor works with the Secretary of Health and other agencies in a variety of ways. The Department of Labor conducts short-term training sessions for their personnel and for personnel from various states.

Under the OSHA Act, the Department of Health and Human Services (HHS) has primary responsibility for conducting health and safety research, evaluating hazards, determining and listing the toxicity of various substances, and workforce development and training. The National Institute of Occupational Safety and Health (NIOSH) is the unit that is responsible for related research and education and training programs. The data from research help develop or modify criteria used for recommending health or safety standards. Surveillance and technical service programs, including hazard evaluation, also are conducted by NIOSH.

Over 117 million men and women are employed in the U.S. All of these individuals are exposed to occupational hazards to some degree. Although OSHA requires that occupational illnesses be reported, reporting has been poor. One reason is the inability and/or failure to identify and correlate potential occupational causes with adverse effects or illnesses. One hundred and twenty-five thousand new cases of occupational disease occur each year according to NIOSH, but the figure may be an underestimate.

Economic Impact

The cost of occupationally related disease and accidents not only must include the loss of dollars due to death, disease, or injury, but also the insurance, effect on society, and the cost of loss of financial stability for the families involved. An additional cost involves the economic impact of OSHA standards and the corrections that must be made in equipment, facilities, and workers' procedures. OSHA must consider the technical feasibility and economic cost of each of its major proposed standards. The cost of implementation can be determined with some accuracy, but the resulting benefits cannot always be estimated. Overall, the loss of workers' time and production, as well as the securing of new equipment and facilities, has cost billions of dollars. Additionally, the cost of operating government is extremely high. It has also been estimated by the commission on federal paperwork that the government generates enough documents each year to fill 51 major-league baseball stadiums.

Occupational Health and Safety Professionals

Although many occupational health and safety principles and practices overlap, there is a distinction between the two areas. Whereas occupational health focuses mainly on illness prevention, safety emphasizes injury prevention. Occupational health professionals are traditionally called "industrial hygienists" and occupational safety professionals are commonly, but not exclusively, called "safety engineers". The traditional industrial hygiene tetrad of anticipation, recognition, evaluation, and control of hazardous agents and factors that can cause ilness and injury among workers is applicable to both occupational health and safety. Indeed, ideally both occupational health and safety issues must be simultaneously addressed to efficiently and effectively minimize risk of illness and injury among workers.

INTERRELATED PHASES OF OCCUPATIONAL HEALTH AND SAFETY

Anticipation and Recognition

Anticipation refers to a proactive estimation of health and safety concerns that are commonly, or at least potentially, associated with a given occupational setting or operation. Recognition closely parallels anticipation and involves actual identification of potential and actual hazards. Indeed, both phases can commence prior to entering a facility or site and continue throughout investigations, evaluations, and general operations as additional data are compiled.

Investigation and characterization involve gathering qualitative and quantitative data regarding a given facility, site, and/or operation. The qualitative and quantitative data include combinations of information, such as:

1. The identification and hazardous characteristics of physical, chemical, and biologic agents and ergonomic, mechanical, and psychological factors
2. The types of unit processes and standard operating procedures
3. Number of operating personnel actually and potentially exposed
4. Proximity and distance of sources of exposure relative to workers, including layout of a facility or site
5. Review of records and documented health and safety programs.

Knowledge of several unit processes in occupational settings ranging from office environments to the traditionally more hazardous manufacturing settings can provide insight about potential exposures. As much information as feasible is gathered during the anticipation and recognition phases and is reviewed and evaluated in order to design strategy necessary for preliminary on-premise survey and evaluation. In addition, the information is useful in developing preliminary health and safety plans to protect occupational health and safety personnel and other workers during specific operations and/or at a given location.

Evaluation

Anticipation and recognition continue during the evaluation phase, which involves a preliminary survey, including visual and instrumental monitoring, of a facility or site. Visual monitoring refers to the observation of processes and/or conditions at the facility or site and, subjectively and objectively identifying potential and actual hazards. Instrumental monitoring involves the use of monitoring equipment to record qualitative and quantitative environmental surveillance data. Chapter 11 provides a summarry of major instrumentation used for identifying, detecting, and measuring various physical, chemical, and biological agents in occupational and non-occupational environments.

Common instrumental monitoring focuses on atmospheric screening for combustible gases and vapors, molecular oxygen, toxic inorganic and organic gases and vapors, toxic inorganic and organic particulates, ionizing radiation, illumination, and sound levels. Meteorological data, such as temperature, relative humidity, barometric pressure, wind speed, and wind direction, also are commonly collected. Mechanical and electrical hazards that may contribute to worker injury also are identified and evaluated. In addition, evaluation of control measures, including personal protective equipment and ventilation systems, is conducted to determine if the controls effectively reduce the potential for worker exposure. Occupational health and safety professionals also may be involved with the ergonomic evaluation of workers to determine if there is an efficient match between them and their physical workplace environment. Relatively recently, psychological factors, including stress encountered by workers, has received increased attention. Indeed, psychological issues are commonly addressed as behavioral aspects of ergonomics.

Qualitative and quantitative environmental surveillance data are augmented by medical surveillance data. Environmental surveillance can provide some data regarding external exposures and/or workplace conditions. Medical surveillance can possibly provide data regarding the impact or adverse effects due to external exposures to various agents or factors. The combination of environmental and medical surveillance data increases the probability of minimizing additional external and internal exposures, as well as, establishing a correlation between cause and effect. The comprehensive environmental surveillance data, plus related medical surveillance data, are used to detect and measure hazardous agents and factors encountered by workers and to design strategy for control and remediation.

Data derived from evaluation are typically summarized in a technical report. The reports are important for documenting exposures and provide insight for future exposure prevention and remediation. The reporting of occupational illnesses, accidents, and incidents on appropriate forms to the safety supervisor, industrial hygienist, or other environmental health professional also is extremely important. Reporting of monitoring data and injury and illness statistics also is necessary to comply with various federal and state laws and to obtain worker's compensation, where necessary, for the victim of an accident or disease.

Control

Measures for controlling exposures to workers engaged in hazardous operations and activities involve implementation of administrative, engineering, and personal protective equipment controls. The implementation of administrative controls, such as development of a health and safety plan and standard operating procedures, should be a primary focus. Indeed, even if engineering and personal protective equipment controls are utilized, their full protective benefit is often compromised due to poor work practices. Engineering controls include appropriate process design, automation, guarding, shielding, grounding, and ventilation. When engineering controls are not adequate or feasible, however, personal protective equipment is frequently warranted. Most common personal protective equipment provides barriers between a worker and contaminated surroundings and sharp or heavy objects. A variety of equipment is available, including respirators, clothing and gloves, ear plugs and muffs, glasses, goggles, shields, and rigid hard hats and shoes.

SCIENTIFIC, TECHNOLOGICAL, AND GENERAL INFORMATION

Overview

Occupational exposure limits have been established and recommended for many of the physical, chemical, and biological agents to limit the degree and duration of worker exposure. Monitoring and analysis are conducted to determine the levels of various physical, chemical, and biological agents present in the workplace, especially within the air matrix. Guidelines also have been set relative to ergonomic, electrical, and mechnical factors other than exposure to airborne physical, chemical, and biological agents that can cause adverse impact to human health.

The general characterisitics and/or subcategroies of the numerous agents and factors present in the occupational environment that can pose health and safety problems to workers that are exposed to elevated levels and/or unacceptable conditions were outlined above. A more detailed summary of three major categories of concern, however, namely physical, chemical, and biological environmental agents and factors, follows.

Physical Agents

Physical agents include energy sources such as sound and vibration, nonionizing and ionizing radiation, and extremes of heat and cold stress. The hazard associated with these agents is typically due to the impact of a form of physical energy interacting with the human system.

Sound

Sound waves are disturbances in the uniform pressure of air molecules, detectable by the human ear, caused by vibrating objects. As objects vibrate back and forth, they alternately cause compressions and rarefractions (partial vacuums) of molecules in the air. Thus, sound is an energy form of vibration which may be conducted through various media including solids, liquids, or gases. The velocity of sound wave travel varies depending upon the mass (density of the medium) and elastic reactions (pressure) of the molecules.

The amplitude or intensity of sound is related to sound pressure. Sound pressure is the deviation of air pressure from normal atmospheric pressure and is related to the amplitude of sound. The best value to use to determine sound pressure amplifications would be average pressure changes. One problem would be created if average pressure were used; average compression and rarefaction pressure change always equal zero or atmospheric pressure. In place of a simple average, a root mean square (RMS) of instantaneous pressures is calculated. RMS sound pressures are measured in dynes per square centimeter, newtons per square meter, microbars, and more commonly, micropascals. Sound power is another measure associated with the amplitude of a sound wave. It refers to the total sound energy radiated by the source. Sound power is typically expressed in the unit watts.

Sound wave amplitude may be defined either as the quantity of sound produced at a given location away from the source, or in terms of its overall ability to emit sound. It is the amount of pressure change from the average atmospheric pressure, which is generally described by sound pressure or sound intensity. The louder the noise, the higher the intensity. Loudness is a term that refers to an observer's perception or impression of loud versus soft sounds.

Associated with the measurement of sound pressure and sound power is the logarithmic unit decibel (dB). By definition, the dB is a dimensionless logarithmic unit relating a measured quantity such as sound pressure in micropascals ($\mu°$) or sound power in watts (W) to a reference quantity. Decibels are commonly used to describe levels of acoustic intensity, acoustic power, and hearing thresholds. Sound pressure levels are measured in decibels referenced to 20 $\mu°$. The reference value corresponds to the lowest sound pressure that will produce an audible sound that is detectable by humans.

$$SPL = 20 \log \frac{P_{meas}}{P_{ref}}$$

$$= 20 \log \frac{P_{meas}}{20 \, \mu P} \qquad \text{(Eq. 10.1)}$$

In addition to amplitude or intensity of sound waves, three other important characteristics of sound waves that are useful in understanding and evaluating sound in the occupational environment are frequency, wavelength, and speed (velocity). Frequency of sound waves is an important attribute that must be determined. Frequency is the number of complete vibrations within a given period of time. It is the total number of completed cycles (compressions and rarefactions) of a sound wave per second (cycles/sec) and is measured in units of hertz (Hz). The A-scale refers to a frequency range of 20 to 20,000 Hz and is considered the frequency range for human hearing. Accordingly, sound measurements conducted for human exposure assessment are reported as decibels A-scale or simply dBA. Wavelength refers to the distance required for one complete pressure cycle to be completed. This attribute is important when considering the effects of sound waves. For example, sound waves that are larger than surrounding objects will diffract or bend around these obstacles. Small sound waves, on the other hand, will refract or be dispersed and surrounding obstacles can function as effective sound barriers. Thus, wavelength is an important variable in noise control activities. The speed of sound or sound wave velocity is dependent upon the ambient temperature of the medium and its environment and at about 22°C is approximately 344 m/sec. The speed (c) of a sound wave is determined by the product of its wavelength ($°$) and frequency (f):

$$c = f \lambda$$
$$344 \text{ m/sec} = f \lambda \qquad \text{(Eq. 10.2)}$$

Sound waves propagate in a spherical form from the source of generation. If sound measurements are taken at the same distance in any direction from a source under free-field conditions, the sound pressure levels should be the same. Free-field refers to a location that is an open area, free of barriers or interferences from ceilings, floors, walls, adjacent equipment, and a wide variety of other items that can diffract, absorb, or transmit sound energy and either increase or decrease sound pressure levels. Under free-field conditions, the longer the distance from the source the lower the sound pressure. Sound pressure changes abide by the inverse square law which states that as the distance increases from the source, there is a corresponding decrease in intensity based on sound level meter readings in decibels. If we are monitoring free-field sound, ideally there should be a 6 dB reduction when the distance doubles. For example, if sound pressure readings are equal to 90 dB when taken 5 feet from a source, when the distance is doubled to 10 feet the corresponding decreased sound pressure level should be 84 dB. The 6 dB reduction in sound pressure, however, is the expected pressure change under ideal conditions. Sound pressure measures failing to decrease by approximately 6 dB as the distance doubles from the source indicates that free-field conditions are not present.

Excessive or unwanted sound is referred to as "noise". Sound energy travels as waves through air and is generated as a result of the oscillation of sound pressure above and below atmospheric pressure. The sound energy interacts with the human system by initially vibrating the tympanic membrane, or ear drum, which conducts the energy to the middle ear where the energy is conducted further via vibration of the ossicles. Ultimately the energy enters the inner ear where pressure changes in the cochlea and movement of hair cells cause stimulation of the auditory nerve which transmits the signal to the brain. Excessive levels of, and exposure to, sound at a given range of frequencies combined with a related excess in duration of exposure can result in damage to the auditory system and concurrent conductive or sensorineural hearing loss. Hearing loss associated only with aging is called presbycussis. Damage to the tympanic membrane (ear drum) or middle ear ossicles (bones) can result in conductive hearing loss. Sensorineural hearing loss results when the inner ear or auditory nerve are damaged. Noise also may cause physical and psychological disorders, interference with voice communications, disruption in job performance, and disruption of rest, relaxation, and sleep. Physical and psychological disorders are due to a change in the response of the nervous system. These effects are annoying and may lead to accidents, but basically are not due to physical changes in the body. Noise is responsible for an increase in muscular activity, constriction in the peripheral blood vessels, acceleration of the heartbeat, changes in the secretion of saliva and gastric juices, an increased incidence of cardiovascular disease, an increase in ear, nose, and throat problems, and an increase in equilibrium disorders. A noise level of 100 dB will interfere with speech. One hundred and thirty decibels cause vibration of the viscera; 133 dB cause a loss of balance. The maximum 8-hr exposure is 90 dB. However, if this is maintained for long period of time, loss of hearing can result. Any level above 130 dB is damaging. At 140 dB, the individual will start to experience pain. Annoyance due to noise is related to its loudness, frequency, and intermittency. As an example, if one were to sit in a quiet room and listen to water drip, it would eventually cause considerable annoyance and discomfort.

Vibration is an oscillating motion of a system ranging from simple harmonic motion to extremely complex motion. The system may be gaseous, liquid, or solid; periodic or continuous. Vibration, which in effect is the motion of a particle, is characterized by the displacement from the equilibrium position, velocity, or acceleration. Some common sources of vibration include tractor operation, pneumatic tools, heavy equipment vehicles, vibrating hand tools, chain saws, stamping equipment, sewing machines, and looms.

Overexposure to vibration while using hand tools may cause neuritis, decalcification of the carpal and metacarpal bones, fragmentation, and muscle atrophy. An extreme condition resulting from the use of power tools is called Raynaud's syndrome or dead fingers. The circulation in the hand is impaired because of extended periods of using vibrating tools. The fingers become white, devoid of sensation, and mildly frosted. The condition usually disappears when the fingers are warmed for some time, but in some cases the condition is so disabling that the individual must seek new work. At present, there is no generally accepted limits for a safe vibration level. This is an area where additional research is needed.

Nonionizing Radiation

Nonionizing radiation refers to waves of electromagnetic energy. In general and as shown above for sound, waves possess frequency, wavelength, and velocity. Frequency (f) is the number of vibrations per unit second, wavelength (λ) is one complete sinusoidal oscillation, and velocity (c) is the time rate change of displacement or the speed at which a wave travels a given distance. The three factors are related by velocity, often measured as the speed of light (3×10^8 m/sec), which is equal to the frequency muliplied by the wavelength:

$$c = f\lambda$$
$$3 \times 10^8 \, \text{m/sec} = f\lambda \qquad \text{(Eq. 10.3)}$$

Electromagnetic waves or radiation consist of oscillating electric and magnetic fields which are perpendicular to each other and each at 90° angles relative to the speed and direction of the traveling wave. Electromagnetic radiation energies make up a spectrum of categories that differ based on ranges of wavelengths (°) and frequencies (f). In addition, the categories of the electromagnetic spectrum may be differentiated by quantity of photon energy. Photon energy is measured in units of joules (j), which refers to the work associated with a one newton force moving an object a distance of one meter. The photon energy (E) is related to the product of Planck's constant (h; 6.624×10^{-34} Joule-sec) and wave electromagnetic wave frequency (f):

$$E = hf$$
$$= (6.624 \times 10^{-34} \, J \, sec) f \qquad \text{(Eq. 10.4)}$$

The photon energy of electromagnetic radiation also can be expressed in units of electron volts (eV), which refers to energy gained by an electron when passing through a potential difference of one volt. One electron volt is equivalent to 1.602 x 10^{-19} J. The power of electromagnetic radiation is related to energy over time and is measured in units of watts (W). One watt represents power over time and is equivalent to 1 J/sec.

Examples of nonionizing electromagnetic radiation in order of increased wavelengths and inversely related decreased frequencies include ultraviolet (UV), visible light, infrared (IR), microwaves (MW), radiowaves (RW), and extremely low frequency (ELF) electromagnetic field (EMF) energies.

The primary difference between ionizing radiation and nonionizing radiation is that ionizing radiation has sufficient energy (>10 eV) to dislodge subatomic orbital electrons upon impact, thus causing ions to form. Nonionizing radiation has lower energies and longer wavelengths (>1.24 x 10^{-7} m) and, therefore, is incapable of dislodging orbital electrons, but may deposit thermal and vibrational energies leaving atoms in an "excited" state.

Ultraviolet radiation is generated from natural and anthropogenic sources and is associated with potential damage to the eyes and skin. Indeed, excessive exposure to UV radiation from welding operations is associated with eye cataracts and exposure to sunlight can damage the eyes and increase the risk of developing dermal cancer.

Visible light is necessary for proper surface illumination and brightness. Too much or too little visible light can pose health and safety problems. Photons of visible electromagnetic light energy are transmitted linearly through air. The direction of light can be changed, however, depending on interaction with various media and incident light will reflect from some surfaces. When light passes through different transparent media, it will bend or refract. In addition, when light passes through and exits a transparent medium it will diffuse. Lighting is important in various environmental settings to increase comfort and safety. Several factors influence lighting, including the quantity and quality. Illumination is related to the quantity of light or the light energy directed on a given area. Quality of light, however, refers to additional factors such as brightness, glare, and color. Glare causes discomfort and interferes with the successful completion of some work. The amount of glare is determined by the design of the lighting fixtures, the types of surfaces on which the light falls, the color of the surfaces and the environment, and the amount of outside light coming into the industrial facility. Both the quantity and quality of illumination are evaluated in various environmental settings to determine if lighting is adequate for specific and general tasks. In general, a minimum of 30 footcandles should be provided in all work areas, including receiving, opening, storing, and shipping areas. Footcandle levels may rise to as much as 2000 in certain examining areas. Poor qualitative and quantitative lighting conditions are associated with increased fatigue.

IR can occur from any surface that is at a higher temperature than the receiving surface. The transfer of radiant energy or heat occurs whenever the radiant heat emitted from one body is absorbed by another body. The organs of greatest concern associated with exposure to IR radiation are the eyes and skin. Radiant heat contributes to heat stress in exposed individuals. Skin erythema and burning, as well as lens and corneal cataracts, have resulted from exposure to specific wavelengths. Glass blowing, foundry, and boiler operations are common occupational sources of infrared radiation.

Microwaves are electromagnetic nonionizing radiation with frequencies and wavelengths ranging from 1 MHz to 300 GHz and 0.1 cm to 10 m, respectively. The energy is generated anthropogenically via devices that cause deceleration of electrons in an electric field which, in turn, generate kinetic energy as microwaves. Depending on the frequency and wavelength, microwaves are transmitted for long distances through air. A major characteristic of microwave energy is that it can be readily absorbed by numerous materials, including human tissue, resulting in the generation of elevated temperatures almost uniformly within a given volume. Relative to human tissue, microwaves are more penetrating in relation to areas of higher water content, such as the eyes, skin, muscles, and visceral organs. In addition, higher frequencies and shorter wavelengths can increase the depth of absorption.

Microwaves cause an increase in heat in the whole body. Microwave absorption and heat generation is greatest in tissues having a high water content. Microwaves change polarity every half cycle, with the tissue molecules doing the same and thereby vibrating rapidly, which produces heat. They also cause damage to the lens of the eye and possible damage to the gonads.

Common uses and sources of microwave radiation include dielectric heaters, radio frequency sealers, radio and television broadcast stations, induction furnaces, communication systems (i.e., radar), and ovens. Microwave ovens are a common use and source of microwaves in both occupational and non-occupational environments. Poorly aligned doors, accumulated food residues, and faulty door gaskets, hinges and latches, however, can result in release of microwave radiation from inside to outside the microwave oven. Leaking microwave ovens can generate significant and dangerous amounts of microwave radiation into the external air. Accordingly, the ovens should be monitored periodically to determine if microwave radiation is detected and measured outside of a closed and operating unit. Microwave oven performance standards specify less than 1 mW/cm^2 at any point 5 cm from the oven prior to purchase and less than 5 mW/cm^2 at any point 5 cm from the oven after purchase.

Specific ranges of wavelengths (0.1 cm to 300 meters) and frequencies (1 MHz to 300 GHz) of microwave and radiofrequency radiation can penetrate the human body and cause adverse effects to internal organs. Thermal impact including heat stress, teratogenesis, and ocular effects (cataracts and cornea damage) have been suggested as possible adverse effects. These thermal effects may include potentially damaging alterations in cells caused by localized increases in tissue temperature. The body's heat sensors, which are located in the skin, are not activated when the RF energy is absorbed deep within the body tissues.

Occupational and even non-occupational exposures to very low and extremely low frequency electromagnetic fields (EMFs) has received increased attention. The nonionizing radiation energy consists of electric and magnetic fields. Electric fields are created by the presence of electric charges associated with operating and non-operating electrical appliances and machinery connected to an electrical power source. Magnetic fields also are associated with electrical devices, but are generated in combination with electric fields when the devices are operating. EMFs are characterized by very low and extremely low frequencies and longer wavelengths. Although there is no specific demarcation, extremely low frequencies (ELFs) are below 10^3 Hz (1 kHz) and very low frequencies (VLFs) range between 10^3 to 10^7 Hz. Major sources of low frequency nonionizing radiation are electrical transmission lines (60 Hz) and numerous electrical devices present in the occupational environment. Occupations requiring installation and maintenance of electrical power lines and devices present some obvious potential sources of EMFs. Relatively recently, however, common office sources such as office computers and video display terminals (VDTs) have received more attention relative to very low frequency EMFs. Accordingly, monitoring is becoming more common to evaluate the levels of nonionizing EMF generated by this equipment in office settings.

Extremely low frequency (ELF) radiation has elicited increased concern especially frequencies of 60 Hz. Sources of ELF include high voltage power lines, VDTs, and close proximity to electrical wiring such as electric appliances. Cancer, such as leukemia, and altered blood chemistry are some of the potential biological effects according to preliminary research of ELF exposure. External exposures and the impact to health, however, remain in need of further investigation.

Light amplification by stimulated emission of radiation (laser) energy is an anthropogenic form of electomagnetic radiation consisting of intense light amplification and concentration in one wavelength and direction. Lasers have been used in the construction industry for measuring distances and welding, and by some surgeons to repair torn retinas. The hazard to the individual is due to the point source of great brightness. The laser may also injure the eye, skin, or possibly internal organs. The laser also may oxidize chemicals in the ambient air.

Ionizing Radiation

Atoms are comprised of three basic building blocks: the electron, which has an extremely small mass and is negative; the proton, which has a mass approximately 1800 times as large as the electron and is positive; and, the neutron, which has about the mass of a proton and electron combined. Nonetheless, the mass of a proton and a neutron are considered approximately equal and the mass of an electron essentially negligible. The protons and neutrons form the central core or nucleus of the atom. The electrons orbit the nucleus. The orbiting electrons do not have exact positions and, therefore, they are considered to be distributed as an electron cloud.

Elements differ based on the number of protons in their nuclei. Atoms of the same element have the same number of protons (the same atomic number), but may have different numbers of neutrons (different atomic weights). These atoms are called isotopes. For instance, the most common form of carbon (C-12) has 6 neutrons and 6 protons in the nucleus. However, an isotope of carbon (C-14) has 8 neutrons and 6 protons. Carbon-14 is an unstable isotope of carbon and, therefore, is considered a radioisotope. Radioisotopes are atoms that have an unstable nucleus that emits energy is the form of ionizing radiation to become more stable. This often brings about a change from one element to another, which causes a change in the chemistry of the substance. For example, fission is the process occurring in nuclear reactors, and refers to the splitting of a uranium atom into two smaller atoms. The two smaller atoms are called fission products. These atoms have a greater number of neutrons compared to protons than the stable atom with the same number of protons. Therefore, in order for the atom to become stable, the neutron spontaneously ejects an electron and becomes a proton. The ejected electron is called a beta particle and travels at extreme velocities. The atom is said to have decayed. The atom has now become a new element.

Thus, chemical elements which are electrochemically unstable will decay due to an imbalance of protons (p^+) or neutrons (n^o) in the atomic nuclei and emanate energy in the form of ionizing radiation. The elements and compounds undergo natural reactions to achieve stability via emanation of atomic energy in the form of particulates and electromagnetic photons. The particles and photons impart excessive energy to matter with which they interact, resulting in ionization.

Ionizing particulate radiation consists of alpha particles ($2 n^o + 2 p^+$ in the form of a charged Helium nucleus, $^4_2He^{+2}$); beta particles (negatron as e^- and positron as e^+); and, neutron particles (n^o). By emission of alpha particles, uranium becomes thorium and radium becomes radon. Ionizing electromagnetic radiation consists of photon energies as x-rays and gamma rays. Alpha or beta particles may be accompanied by gamma radiation, which is similar to X-ray. X-rays are of short wavelength, high frequency, and high energy. X-rays travel through space at the speed of light. They possess no real mass charge. X-rays are produced by bombarding a material with electrons which contain energy. Only about 1% of the electrons in the diagnostic region will produce X-radiation. Most of the electrons give up their energy as they hit solid material and increase the heat of the material. The difference between the gamma ray and the X-ray is in their origin. A gamma ray is emitted from the nucleus of the atom when the nucleus goes at a higher or lower energy level.

The particulate and electromagnetic wave forms of ionizing radiation can interact via direct or indirect ionization of macromolecular or cellular components of the human body which, like those materials classified as toxic, may result in adverse biochemical and physiological changes manifested as abnormalities, illnesses, or premature deaths among those exposed or their offspring.

Ionization energies are measured in electron volts (eV). The rate at which an element undergoes radioactive decay is referred to as "activity" and measured in units of curies (Ci) or bequerals (Bq). One curie is equivalent to 3.7 x 10^{10} disintegrations per second (dps). The unit bequeral (Bq) is an alternative to curies, where 1 Bq equals 1 dps. Half-life refers to the time required for 50% of a radioactive element to decay (disintegrate) and, accordingly, the activity to decrease. Each radionuclide decays with a specific half-life. During decay, it generally produces other radioactive substances. Radium-226 decays with a half-life of 1620 years. Iodine-131 decays with a half-life of 8 days. Sodium-24 decays with a half-life of 15 hr. The rate of decay is important to environmental health personnel.

Activity does not specify the type of ionizing radiation and the degree of hazard. "Exposure" is the amount of ionization (electrical charge produced by X-rays or gamma rays) per mass or volume of air. Measured in units of "roentgens" (R = 2.58 coulomb/kg = 2.083 x 10^9 ion pairs/cc air). Since 34 eV of energy is required to produce one ion pair in air, 1 R= 2.083 x 10^9 ion pairs/cc air x 34 eV/ion pair = 7.08 x 10^{10} eV/cc. In relation, 7.08 x 10^{10} eV/cc multiplied by 1.6 x 10^{-2}ergs/eV = 0.113 ergs/cc air. As a result, since 1 cc air has a density equal to 0.001293 g/cc, 1 R = 0.113 ergs/ 0.001293 g air = 87.6 ergs/g air. An erg, like the unit joule, is unit of measure for work or force through distance. One joule is equivalent to 10^7 ergs and 1 electron volt (eV) = 1.6 x 10^{-12}ergs.

Exposure measured in units of roentgens has been replaced with the term "absorbed dose". Absorbed dose is the amount of radiation energy (i.e. energy departed from any ionizing source) absorbed by the entire body or a specific body site. This is equivalent to the quantity of ionization energy deposited per unit mass of tissue (or other matter) and is measured in "rads" (r = 100 ergs energy/g tissue). An alternative unit, "gray" is equivalent to 100 rads which equals 1 joule/kg. "Equivalent dose" refers to the quantity of absorbed dose multiplied by factors associated with a given form of ionizing radiation, such as the tissue weighting factor. The dose is measured in units of "rem" and, more recently, "sieverts" where 1 Sv = 100 rems.

Ionizing radiation is an important human hazard, since absorption of energy in the body tissues causes physiological injury by means of ionization or excitation. Ionization is a process involving removal of an electron from an atom or molecule leaving the system with a net positive charge. Excitation is the addition of energy to an atomic or molecular system, changing the system from a stable to an excited state. Radiation causes cell damage by upsetting the cell chemistry or physics, and by disrupting the ability of the cell to repair itself. Radiation causes somatic damage, which means that it affects the cells and organs during the lifetime of the human. It also causes genetic damage, which affects future generations. The amount of damage depends on the amount of radiation received, the extent of body exposure, the part of the body exposed, the amount of exposure, and the type of exposure, whether external or internal.

External radiation occurs when radioactive materials pass through the body. Internal radiation occurs when radioactive material is inhaled, swallowed, or absorbed through breaks in the skin. A large dose of acute radiation exposure causes nausea, fatigue, blood and intestinal disorders, temporary loss of hair, and potentially serious damage to the central nervous system. This might happen in the event of a nuclear war or a massive nuclear accident. Long-term effects include increased incidence of lung cancer suffered by miners of uranium and pitchblende; bone cancer, which occurred in luminous watch painters years ago who had swallowed small quantities of radium; skin cancer and leukemia, which occurs in physicians and dentists from repeated exposure to radiation. Radiation also causes mutations in the genes of reproductive cells.

The extent of the hazard to humans is based on the dose, the body part, and the time of exposure. Large amounts of radiation to the whole body cause acute exposure. Large amounts of radiation to small parts of the body cause acute and chronic exposures. Small amounts of radiation to the whole body causes chronic exposure. Small amounts of radiation to limited portions of the body causes acute and chronic exposures.

Humans are exposed to radiation from the natural environment, including cosmic rays, the earth, and the atmosphere. Radiation from artificial sources include medical and dental sources, occupational sources, nuclear energy plants, radioactive fallout, radioactive dial watches, clocks, meters, radioactive static eliminators, television sets, isotope-tagged products, radioactive luminous markers, and certain household goods. Presently, the largest artificial source of radiation exposure is medical and dental X-ray machines. The possibility of acute radiation injury due to X-rays is minimal, however, long-term effects are not fully understood.

Industrial use of radiation produces additional sources of radioactive exposure. This includes the inspection of materials by means of radiographic inspection, fluoroscopic inspection, and continuous automatic inspection by X-rays to measure the thickness of material and the height at which something is filled. Personnel are checked for stolen articles, contraband, or weapons by whole-body fluoroscopes. The shoe-fitting fluoroscope has fortunately been eliminated in most states, since it was an important source of radiation, especially for small children. X-rays are also used to determine the diffraction pattern of crystalline materials.

Ionizing radiation has been used in industry for sterilization of foods and drugs, killing insects in seeds, toughening polyethylene materials, and activation of chemical reactions in the petroleum process. One device used is the Van de Graaf apparatus, which is an electron-beam generator. In the past, radium was used as a source for industrial radiography. It was also used in luminous paints and in the removal of static electricity generated by the movement of nonconducting materials at high speeds through machinery. The radium produced ionized air which removed the static charge. Today, other artificial radionuclides are used in industry, including cobalt-60, iridium-192, and cesium-137. Strontium-90 is used in some luminous paints and in the krypton switchlamp used by railroads in remote locations.

Atomic batteries are used to produce high-voltage, low-current sources in special instruments. Radionuclides are used to determine thickness, density, and depth of materials, and are used as tracers. Sources of radiation also include nuclear reactors, ventilation air, and cooling wastes. The latter come from nuclear power plants or nuclear ships and submarines. Thousands of mine workers are exposed to radiation in uranium mines and through milling.

Hot and Cold Temperature Extremes

Heat and cold stress are physiological responses to extremes of hot and cold temperatures, respectively. Accordingly, thermal agents must be considered as part of the evaluation of the occupational environment. The body generates metabolic heat via basal and activity processes. If the environment is too cold, the body loses heat faster than it can produce heat and a condition known as hypothermia can develop. Alternatively, if the environment is too hot and the body is not able to cool fast enough, heat stress can result. The potential for heat stress is exacerbated if an individual is also very active and simultaneously exposed to elevated temperatures.

There are four factors that influence the interchange of heat between people and the environment:

1   air temperature
2.   air velocity
3.   moisture content of the air
4.   radiant temperature

The industrial heat problem is created when a combination of these factors produce a working environment that may be uncomfortable or hazardous to individuals because of an imbalance of metabolic heat production and heat loss.

Humans regulate internal body temperature within certain narrow limits (i.e., 97 to 99.5°F). Blood flows from the site of heat production in the muscles and deep tissues, which is the deep region or core, to the body surface, which is the superficial regions where it is cooled. The body produces sweat to help cool itself. Normally, the human body automatically controls the internal heat load. External heat load, which is produced by the environment, affects humans in different ways, based on physical fitness, work capacity, age, health, living habits, and degree of acclimation to heat.

Humans exchange heat with the environment through conduction and convection, radiation, and evaporation. Conduction is the transfer of heat through direct contact. If the substance in contact with humans, such as air, water, clothing, or other objects, is higher in temperature than the skin, the body will gain heat; if the substance is lower in temperature than the skin, heat will be lost. The rate of transfer is based mostly on the difference between the two temperatures. Air and water accelerate the transfer. Convection is the transmission of heat by the motion of air or water. Thermal or radiant heat is the transmission of energy by means of electromagnetic waves of infrared radiation. Radiant energy, when absorbed, becomes thermal energy and results in an increase in the temperature of the absorbing body. Evaporation is based on air speed and the difference in vapor pressure between perspiration on the skin and the air. In hot, moist environments, evaporation may be limited or stopped. Cold is the absence of heat.

Thus, heat stress is influenced by several environmental factors, including air temperature measured in degrees celsius or farenheit, air movement measured in feet per minute, humidity measured in percent moisture, and radiant heat also measured in degrees celsius or farenheit. Elevated air temperature increases the heat burden on the human body. The adverse influence of air temperature is exacerbated by the presence of excessive water vapor in the air due to high humidity. Elevated humidity decreases the evaporation rate of perspiration from the skin. Perspiration and its subsequent evaporation is a natural defense mechanism to transport excessive heat from the body. Excessive water vapor in the air hinders the evaporation process and elimination of heat via evaporative cooling. The evaporative cooling is facilitated by increased air movement and, accordingly, also hindered by lack of sufficient wind or breeze. Finally, radiant heat from various sources penetrates through the air and is absorbed by the body. In turn, the absorbed radiant heat increases the heat burden on the body. These thermal factors in combination with metabolic factors related to the worker based on his/her physiology and degree of exertion influence the potential for an individual developing heat stress. In addition, the potential for heat related disorders is increased by such factors as physical conditioning, age, weight, gender, and impermeable protective clothing.

The following are heat stress illnesses: heat stroke, caused by considerable exertion in a hot environment, lack of physical fitness, obesity, dehydration, and so forth; heat exhaustion, caused by extreme exertion and heat, lack of water, and loss of salt; heat cramps, caused by heavy perspiration during hot work, large intake of water, loss of salt; heat syncope, due to an unacclimated worker standing erect and immobile in the heat, causing blood to accumulate in the lower part of the body, which results in inadequate venous blood return to the heart; heat rash, due to unevaporated sweat from the skin; heat fatigue, caused by extreme heat for long periods of time; interference with the heat-regulating center in the brain, causing body temperatures to rise to 108–112°F , which may result in brain damage or death. Certain safety problems are common in hot environments. Heat tends to promote accidents due to the slipperiness of sweaty palms, dizziness, or the fogging of safety glasses. Also molten metal, hot surfaces, and steam create serious contact burns. Heat lowers an individual's mental alertness and physical performance. It also promotes irritability, anger, and overt emotional reactions to situations, which may lead to accidents.

Heat stroke symptoms include hot skin, usually dry, red, or spotted. The body temperature rises to 105°F or higher. The victim is mentally confused, delirious, possibly in convulsions, or unconscious. Unless the person receives quick and appropriate treatment, the individual will die. Individuals with these signs or symptoms of heat stroke need to be immediately taken to a hospital. Any emergency medical care prior to hospitalization should include removing the person to a cool area, soaking the clothing with water, and vigorously fanning the body to increase cooling.

Heat exhaustion includes several clinical disorders which have symptoms that may resemble the early symptoms of heat stroke. The person suffering from heat exhaustion still is able to sweat but experiences extreme weakness or fatigue, giddiness, nausea, or headache. In more serious cases the person may vomit or lose consciousness. The skin is clammy and moist, the complexion is pale or flushed, and the body temperature is normal or slightly elevated. In most cases, have the person rest in a cool place and drink plenty of liquids. If the case of heat exhaustion is mild, the individual will recover quickly. If the problem is severe take the individual to a hospital. Extreme caution must be taken concerning individuals who are on low sodium diets and who work in hot environments.

Heat cramps are painful spasms of the muscles that occur in individuals who sweat profusely because of heat exposure. They may be drinking large quantities of water but not replacing the body's salt loss. The water tends to dilute the body's fluids while the body is still losing salt. The low salt level in the muscles causes cramps, which may occur in the arms, legs, or abdomen. Again, be most cautious concerning individuals with heart problems or those on low sodium diets.

Heat rash, which is also known as prickly heat, occurs in hot, humid environments, where sweat is not easily removed from the surface of the skin by evaporation and the skin stays moist. The sweat ducts become plugged and a skin rash then appears. Extensive heat rashes may become complicated by infection. An individual can prevent the problem from occurring by resting in a cool place each day and by bathing regularly and drying the skin.

Transient heat fatigue is the temporary state of discomfort and mental or psychological strain coming from prolonged heat exposure. If the person is unaccustomed to heat, he or she is particularly susceptible to the problem. The individual suffers a decline in test performance, coordination, alertness, and vigilance, which may lead to accidents. Transient heat fatigue can be lessened or eliminated by using proper techniques of heat acclimatization.

Cold related stress is initiated by a peripheral vasoconstriction, especially in the extremities, which results in a sharp drop in skin temperature. This helps reduce the body heat loss to the environment. However, this causes a chilling of the extremities, which results in the toes and fingers approaching freezing temperatures very rapidly. When the temperature of hands and fingers drop below 15°C, they become insensitive, malfunction, and the chance for accidents increases. The cooling stress is proportional to the total thermal gradient between the skin and the environment. The loss of heat through evaporation of perspiration is not significant at temperatures lower than 15–20°C. When vasoconstriction can no longer adequately maintain the body heat balance, shivering becomes an important mechanism for increasing the body temperature by causing the metabolic heat production to increase several time the resting rate. When proper clothing is used and the clothing does not become wet with water or sweating due to excessive work, then all parts that are protected by the clothing will help maintain the body temperature. Exposed parts such as the face or fingers, may become excessively chilled and frostbite may occur.

Frostbite occurs when there is an actual freezing of the tissues with the mechanical disruption of the cell structure. The freezing point of the skin is –1°C. Increasing wind velocity will increase heat loss and cause frostbite to occur more rapidly. When the velocity reaches 20 mi/hr, the exposed flesh will freeze within about 1 min at –10°C. If the skin comes in contact with below-freezing-temperature objects, frostbite may occur, even though the environmental temperatures are warmer. The first symptom of frostbite is often a sharp, pricking sensation. Cold, however, produces numbness and anesthesia, which may allow serious freezing to develop without warning of acute discomfort. The injury from frostbite may range from a single superficial injury with redness of the skin, to deep cyanosis and gangrene.

Trench foot may be caused by long, continuous exposure to cold without freezing, combined with a persistent dampness or immersion in water. Edema, tingling, itching, and severe pain may occur. This is followed by blistering, superficial skin necrosis, and ulceration.

General hypothermia is an extremely acute problem that results from prolonged cold exposure and heat loss. If a person become fatigued during physical activity, he or she will be more apt to have heat loss. As exhaustion approaches, the vasoconstrictor mechanism is overwhelmed and suddenly vasodilation occurs, with a rapid loss of heat and critical cooling. Alcohol consumption and sedative drugs increase the danger of hypothermia.

Chemical Agents

Chemical agents are classified as inorganic and organic and most notable for toxicity and flammability from an occupational health perspective. Aerosols or particulates are divided into dusts, mists, fumes, fibers, and smoke. Dusts are generated from mechanical actions such as crushing, grinding, or sawing solid materials. Mists are finely divided liquid droplets generated via agitation of liquids or certain operations such as spraying or dipping. Fumes result when a metal is heated to its sublimation point. The metal forms a vapor which rises into the air where it often reacts with molecular oxygen to form metal oxides. The vapor then condenses and forms a very small particle (<1 μm) called a fume. Fibers are solids generated in much the same way as dusts. Fibers differ from dusts in their form. Dusts are generally regarded as spherical or round, whereas fibers have a defined length to width ratio (e.g., 3:1) making them appear more columnar. Smoke consists of small particles of carbon and other organic matter generated from incomplete combustion reactions. All aerosols, except fibers, are measured as weight per volume in units of milligrams per cubic meter of air (mg/m³). Fibers are measured as counts per volume in units of fibers per cubic centimeter of air (f/cc).

Unlike gases and vapors, aerosols do not readily diffuse and uniformly distribute throughout their area of generation. In addition, particulates are classified as respirable and non-respirable depending on particle size or diameter. Particulates with diameters <10 μm are considered respirable, that is, they can by-pass the passages of upper respiratory system. Particle diameters <5 μm can reach the lower regions and those <2 μm can travel to the terminal points of the bronchial system known as the alveolar region.

Gases and vapors are frequently generated in the occupational environment. True gases exists in the gaseous state at normal temperature (25°C) and pressure (760 mmHg), while sources of vapors exist as liquids or solids under these conditions. Gases and vapors form true solutions in the atmosphere, which means that they expand and mix completely with ambient air.

Gases and vapors are monodisperse in terms of molecular size. This means that all of the molecules of a particular gas or vapor will be the same size. Indeed, gaseous substances are not classified by size like aerosols. Unlike aerosols, gases and vapors are of molecular size and, accordingly, means that they can not be removed from the air via physical filtration methods. The gas or vapor must somehow be separated from the surrounding air molecules. Toxic gases and vapors are measured as volume per volume in units of parts per million (ppm). Flammable gases and vapors and oxygen gas are also measured as volume per volume, but in units of percent (%) gas or vapor in air. Note that 10,000 ppm equals 1%.

Toxic Chemicals

Toxic substances include metals, aqueous-based acids and alkalies, and petroleum-based solvents, among others. The human toxicity of chemical agents is related to several factors including the duration of exposure, concentration, and mode of contact. Inhalation and dermal contact are the primary and secondary modes of exposure, respectively, to chemical agents in the occupational environment. Accordingly, the occupational environment is commonly evaluated for airborne toxic chemical agents. Thus, exposure to chemical agents can occur directly via dermal contact with the material during handling and indirectly via respiratory and dermal contact with airborne agents. Materials also can be ingested via handling food with contaminated hands or as a result of mucociliary escalation of materials that were originally inhaled into the respiratory system and subsequently transported to the throat and mouth where they are swallowed.

Chemical agents are either organic or inorganic. The organic materials are either hydrocarbons or substituted hydrocarbons. The hydrocarbons are composed solely of carbon and hydrogen while substituted hydrocarbons are composed of carbon and hydrogen plus functional groups composed of elements such as chlorine, nitrogen, phosphorous, sulfur or oxygen. The hazards associated with organic compounds and the fate of the compounds in the environment are mainly dependent on their chemical compositions and associated physical properties. Toxic inorganic materials are either metallic or non-metallic elements, commonly in the form of salts, hydrides, and oxides.

Although distinctions are made between the subclasses of chemical agents, it should be noted that a single chemical component may exhibit a combination of characteristics. Indeed, in view of the fact that any chemical is toxic at an appropriate dose or concentration, then all flammable, corrosive, and reactive materials, as well as radiological and some biological agents are also toxic. The reverse, however, is not necessarily true. That is, all toxic materials do not exhibit other hazardous characteristics.

Toxic agents may induce biochemical and physiological changes in human systems following either systemic contact, via absorption into blood and tissues, or local contact. The changes may be ultimately manifested as adverse effects such as morphological and functional abnormalities, illnesses and premature deaths among those exposed or their offspring. As suggested above, toxicity is inherent in all compounds. The toxicity of a given agent, however, may be directly attributable to an original parent compound or indirectly attributable to an active metabolite formed via biotransformation in a human system. In addition, it should also be recognized that secondary toxicants can be generated by flammable, corrosive, and reactive chemicals in the form of toxic by-products released from reactions, fires, and explosions.

Toxic conditions also can be present when levels of atmospheric oxygen are not normal. Oxygen gas is a normal and obviously essential component in air at a concentration of approximately 20.8%. When levels decrease significantly below 20.8% the atmosphere is designated "oxygen deficient" (e.g., <19.5% O_2) and when levels increase significantly, the atmosphere is "oxygen enriched" (e.g., >22% O_2). Deficient and excessively enriched atmospheres can result in toxic effects to individuals present due to hypoxia and hyperoxia, respectively. Enriched atmospheres also increase the potential for initiation of combustion and explosion reactions.

A summary of some representative toxic chemical agents encountered in the environment, including occupational settings, follows.

Crystalline Silica is a form of silicon oxide and is a natural component of the earth. Exposure is associated with underground mining, quartz-bearing rock, cutting of granite, manufacture of pottery, porcelain, abrasives, sand-blasting operations, stripping, mining, and steel and furnace making. The degree of hazard is based on the concentration of the dust, the percentage of free crystalline silica in the dust, and the duration of exposure. *Silicosis* is a fibrogenic lung disease caused by the inhalation of crystalline silica. The disease has a relatively long latency period and frequently causes death. It also may be complicated by tuberculosis and acute lung infections.

Asbestos refers to a group of inorganic silicates occurring in nature as rocks. Asbestos particulates are characterized by a fibrous morphology instead of a common spheroid shape. Several types exist including chrysotile, crocidolite, amosite, anthophyllite, tremolite, and actinolite. Chrysotile represents about 90% of all asbestos processed in the U.S. Asbestos has been used in the construction industry for roofing, plastics, insulation, floor tiling, and cement production. It also has been used in the automotive industry as friction material in brake and clutch linings and the textile industry for fire-resistant clothing, safety equipment, and curtains. Inhalation of asbestos fibers has been associated with a fibrogenic lung disease called *asbestosis.* In addition, asbestos is associated with lung cancer and a cancer of the pleural lining called *mesothelioma.*

Metals are components of many inorganic compounds, commonly metallic oxides and salts, used in a variety of industrial and commercial areas. Metals bound to organic compounds are called organometallics. Organometallics are more lipid soluble than the inorganic metallic oxides and salts and, as a result, can more easily be absorbed into the blood. The blood, cardiopulmonary system, gastrointestinal system, kidney, liver, lungs, central nervous system, and skin are potential target sites for metals. The kidneys are the major target site and virtually all metals can accumulate and damage the renal system. Metals have a high affinity for indigenous protein groups, including enzymes. Many metals, including lead, arsenic, and mercury bond to enzymes and inhibit essential catalytic activity. For example, lead can bond to and inhibit two different enzymes which catalyze reactions essential for formation of normal molecular components of red blood cells. If inhibited, levels of zinc protoporphyrin (ZPP) and aminolevulinic acid (ALA) increase. Indeed, both ZPP and ALA are biomarkers of lead toxicity.

Arsenic, cadmium, chromium, nickel, and *uranium* are respiratory carcinogens. *Berrylium* and chromium are associated with ulceration of respiratory and dermal tissues. Berrylium also can cause a debilitating lung disease called berrylliosis. *Cobolt* can damage the heart. Many metals, including *arsenic, mercury, manganese*, and *lead* are toxic to the nervous system causing adverse behavioral and neuromuscular effects. Metal oxide fumes generated from welding and foundry operations can cause an acute flu-like illness called *metal fume fever.*

Organic *solvents* are pure or substituted hydrocarbons derived from petroleum and are used to dissolve other organic materials. The organic solvents used in industry are divided into several chemical groups including hydrocarbons (aliphatic and aromatic), halogenated hydrocarbons, alcohols, ethers, glycol derivatives, esters, and ketones. Solvents are used in industry for extracting oils and fats, degreasing, dry cleaning, and printing inks. They are also present in a variety of products, including paints, varnishes, lacquers, paint removers, adhesive, plastics, textiles, polishes, waxes, thinners, chemical reagents, and drying agents. Direct dermal contact with liquid organic solvents can cause dermatitis. In addition, nonpolar solvents can absorb through the skin and enter the blood. Solvents also pose an inhalation hazard since most volatilize and form airborn-solvent vapors. Virtually all organic solvents have some effect on the central nervous system and the skin. The narcotic effects of these solvents are extremely important since the worker becomes unable to use proper judgement and, as a result, may be involved in serious accidents. *Hydrocarbons* are powerful narcotic agents causing drowsiness and impaired cognitive and motor functions. In addition, many solvents cause damage to the blood, liver, kidneys, gastrointestinal system, eyes, and respiratory system. *Aromatic hydrocarbons*, which include benzene, ethyl benzene, toluene, and xylene, are commercial solvents and intermediates in the chemical and pharmaceutical industries. All of the aromatic hydrocarbons cause a central nervous system depression marked by decreased alertness, headaches, sleepiness, and loss of consciousness. They cause a defatting, contact dermatitis. *Benzene* is extremely toxic to the blood-forming organs and causes anemia and leukemia. *Halogenated aliphatic hydrocarbons* include carbon tetrachloride, chloroform, ethyl bromide, ethyl chloride, ethylene dibromide, ethylene dichloride, methyl chloride, methyl chloroform, methylene chloride, tetrachloroethane, tetrachloroethylene (perchloroethylene), and trichloroethylene. These chemicals are commercial solvents and intermediates in organic synthesis. They affect the central nervous system, kidney, liver, and skin. All of these chemicals cause a central nervous system depression, decreased alertness, headaches, sleepiness, and loss of consciousness. They also can damage the kidneys. These chemicals may cause anemia. The halogenated aliphatic hydrocarbons cause liver changes that lead to fatigue, malaise, dark urine, liver enlargement, and jaundice. Tetrachloroethane or *carbon tetrachloride*, causes serious damage to the liver and kidneys and is considered carcinogenic. *Alcohols* such as methyl or *n*-butyl are mildly narcotic and have low flashpoints and therefore are highly flammable. Ethers are highly flammable and have strong narcotic properties. *Glycol derivatives*, such as ethylene glycol, are highly flammable and have a toxic effect on the nervous system and the blood. *Esters* are flammable and cause irritation to the eyes, nose, and upper respiratory tract. Ketones, such as acetone, are quite flammable. *Carbon disulfide* is a very hazardous solvent used in industry. It is highly flammable and highly toxic, acting mainly on the central and peripheral nervous systems.

Many inorganic and organic gases cause toxic effects. Oxides of *nitrogen* and *sulfur, ozone, chlorine, formaldehyde, hydrogen chloride, hydrogen sulfide*, and *hydrogen cyanide* are respiratory and ocular irritants. *Hydrogen cyanide* and *hydrogen sulfide* are chemical asphyxiants that can cause suffocation due to inhibition of a cellular enzyme causing inhibited utilization of inhaled molecular oxygen. *Carbon monoxide*, produced by the incomplete combustion of fuel, also is a chemical asphyxiant, but inhibits transport of inhaled oxygen by bonding to hemoglobin in red blood cells. *Vinyl chloride* is a flammable, compressed gas used as a raw material for the production of polyvinyl chloride and causes damage to the liver (including cancer), skin, vascular system, nervous system, and kidneys.

Insecticide manufacture and use is potentially hazardous to the workers. The *organochlorine* insecticides are chlorinated ethanes, which include dichloro dibenzo trichloroethane (DDT), and the cyclodienes, which include aldrin, chlordane, dieldrin, and endrin. They also include the chlorocyclohexanes such as lindane. They are used for pest control. They affect the kidney, liver, and central nervous system. The potential health effects include acute symptoms of apprehension, irritability, dizziness, disturbed equilibrium, tremor, and convulsions. The cyclodines may cause convulsions without any other initial symptoms. The chlorocyclohexanes can cause anemia. The cyclodienes and chldorocyclohexanes cause liver toxicity and can cause permanent kidney damage. The *organophosphate* insecticides include diazinon, dichlorovos, dimethoate, trichlorfon, malathion, methyl parathion, and parathion. These chemicals are used for pest control. They affect the central nervous system, liver, and kidney. The major neurotoxic mechnism involves inhibition of the enzyme acetylcholine esterase. Depending on the amount of insecticide that has been absorbed by the individual, the acute symptoms range from headaches, fatigue, dizziness, increased salivation, crying, profuse sweating, nausea, vomiting, cramps, and diarrhea, to tightness in the chest, muscle twitching, and slowing of the heartbeat. Severe cases can bring about a rapid onset of unconsciousness and seizures, leading to death. A delayed effect may be weakness and numbness in the feet and hands, with long-term permanent nerve damage possible. The *carbamate* insecticides include aldicarb, baygon, and zectran. They are used for pest control. They affect the central nervous system, liver, and kidney. Their potential neurotoxic effects involve inhibition of the the enzyme acetylcholine esterase as discussed for the organophosphates.

Tobacco emissions consist of smoke particles and gaseous organic and inorganic molecules. Employees exposed in the workplace to toxic chemicals can receive additional exposure because of the presence of toxic chemicals in tobacco products. Cigarette smoking causes increased exposure to carbon monoxide. Workers are frequently exposed to carbon monoxide as part of their job. Therefore, the additional levels of carbon monoxide related to smoking can cause cardiovascular changes that are dangerous, especially to people who have coronary heart disease. Other chemicals found in tobacco that workers might be exposed to at their jobs include acetone, acrolein, aldehydes, arsenic, cadmium, hydrogen cyanide, hydrogen sulfide, ketone, lead, methyl nitrate, nicotine, nitrogen dioxide, phenol, and polycylic aromatic compounds.

Tobacco products may become contaminated by chemicals used in the workplace and therefore increase the amount of toxic chemicals entering the workers' bodies. Smoking may contribute to an effect comparable to that which can result from exposure to toxic agents found in the workplace and therefore cause an additive biological effect. Combined worker exposure to chlorine and cigarette smoke can cause a more damaging biological effect than exposure to chlorine alone. Smoking may act synergistically with toxic agents found in the workplace and cause a more profound effect than the simple exposure that the individual is subjected to. Workers who were heavy smokers and installed asbestos had a higher level of cancer than those who were nonsmokers. Radon daughters also act synergistically with tobacco smoke. Smoking may contribute to accidents in the workplace. The accidents apparently are due to a lack of attention, preoccupation for the hand for smoking, irritation of the eyes, and coughing. Smoking can also contribute to fires and explosions where flammable or explosive chemicals are being used or stored. Smokers have poor lung construction and a higher incidence of urinary abnormalities when they are exposed to cadmium. A chronic cough and expectoration occurs when individuals are exposed to both smoking and various ethers.

Flammable Chemicals

Flammable chemical agents include materials which serve as fuels (reducing agents) that can ignite and sustain a chain reaction when combined in a suitable ratio with oxygen (oxidizing agent) in the presence of an ignition source (heat; spark). Flammable agents are characterized as having a low flash point (i.e., <60°C) which is directly related to vapor pressure. In general, organic compounds such as organic solvents vaporize at relatively lower temperatures and, accordingly, are much more sensitive to heat than inorganic compounds. Flammable materials pose an obvious hazard from the standpoint of potential burns to human tissue. Indirectly during combustion, however, flammable materials can contribute to the formation of toxic atmospheres due to generation of by-products such as strong irritants and chemical asphyxiants, as well as, via consumption of molecular oxygen during combustion.

Numerous organic gases (e.g., methane; propane) and solvent vapors act as combustible "fuels" if there is an adequate ratio of fuel to air. The combustible gas or vapor is a "reducing" agent and oxygen in air is an "oxidizing agent". In the presence of an ignition source, an oxidation-reduction reaction involving the fuel and air can be initiated and result in propagation of flame and/or an explosion. Most organic gases and vapors will combust or explode within a range of ratios of "fuel:air" known as the "flammability range". The extremes of the flammability range are referred to as the "lower flammability or explosive limit" (LFL or LEL) and "upper flammability or explosive limit" (UFL or UEL). The LFL or LEL is the minimum ratio of fuel to air that will support propagation of flame; ratios less than the LFL are considered too "lean" to support combustion. The UFL or UEL is the maximum ratio of fuel to air that will support propagation of flame; ratios greater than the UFL are considered too "rich" to support combustion.

Corrosive Chemicals

Corrosive chemical agents include those materials which upon contact with human tissue can induce severe irritation and destruction of tissue due to accelerated dehydration reactions. Corrosive agents also include materials which can dissolve metal in a relatively short period of time. Typical examples of corrosives are organic and inorganic acids and bases. The strengths of acids and bases and the extremes of pH (i.e., <pH 2 and >pH 12.5, respectively) are directly correlated with the degree of corrosiveness.

Inorganic acids are compounds of hydrogen and one or more other elements that dissociate in water or other solvents to produce hydrogen ions. Inorganic acids include chromic, hydrochloric, nitric, and sulfuric. Inorganic acids, when mixed with other chemicals or combustible materials, cause fires or explosions. Organic acids include formic, acetic, propionic, chloracetic, and trifluoracetic. Acids are corrosive in high concentrations and will destroy body tissue and cause chemical burns to the skin and eyes. Inhalation exposure to acidic mists, vapors, and powders can irritate the respiratory system.

Alkalies are caustic substances that dissolve in water and accept hydrogen ions to form a solution with a pH higher than 7. These include ammonia, ammonium hydroxide, calcium hydroxide, potassium hydroxide, and sodium hydroxide. The alkalies in solid form or concentrated liquid solutions are more hazardous and destructive to tissues than most acids. They cause irritations of the eyes and respiratory tract and lesions in the nose. Potassium and sodium hydroxide, even in dilute solutions, soften the skin and dissolve skin fats.

Alkaline ammonia gas is an important part of many compounds containing nitrogen. It is used in making fertilizers, nitric acid, acetylene, urea, explosives, synthetic fibers, and synthetic resins. It is also used in producing paper products, photographic film, dyes, inks, glues, medicine, blueprints, some cleaning solutions, and in ice-making machines and commercial refrigeration. Ammonia can be extremely irritating to the eyes, throat, and upper respiratory system. Large concentrations of ammonia can produce convulsive coughing by preventing breathing and can therefore cause suffocation in a short time. It also dissolves in the skin's moisture, causing a corrosive effect on the skin. Contact with anhydrous and aqueous ammonia causes first- and second-degree burns of the skin and eyes. Blindness can be the ultimate result of this contact. Ammonia gas in the presence of oil or other combustible materials causes fires.

Reactive Chemicals

Reactive chemical agents consist of chemically unstable materials which are typically characterized as either strong oxidizing or reducing agents. Chemical instability results in increased sensitivity to violent reactions which may result in extremely rapid generation of heat and gases which, in turn, may culminate in ignition, explosion or emission of toxic by-products. Some unstable compounds can react with air or water. Ethers can oxidize forming explosive peroxides which are sensitive to shock.

Biological Agents

Biological agents are commonly referred to as pathogenic organisms. There are numerous identified pathogenic biologic agents which may be encountered in the environment. They consist of agents which if introduced into the human body, may disrupt biochemical and physiological function via infectivity or induction of toxicity. The disruption can result in illness and death if the immune system is not able to destroy the biological agents. Infectivity is related to the virulence and the population density of organisms present at a given target site. Toxicity can be induced by microbiological agents which synthesize and release a chemical toxin. Examples of microbial pathogenic agents include bacteria, actinomycetes, rickettsia, fungi, protozoans, helminths, nematodes, and viruses. Unlike radiological and chemical substances, all biological agents, except viruses, are examples of biotic or living organisms. Viruses are abiotic or non-living agents composed of biochemicals (i.e., proteins and nucleic acids) which may insert into and disrupt human cells. Microbiological agents are measured as counts per volume in units of colony forming units per cubic meter of air (cfu/m^3).

Biological hazards vary with the type of occupation. The medical profession is exposed to staph and strep infections, viral infections, and many other types of organisms that their patients carry. Some healthcare workers and biomedical researchers are subjected to infectious hepatitis and other diseases, such as Acquired Immune Deficiency Syndrome (AIDS). AIDS is caused by the Human Immunodeficeincy Virus (HIV) and has an insidious onset with an uncertain incubation period. People at greatest risk, besides those involved in sexual contact, purposeful needle contact, or necessary blood transfusions, are medical and other health-related personnel, as well as safety and security personnel. Individuals working with solid waste removal and disposal are also exposed to the same type of biological hazards. Research scientists are constantly working with bacteria, viruses, and other pathogenic microorganisms and are potentially exposed.

Laboratory exposures can result from the use of biological materials or laboratory work done on biological agents of unknown epidemiology or etiology. The worldwide literature reports an estimated 6000 cases of accidental infection have occurred in the laboratory. They include bacterial, fungal, parasitic, rickettsial, and viral infections. Among the causes of accidents that result in laboratory-acquired infection are oral aspiration through pipettes, accidental syringe inoculation, animal bites, spray from syringes, centrifuge accidents, cuts or scratches from contaminated glassware, cuts from instruments used during animal autopsy, and spilling or splattering of pathogenic cultures on floors and table tops. A large number of these infections have not been traced, and therefore whenever hazardous infectious material is used, extreme caution must be taken.

The zoonoses, which are diseases transmitted from animals to humans, are another biological hazard to which the worker is subjected. These zoonoses include anthrax, brucellosis, tetanus, encephalitis, leptospirosis, Q fever, rabies, salmonellosis, trichinosis, bovine tuberculosis, tularemia, ringworm, and other common fungus infections. Workers also acquire roundworm, hookworm, and many diseases transmitted by fleas, ticks, and flies. Many workers, in addition, are subjected to various dermatoses.

Summaries of example biological agents and related health effects that can be related to occupational exposures follow.

Cotton dust is generated from the finely pulverized part of the cotton plant. The hazard is greatest in handling the raw cotton during the carding stage, but hazards also exist during other stages in the cotton manufacturing process. Workers who inhale this dust may acquire a respiratory disease called *byssinosis*.

The bacteria *Bacillus anthracis* causes an infectious disease called *Anthrax*. The disease is an acute systemic disease involving the skin, gastrointestinal tract, or lungs. In most cases the organism enters the body through a cut or abrasion. The organisms are found in the carpet and leather-goods industries. They are present in hair, wool, hides, and skins. The incubation period is within 7 days, usually 2–5 days.

Large reservoirs of *Salmonella* bacteria are present in chickens, human and animal foods, uncooked meat, vegetable products, and animal feeds. Ingestion of these organisms can cause serious gastrointestinal illness.

Tularemia is caused by the bacteria *Francisella tularensis* and occurs when an individual is scratched by a piece of bone from a wild rabbit. Other animals, such as muskrats, woodchucks, squirrels, and skunks harbor and transmit this organism to humans. Tularemia is also spread from animals to humans by wood ticks, dog ticks, and the blood-sucking deer fly. The incubation period is 2–10 days, usually 3 days, and is related to the virulence of the infecting strain and the number of organisms.

Q fever is a rickettsial disease of rats, cattle, horses, sheep, goats, and dogs. Cattle barns and holding areas for goats and sheep become heavily contaminated with these organisms. The organism, *Coxiella burnetti*, is thought to be inhaled by the victim. Raw milk from infected cows has caused Q fever. The incubation period is 2–3 weeks, depending on the amount of organisms.

Aspergillosis is a fungus transmitted to humans through dust contaminated with *Aspergillus fumigatus, A. niger*, or *A. flavus*. It is a fungus disease of birds, ducks, chickens, and cattle. The incubation period is probably a few days to weeks.

Brucellosis comes from infected cattle and swine. The disease is transmitted by direct contact with the diseased animals and by drinking unpasteurized milk or milk products contaminated with *Brucella sp.* bacteria. The incubation period is highly variable, usually 5–30 days, occasionally several months.

Bacterial, fungal, or parasitic agents in the environment may cause various types of *dermatitis*. Fungus infections are found among kitchen workers, bakers, food handlers, fur, hide, wool handlers and sorters, barbers, and beauticians. Workers who handle grain or straw may also acquire parasites that cause grain itch and ground itch.

Viral hepatitis is insidious with anorexia, vague, abdominal discomfort, nausea, and vomiting, often progressing to jaundice. The occurrence is worldwide among drug abusers, homosexual men, and patients and employees of various medical care institutions. The reservoir is people and the transmission is through blood, semen, saliva, and vaginal fluids. The incubation period is usually 15–180 days with an average of 60–90 days.

Ergonomic and Psychological Factors

Ergonomic factors include human attributes, abilities and limitations as applied to the living and occupational environments. Ideally, optimal conditions will exist to maximize human health, comfort, and well-being while promoting performance efficiency and effectiveness through the appropriate design of tools, machines, tasks, and environments.

Specific human factors include psychological capabilities, physiological dimensions (anthropometrics) and capabilities (biomechanics), and psychosocial issues. Fatigue, boredom, occupational stress, vigilance, mental (memory/recall) abilities/limitations, circadian rhythms, sensory capabilities, anthropometric and biomechanical attributes, and peer, labor-management, or organizational climate variables are just a few of the many factors that can contribute to unfavorable conditions in the occupational setting.

Some of the workplace factors of concern could include hazards associated with the machines, equipment, tools, and layout and design variables found in the occupational environment. Confusing displays (gauges), controls which require excessive force to operate, handtools requiring users to assume awkward body postures or positions, materials that are excessively heavy, or manual materials handling with unreasonable frequency, duration, pace, or transfer distance requirements are just some of the many variables of concern when assessing the causes of occupational-related ergonomic problems. Light and vision adaptation when moving from dark to well lighted environments could contribute to safety-related problems.

Work strain due to unusual postures or improper postural adjustment to the work results in back pain, headaches, nervousness, aching muscles, excessive perspiration, and depression. The most serious results of work stress are serious traumatic injuries, accumulative injuries, or death. Stress is due to excessive heat, improper chairs, poor work space, or any other type of situation in which the individual performs his or her work in an unusual situation or in a continuous manner. An example of this would be writer's cramp, caused by continued use of the same muscles for long periods of time. Other environmental factors, such as illumination, climate, and noise increase stress. Restlessness and discomfort are often associated with sitting in certain hard-back chairs for extended periods.

Occupational stress can increase susceptibility to illness and injury. Stress is the nonspecific response of the body to any demand made upon it. Stress may result in a fight-or-flight response to potentially dangerous situations. The body reacts through elevated heart rate and blood pressure, and a redistribution of blood flow to the brain and major muscle groups and away from the distal body parts. Subjective factors may play a much larger role in the experience of stress than objective factors. Job conditions that may result in stress include undue task demand, work conditions, or work situations, exposures to chemical and physical hazards, pressure to accomplish tasks, and suppressed anger at inappropriate supervision and management decisions. Job stress may be a combination of conditions at work and at home that interact with the worker and results in acute disruption of psychological or behavioral homeostasis. If these reactions or disruptions are prolonged, it may lead to a variety of illnesses that are stress related. These illnesses include hypertension, coronary heart disease, alcoholism, drug use, and mental illness. Lower resistance to infectious diseases may also occur. The individual may be more apt to have accidents that may result in injuries.

The stressors (those things that cause stress) may be sociocultural, such as racism, sexism, and economic variables; organizational, related to hiring policies, plant closing, and automation; the work setting, such as the time and speed necessary to complete tasks, supervision and management problems, and ergonomics; interpersonal, such as marital problems, death, and illness; psychological, such as improper coping skills, poor self-image, poor communication, addictive behavior, and neurosis; biological, such as disease, poor sleep, poor appetite, chemical dependency, and biochemical imbalance; and physical or environmental, such as air problems, climate, noise, poor lighting, poor equipment design, exposure to radiation, and exposure to toxic substances. The distress cycle includes job stress, job dissatisfaction, organizational distress, stressful life changes, life and health risks, accident risks, and stress due to new technology.

Surveys suggest that 75–90% of all visits to primary care physicians are due to stress-related problems. These problems include backache, headache, insomnia, anxiety, depression, chest pain, hypertension, gastrointestinal problems, and dermatological problems. Stress also may lower your resistance to a variety of diseases and contribute to a slow healing process. Uncertainty, doubt, lack of recognition, pressure related to time, and insecurity all are causes of health problems, and all may be related to an inordinate amount of stress created on the job.

OCCUPATIONAL HEALTH CONTROLS

General Measures for Controlling Workplace Agents and Factors

Substitution

Sometimes it is possible to substitute less hazardous materials, equipment, or even processes for more hazardous ones. Solvents such as dichloromethane and methylchloroform have been substituted for carbon tetrachloride. Toluene or xylene have been substituted for benzene. It is safer to dip an object into paint than to spray paint. This is an example of a process change.

Safety cans are much safer for storing flammable solvents than glass bottles. Another example of change in equipment would be the use of safety glass as opposed to regular glass in the fume hood and in glasses used by workers. The substitution of a wet method of operation for a dry method, where dust becomes a serious hazard, is another technique in good environmental control. An example of this would be the wetting of floors before sweeping up dust or the use of treated mops rather than regular dry mops in the hospital. The use of water drills by dentists has not only reduced the heat generated by the drills and the potential pain, but has also reduced the number of particles that could fly into the dentist's eyes or injure his or her hands.

Isolation

Isolation is the physical separation of the worker from the hazard by use of a barrier, by increasing the distance from, or decreasing the time of exposure to, the hazardous material. Stored materials may be isolated physically from other materials and the worker or they may be stored in small units in specially ventilated rooms or under protective coverings. Radioactive materials may be stored in concrete vaults or in lead pigs. When the material to be mixed or processed is extremely hazardous and must be mixed by hand, either remote guidance controls should be used with mechanical hands within a specially protected container, or the individual might use a glove box with his or her arms and hands in protective gloves and the rest of his or her body shielded from the area of material preparation.

Extremely hazardous equipment, such as that which is under high pressure, may be separated from the worker by reinforced concrete, steel, or armor plate. The work is performed by remote control and the worker observes the process through television cameras, mirrors, or periscopes. Remote-process control is used when the process itself is extremely hazardous. An example of this is automatic sampling analysis in petroleum processing plants, censors used on the production line, and processed equipment controlled through the use of a computer. Enclosures are another method of isolation. Enclosures are used in such operations as sandblasting, heat-treating, mixing, grinding, and screening. Enclosures also are a useful tool in noise control.

Ventilation

Ventilation is used to provide a comfortable workplace environment and to keep dangerous materials from accumulating in the air. Ventilation is brought about when air is removed from a given area and fresh air replaces it. The exhausted air carries with it heat and possible odors that have been generated by people, processes, or equipment. The quantity and type of distribution of air by means of ventilation systems depends on the amount of heat produced, the kinds of contaminants getting into the air, the quantity of contaminants, and the ultimate air quality desired. Proper ventilation dilutes the concentration of a hazardous material and prevents fires and explosions. The various types include natural ventilation, forced general ventilation or dilution ventilation, general exhaust ventilation, and local exhaust ventilation.

Natural ventilation refers to air flow into the room through cracks, crevices, and open windows. Since it is difficult to control natural ventilation because of variations in temperature, pressure, wind velocity, and general weather conditions, this process is not very practical for an industrial environment.

Forced general ventilation or dilution ventilation is accomplished by blowing fresh air into a workroom. This may be used to control vapors given off by organic solvents at room temperature or to reduce the heat load within the working area. Exhaust ventilation is accomplished by mechanically withdrawing large quantities of air from the work area. The air is then replaced by fresh air being sucked or blown into the work area. However, the movement of large quantities of air can cause drafts and may be uncomfortable for the worker. When this technique is used, it is preferable to have several low velocity exhaust fans. Diluting the contaminant in the air, either by forcing in large quantities of air or exhausting large quantities of air in the work area, does not reduce or eliminate the source of hazardous material introduced into the workroom. This is an important consideration when the hazardous material can cause a serious health problem for the worker. General dilution ventilation is acceptable if the toxicity of the contaminant is low, the amount of material is produced at a low rate, the kind of material at low concentrations does not have serious effects on the individuals, and the material can be diluted adequately with a large volume of air. Local exhaust ventilation utilizes less make-up air, and therefore reduces costs due to heating or cooling. It helps control the generation and accumulation of particulates and vapors at the source. As a result, a lower if not neglible amount of contaminant is discharged into the air helping to reduce worker exposures and housekeeping problems. Since local exhaust ventilation controls the contaminant prior to disharge into the air, it is the preferred method for controlling worker exposure to toxic agents.

Personal Protective Equipment

Personal protective devices consist of respirators, protective clothing, gloves, shoes, helmets, eye and face shields and goggles, earmuffs, plugs. Respiratory equipment is used to protect against inadequate oxygen, toxic gases, toxic vapors, toxic particulate matter, or any combination of these hazards. The type of respiratory equipment used is based on the kind of hazardous situation in which the individual will work. It must be determined if there is adequate oxygen in the atmosphere, the types of contaminants in the air, the effects of the contaminant on the body, how quickly the effect will occur, the concentration of the material, and the length of time the worker will be present within that environment.

There are three basic types of respiratory equipment:

1. air-purifying respirators contain special filters or sorbents for separating the contaminant from the inhaled air prior to entry into the respiratory system
2. air-supplied equipment have hoses and facepieces or hoods and are connected to a central system that provides breathing air via a stationary compressor or air cylinders
3 self-contained equipment that provides breathing air from a cylinder carried by the individual

Protective clothing includes the use of gloves, aprons, boots, coveralls, and other articles made of impervious materials. Their function is to prevent chemicals from coming in contact with the skin. Special types of reflective aluminum clothing are used in areas where there is considerable radiant heat. Helmets or hard hats are needed in any type of construction or in such industries as the steel industry to absorb impact from heavy materials.

The eyes and face are protected against contact with corrosive solids, liquids, and vapors or impact with projectiles by using safety goggles, face shields, and glasses. It is important to protect the eyes against numerous hazards, such as glare, ultraviolet light, and other kinds of physical or chemical hazards. Such industries as grinding are extremely dangerous, since the particles may easily penetrate the eye or cornea, or cut the eyelid. Chemical burns in the eyes due to splash may cause temporary or permanent loss of vision.

Ear protection is provided by the use of earplugs or earmuffs. The devices attenuate sound levels so that a reduced level actually enters the auditory system. It is essential that the plugs and muffs are kept very clean to prevent ear infection, and the process should be evaluated, since the noise level may exceed the protection afforded by the muffs or plugs.

Housekeeping

The satisfactory environment is a complex interrelationship between the physical facilities, color, light, ventilation, aesthetics, dirt, chemicals, and microbiological hazards. The dark, shabby, dirty, littered environment is depressing and leads to accidents and poor occupational health practices.

The procedures and techniques of cleaning are extremely essential within the industrial environment. Attention must be paid to dustless removal, by the use of treated cloths, mops and vacuum cleaners with filters, of inert or hazardous dusts that may fall on the floors, ledges, equipment, and partitions. Wet or damp sweeping is an important technique in the removal of this dust, provided it does not cause a slipping hazard or the chemicals present do not react in the presence of water.

It is essential to use proper detergents, disinfectants where needed, and equipment capable of performing a good housekeeping job. The personnel must have proper training in the use of the equipment. They must also have an understanding of the types of materials being removed from surfaces. It is extremely important that housekeeping personnel be well supervised in order that a minimum of contaminants may be spread and a maximum of contaminants removed during the housekeeping process.

Where spills of toxic materials have occurred, previously established techniques for removal of the toxic material must be used. The workers must be trained to utilize these techniques and to know which techniques to use to protect themselves during the clean-up procedure. The housekeeping must be performed on a daily or shift schedule and also on special schedules depending on what must be done.

The types of equipment and floors to be cleaned vary with the industrial operation, making it difficult to list specific cleaning procedures. However, in all cases, materials used for cleaning must be evaluated so that they do not in some way combine with the hazardous or inert material to cause new hazards. Each industrial plant should be evaluated for its particular housekeeping problem by environmental or industrial hygiene consultants. Specialized programs should be established for each plant.

Education

The educational process is used in the occupational environment for training workers in job performance, retraining workers in their job assignments, or teaching new procedures, and hopefully, changing attitude toward the utilization of good personal practices. Although the machinery may be the best engineered, contain all available safeguards, and the process and materials may not be causing a problem within the environment, lack of good attitudes on the part of the employees may lead to hazardous situations. Employees may improperly utilize equipment, remove guards, wear unsafe clothing, or allow their minds to wander during the work process. Good educational procedures should help workers develop a significant understanding of accident and health hazards. They should help workers to be flexible in their work habits. They should learn to work at a normal rate of speed, not intentionally violate existing rules and regulations, and carefully assess the potential problems of their unique environments.

Control of Physical Agents

Noise and Vibration

Noise is controlled by use of proper engineering, proper administrative techniques, and personal protective equipment. When replacing equipment, total noise levels for the area should be considered and pieces of equipment with lower noise levels should be utilized wherever possible. The proper maintenance of equipment is essential, since work or imbalanced parts, improper adjustments, inadequate lubrication, and improperly shaped tools contribute to the noise problem. Belt drives on machines should be substituted for gears wherever possible. Vibrations may be dampened, or decreased, by increasing the mass or stiffness of the equipment or material and by using rubber or plastic linings. The support of machines should be strengthened. Flexible mounts, hoses, or pipes should be used, if possible. Mufflers at either the intake or exhaust on internal combustion engines and compressors should be checked regularly and replaced when needed. The noise source and the operator should be isolated when possible.

Proper administrative controls reduce excessive noise exposure by adequate arrangement of work schedules, removing the worker to a low noise level when he or she has worked a high level during the day, dividing the work at high noise levels into several days or among several different people, running a noisy machine a small portion of the day, and running high-level noise machines when a minimum number of employees are present.

Personal protective equipment includes the use of plugs and earmuffs that reduce the sound level from 25–40 dB. Further, a hearing conservation program should be implemented. The individuals subjected regularly or frequently to 90 dBA or above should be tested. A complete medical examination and previous work history should be completed for each worker and maintained on a regular basis. Individuals showing signs of physical damage or having spent long periods of time at high noise levels in the past should be assigned to a much quieter environment.

Vibration is reduced by isolating the disturbance from the radiating surface, reducing the response of the radiating surface, reducing the mechanical disturbance causing the vibration, using materials that will deaden the vibration, reducing friction, and absorbing the energy created by the vibration. Proper design of equipment, adequate balancing of rotating machinery, and proper maintenance of machinery also act as effective controls.

Nonionizing Radiation

Individuals should be protected against ultraviolet rays by minimizing the time of exposure, keeping a maximum distance from the source, and using proper shielding. Glasses should be used when needed. Since ventilation does not control radiant heat, other techniques must be used. There techniques include installation of shields that reflect the radiant heat or absorb it without reradiating it to the individual and the use of reflective clothing, such as aluminized coats, pants, coveralls, and helmets. An individual should only be permitted to work within a hot area for short periods of time, wearing a breathing apparatus that prevents inhalation of hot vapors. Another means of protection is maintenance of proper distance between the source of heat and the worker.

The controls for microwave hazards include limiting the time and amount of exposure, and providing proper shielding and distance from the source. The controls for lasers include proper engineering to prevent exposure to the laser beam, proper procedures to avoid errors, and proper personnel equipment coupled with periodic eye examinations.

A pleasant visual environment is the result of adequate quantities and proper quality of illumination. Since the most common faults of lighting systems are low levels of illumination, glare, shadows, and poor brightness ratios, these factors must be carefully considered. The size, contrast, brightness, and time involved in visual activities must also be considered. Any glare present within the field of vision may cause visual discomfort. This glare may either be direct or indirect. Shadows should be eliminated, since they interfere with effective vision and are very annoying. Care must be taken to provide proper brightness ratios between the visual task and the immediate surrounding environment. For example, a student should never sit in a darkened room with only a single source of light on his or her textbook. Whenever he or she looks up his or her eyes must adjust from the brightness of the task to the darkened environment. This causes fatigue and reduces the student's ability to concentrate and absorb the material he or she is studying. The environment itself must be clean and pleasant. It is always better, if possible, to use pastel color on walls and ceilings. Where it is impossible to eliminate a source of glare, the worker should be shielded from it and, if necessary, should wear protective goggles.

Ionizing Radiation

The type of controls used for ionizing radiation are based on the intensity and energy of the radiation source and whether it is an external or an internal hazard. X-rays and gamma rays are the most common type of external radiation hazard, for they can penetrate the entire body. Beta rays may or may not be an external hazard, depending upon their energy. If the beta rays penetrate to the basal layer of the epidermis, they are hazardous. Neutrons, because of their high penetrating powers, are also external radiation hazards. Prevention of these hazards is accomplished through distance, proper shielding, and exposure time. The inverse square law applies to the reduction of radiation intensity due to point sources of X-ray, gamma ray, and neutron radiation. Good shielding includes proper enclosure of the radiation source and the use of diaphragms and cones to limit the beam to avoid stray radiation. Radiation will scatter, so it is necessary to limit this beam and have proper absorptive shielding. The exposure time should be kept to an absolute minimum. Other controls include the use of protective aprons and gloves, wearing of film badges, carrying of dosimeters, and evaluation of machines to ensure that optimum voltage is used in the operation of the machine and the fastest film is utilized for best exposure. Since the radioactive material may enter the body through inhalation, it is important to wear proper masks in potentially dangerous areas and to contain the radioactive material in hoods with special exhaust systems. Food and water must be thoroughly monitored to ensure that individuals do not ingest radioactive material. The skin should be protected by the use of protective clothing. Absolute cleanliness must be practiced at all times and double containers should be used for all waste or cleaning materials.

Heat

There are numerous controls recommended for workers in hot environments, including acclimatization, proper work and rest periods, distribution of the work load, assigning younger and better equipped personnel to hot work, scheduling hot jobs for the coolest part of the day, regular breaks, frequent physical examinations, adequate quantities of drinking water, replacement of salt, personnel education concerning work problems related to heat and the effect of alcohol and drugs, protective clothing, shielding from the sun, adequate ventilation, and shielding from sources of radiant heat. Acclimatization involves a progressively increasing exposure to a given temperature based on a known activity over a period of typically at least a week.

Control of Chemical Agents

Exposure to chemical agents also is controlled by implementing administrative, engineering, and personal protective equipment controls. Toxic and flammable atmospheres are best controlled by minimizing generation by using less hazardous substitutes and controlling generation via process design and local and general dilution ventilation. When administrative and engineering controls are not feasible or they do not adequately control the hazard, use of personal protective equipment is warranted.

Control of Biological Agents

Special controls are needed when working with biological hazards. These controls include air-pressure differentials, filtered supply air, filtered exhaust air, special airlock and pass-through autoclaves, ultraviolet barriers at throughways and in special laboratory areas, special treatment of contaminated sewage and waste, and isolation and separation of all of the water, steam, natural gas, and other utilities. Additional controls are discussed in the section on controls of hazardous and toxic materials.

Control of Ergonomic and Psychological Factors

Tools and work situations must be selected to reduce as much work stress as possible. If the individual has difficulty reaching operating equipment, or suffers from poor posture, discomfort, or emotional strain, then physical illness or accidents may result. Proper design of the employee's work space is also essential. Care must be taken to eliminate fatigue. An evaluation must be made of the individual to determine if he or she can effectively handle the physical and mental loads of the assignment. The worker should be trained for the task and then checked by a knowledgeable supervisor who has the ability to communicate with him or her.

An occupational stress management program should help the employees cope with the occupational stresses, as well as the personal and societal stresses. Many companies have not only developed stress management training programs, but further have developed an "employee assistance program." The function of these programs is to have individuals establish a mechanism to reduce stress, plan alternatives to problems, and assist individuals in planning techniques to deal with stress when the problems cannot be relieved. The actual program could consist of lectures, TV tapes, small group dynamics, and referrals to physicians when specific problems of a personal nature need to be determined and treated. All stress management programs should include techniques of relaxation therapy, good nutrition, preventive measures, and control of habits that may lead to stress. Above all else, a good supervision and management program to train the supervisors and managers would be extremely useful, since good management can help reduce stress levels among employees.

OCCUPATIONAL HEALTH AND SAFETY PROGRAM

General Aspects

The occupational health and safety program has as its major objectives protection of workers against health and safety hazards in the work environment; proper placement of workers according to their physical, mental, and emotional abilities; maintenance of a pleasant, healthy work environment; establishment of preplacement health examinations; establishment of regular, periodic health examinations; diagnosis and treatment of occupational injuries or diseases; consultation with the worker's personal physician, with the worker's consent, of other related health problems, such as heart disease; health education and counseling for employees; safety education for employees; establishment of research to identify hazardous situations or find means of preventing hazardous situations; and establishment of necessary surveys and studies of the industrial environment for the protection of the worker, his or her family, and the community.

A good occupational health program not only benefits the worker but is also a value to the industry. It reduces the cost of worker's compensation insurance, reduces the cost of hospital insurance, reduces absenteeism and labor turnover, helps satisfy the legal requirements set forth in the OSHA laws, and other federal laws and state laws, and creates a good working relationship between management and workers.

Hazard Communication Program

The Hazard Communication Program is required by the OSHA standards and by standards established for Federal Employee Occupational Safety and Health Programs. The program consists of three elements. They are hazard warning labels, material safety data sheets (MSDSs) and employee training programs. The standard which was issued in 1983 by OSHA, requires chemical manufacturers and importers to make a comprehensive hazard determination for the chemical products they sell. The manufacturers, importers, and distributors must provide information concerning the health and physical hazards of these products. This is accomplished by means of warning labels. The manufacturing sector that uses the chemicals to turn out additional products must protect their employees by meeting the following standards:

1. Preparing a written Hazard Communication Program, which includes an inventory of all hazardous chemicals used in their facilities.
2. Obtaining an MSDS for each hazardous chemical they use.
3. Displaying appropriate facility placards and warnings.
4. Preparing a hazard communication training plan.
5. Providing training to employees who are potentially exposed to hazardous chemicals in their facilities.

The elements of the Hazard Communication Program should include preparation of a written hazard communication plan; identification and evaluation of chemical hazards in the work place; preparation of a hazardous substance inventory; development of a file of MSDSs; provision for access to MSDSs for employees; ensurance that incoming products have proper labels; development of a system of labeling within the facility; development of the training program; identifying and training employees who are potentially exposed to the hazardous chemicals; and evaluation of the programs and redirection where necessary. The definition of hazardous chemicals will need to be clearly stated in order to properly identify the potential problems within the workplace. The written plan must describe how complaints will be handled, labeling, how MSDSs will be obtained and made available, and how information and training will be provided. The inventory of all the toxic chemicals in the workplace will have to be cross-referenced to the MSDSs. Provision will be made for employees to deal with accidental spills and leaks, and for making employees aware of the types of hazards that may occur as a result of the chemicals or the pipes carrying the chemicals.

A hazard assessment should be conducted by trained environmental health personnel such as an industrial hygienist. The survey itself involves identifying the chemicals, how they are being used or produced, and in what quantities. Determine if warning labels are placed in readily observable places on drums or containers. Next set up a flow chart of each operation. The flow chart should start at the place where the raw materials enter the building, is taken to storage, and then is introduced into the actual manufacturing operation. From this point, each step along the way in the manufacturing of the product should be shown and the potential problems listed under the particular step. The flow chart will proceed to the final step of production packaging and storage, and also to byproducts, waste products, waste product storage, and disposal. By using the flow chart and diagramming the actual flow, it is easier to determine who may come in contact with the hazardous chemical, where it may occur, how it may happen, when it may happen, why it may happen, what the final results may be, and the kind of protective measures that are being utilized by the employee to avoid contact with the chemical.

An air sampling survey should be conducted in those areas that appear to have potential chemical hazard problems. The air sampling survey should consist of a general air survey, specific air samples in high concentration areas, and the monitoring of personnel with personal monitors at the breathing level.

An important part of the study includes ventilation controls related to local exhaust ventilation as well as general ventilation. Since the chemicals vary so, the requirement for respirators and protective equipment varies considerably.

Once the survey has been concluded, a hazardous substances inventory should be developed and a hazardous chemical MSDS file should be provided. When hazardous chemicals are utilized, they should be clearly identified by chemical name, trade name and common name. All of the hazardous ingredients in the substance should be listed. The physical and chemical characteristics of the substance, such as boiling and freezing points, density, vapor pressure, specific gravity, solubility, volatility, and the product's general appearance and odor, should be included. All physical hazards, such as carcinogenicity, corrosive, toxic, irritant, and other hazards, should be included. The route of entry should be clearly described. Special precautions and procedures should be established for spills, leaks, cleanup controls, emergencies and first aid. The program must provide access to the MSDSs for employees. All incoming labels and existent labels need to be checked.

Employee education and training is an important requirement of the program. The employees need to clearly understand the operations of their work areas, the hazardous chemicals present, the means of prevention and control, the physical and health hazards associated with the chemicals, and the types of protective clothing or respirators that they should wear in order to avoid problems. Inherent to this is hazard recognition or the ability to recognize when conditions are not acceptable from health and safety perspectives. Finally, workers should be well aware of any symptoms of illness that may be attributed to the chemicals or other substances in the workplace.

SUMMARY

Occupational disease, injury, and death is of vital concern to our society because the costs and physical and emotional anguish are great. Occupational illnesses and injuries may be caused by physical, chemical, biological, ergonomic, and psychological factors. Although there are numerous techniques available for hazard reduction and the potential for intervention is excellent in certain areas, it is extremely poor in others. Environmental health and clinical practitioners must work together for ongoing improvement.

The basic principles of occupational disease and injury control involve prevention primarily and treatment secondarily. To have proper prevention and treatment, it is necessary to not only study the individual employee and ensure that job work activties are carried out in an effective manner. It is also important to evaluate and control the level of anticipated and recognized hazards associated with the various agents in the workplace encountered by workers. Once the controls are established, both the environment and the worker must be constantly monitored to ensure that exposure levels to a given set of occupational hazards will not exceed the appropriate established standards.

Chapter 11

MAJOR INSTRUMENTATION FOR ENVIRONMENTAL EVALUATION OF OCCUPATIONAL, RESIDENTIAL, AND PUBLIC INDOOR SETTINGS

BACKGROUND AND STATUS

Environmental evaluation includes the collection, detection, and measurement of representative samples from an environmental matrix such as air, water, soil, and even food and beverages. This chapter provides a summary of the major forms of major monitoring and analytical instrumentation that are used to evaluate several types of physical, chemical, and biological agents associated with occupational, residential, and public indoor settings. The air is a common matrix sampled and analyzed in these environmental settings since inhalation of contaminated air is considered a primary mode of toxic or pathogenic agent entry. In addition, the air serves as a major matrix for elevated sound levels, extremes of temperature and humidity, and transfer of radiation energies. Accordingly, this chapter only focuses on related instrumentation for detecting and measuring contaminants and energies in the air matrix. It must be noted, however, that although the majority of the instrumentation summarized in this chapter are used mainly to evaluate occupational and non-occupational exposures in indoor environments, some also have application for evaluation of outdoor air contaminants and energies.

The data collected and analyzed are used to evaluate both actual and potential external exposure to agents encountered by humans. In turn, the levels are compared to what are considered acceptable limits of exposure. These limits include the Permissible Exposure Limits (PELs) enforced by the Occupational Safety and Health Administration (OSHA), recommended Threshold Limit Values (TLVs) by the American Conference of Governmental Industrial Hygienists (ACGIH), and Recommended Exposure Limits (RELs) by the National Institute for Occupational

Safety and Health (NIOSH), among others. In some cases, a parameter will be monitored without concern necessarily for excessive exposure. For example, illumination is evaluated and compared to recommend guidelines to assure that there is an appropriate, neither inadequate nor excessive, quantity of lighting.

Environmental and occupational health and safety specialists also must evaluate engineering controls designed to reduce or eliminate airborne contaminants and to provide atmospheric comfort in indoor environments. Thus, both local, general, or dilution ventilation systems are evaluated to assure that the design and operation is efficient relative to the characteristics of the setting, including the type and levels of contaminants and the number of occupants present.

EVALUATION CATEGORIES

Monitoring and related analytical activities are divided into several categories to reflect the type of monitoring that is conducted. Categories are based on factors including time, location, and method of collection and analysis. Each serves a purpose in evaluating the occupational, residential, and public indoor environments to determine the degree of external exposure of occupants to various agents.

Instantaneous or Real-Time Monitoring

Instantaneous monitoring refers to the collection of a sample for a relatively short period ranging from seconds to typically less than 10 minutes. A major advantage of instantaneous monitoring is that both sample collection and analysis are provided immediately via direct read-out from the monitoring device. The data represents the level of an agent at the specific time of monitoring. Accordingly, instantaneous monitoring also is referred to as "direct reading" and "real-time" monitoring. Indeed, real-time monitoring is perhaps a more appropriate designation since there are some devices already developed and some being designed for integrated or continuous monitoring that provide a direct readout or instantaneous result without need for laboratory analysis and the associated delays. In addition, the main purpose of real-time monitoring is to reveal what a level of an agent is at an immediate point of time or during "real-time".

The application of real-time monitoring varies. The strategy is used when preliminary information regarding the level of an agent is needed at a specific time and location. For example, real-time monitoring is commonly used for screening to identify agents and measure related levels during the initial stages of an indoor air quality investigation. This is important for developing follow-up monitoring strategy and to determine if integrated monitoring (see the following section) is warranted. Real-time monitoring also is beneficial for determining levels of agents during short-term operations or specific isolated processes when peak levels are anticipated or suspected.

Integrated or Continuous Monitoring

Integrated monitoring refers to the collection of a sample continuously over a prolonged period ranging from more than 10 or 15 minutes to typically several hours. Thus, integrated monitoring also is referred to as "continuous" monitoring reflective of the extended period of sample collection. Most work shifts are 8 hours, therefore occupational exposure limits are most commonly based on an 8 hour exposure period. Accordingly, it is very common as well for sampling to cover the duration of the shift. Sampling for longer periods, however, is not uncommon. Several strategies can be followed. For example, a sample run could be started immediately at the beginning of an 8 hour period and allowed to run until the end. Analysis of the sample would provide a single value representative of the level of a particular agent during the 8 hour period. The single value represents an "integration" of all the levels during the 8 hour period. The single value, however, does not provide information regarding fluctuations of levels that were higher, lower, or not detectable during shorter periods within the 8 hours. In addition, there is no indication as to the levels at specific times and locations during the 8 hours. As a result, an alternative strategy involves collection of several samples of shorter duration during the entire sampling period. In turn, analysis of the individual samples provides levels associated with specific times, tasks, and locations during the total (i.e., 8 hour) sampling period. Concentration (C) and corresponding sample time (T) data from one sample (C_1) or several individual samples (C_1 to C_n) can be time-weighted (C x T) and averaged to provide a single overall time-weighted average (TWA) for the 8 hour period.

$$8 \ hr \ TWA = \frac{\left(C_1 T_1\right)}{8 \ hr}$$

$$or, \ 8 \ hr \ TWA = \frac{\left(C_1 T_1\right) + \left(C_2 T_2\right) + \ ... \ + \left(C_n T_n\right)}{8 \ hr} \qquad (eq. \ 11.1)$$

The major advantage of integrated monitoring is that it provides a single value for the level of an agent over a prolonged period. The level of an agent can be determined during discrete times and locations within a workshift to assist in identifying factors which influence elevated values of exposure or external exposure. A major disadvantage associated with integrated monitoring is that in most cases samples must be submitted to a laboratory for analysis prior to knowing the measurement. This frequently results in a delay between sample collection and data reporting; a delay that potentially exposed occupants are not always readily willing to understand or accept.

Personal Monitoring and Area Monitoring

Personal monitoring involves direct connection of an integrated monitoring device to an individual. The device, in turn, will collect a sample and/or record the intensity of an agent in the specific areas and during specific tasks conducted by an individual. Indeed, personal monitoring is frequently a form of mobile monitoring since the sampling device travels to the same areas at the same times as the individual that wears it.

If inhalation is the mode and the respiratory system the route of entry of an agent, the monitoring device is positioned in the individual's "breathing zone", that is, an area within a one foot radius of the individual's nose and mouth. Typically, an integrated monitoring device for personal sampling is attached near the individual's scapula or collar bone. Relative to evaluation and impact of sound levels, however, hearing is the major mode and the auditory system the route of entry. When conducting personal monitoring for sound, therefore, the sampling device can be connected to the individual's "hearing zone"; a region within a one foot radius of the ear. This zone is ideally the ear itself, or more commonly, the trapezius region of the shoulder.

Instantaneous or real-time monitors also can be used to determine levels of agents in an individual's breathing and hearing zones. For example, a sound level meter can be held by the professional or technician conducting the monitoring in the auditory region of an individual whose exposure is being evaluated. This would provide an instantaneous assessment of sound levels in the individual's hearing zone at the specific time of monitoring.

Area monitoring is conducted to evaluate the levels of agents in a specific location, instead of evaluating levels encountered by a specific individual. Integrated monitoring devices are typically positioned in a stationary location. The data represents the level of an agent in the specific area for the sampled time. Instantaneous or real-time monitoring, however, involves area sampling in either a stationary or mobile mode. Instantaneous monitoring may be conducted while standing still in a given location or the monitor can be transported to various locations while intermittently checking the indicator or read-out on the device.

EVALUATION OF AIRBORNE PARTICULATE AND GASEOUS AGENTS

Integrated or Continuous Monitoring for Particulates

As defined in Chapter 10, there are several subclasses or forms of airborne particulates including dusts, fibers, and fumes. Exposure limits for airborne dusts, fibers, and fumes have been established for many individual elements and compounds based on their inherent toxicities. There are situations when it is important to measure and know the concentration of a specific particulate that is either highly, moderately, or slightly toxic. Other situations arise, however, when the concentration of a more general composition of particulates, including a combination of highly to only slightly toxic, is warranted. In addition, situations also exist in which there are no established exposure limits for a specific type of particulate air contaminant, but measurement of that material as "total particulate" will provide important data during exposure assessment.

"Total particulate" refers to generic particulate mixtures consisting of various diameters, compositions, and inherent toxicities. These materials may be generically classified as "nuisance dust" or "particulate not otherwise classified". The fraction of total particulate that consists of particles with diameters <10 microns is referred to as "respirable particulate". The respirable fraction of particulates also can consist of mixtures of various compositions of particulates, but has a higher probability of entering the lower respiratory system because the smaller particle diameters (<10 microns) potentially permit the upper respiratory defense mechanisms to be by-passed.

Fequently, however, only knowing the concentration of more generic total or respirable particulates in the air is insufficient. For example, if there is a real or potential source of specific fibrous dusts, such as asbestos, it is important to know what the concentration of these specific particulates are due to their inherent toxicity. The morphology of fibers, unlike other classes of particles, are more columnar than spherical. As a result, the sampling and analysis involves methods that will assure accurate collection and quantification of fibers, such as asbestos. Specific metallic dusts and fumes are common particulates present in the environment for which many specific exposure limits for airborne levels have been established. If there is a real or potential source of metallic dusts or fumes, it is important to know what the concentration of a specific metallic particulate is due to the associated toxicity. Indeed, most metallic dusts and fumes are considered highly toxic relative to "nuisance" dust or "particulates not otherwise classified" and, accordingly, occupational exposure limits are much lower.

Specific sampling and analytical methods are available for dusts, fibers, and fumes. Sampling involves moving an airstream through a filter collection medium, selected in accordance to particle size (diameter) and/or composition of the particulates. Airborne particulates are electrostatically attracted to, intercepted by, and impacted on the filter medium, resulting in separation of the particulates from the air (Figure 11.1). The filter medium is contained in a plastic holder or cassette which, in turn, is connected to an air-moving device or sampling pump via a short length of flexible tubing (Figure 11.2). The two most common filter media are polyvinyl chloride (PVC) for collecting total and respirable dusts, mixed-cellulose ester fiber (CEF) for collecting metallic dusts, fumes, and asbestos fibers. The common sample filter media range from 25–37 mm diameter and 0.8–5.0 μm pore sizes. The diameter of the filter provides a cross-sectional surface area for deposi-

Figure 11.1.   Particulates or aerosols are separated and collected from contaminated air using a sampling filter. Polyvinyl chloride (PVC) and cellulose ester fiber (CEF) filter diameters are commonly 25 and 37 mm and specific pore sizes often range from 0.8 to 5.0 μm.

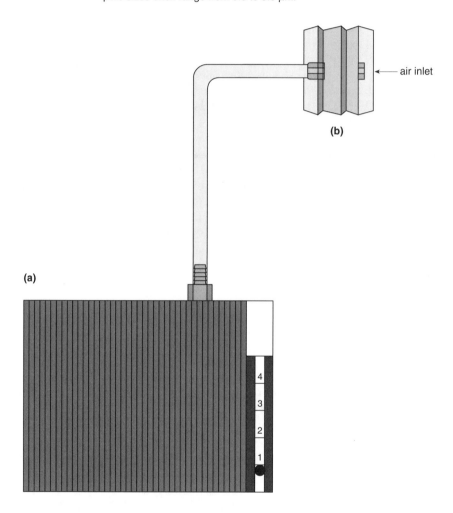

Figure 11.2.   Airborne particulates such as total dust, metal dust, metal fumes, and asbestos fibers are sampled using (a) a high-flow or multi-flow battery-operated air sampling pump connected via flexible tubing to (b) a membrane filter contained within a plastic cassette.

tion and collection of particulate. The pore size causes particulates with diameters greater than the pore diameter to collect on the surface of the filter. Particulates with diameters less than the pore size, however, are collected within the pores due to an electrostatic attraction between the particles and the filter. In addition, the pores are tapered and allow interception and impaction of particulates within the medium. Indeed, collection efficiency increases as particulates accumulate on the filter.

The "high-flow" or "multi-flow" air sampling pumps used to collect air containing suspended particulates are typically operated at flow rates between 1.5 and 3 liters per minute (L/min). "Vaccum pumps" also are used, however, to collect very high volumes of air in a relatively short period and operate at flow rates reaching 1 ft³/min (28.3 L/min). The total volume of air sampled is calculated by multiplying the flow rate of the air sampling pump (Liters/min) and sampling time (minutes) together. The volume collected is expressed in liters (L) which, in turn, is commonly converted to cubic meters (m³) or cubic centimeters (cc). The volume actually represents how much air passed through the filter sampling medium.

$$Volume\ L\ =\ Flow\ Rate,\ \frac{L}{min}\ \times\ Time,\ min \qquad (eq.\ 11.2)$$

The sample collection medium for *total dust* is prepared by pre-weighing a PVC filter for each field sample that is going to be collected. A filter is pre-weighed using a mechanical or electrobalance to determine "pre-weight". The pre-weighed filter is subsequently positioned on a cellulose support pad in a three-stage plastic cassette and attached to an air sampling pump using a short length of flexible hose.

A PVC filter medium is prepared the same way for sampling *respirable dust*, however, the filter and a two-stage cassette is used in combination with a particle size-selective device or cyclone (Figure 11.3) to separate and collect the respirable fraction from total particulates. Only particles with diameters <10 μm flow through the upper cyclone, into the cassette, and on to the filter. Air enters the cyclone which creates a vortex, and due to centrifugal force, the larger and denser particles flow downward and settle into the bottom of the device, called a grit cap. Simultaneously, particles with smaller diameters (<10 μm) flow upward in the opposite direction and enter the inlet port of a two-stage cassette and deposit and collect on a pre-weighed PVC filter.

Gravimetric analysis involving a mechanical or electrobalance is used to measure amount of either total or respirable particulate (Figure 11.4). Unlike specific particulate air contaminants such as asbestos fibers, metallic dusts, and fumes that require an analytical method that will both identify the specific analyte and measure the amount collected, grametric analysis of either total or respirable particulate only involves measurement of the amount (weight) collected on the filter. The gravimetric procedure involves comparison of post-weight and pre-weight of filter. Theoretically, the pre-weight is only the weight of the filter and the post-weight is the weight of the filter plus the collected particulate. The difference in weights (post-weight minus

(a)

(b)

air inlet

Figure 11.3.   Airborne respirable particulates are sampled using (a) a high-flow or multi-flow battery-operated air sampling pump connected via flexible tubing to (b) a membrane filter contained within a plastic cassette positioned in a cyclone assembly.

| **Sample Chamber:** shields sample filter from air current | **Weighing Pan:** stage where sample filter is placed | **Balance Mechanism:** responds to weight of sample filter | **Read-Out:** displays weight of sample filter |

Figure 11.4.   Major components of an electrobalance used for conducting gravimetric analyses of either sampled total or respirable particulates collected on a PVC filter.

pre-weight) is the mass of particulate collected. Gravimetric analysis of total and respirable particulate is the easiest of the major analytical methods. Both pre-and post-weights of the PVC filters are determined as follows:

1.   filter is placed on a wieghing pan inside the transparent glass sample enclosure with sliding door or/sash to shield it from moving airstreams and related turbulence during weighing
2.   the mechanical or electrical balance device responds to the weight of filter
3.   a digital read-out is displayed in response to the mechanical or electrical signal from the balance mechanism and provides weight of filter in milligrams (mg).

Dividing milligrams of particulate by cubic meters of air sampled provides results in mg/m^3

Sampling for *asbestos fibers* involves the same type of air sampling pump, but use of a different filter and cassette than those used to collect the more generic total or respirable particulate. A sample is collected using a 25mm diameter cellulose ester fiber (CEF) filter with 0.8 μm pore size positioned within a longer two-stage open-faced cassette. The open-faced, longer (50mm extended cowl) permits more uniform collection and distribution of fibers across the face of the filter.

The method of fiber analysis commonly involves *phase contrast microscopy*. The CEF filters are chemically digested and made transparent on a microscope slide so that individual fibers can be counted via microscopy. The method allows both identification of the particulate analyte as fibers and the count of the number of fibers collected.

Although transmission electron microscopy (TEM) is better for identifying and measuring specific types of fibers, phase contrast microscopy (PCM) still is the most common analytical method for counting fibers (Figure 11.5). Unlike TEM, PCM analysis is not generally used to identify the type of fiber. The number of fibers collected on the CEF filter are determined using PCM as follows:

1.   a segment (ca. 25%) of the CEF sample filter is cut and mounted on a glass microscopy slide
2.   the filter section is digested and collapsed on a microscope slide via acetone vapor and following mounting, is eventually positioned on the microscope stage for viewing
3.   while looking through the ocular of the microscope, a Walton-Beckett (W-B) graticule positioned in the ocular appears as a grid superimposed over the microscope viewing field of the sample
4.   the calibrated graticule grid is used for measuring length, width, and number of fibers.

Figure 11.5.   Major components of a phase contrast microscope used for conducting analysis of sampled fiber particulates collected on a CEF filter.

Any particle having a length to width ratio greater than 3:1 and a length of 5 μm or greater, is counted as a fiber. Samples are veiwed at 400x or 450x magnification power and results are presented as the number of fibers per cubic centimeter of air collected (f/cc).

It should be noted that phase contrast microscopy is a counting method and does not measure the specific properties of a substance. It is not inherently specific for asbestos. All particles meeting the criterion described (length-to-width ratio and length) are counted as fibers. Smaller fibers that may be present are not counted.

Metallic dusts and fumes also are separated from a contaminated airstream using filtration that involves electrostatic attraction, interception, and impaction of metal particles with a filter medium. The sample medium is a 37 mm diameter cellulose ester fiber (CEF) or paper filter with 0.8 μm pore size. CEF filters are used because, relative to PVC filters, they hydrolyze (dissolve) in acid and ash easily; characteristics which are important during post-sampling analysis of samples for metals. The smaller pore size (0.8 μm) filter is warranted since typically the variety of particle sizes is lower and the particle diameters smaller for metal dusts and especially fumes relative to total dusts.

The type of air sampling pump and the assembly of the CEF filter medium and cassette for metal dust and fume sampling is the same procedure described for total particulate sampling, except for the differences in filters.

Metallic dusts and fumes are another example of specific particulate air contaminants that require an analytical method that will both measure the amount of particulate collected and identify the specific metallic analytes (e.g., Pb; Cd; Be). Analysis of the sample typically involves either atomic absorption spectroscopy (Figure 11.6) or inductively coupled plasma emission spectroscopy. The identity and amount of metal dust or fume collected on the CEF filter are determined using an *atomic absorption spectrophotometer* as follows:

1. samples are prepared for analysis by digesting and/or ashing the CEF filter so that the metal air contaminant is present as a dissolved analyte
2. the solution of dissolved analyte(s) is injected or aspirated into the analytical instrument for qualitative and quantitative analysis
3. the sample enters the chamber where extremely high temperatures fueled by gases such as oxygen and acetylene converts metal *ions* in the sample solution to ground state metal *atoms* via process called "atomization"
4. a cathode-ray lamp generates a specific wavelength of light directed at the sample chamber that is required for detection of specific metallic analyte
5. a monochromator lens positioned between the sample chamber and detector is adjusted to wavelength absorbed by analyte
6. the photodetector measures the intensity of light transmitted through the sample and monochromator lens, that is, the light not absorbed by the sample
7. a direct read-out receives electronic signal from detector and displays data as Absorbance

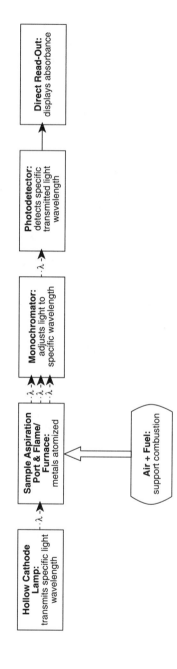

Figure 11.6.   Major components of an atomic absorption spectrophotometer used for conducting analysis of sampled metallic particulates collected on a CEF filter.

8. the amount of metal detected and measured is determined based on comparison of measured Absorbance to a Standard Curve of Absorbance versus Known Concentrations.

The concentration is expressed in units of milligrams per cubic meter (mg/m^3) based on the amount (mg) of metal detected and measured divided by the volume of air sampled (m^3).

Bioaerosols represent a group of viable or living particulates, such as bacteria, actinomycetes, fungi, and protozoa. Although specific exposure limits are not well established for airborne viable organisms, monitoring is often conducted to determine if airborne organisms, as bioaerosols, can be detected and measured for comparison to general recommended guidelines and/or background levels. The organisms are frequently combined with airborne mists, sprays, and/or dusts and the particle size distribution can be classified as non-respirable and respirable fractions.

The collection of airborne bioaerosols can be achieved through use of various methods, including filtration and impingement. A relatively accurate method, however, involves use of specialized sieve-like devices called impactors connected with a short length of heavy gauge flexible hose to a high-flow vacuum air sampling pump (Figures 11.7 and 11.8). The sampling train is commonly operated at flow rate of 1 ft^3/ min which is equivalent to 28.3 L/min. To detect the presence of organisms, they are collected on media called nutrient agar which enhances growth of colonies that are visible for counting. Dishes of agar are positioned within the impactor during monitoring so that the bioaerosols are initially separated based on particle sizes by the impactor and deposited on respective dishes of semi-solid agar for growth, detection, and measurement. Common impactors consist of two and six stages.

Agar serves as a growth media for numerous microorganisms, such as bacteria and fungi. The agar is prepared and sterilized using an autoclave prior to each use. The culture dishes also must be sterilized. The purpose of sterilizing the materials is to minimize the erroneous growth of microorganisms not associated with the actual air monitoring.

A two-stage cascade impactor can separate respirable from non-respirable fractions. A six-stage cascade impactor, however, permits even more specific separation based on particle sizes. Prior to and in between use, the cascade impactor should be disinfected using a chemical disinfectant or autoclave to minimize cross-contamination of samples.

Bioaersol analysis involves a preliminary incubation period to allow for growth of detectable and measurable colonies of microorganisms (Figure 11.9). The incubation time, temperature, humidity, and lighting varies depending on the type of organisms that were collected and are being selected for analysis. Following a prescribed incubation, the culture dishes are observed for detectable and countable microbial colonies. Dishes can be positioned under a colony counter, which is basically an illuminated grid surface with a magnification lens, to facilitate

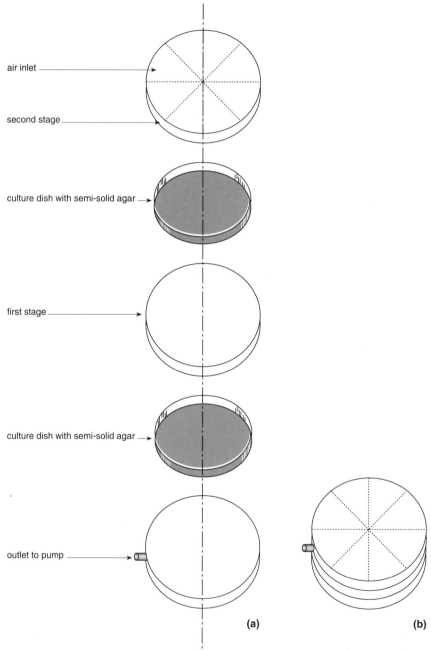

air inlet

second stage

culture dish with semi-solid agar →

first stage

culture dish with semi-solid agar →

outlet to pump →

(a)

(b)

Figure 11.7.   An (a) exploded view and (b) assembled view of a two-stage cascade impactor containing two culture dishes filled with semi-solid nutrient growth agar for sampling bioaerosols.

Figure 11.8.   Airborne bioaerosals can be sampled using (a) a high-flow vacuum air sampling pump connected via flexible tubing to (b) a two-stage cascade impactor containing two culture dishes filled with semi-solid nutrient growth agar.

Figure 11.9.   Following sample collection of bioaerosols onto culture dishes filled with nutrient growth agar, the dishes are incubated at a specific temperature for a specific duration prior to counting colony forming units.

detection and counting. The concentration of airborne bioaerosol is based on the number of colonies counted per volume of air sampled and is typically expressed as colony forming units per cubic meter of air (cfu/m^3).

Instantaneous or Real-time Monitoring of Particulates

Instrumentation and methods for conducting integrated or continuous personal and area monitoring and analysis of classes of airborne particulates were summarized above. Instrumentation and methods also are available, however, for conducting instantaneous or real-time area monitoring of generic total and respirable particulates as well as more specific airborne particulates such as fibers. The instrumentation typically involves battery-operated electronic meters that will actively pump or blow air into the device where airborne particulates are detected and measured. The concentration of airborne particulate is indicated on a direct read-out display (Figure 11.10).

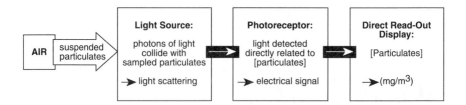

Figure 11.10.   Various types of direct-reading aerosol/particulate analyzers can be used for instantaneous or real-time monitoring of airborne dusts and fibers. The schematic represents the mechanism of photometric detection and measurement of airborne particles.

Direct-reading instruments for monitoring airborne particulates are generally less accurate, but more convenient than integrated air monitoring techniques which require time for laboratory analysis. Their speed of operation allows rapidly changing conditions to be assessed at the specific time of monitoring.

Typically, the monitor is placed in a stationary position four to six feet above the working surface to collect area samples. Area samples also are collected using hand-held devices which permit movement to various locations while simultaneously monitoring for airborne aerosols. Instantaneous area monitoring is important to determine the concentration of particulates within a given area, the sources of emission, the effectiveness of ventilation controls, and/or the peak concentrations of particulates during specific operations and/or periods.

Several types of *direct-reading aerosol meters* are available. The devices differ based on application and method of detection and measurement. Two major principles of particulate detection and measurement are light scattering and electrical precipitation and oscillation. In addition, based on application, some devices differentiate between concentrations of spherical (mg/m^3) and fibrous (fibers/cc) particulates.

Integrated or Continuous Monitoring of Gases and Vapors

Exposure limits and specific monitoring and analytical methods have been established for non-particulate air contaminants known as *gases* and *vapors*. Airborne gases and vapors can be separated from a contaminated airstream via adsorption on to a solid medium or absorption into a liquid medium following contact. Accordingly, some integrated sampling methods for gases and vapors involve sorptive techniques using a solid medium such as activated carbon or silica gel as adsorbents. *Adsorption* (Figure 11.11) refers to the immediate bonding of a gas or vapor to the surface of a solid under normal conditions of temperature and pressure. Indeed, adsorption specifically refers to bonding at the surface interface where contact occurs between the respective surfaces of the solid adsorbent and the

Figure 11.11.   Organic gases and vapors are separated and collected from contaminated air via adsorption onto a solid adsorbent medium.

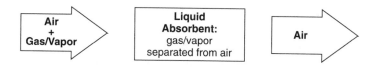

Figure 11.12.   Organic gases and vapors are separated and collected from contaminated air via absorption into a liquid absorbent medium.

flowing gas or vapor molecules. Although adsorption of gases and vapors on to *solid adsorbent media* is preferred for conducting integrated monitoring, some compounds are not collected efficiently using these media. As a result, *liquid absorbent media* are used to collect relatively more soluble and reactive organic gases and vapors, as well as, some inorganic gases and mists. Unlike adsorption where gas and vapor molecules bond to the surface of a solid medium, *absorption* (Figure 11.12) involves dissolution of molecules of gases and vapors into a liquid medium.

The solid adsorbents such as *activated charcoal* and *silica gel* are used most frequently to collect relatively water-insoluble and nonreactive gases or vapors. Solid adsorbent media are characterized as small porous spheroid particles. This characteristic provides high specific surface area (i.e. mm^2 surface area/g media) for contact between airborne gases and vapors and the surface of the adsorbent. In turn, adsorption efficiency is directly related to the amount of specific surface area that is exposed. Other factors which affect collection efficiency include sample flow rate, concentration of the gas or vapor, temperature, and humidity. If flow rates are too high, not enough time is allowed for the contaminant to react with the adsorbent. The lowest flow rate which will give a sample of sufficient size for analysis should be utilized, but flow rates are rarely less than 10 cc per minute (cc/min). To compensate for high concentrations of gases or vapors, sampling time and or flow rates can be reduced, or a larger sorbent tube containing more adsorptive media can be utilized. Elevated temperature and humidity can increase the rate at which the surface of adsorbent media become saturated.

Activated charcoal is used for the collection of nonpolar organic vapors with boiling points greater than 100°C. Silica gel is generally used to collect polar substances that do not efficiently adsorb to activated carbon. Silica gel can be used to collect organics with a minimum of three carbons, compounds containing hydroxyl groups, halogenated hydrocarbons, and inorganic compounds. The silica medium is very hygroscopic and, accordingly, collection efficiency decreases under conditions of high relative humidity.

The adsorbent media are contained within flame-sealed glass tubes tapered at each end. Prior to sampling, both ends of the glass sampling tube are carefully snapped open for air to flow through during sample collection. Integrated monitoring for airborne gases and vapors involves a low-flow or multi-flow pump set at a flow rate less than 1000 cc/min and commonly lower than 100 cc/min. The pump is connected to a short length of flexible hose and tube holder assembly, which in turn, contains the glass tube packed with the solid adsorbent (Figure 11.13). The sampling pump creates a vacuum which causes air to be pulled into the tube and through the front then the back-up sections prior to eventual exhaust from an outlet port on the pump. In theory, if gases or vapors are present in the air as it passes though the column of solid adsorbent particles, they would be immediately removed from the airstream and collected via adsorption onto the solid medium. Subsequent to sample collection, the sampling tube is removed and plastic caps are placed on each end prior to laboratory analysis.

Gas chromatography (Figure 11.14) is the method of analysis most often utilized for contaminants collected on solid adsorbent tubes. Analysis consists of the following:

1. solid adsorbent is removed from the glass sampling tube and placed into a vial to which a solvent such as carbon disulfide for charcoal and methanol for silica gel is added to desorb the gas or vapor contaminant from the solid adsorbent medium
2. after a short incubation period for desorption, an aliquot of desorption solvent containing dissolved analytes is manually or automatically withdrawn from the sample vial and injected into a heated injection port of a gas chromatograph for vaporization
3   vaporized desorption solvent and analytes are transported (carried) to and through a heated column via an inert carrier gas ("mobile phase")
4. the desorption solvent and each analyte interact differently with the medium ("stationary phase") coated or packed inside the column and, as a result, separate (partition) from each other
5. the analytes elute off the column to a detector in the order of shortest to longest retention time on the column and are detected and plotted on a recorder as curves/peaks on a graph called a "chromatogram", "retention time", and "area under the curve" are indicated for each analyte.

The analytes are identified by comparing the retention time for the unknown sample analyte to the retention time for a known standard analyte. Analytes are quantified based on the area under the curve or "peak area" (area ratio).

The alternative and typically less preferred sampling media, liquid absorbents, are contained in specialized sampling devices called *midget impingers* or *midget*

Figure 11.13.   Organic gases and vapors are separated and collected from contaminated air using (a) a low-flow or multi-flow battery-operated air sampling pump connected with flexible tubing to (b) a glass tube containing a solid adsorbent.

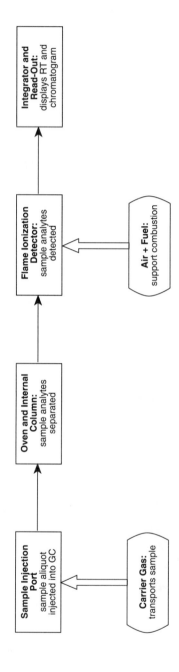

Figure 11.14.   Major components of a gas chromatograph used for conducting analysis of sampled organic gases and vapors collected onto a solid adsorbent.

Figure 11.15.   Organic gases and vapors are separated and collected from contaminated air using (a) a high-flow or multi-flow battery-operated air sampling pump connected via flexible tubing to (b) a glass midget impinger containing a liquid absorbent.

bubblers connected via flexible tubing to a high-flow sampling pump during monitoring (Figure 11.15). The devices enhance aeration of the liquid absorbent with the sampled air so that increased surface area and mixing is available. In turn, increased contact between the sampled gas and vapor molecules and the liquid absorbent enhances the efficiency of collection of the contaminant.

Midget impingers and bubblers are columnar glass devices that provide a reservoir for liquid absorption reagents. Inserted in the reservoir is a hollow glass inlet tube open at each end and tapered (impingers) or fritted (bubblers) at the bottom. The tube is attached and extends through the top of the reservoir. When the reservoir is filled with 10–20 ml of a liquid medium, the tapered or fritted end of the tube is immersed in the fluid. The tube serves as a conduit for the air to enter the impinger or bubbler and mix with and dissolve into the liquid absorbent.

Ultraviolet (UV) or *visible (Vis) spectrophotometry* is the method of analysis most often utilized for contaminants collected in liquid absorbents (Figure 11.16). The analyisis proceeds as follows:

1. liquid absorbent plus dissolved contaminant is removed from the impinger or bubbler and mixed with reagents yielding a detectable analyte that will maximally absorb a wavelength of light in either UV or visible spectrum
2. a specific spectrum (UV or Vis) and wavelength of light is selected relative to the contaminant or analyte in the sample by adjusting the monochromator positioned between the light source and sample chamber
3. a sample solution is placed in a special cuvette into the sample chamber where a specific wavelength of light is directed to the sample chamber where analyte molecules can absorb the light energy in direct relation (linearly) to its concentration ("Beer's Law")
4. light not absorbed by the sample analyte exits the sample chamber to the detector which generates an electrical signal to the read-out
5. intensity of the light detected is converted to display as Absorbance (Abs) on a direct read-out.

Absorbance reading for the analyzed sample (unknown concentration) compared to a Standard Curve consisting of Absorbance readings corresponding to known concentrations of analyzed standards. The amount of analyte in the sample solution that was detected and measured by the UV/Vis spectrophotometer is extrapolated from the Standard Curve and divided by sampled volume of air to give concentration.

Instantaneous or Real-time Monitoring of Gases and Vapors

Instantaneous, or real-time direct-reading instruments, also are used to detect and measure the concentrations of airborne gases and vapors. The gaseous agents detected and measured include combustible and/or toxic gases or vapors and oxygen gas.

The presence of a combustible or explosive atmosphere and either an oxygen deficient or enriched atmosphere can pose immediate hazards to individuals. Monitoring devices are available that instantaneously detect and measure the level of airborne combustible gas or vapor and oxygen gas. The most common devices for detecting and measuring combustible and oxygen gases are used for area monitoring. Individual instruments are available to measure either combustible gases or vapors or oxygen gas. It is more common, however, to use a combination

Figure 11.16.   Major components of a UV/VIS spectrophotometer used for conducting analysis of sampled organic gases and vapors collected into a liquid absorbent.

Figure 11.17.   Various types of direct-reading combined combustible gas and vapor and oxygen analyzers can be used for instantaneous or real-time monitoring of combustible and oxygen enriched or deficient atmospheres. The schematic represents two respective mechanisms of detection and measurement of airborne combustible gases and vapors and oxygen gas.

unit that is designed to detect and measure both parameters. These combined devices are called combination combustible and oxygen gas meters or, more simply, combustible gas meters (Figure 11.17). The later name, however, does not reflect the potential to monitor oxygen. Many of the combination meters also measure a toxic parameter such as carbon monoxide or hydrogen sulfide gases.

Monitoring of combustible gas or vapor and oxygen gas is instantaneous and represents real-time measurements of the analytes. The devices are typically electronic active-flow monitoring instruments that collect an air sample by automatically pumping or blowing air into the device. As the air flows through the instrument, combustible gas or vapor and oxygen gas interact with sensors or detectors for the respective analytes. The meters are generally calibrated for combustible gas or vapor using a known concentration of reference gas, such as pentane. Responses to other combustible gases or vapors actually detected and measured during monitoring, however, must be adjusted via conversion factors to determine the exact concentration of a specific gas. For oxygen, some meters are calibrated using a known concentration of molecular oxygen gas.

Common *combustible gas meters* contain a heated platinum filament which internally ignites the combustible gas or vapor. The ignition of a combustible gas or vapor generates a detectable and measurable quantity of heat. This heat of combustion changes the resistance of the filament in proportion to the concentration of the combustible gas or vapor present. The read-out of the meter indicates the concentration of combustible gases or vapors as "percent of the LEL". This actually represents a percentage of the LFL or LEL of the calibration gas. Various makes and models are available.

Combustible gas meters are intended for use only in normal atmospheres, not atmospheres that are oxygen enriched or deficient. Oxygen concentrations significantly less than or greater than normal (ca. 20.8%) may cause erroneous readings. Accordingly, the level of airborne oxygen gas is usually measured simultaneously. Leaded gasoline vapors, halogens, and sulfur compounds interfere with the filament and decrease its sensitivity. Silicone-containing compounds can damage the filament.

Oxygen gas meters typically involve an oxygen sensing device and a meter readout. In some units, air is drawn to the oxygen detector with an aspirator bulb or pump, while in other units, equalization due to diffusion is allowed to occur between the ambient air and the sensor. An electrochemical sensor is used to determine the oxygen concentration in air. Components of a typical sensor include a counting and sensing electrode; a housing containing a basic electrolytic solution; and a semipermeable Teflon membrane. Oxygen molecules diffuse through the membrane into the electrolytic solution. A minute electric current is produced by the reaction between the oxygen and the electrodes. This current is directly proportional to the oxygen content of the sensor. The current passes through the electronic circuit, resulting in a needle deflection on the meter. The meter is calibrated to read 0–10%, 0–25%, or 0–100% oxygen. When evaluating an unknown environment, the oxygen indicator should be used first to determine if an oxygen deficient or oxygen enriched atmosphere exists.

It also is frequently useful and necessary to instantaneously detect and measure the concentration of an airborne *toxic inorganic* or *organic gas* or *vapor*. Devices are available that consist of a manual air sampling pump with *detector tube media* to serve this purpose. An air sample can be collected, and within a minute or two, the presence of an individual contaminant or a class of contaminants can be determined qualitatively and quantitatively. Various detector tubes are available for the numerous contaminants of concern.

Detector tubes are very useful for intital screening of areas and to develop strategies for more comprehensive integrated or continuous monitoring techniques for specific gaseous contaminants of concern. Detector tubes for real-time monitoring also are used concurrently during integrated sampling to trace sources of exposure and to track variations in exposure levels throughout the sampling area and period.

There are two classes of detector tubes, length of stain and colorimetric. In both cases, a sealed glass tube is filled with a chemically coated solid sorbent that has been impregnated with an appropriate chemical reagent to react specifically with the contaminant of interest. Prior to sampling, both ends of a glass detector tube are carefully snapped open for air to flow through during sample collection. The detector tube is inserted into the inlet orifice of the air sampling pump. Two major types of air sampling pumps are available for use with detector tubes. Both types are manual and are referred to as "piston pumps" and "bellows pumps" (Figure 11.18).

Figure 11.18. A direct-reading (a) piston pump or (b) bellows pump with an attached detector tube can be used for instantaneous or real-time monitoring of inorganic and organic gases and vapors.

As contaminated air is drawn through a detector tube using a manual pump, molecules of the gaseous contaminant contact the solid medium and react. The reaction results in either a dark stain or color development, depending on the type of tube (i.e., "length of stain" or "colorimetric"). The length of stain or the intensity and shade of color development corresponds with the concentration of the gas or vapor contaminant in the sampled air. Length of stain tubes either have the concentration scale printed on them, or the ratio of the length of stain to the total media length is compared against a chart of corresponding concentrations. When contaminated air is drawn through a colorimetric tube, the reagent-impregnated sorbent yields a progressive change in color intensity. At the completion of sampling, the sample tube is compared with a chart of tinted colors that correlate to the concentration of the gas or vapor present in the atmosphere.

Detector tubes are one of the easiest monitoring devices to use, but data must be interpreted with caution since the tubes are known to be relatively inaccurate. Accuracy is only ± 25% in many cases and for some contaminants error is greater. Therefore, a sample detector tube may reveal a concentration of a contaminant equal to 200 ppm for example, but the actual concentration may be 150 to 250 ppm (200 ppm ± 25% or 200 ± 50 ppm).

Electronic devices also are available for instantaneous or real-time area monitoring of organic gases and vapors. These devices operate based on different principles, but also share several similar characteristics. The devices consist of air sampling pumps that cause active flow of air and airborne contaminants into a chamber. The air and related organic gas and vapor molecules flow across a detector that qualitatively detects and identifies the contaminant as organic and quantitatively measures the concentration in parts per million on a direct read-out display (Figure 11.19). In most cases, the identity of the specific components of the airborne organic gas and vapor may be unkown or uncertain. Accordingly, data are commonly reported as total organic vapor (TOV).

Meters with *flame ionization detectors (FID)* measure the concentration of total ionizable organic gases and vapors and, if the identity of the gas or vapor is known, also measure the concentration of specific organics in the atmosphere. Common components of a meter with a FID are battery-operated air sampling pump, FID, cylinder of compressed hydrogen gas, read-out, and sampling probe.

Response will vary depending upon the number of carbon atoms in the molecules. A gas or vapor is pumped into the instrument and passed through a hydrogen flame at the detector which ionizes the organic vapors. When most organic vapors burn, positively charged carbon-containing ions are produced, which are collected by a negatively charged collecting electrode in the chamber. An electric field exists between the conductors surrounding the flame and a collecting electrode. As the positive ions are collected, a current proportional to the hydrocarbon concentration is generated on the input electrode. This current is measured with a preamplifier which has an output signal proportional to the ionization.

Figure 11.19.   Various types of direct-reading organic gas and vapor analyzers can be used for instantaneous or real-time monitoring of toxic gases and vapors. The schematic represents the mechanism of flame ionization detection and measurement of airborne organic vapors.

Meters with a *photoionization detector (PID)* also measure the concentration of total ionizable organic gases and vapors. Meters with a PID consist of a readout unit consisting of an analog meter, a rechargeable battery, an amplifier, an ultraviolet (UV) lamp, and a sensor ionization chamber.

An electrical pump pulls the gas or vapor sample past a UV source. Constituents of a sample are ionized, producing an instrument response, if their ionization potential is equal to or less than the ionizing energy supplied by the instrument UV lamp being utilized. The radiation produces an ion pair for each molecule of contaminant ionized. The free electrons produce a current directly proportional to the number of ions produced. The current is amplified, detected, and displayed on the meter. By varying the electron volts of UV light, a wide range of organic compounds can be detected and quantified.

Infrared spectrophotometric meters can detect and measure concentrations of specific organic gases and vapors. The instruments have direct reading scales, recorder output, and a built-in sampling pump. Compounds are detected by an infrared spectrophotometer-based absorption of a discrete wavelength of infrared light. In turn, absorption is proportional to the concentration of gas or vapor.

Portable gas chromatographs are versatile instruments for instantaneous monitoring. The principle of detection and measurement is similar to bench-top laboratory gas chromatographs described earlier. The instruments separate compounds and permit detection and quantification of specific organic compounds. Sampled gas or vapor is passed through a column packed with a liquid phase on solid support or a solid phase support. Components of the sampled airstream separate as they pass through the column based on their inherent chemical characteristics. The individual components ("analytes") are identified based on their retention time on the column and detected via a variety of commercially available detectors, typically a FID.

EVALUATION OF AIRBORNE SOUND LEVELS

Integrated or Continuous Monitoring of Sound

Integrated or continuous personal monitoring using *audio dosimeters* is frequently conducted to detect and measure *sound levels* when evaluating the occupational environment. Following monitoring, data are calculated to determine whether levels are within standard or recommended criterion levels. A criterion level is the continuous A-weighted sound level which constitutes 100% of an allowable exposure. For example, a standard criterion level is 90 dBA for an eight hour day. If noise exposures greater than 90 dBA are based upon a 5 dB "doubling exchange rate", then workers are permitted to be exposed to 95 dBA for four hours, 100 dBA for two hours, 105 dBA for one hour and so on. Allowable time (T) can calculated for exposure to a given sound pressure level (L_a) based on a criterion level of 90 dBA and a doubling exchange rate of 5 as follows:

$$T = \frac{480 \text{ min}}{2^{(L_a - 90)/5}} \qquad \text{(eq. 11.3)}$$

To calculate noise exposure in locations or job classifications where sound pressure levels vary, the total sound pressure exposure or dose must be determined. Dose (D) is calculated based on the actual exposure time at a given sound pressure level (C) divided by the allowable exposure time at that level:

$$D\% = \left(\frac{C_1}{T_1} + \frac{C_2}{T_2} + \ldots + \frac{C_n}{T_n} \right) \times 100 \qquad \text{(eq. 11.4)}$$

If the total dose exceeds 100% or unity, then individuals exposed to the various sound sources at the measured levels would be excessively exposed. To determine the equivalent sound pressure level (L_{eq}) in dBA for an eight hour day based upon dose (D) values, the following formula is used:

$$L_{eq} = 90 + 16.61 \ \log \frac{D\%}{100} \qquad \text{(eq. 11.5)}$$

To determine the equivalent sound pressure level (L_{eq}) in dBA for dose (D) exposure durations not equal to an eight hour shift, the following formula is used where T is the actual time in hours:

$$L_{eq} = 90 + 16.61 \ \log \frac{D\%}{12.5 \ T} \qquad \text{(eq. 11.6)}$$

Audio dosimeters are special SLMs that integrate sound pressure levels over time and are used to measure personal noise exposures. Audio dosimeters (Figure 11.20) are composed of a microphone attached to a meter that is similar to a sound level meter (see below). The detected sound levels, however, are integrated into memory and measurements are indicated via direct read-out. Internal and external computerized read-out devices are available. The audio dosimeter microphone is typically clipped to an individual's clavicle (collar) or trapezius (upper shoulder) region provided it is within 12 to 18 in. of the ear (i.e., "hearing zone"). Ideally, the microphone of the device should be connected to the worker's ear, but this is not always practical nor desired. The meter is either placed in the shirt pocket or attached to the belt. Audio dosimeters integrate the A-scale sound pressures that the worker is exposed to over an entire shift (or monitoring period) and total exposure or dose is then determined.

By calculating the integrated dose obtained with the audio dosimeter and inserting this value into one of the equivalent sound pressure level formulas presented above, it is possible to determine level of exposure. Many dosimeter models may be down-loaded to a computer and printouts can be generated recording a wide variety of measures obtained throughout the monitoring activity.

Instantaneous or Real-time Monitoring of Sound

The *sound level meter (SLM)* and *octave band analyzer (OBA)* are two generic types of sound monitoring instruments typically used when assessing noise problems in various environmental settings. The SLM is an instrument used to monitor air pressure changes resulting from environmental sound/noise. The octave band analyzer (OBA) is a special SLM for measuring pressure levels of sound for octave frequency bands and is referred to as a sound level meter octave band analyzer (Figure 11.21). The octave band analyzer is actually an electronic filter attached to or incorporated with a SLM. There are a wide variety of SLMs and SLM/OBAs commercially available. These instruments are used in the occupational environment to assist in detecting and measuring sound levels and determining whether a noise problem exists. SLMs consist of three major generic components; a microphone, a meter, and a connection for an octave band analyzer.

While each sound level meter is slightly different each may have one or more sound measurement "scales" corresponding to a range of frequencies. Three more common scales are labelled A, B, and C. The A-scale consists of frequencies (20 to 20,000 Hz) closely associated with human hearing and is the mandated scale required for monitoring activities involving human exposure assessment. The B-scale is used in sound research activities, and the C-scale is used for engineering and maintenance or trouble-shooting activities.

There are four types of approved SLMs commercially available that differ by their certified accuracy. Type "0" has an accuracy of ± 0 dB and is the most accurate SLM available. It is predominantly used for laboratory research applications. Type

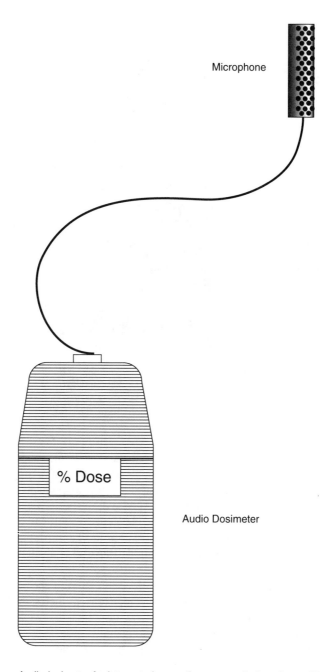

Figure 11.20.   Audiodosimeter for integrated or continuous monitoring of sound levels.

Figure 11.21.   Combination sound level meter and octave band analyzer for instantaneous or real-time monitoring of sound levels at specific frequency
ranges.

"1" SLMs are instruments that have an accuracy of ±1 dB. These are precision units
intended for extremely accurate field and laboratory measurements. Type "2"
SLMs are general purpose sound level meters with an accuracy of ± 2 dB. Type 2
SLMs better set at A-scale are mandated for use to evaluate the occupational
environment for noise. The last type of SLM is the Type "S" or "Special Purpose"
sound level meter. It is equivalent to the old Type "3" unit with design tolerances
similar to those listed for the Type 1 unit. The Type "S" SLM differs from the type
1 because these instruments are not required to have all of the functions of the units
previously listed.

While the SLM without octave band electronic filter can be set at a scale (A, B,
or C) which represents a relatively broad range of frequencies, the SLM/OBA can
be adjusted to specific frequencies which represent relatively narrow octave band
frequencies. The specific "center band frequencies" are 31.5, 63, 125, 250, 500,
1000, 2000, 4000, 8000, and 16,000 Hz. Each center band frequency represents an
individual range or band of frequencies.

EVALUATION OF AIR TEMPERATURE FACTORS

Instantaneous or Real-time Monitoring of Elevated Temperature Factors

Several devices and procedures are available for evaluating the occupational environment for conditions that may contribute to heat stress. The most common method, however, involves determination of the *Wet-Bulb Globe Temperature (WBGT) Index*. The method involves collection of two to three major measurements that reflect the contribution of environmental factors to heat stress.

The measurements include the dry-bulb temperature, natural wet-bulb temperature, and the globe temperature. The dry-bulb temperature measures the ambient temperature. The natural wet-bulb temperature records the ambient temperature, but reflects the influence of humidity and air movement. High humidity and/or low air movement result in an elevated wet-bulb temperature. The globe temperature reflects the contribution of radiant heat. An integration of all three measurements yields a WBGT Index. For indoor and outdoor settings where there is no solar load, only two measurements are needed, natural wet-bulb (WB) and globe temperature (GT):

$$WBGT_{in} = 0.7WB + 0.3GT \qquad \text{(eq. 11.7)}$$

For outdoors in sunshine, the air temperature based on the dry-bulb (DB) temperature must also be measured:

$$WBGT_{out} = 0.7WB + 0.2GT + 0.1DB \qquad \text{(eq. 11.8)}$$

Although the WBGT is easy and simple to use, it does not in itself include a factor for the activity or excertion of an individual. It is not possible to determine an allowable exposure time directly from the WBGT. Guidelines have been developed, however, which eliminate these difficulties. The work load for an individual is determined by estimating the worker's metabolic rate. A chart has been developed to aid in this process. An occupational exposure limit has been presented based on the amount of time a worker is involved in continuous work and the level of work that the worker is performing.

Although measurement of the ambient air temperature by itself is insufficient to determine heat stress, it is an integral part of heat stress determinations. There are various methods to assess ambient air temperature. The most common method of measurement is the dry-bulb thermometer. A dry-bulb thermometer consists of a hollow glass tube with bottom reservoir of mercury. The range of the thermometer should be –5° to +50°C and accurate to ± 0.5°C. Temperatures beyond the range of a dry-bulb thermometer can break the thermometer. When the measurements are taken, the dry-bulb thermometer must be shielded from radiant heat sources so that only the temperature of the ambient air will be detected. Thermoelectric thermometers or thermocouples are also used to measure the ambient air temperature.

The black globe thermometer, or Vernon Globe, is the standard method for measuring radiant heat. A black globe thermometer consists of a 6-inch diameter hollow copper sphere that is painted matte black. A thermometer is inserted so that its bulb is centered inside the globe. The range of the thermometer should be $-5°$ to $+100°C$ and accurate to $\pm 0.5°C$. The black globe absorbs radiant heat. The thermometer inside the globe is allowed to reach equilibrium. The energy measured is generally termed the mean radiant temperature, because some heat is lost from the globe via convection.

As the temperature of air rises, it will hold more water vapor. When the air at a given temperature holds all the water vapor that it can, then is said to be saturated. The water in the air exerts a vapor pressure which is measured in millimeters of mercury. Humidity in the air is usually measured with a wet bulb thermometer or a sling psychrometer. The wet-bulb thermometer consists of a standard dry-bulb thermometer covered with a cotton sleeve or wick moistened with distilled or deionized water. The covered thermometer bulb is partially immersed in water to keep the wick wet. As the wick evaporates water, a certain amount of heat energy is dissipated through evaporative cooling. The bulb is then cooled by the heat absorbed during evaporation of the water, and an equilibrium is reached between the evaporation rate and the water vapor pressure in the air. The temperature is indicated on the thermometer. At full saturation, the wet bulb thermometer temperature will equal the dry-bulb temperature since no evaporative cooling will be experienced.

The thermometers and accessories needed to measure wet-bulb temperature, globe temperature, and dry-bulb temperature can be assembled and positioned in a ring stand (Figure 11.22). Commercially available electronic WBGT meters are also available (Figure 11.23). The electronic devices integrate the temperature measurements and provide a direct read-out for individual parameters and the combined WBGT.

EVALUATION OF ILLUMINATION

Instantaneous or Real-time Monitoring of Illumination

Visible light is generated from both natural and anthropogenic sources. Luminous flux refers to the total energy from a source of light that is emitted and can influence sight. The luminous flux is measured in standard units called lumens which refer to the total radiant power or flux shining on a 1 ft^2 surface. llumination is the density of the luminous flux on a given area of a surface. Thus, illumination equals lumens per area. Relative to an area of 1 ft^2, illumination is measured in units of footcandles or Lux.

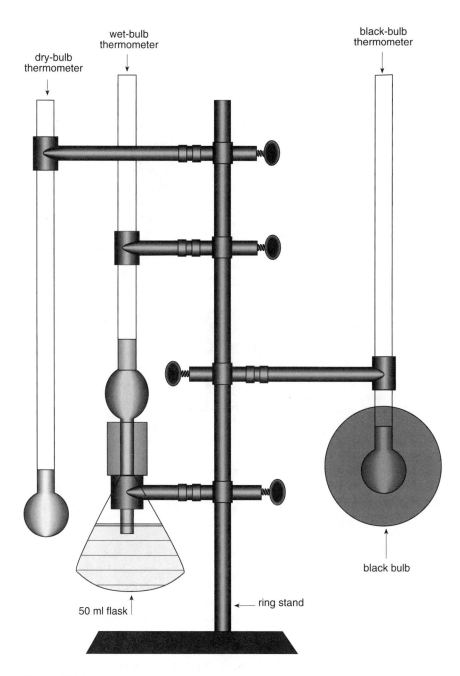

Figure 11.22.   Assembly for measuring wet-bulb globe temperature (WGBT) indices for heat stress.

Figure 11.23.   Electronic meter for measuring wet-bulb globe temperature (WBGT) indices for heat stress

Light detection and measurement is called *photometry*. An electronic *illumination meter* is used to measure lighting. The devices consist of a relatively flat photosensitive sensor with a color attenuator or filter connected to a detector (Figure 11.24). Light energy focused directly on the sensor or photocell causes emission of electrons and related generation of increased current. The current generated is directly related to the intensity of light. Measurement of illumination, in foot candles or Lux, is indicated on a direct read-out display.

Illumination levels are measured and recorded at various points within a given area in relation to the activity that is conducted at the location. The distances are

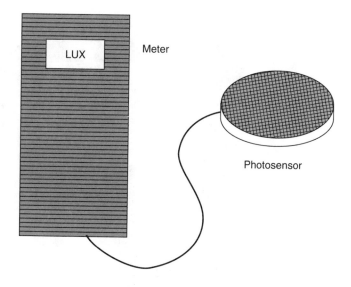

Figure 11.24. Illumination meter for instantaneous or real-time detection and measurement of light.

measured between the light source and the work surfaces, such as assembly lines and desk tops. Illumination is then monitored at the specific work surfaces and general areas of a given area.

EVALUATION OF MICROWAVE RADIATION

Instantaneous or Real-time Monitoring of Microwave Radiation

Several types of microwave detectors are available, including thermal devices that respond to increased temperatures from absorbed radiation by reflecting a change in resistance or voltage. Electronic *microwave meters* that convert microwave frequency into direct current are commonly used for instantaneous area monitoring of nonionizing radiation from microwave ovens. The meters consist of an antenna or probe that is connected to a detector and amplifier (Figure 11.25). A spacer may be present on the probe so that a uniform distance is maintained between the source and the antenna. When evaluating a microwave oven, a 5 cm distance is required between the oven door and the probe during monitoring.

A thorough inspection of the microwave oven is conducted as part of the evaluation to check the electrical wiring and plug, door alignment, door seal condition and cleanliness, and door hinge and latch condition. Following inspection, materials used for the "standard loading" of the unit must be placed in the

Microwave Oven

Figure 11.25.   Evaluation of a microwave oven for instantaneous or real-time detection and measurement of microwave radiation using (a) a survey probe attached to (b) a direct-reading meter.

microwave oven. Operation of ovens with no load may cause damage to the unit and must be avoided. Accordingly, a standard load typically consists of a 500 ml nonmetal (i.e., plastic, glass) beaker filled with approximately 275 ml of tap water.

Monitoring is performed while the door is closed and the oven is operating by keeping the probe 5 cm from the surface of the unit and scanning the door area. Following the evaluation, a thermometer is immersed in the beaker of water to verify an increase in the temperature of the fluid due to microwave energy.

EVALUATION OF SOURCES OF ELECTRIC AND MAGNETIC FIELDS (EMFS)

Instantaneous or Real-time Monitoring of EMFs

Monitoring requires use of a specialized electronic meter with antenna that can detect and measure low frequency *electric and magnetic fields* (Figure 11.26). The antenna is typically a diode-dipole design and serves as an electronic sensor of electric and magnetic field energies. The root-mean-square (RMS) of the detected signal is computed and the measurement is displayed on a direct read-out meter. Magnetic fields are commonly measured in units of milligauss (mG) and electric fields in units volt/meter (V/m).

Distances and heights relative to the sources of electromagnetic radiation must be considred during monitoring. Measurements are recorded at various vertical and horizontal distances from sources such as electrical appliances and devices. In addition, it is useful to conduct measurments when the suspected sources are turned "OFF" and then when turned "ON". The contribution of electric and magnetic energies to the area then can be attributed to the operating source.

EVALUATION OF IONIZING RADIATION

Integrated or Continuous Monitoring of Ionizing Radiation

Film badges and pocket dosimeters are two common devices used for conducting personal integrated monitoring of ionizing radiation (Figure 11.27). *Film badges* consist of photographic film coated with a thin layer of a special emulsion containing silver halide (i.e., silver bromide). The film is positioned within a special badge that can be modified depending on the use. Interaction with ionizing radiation causes the silver halide to ionize forming silver cations. The silver cations are attracted to negatively regions of the film yielding free-silver compounds. After use, the film is removed from the badge and submitted for processing. Development of the film rinses away silver halide and silver ions. Free-silver, however, remains and its present causes the film to exhibit increased opacity in proportion to the amount of ionizing radiation exposure. The measurement of exposure is determined using a densitometer that measures the degree of film opacity.

Pocket dosimeters, like the film badge, are typically clipped to an individual's lapel or shirt pocket. They are similar to ionization chamber technology. The

Figure 11.26.   Combined electric and magnetic field meter for detection and measurement of nonionizing electric and magnetic fields.

film badge

pocket ionization chamber

Figure 11.27.   Integrated or continuous personal sampling (a) film badge and (b) pocket
dosimeter for detection and measurement of ionizing radiation.

devices reveal accumulated exposure based on a direct read-out visible in the
dosimeter. A similar type of pen-like dosimeter, however, requires a separate read-
out device to determine the measurement of absorbed dose.

Another common personal monitoring device for ionizing radiation is the
thermoluminescence dosimeter (TLD). Energized electrons resulting from ionizing
radiation interact and are trapped within a phosphor crystal. Thermal analysis of the
dosimeter causes generation of detectable light. The measured light emission
reflects the level of absorbed dose.

Instantaneous or Real-time Monitoring of Ionizing Radiation

Instruments also are available for conducting instantaneous or real-time area
monitoring of ionizing radiation. Levels of ionizing radiation that are detected and
measured vary depending on several factors, including the source and the distance

Figure 11.28.   Geiger-Müller meter for instantaneous or real-time detection and measurement of ionizing radiation.

from it. Instantaneous area monitoring devices commonly consist of a probe connected to a meter with a direct read-out display (Figure 11.28). Selection of monitoring devices depends on the type and level of ionizing radiation that is present.

An *ionization chamber* is a real-time electronic monitoring device that consists of a chamber filled with a known volume of gas. Within the chamber is a positively charged anode and negatively charged cathode. When airborne ionizing radiation enters the chamber it interacts with the gas molecules and forms primary ion pairs. The ions are attracted to the respective electrodes which, in turn, creates a change in voltage. The change in voltage and accompanying current resulting from the ions is proportional to the amount of ionizing radiation detected. The measurement is displayed on a direct read-out. Ionization chambers are useful for area measurements of relatively high levels of ionizing radiation. The meters do not, however, differentiate between or identify the types of ionizing radiation detected.

Proportional counters are electronic meters that function very similarly as ionization chambers and are used for real-time area measurements of ionizing radiation. Indeed, they are a special type of ionization chamber. The major difference, however, is that these meters cause a magnification of ionization when ionizing radiation interacts with the gas molecules in the chamber. A higher voltage is generated and produces secondary ions. As a result of secondary ion formation, the current is amplified and corresponds to increased sensitivity of the instrument. Accordingly, proportional counters respond to lower levels of ionizing radiation than standard ionization chambers. The instrument is especially useful for detecting, measuring, and identifying alpha and beta particles.

Geiger-Mueller (G-M) meters also operate based on the principal of ionization as discussed for the preceding two devices. The instrument, however, exhibits even greater secondary ionization than described for the proportional counter. Accordingly, the sensitivity to detect low-level ionizing radiation is even greater. Unfortunately, the electrical current that is generated as a result of the interaction between the detected ionizing radiation and the gas molecules within the chamber and the secondary ions formed and attracted to the electrodes is not typically proportional to the ionization energy present. The G-M meter, therefore, is most useful as a screening device to detect low-levels of ionizing radiation, but is not as effective for measuring accurate doses. High level radiation can cause the meter to block resulting in no response when levels of ionizing radiation are actually present. Major application is for area measurements of low-levels of beta and gamma radiation.

EVALUATION OF AIR PRESSURE, VELOCITY, AND FLOW RATE

Instantaneous or Real-time Monitoring Within Ventilation Systems

Inside measurements are collected within ducts of local and general ventilation systems to determine pressures such as velocity pressure. In turn, air transport velocities and flow within the systems can be calculated.

To measure air *velocity pressures* as well as static and total pressures, the most common devices used are the *pitot tube* connected to an *inclined fluid manometer* (Figure 11.29). A pitot tube is a concentric tube that is designed to simultaneously account for total and static pressures and via differential pressure, measure velocity pressure. The inclined manometer contains a fluid such as water or oil. The pitot tube is connected to the manometer and inserted into a duct. Air pressures within the duct cause the fluid in the manometer to deflect. The deflection is measured in "in. of water" ($''H_2O$). Several velocity pressure measurements are taken horizontally and vertically across the area of the duct lumen. In turn, the velocity pressures (VP) are converted to velocity (V) and averaged to yield "average transport velocity" (V) in "feet per minute" (ft/min). Average velocity multiplied by the cross-sectional area (A) of the duct in "square feet" (ft^2) yields "volumetric flow rate" in "cubic feet per minute" (ft^3/min).

$$Q \frac{ft^3}{min} = V \frac{ft}{min} \times A \; ft^2 \qquad \text{(eq. 11.9)}$$

Instantaneous or Real-time Monitoring Outside Ventilation Systems

Measurements outside the system include face and capture velocity measurements. Face velocity refers to the movement of air in feet per minute at the immediate opening of a hood, vent, or diffuser. Capture velocity is a measurement of velocity at a specific distance from the source of contaminant generation to the face of the hood.

Figure 11.29.   Pitot tube and inclined manometer for measuring velocity pressures at designated cross-sectional traverse points inside a duct of a local or general ventilation system.

Measurements of face and capture velocities outside of a ventilation system are commonly determined using a swinging vane anemometer or a thermal anemometer (Figure 11.30). The swinging vane anemometer is a mechanical device that involves a pendulum mechanism that deflects in response to moving air. The degree of deflection is related to the velocity of the air. A read-out indicates the quantitative velocity. Thermal velometers are battery operated and consist of a probe containing a thin wire that is heated to a stable temperature. Movement of air across the wire will cool it. The degree of cooling is related to the air velocity shown quantitatively on the meter read-out.

Velocity measurements are taken at several points at the face of a hood, vent, or diffuser and averaged to give "average face velocity" in "feet per minute" (fpm). For capture velocity, several measurements are taken at a distance from the source to the exhaust hood opening and averaged to give "average capture velocity" also in units of fpm. In addition to these instruments used for collecting quantitative data, a qualitative estimate of airflow can be made using smoke tubes containing titanium tetrachloride to determine if air contaminants, based on the movement of the visible cloud of generated smoke, are being captured by an exhaust system.

Figure 11.30.   (a) Swinging-vane anemometer and (b) thermal or hot wire velometer for measuring face and capture velocities at designated points outside of a local or general ventilation hood, vent, or diffuser.

REFERENCES

Chapter 1

Abercrombie, M. *Dictionary of Biology* (Baltimore: Penguin Books, 1964).

Adams, J. "Clinical Relevance of Experimental Behavioral Teratology", *Neurotoxicology* 7(2):44–90 (1980).

American Cancer Society. *Cancer Facts and Figures, 1983* (New York: American Cancer Society, 1983).

American Gas Association. *Total Energy Resource Analysis* (Arlington, VA: January 1988).

Ames, B. N. "Identifying Environmental Chemicals Causing Mutation and Cancer", *Science* 204:587–592 (1979).

Anderson, L. E. *The Mosby Medical Encyclopedia* (New York: New American Library, 1985).

Archer, V. E. *Health Implications of New Energy Technologies* (Ann Arbor, MI: Ann Arbor Science, 1980).

Arthur Anderson & Company. *Oil & Gas Reserve Disclosures* (Houston, TX: 1988).

Atlas, R. M. *Basic and Practical Microbiology* (New York: MacMillan Publishing Co., 1986).

Austin, D. F. *Epidemiology for the Health Sciences* (Springfield, IL: Charles C Thomas, 1982).

Beckman, D. A., and R. L. Brent. "Mechanism of Known Environmental Teratogens: Drugs and Chemicals", *Clin. Perinatol.* 13(3):649–87 (1986).

Billings, C. H. "Health Risk Assessment", *Public Works* 117:107 (1986).

Bloomfield, M. M. *Chemistry and the Living Organism* (New York: John Wiley & Sons, 1984), p. 610.

Boehm, P. D., and D. L. Fiest. "Subsurface Distributions of Petroleum From an Offshore Well Blowout", *Environ. Sci. Technol.* 16(2):67–76 (1982).

Boyland, E. "The Metabolism of Foreign Compounds and the Induction of Cancer", *Xenobiotica* 16:901–915 (1986) .

Brenson, A. *Control of Communicable Diseases in Man* (Washington, DC: American Public Health Association, 1985).

Brent, R. L. "Evaluating the Alleged Teratogenicity of Environmental Agents", *Clin. Perinatol.* 13(3):609–613 (1986).

Broom, B. *Anatomy and Physiology* (New York: Cambridge Communications, 1985).

Brown, T. L. *Chemistry: The Central Science* (Englewood Cliffs, NJ: Prentice Hall, 1984), p. 840.

Butler, G. C. *Principles of Ecotoxicology* (New York: John Wiley & Sons, 1981), p. 350.

Carlo, G. L., and S. Hearn. "Practical Aspects of Conducting Reproductive Epidemiology Studies Within Industry", *Progr. Clin. Biol. Res.* 160:49–65 (1984).

Chen, R. "Detection and Investigation of Subtle Epidemics", *Prog. Clin. Biol. Res.* 163B:39–43 (1985).

Clemmesen, J. "International Commission for Protectoin Against Environmental Mutagens and Carcinogens", *Mutat. Res.* 98(1):97–100 (1982).

Committee on the Institutional Means for Assessment of Risks to Public Health, Commission of Life Sciences, National Research Council. *Risk Assessment in the Federal Government: Managing the Process* (Washington, DC: National Academy Press, 1983).

Coon, M. J., and A. V. Persson. *Enzymatic Basis of Detoxificaton* (New York: Academic Press, 1980).

Covello, V., Von Winterfeldt, Detlof, Slovic and Paul. *Risk Communication: An Assessment of the Literature on Communicating Information about Health, Safety and Environmental Risks,* draft (Los Angeles: Institute of Safety and Systems Management, 1986).

Darmstadter, J. *Energy in America's Future: The Choices Before Us* (Baltimore, MD: John Hopkins Press, 1980).

Data Resources, Inc. *Energy Review,* (Lexington, MA: Data Resources Inc., 1988).

Davies, D.M. *Textbook of Adverse Drug Reactions* (New York: Oxford Medical Publications, 1981).

Department of the Environment. *Heavy Metals in the Environment* (London: H.M.S.O., 1983).

Di Carlo, F. J., P. Bickart and C. M. Auer. "Structure—Metabolism Relationships (SMR) for the Prediction of Health Hazards by the EPA", *Drug Metab. Rev.* 17(3–4):187–220 (1986).

Doll, R. "Relevance of Epidemiology to Policies for the Prevention of Cancer", *Human Toxicol.* 4(1):81–96 (1985).

Doyle, D. K. "Teratology: A Primer", *Neonatal Network* 4:24–29 (1986).

Dufus, J. H. *Environmental Toxicology* (London: Edward Arnold Publishers, 1982), p. 164.

Ecker, M. D., and N. J. Bramesco. *Radiation* (New York: Vintage Books, 1981).

El-Hinnawi, E. E. *The Environmental Impacts of Production and Use of Energy* (Shannon, 1981).

Ellerman, V. "Cancer Carcinogenesis", *Science America* 243:98–105 (1984).

Energy Information Administration. *Annual Energy Outlook 1987* DOEEIA-0383(87) (Washington, DC: Dept. of Energy, 1988).

Energy Information Administration. *Manufacturing Energy Consumption Survey: Consumption of Energy, 1985* DOE/EIA-000512(85) (Washington, DC: Dept. of Energy, 1988).

Evans, D. W. "Effects of Ionizing Radiation", *Cana. Med. Assoc. J.* 129:10–11 (1983).

Evans, W. E. *Applied Pharmacokinetics* (San Francisco: Applied Therapeutics Inc., 1980).

Fawcett, H. H. *Hazardous and Toxic Materials* (New York: John Wiley & Sons, 1984).

Fein, G. G. et al. "Environmental Toxins and Behavioral Development", *Am. Psychol.* 38(11):1188–1197 (1983).

Fein, G. *Intrauterin Exposure of Humans to PCB's: Newborn Effects,* (Duluth, MN: U.S. EPA, 1984).

Fochtman, E. G. *Biodegration and Carbon Adsorption of Carcinogenic and Hazardous Organic Compounds* (Cincinnati, OH: U.S. EPA, 1981).

Frederick, G. L. "Assessment of Teratogenic Potential of Therapeutic Agents", *Clin. Invest. Med.* 8(4):323–27 (1985).

Fukino, H., and S. Mimura. "Mutagenicity of Airborne Particles", *Mutat. Res.* 102: 237–241 (1982).

Gas Resources Institute. *1987 GRI Baseline Projection of U.S. Energy Supply and Demand* (Chicago, IL: Gas Resources Institute, 1987).

Gofman, J. W. *Radiation and Human Health* (San Francisco: Sierra Club Books, 1981).

Goldsmith, J. R. "Epidemiology Monitoring in Environmental Health, Introduction, and Overview", *Sci. Total Environ.* 32:211–218 (1984).

Gorbaty, M. L. *Coal Science* (New York: Academic Press, 1982).

Grahame-Smith, D. G., and J. K. Aronson. *Oxford Textbook of Clinical Pharmacology and Drug Therapy* (Oxford: Oxford University Press, 1984).

Green, G. "Ionizing Radiation and its Biological Effects", *Occupat. Health,* 38:354–357 (1986).

Green, L. W. *Community Health* (St. Louis, MO: Times Mirror/Mosby College Publishing, 1986).

Heinonen, O. P., D. Slone and S. Shapiro. *Birth Defects and Drugs in Pregnancy* (Boston, MA: John Wright-PSG Inc., 1982).

Hendee, W. R. "Real and Perceived Risks of Radiation Exposure", *West. J. Med.* 138:380–386 (1983).

Henold, K. L., and F. Walmsley. *Chemical Principles, Properties, and Reactions* (Reading, MA: Addison-Wesley, 1984).

Hodgson, E. *Introduction to Biochemical Toxicology* (New York: Elsevier, 1980).

Inskip, H., V. Beral, P. Fraser and J. Haskey, "Epidemiological Monitoring: Methods for Analyzing Routinely-Collected Data", *Sci. Total Environ.* 32:219–232 (1984).

Jefrey, A. M. "DNA Modification by Chemical Carcinogens", *Pharmacol Terminol.,* 28:237–241 (1985).

Khoury, D. L. *Coal Cleaning Technology* (Parkridge, NJ: Noyes Data Corp., 1981).

Khoury, M. J., and N. A. Holtzman. "On the Ability of Birth Defects Monitoring to Detect New Teratogens", *Am. J. Epidemiol.* 126(1):136–143 (1987).

Kleinbaum, D. G., L. L. Kupper and H. Morgenstern. *Epidemiologic Research* (London: Lifetime Learning Publications, 1982).

Larson, T. E. *Cleaning Our Environment: A Chemical Perspective* (Washington DC: American Chemical Society, 1983).

Legator, M. S., M. Rosenberg and H. Zenick. *Environmental Influences on Fertility, Pregnancy, and Development* (New York: Alan R. Liss, 1984).

Lilienfeld, A. *Foundations of Epidemiology* (New York: Oxford University Press, 1980).

Lippmann, M., and R. B. Schlesinger. *Chemical Contamination in the Human Environment* (New York: Oxford University Press, 1980).

March, J. *Advanced Organic Chemistry* (New York: John Wiley & Sons, 1985).

Marcus, J., and S. L. Hans. "Neurobehavorial Toxicology and Teratology", Department of Psychiatry, University of Chicago, 1982.

Meyers, V. K. "Chemicals Which Cause Birth Defects—Teratogens A Special Concern of Research Chemists", *Sci. Total Environ.* 32:1–12 (1983).

Miller, G. T. *Living in the Environment* (Bellmont, CA: Wadsworth Publishing Company, 1985).

Morgenstern, H. "Uses of Ecologic Analysis in Epidemiological Research", *Am. J. Public Health* 72:1336–1344 (1982).

Mortimer, C. E. *Chemistry* (Belmont, CA: Wadsworth, 1981).

Naeye, R., and H. Tafari. *Risk Factors in Pregnancy and Diseases of the Fetus and Newborn* (Baltimore: Williams and Wilkins, 1983).

Nau, H. "Species Differences in Pharmacokinetics and Drug Teratogenesis", *Environ. Health Perspec.* 70:113–129 (1986).

Neubert, D., et al. "Principles and Problems in Assessing Prenatal Toxicity", *Arch. Toxicol.* 60:238–245 (1987).

NIH. *Cancer Risk and Rates.* NIH Publication No. 85–691 (Washington, DC: NIH, 1985).

Niswander, K. R. *Obstetrics* (Boston: Little Brown and Co., 1981).

Oftedal, P., and A. Brogger. *Risk and Reason—Risk Assessment In Relation to Environmental Mutagens and Carcinogens* (New York: Alan R. Liss, 1986).

Parks, C. F., Jr. *Earthbound: Minerals, Energy, and Man's Future* (San Francisco: Freeman, Cooper and Company, 1975).

Poch, D. I. *Radiation Alert* (Toronto/Ontario: An Energy Prove Project, 1985).

Pochin, E. E. *Nuclear Radiation: Risks and Benefits* (Oxford: Clarendon Press, 1983).

Porter, C. R. *Coal Processing Technology* (American Institute of Chemical Engineers, 1981).

Price, S. A. *Pathophysiology Clinical Concepts of Diseases Processes* (New York: McGraw-Hill, 1986).

Quilligan, E. J., and N. Kretchmer. *Fetal and Maternal Medicine* (New York: John Wiley & Sons, 1980).

Robertson, R., and P. M. MacLeod. "Accutane-Induced Teratogenesis", *Canadian Med. Assoc. J.* 133(11):1147–1148 (1985).

Rosa, F. W., A. L. Wilk and F. O. Kelsey. "Teratogen Update", *Teratology* 33(3):356–364 (1986).

Rosenber, M. J., and W. E. Halperin. "The Role of Surveillance in Monitoring Reproductive Health", *Teratogen. Carcinogen. Mutagen.* 4(1):15–24 (1984).

Rowland, M., and T. Tozer. *Clinical Pharmacokinetics: Concepts and Applications* (Philadelphia: Lea & Febiger, 1980).

SCEP. "Study of Critical Environmental Problems", in *Man's Impact on the Global Environment* (Cambridge, MA: MIT Press, 1983).

Schardein, J. L. "Current Status of Drugs as Teratogens in Man", *Progr. Clin. Biol. Res.* 163C:181–190 (1985).

Schlesselman, J. *Case-Control Studies* (New York: Oxford University Press, 1982).

Schottenfeld, D., and J. Fraumeni, Jr. *Cancer Epidemiology and Prevention* (Philadelphia: W. B. Saunders, 1982).

Shepard, T. H. "Human Teratogens: How Can We Sort Them Out?", *Ann. N. Y. Acad. Sci.* 477:105–115 (1986).

Sherwin, R. P. "What is an Adverse Health Effect?", *Environ. Health Perspec.* 52:177–182 (1983).

Sorensen, H. *Energy Conversion Systems* (New York: John Wiley & Sons (1983).

Swaab, D. F., and M. Mirmiran. "Functional Teratogenic Effects of Chemicals on the Developing Brain", *Monogr. Neural Sci.* 12:45–57 (1986).

Ujino, Y. "Epidemiological Studies on Disturbances of Human Fetal Development", *Arch. Environ. Health* 40:181–184 (1985).

Ulsamer, A. G., P. D. White and P. W. Preuss. "Evaluation of Carcinogens: Perspective of the Consumer Product Safety Commission", in *Handbook of Carcinogen Testing*, H. A. Milman and E. K. Weisburger Eds. (Park Ridge, NJ: Noyes Publications, 1985).

U.S. Department of Health and Human Services. *Determining Risks to Health* (Dover, MA: Auburn Publishing Co., 1986).

U.S. Department of Health and Human Services, Food and Drug Administration. "Identifying and Regulating Carcinogens", Comments on unpublished OTA draft report (1987).

U.S. Department of Health and Human Services, National Toxicology Program. *NTP Tech. Bull.* 1(3):1–2, (1980).

U.S. Department of Health and Human Services, National Toxicology Program. "Explanation of Levels of Evidence of Carcinogenic Activity", 1986.

U.S. Department of Health and Human Services, National Toxicology Program. *Annual Plan, FY 1986*, DHHS Publ. No. NTP-86-086 (Research Triangle Park, NC:1986).

U.S. Department of Health and Human Services, National Toxicology Program. Board of Scientific Counselors *Report of the NTP Ad Hoc Panel on Chemical Carcinogensis Testing and Evaluation* (Research Triangle Park, NC: 1980).

U.S. Department of Health and Human Services, National Toxicology Program. *Fourth Annual Report on Carcinogens* (Research Triangle Park, NC: 1985).

U.S. Department of Health and Human Services, National Toxicology Program. *Review of Current DHHS, DOE, and EPA Research Related to Toxicology, Fiscal Year 1986*, Publ. No. NTP-86-087 (Research Triangle Park, NC: 1987).

U.S. Department of Health and Human Services, National Toxicology Program. *Second Annual Report on Carcinogens* (Research Triangle Park, NC: 1981).

U.S. Department of Health and Human Services, National Toxicology Program. *Third Annual Report on Carcinogens* (Research Triangle Park, NC: 1983).

U.S. Environmental Protection Agency. *Environmental Outlook 1980* (Washington, DC: 1980).

U.S. Environmental Protection Agency. "Guidelines for Carcinogen Risk Assessment", *Fed. Reg.* 51:33992 (1986).

U.S. Environmental Protection Agency. "Guidelines for Mutagenicity Risk Assessment", *Fed. Reg.* 51:34006 (1986).

U.S. Environmental Protection Agency, Office of Policy and Resource Management. *Environmental Mathematical Pollutant Fate Modeling Handbook/Catalogue* (draft), Contract no. 68-01-5146 (Washington, DC: 1982).

U.S. Environmental Protection Agency. "Proposed Guidelines for Carcinogen Risk Assessment", *Fed. Reg.* 49(227):46294–46301 (1984).

U.S. Environmental Protection Agency. "Proposed Guidelines for the Health Risk Assessment of Chemical Mixtures and Request for Comments", *Fed. Reg.* 50(6):1170–1176 (1985).

U.S. Environmental Protection Agency. Proposed Guidelines for the Health Assessment of Suspect Developmental Toxicants and Request for Comments", *Fed. Reg.* 49 (227):46324–46331 (1984).

U.S. Environmental Protection Agency. *Risk Assessment and Management: Framework for Decision Making* (Washington, DC: 1984).

Vainio, H. "Current Trends in the Biological Monitoring of Exposure to Carcinogens", *Scand. J. Environ. Health* 11:1–3 (1985).

Verschueren, K. *Handbook of Environmental Data on Organic Chemicals* (New York: Van Nostrand Reinhold Company 1983).

Vesilind, A. P., and J. Pierce. *Environmental Pollution and Control,* 2nd ed. (Ann Arbor, MI: Ann Arbor Science, 1983).

Vesiland, P. A., and J. J. Peirce. *Environmental Engineering* (Boston: Butterworth Publishers, 1982).

Wagner, P. *Coal Processing Technology* (American Institute of Chemical Engineers, 1980).

Wallace, R. A. *Biology, the Science of Life* (San Francisco: Scott, Foresman and Company, 1983), p. 1140.

Wartak, J. *Clinical Pharmacokinetics* (Baltimore: University Park Press, 1983).

Washington Policy and Analysis, Inc. *Natural Gas: A Strategic Resource for the Future* (Washington, DC: 1988).

Wehrle, P. *Communicable and Infectious Disease* (St. Louis, MO: C.V. Mosby Co., 1981).

Wharton Econometric Forecasting. Associates *Energy Analysis Quarterly* (Bala Cynwyd, PA: 1988).

Whitmore, R. W. *Methodology for Characterization of Uncertainty in Exposure Assessments* (Research Triangle Park, NC: Research Triangle Institute, 1985).

Winter, M. E., *Basic Clinical Pharmacokinetics* (San Francisco: Applied Therapeutics, Inc., 1980).

Yoyn, D., and Blkeslee, Eds. *Advances in Comparative Leukemia* (New York: Elsevier North Holland, 1981), p. 1–16.

Zitko, V. "Graphical Display of Environmental Quality Criteria", *Sci. Total Environ.* 72:217–220 (1988).

Chapter 2

Abbot, J. "Analysis of the response of Saccharomyces cerevisiae cells to Kluyveromyces lactis toxin," *J. Gen. Microbio.* 137(1757): 15-18 (1991).

Akinboade, O. A., J. O. Hassan and A. Adejinmi. "Public Health Importance of Market Meat Exposed to Refuse Flies and Air-borne Microorganisms", *Int. J. Zoonoses* 11 (1):111–114 (1984).

Archer, D. L., and F. E. Young. "Contemporary Issues: Diseases with a Food Vector", *Clin. Microbiol. Rev.* 1(4): 377–398 (1988).

Baddour, L. M., S. M. Gaia, R. Griffin and R. Hudson. "A Hospital Cafeteria-Related Food-borne Outbreak due to *Bacillus Cereus*: Unique Features", *Infect. Control* 7 (9):4652–4655 (1986).

Blumenthal, D. "Food IrridationL Toxic to Bacteria, Safe for Humans" *FDA Consumer*, p. 11-15 (Nov. 1990).

Brown, P., D. Kidd, T. Riordan and A. Barrell. "An Outbreak of Food-Borne *Campylobacter jejuni* Infection and the Possible Role of Cross-Contamination", *J. Infect.* 17 (2):171–176 (1988).

Bull, P. *Prevention of Food-Borne Disease* 22(2):207–208 (1988).

Bush, M. F. "The Symptomless Salmonella Excretor Working in the Food Industry", *Community Med.* 7 (2):133–135 (1985).

Chia, J. K., J. B. Clark, C. A. Ryan and M. Pollack. "Botulism in an Adult Associated With Food-Borne Intestinal Infection With *Clostridium botulinum*", *N. Engl. J. Med.* 315(4):239–241 (1986).

Cohen, M. L., and R. V. Tauxe. "Drug-Resistant Salmonella in the United States: An Epidemiologic Perspective," *Science* 234(4779):964–969 (1986).

Corbe, S., P. Barton, R. C. Nair and C. Dulbert. "Evaluation of the Effect of Frequency of Inspection on the Sanitary Conditions of Eating Establishments ", *Can. J. Public Health* 75(6):434–438 (1984).

Cordle, M. K. "USDA Regulation of Residues in Meat and Poultry Products ", *J. Anim. Sci.* 66(2):413–433 (1988).

de-Wit, J. C., and E. H. Kampelmacher. "Some Aspects of Bacterial Contamination of Hands of Workers in Food Service Establishments", *Zentralbl. Bakteriol. Mikrobiol. Hyg. B.* 186(1):45–54 (1988).

Diel, J.F. *Safety of Irradiated Foods: Potential and Actual Applications* (New York: Marcel Decker Inc., 1990) p. 165–170, 237–242.

Engel, R.E., A.R. Post, and R.C. Post"Implementation of Irradiation of Pork for *Trichina* control," *Food Tech.* 42:71–75 (1988).

Farkas, J. *Irradiation of Dry Food Ingredients* (Boca Raton, FL: CRC Press, Inc., 1988).

Gay, C. T., W. A. Marks, H. D. Riley, J. B. Bodensteiner, M. Hamza, P. A. Noorani and G. B. Bobele. "Infantile Botulism ", *South. Med. J.* 81(4):457–460 (1988).

Gelderblom, W.C.A. "Effects of Temerature and Incubation Period on Production of Fumonisin B1 by Fusarium moniliform," *Appl. Environ. Microbiol.* 56(1733):75–78 (1990).

Greene, J. "FDA Talk Paper ", *Food and Drug Admin.* 25 Feb. 1983.

Greene, J. "FDA Talk Paper ", *Food and Drug Admin.* 5 Apr. 1983.

Greene, J. "FDA Talk Paper ", *Food and Drug Admin.* 26 Sept. 1984.

Guilfoyle, D. E., and J. F. Yager. "Survey of Infant Foods for *Clostridium botulinum* Spores", *J. Assoc. Off. Anal. Chem.* 66(5):1302–1304 (1983).

Harris, N. V., D. Thompson, D. C. Martin and C. M. Nolan. "A Survey of Campylobacter and Other Bacterial Contaminants of Pre-Market Chicken and Retail Poultry and Meats, King County, Washington", *Am. J. Public Health* 76(4):401–406 (1986).

Hayes, A. H. "The Food and Drug Administration's Role in the Canned Salmon Recalls of 1982 ", *Public Health Rep.* 98(4):144–149 (1983).

Hill, G.W. "Levels of Bacteria, Fungi, and Endotoxin in Bulk and Aerosolized Corn SIlage," *Appl. Environ. Microbiol.* 55(1039):200–210 (1989).

Hobbs, G. "Food Poisoning and Fish ", *J. R. Soc. Health* 103(4):144–149 (1983).

Hay, R. N. "Environmental Effects on Mastitis and Milk Quality ", *Vet. Clin. North Am. [Large Anim. Pract.]*, 6(2):371–375 (1984).

Kay, R. N. "Poultry Meat Inspection — A Service to the Community ", *J. R. Soc. Health* 108(2):39–40 (1988).

Kliks, M. M. "Anisakiasis in the Western United States: Four New Case Reports From California ", *Am. J. Trop. Med. Hyg.* 32(3):526–532 (1983).

Knight, K.G., "Effect of Temperature on Aflatoxin Production in Mucuna Pruriens Seeds," *Appl. Environ. Microbiol.* 55(532):70–75 (1989).

Kwee, H. G. and R. L. Sautter. "Anisakiasis ", *Am. Fam. Phys.* 36(2):137–140 (1987).

Lim, Y. S., S. Y. Khor, M. Jegathesan and S. H. Kang. "The Isolation of Salmonella from Raw, Cooked and Dried Foods", *Med. J. Malaysia* 39(3):220–224 (1984).

Macdonald, K. L., and P. M. Griffin. "Foodborne Disease Outbreaks, Annual Summary, 1982", *MMSR. CDC. Surveill. Summ.* 35(1):775–1655 (1986).

Martin, D. L., T. L. Gustafson, J. W. Pelosi, L. Suarez and G. V. Pierce. "Contaminated Produce — A common Source for Two Outbreaks of Shigella Gastroenteritis ", *Am. J. Epidemiol.* 124(2):299–305 (1986).

Mascola, L., L. Lieb, J. Chiu, S. L. Fannin and M. J. Linnan. "Listeriosis: An Uncommon Opportunistic Infection in Patients with Acquired Immunodeficiecy Syndrome. A report of five cases and a review of the Literature ", *Am. J. Med.* 84(1):162–164 (1988).

Meitert, E., A. Vlad and F. Sima. "Food Poisoning Associated with *Pseudomonas aeruginosa"* *Arch. Roum. Pathol. Exp. Microbiol.* 43(2):115–119 (1984).

Miller, S. A. "Toxicology and Food Safety Regulations ", *Arch. Toxicol.* 60(1–3):212–216 (1987).

Nastasi, A., C. Mammina, M. R. Villafrate, G. Scarlata, M. F. Massenti and M. Diquattro. "Multiple Typing of *Salmonella typhimurium* Isolates: An Epidemiological Study", *Microbiologica* 11(3):173–178 (1988).

Nelson, P. "Mycotoxicology: Introduction to the Mucology, Plant Pathology, Chemistry, Toxicology, and Pathology of Naturally Occurring Mycotoxicoses in Animals and Man," *Am. Scientist* 78(471):130–133 (1990).

Prescott, L., J. Harley, and D. Klein *Microbiology* (Dubuque, IA: William C. Brown, 1990).

Reed, D. V. "The FDA Surveillance Index for Pesticides: Establishing Food Monitoring Priorities Based on Potential Health Risk", *J. Assoc. Off. Anal. Chem.* 68(1):122–124 (1985).

Rennels, M. B., and M. M. Levine. "Classical Bacterial Diarrhea: Perspectives and Updates—Salmonella, Shigella, Escherichia coli, Aeromonas, and Plesiomonas", *Pediatr. Infect. Dis.* 5(Suppl. 1):91–100 (1986).

Tartakow, I., and Jackson. *Foodborne and Waterborne Disease* (Westport, CT: AVI Publishing Company, 1981).

U. S. Department of Health and Human Services, Food and Drug Administration. "General Principles for Evaluating the Safety of Compounds Used in Food-Producing Animals", typescript (July 1983).

Wainwright, R. B., W. L. Heyward, J. P. Middaugh, C. L. Hatheway, A. P. Harpster and T. R. Bender. "Food-Borne Botulism in Alaska, 1947–1985: Epidemiology and Clinical Findings", *J. Infect. Dis.* 157(6):1158–1162 (1988).

Wilson, J. "Foodborne Illness: The Problem That Just Won't Go Away", *Can. Nurse* 82(2):14–15 (1986).

Wong, H. D., M. H. Chang and J. Y. Fan. "Incidence and Characteriziation of *Bacillus cereus* Isolates Contaminating Dairy Products", *Appl. Environ. Microbiol.* 54(3)669–702 (1988).

Yamagishi, T., K. Sakamoto, S. Sakurai, K. Konishi, Y. Daimon, M. Matsuda, Y. Gyobu, Y. Kubo and H. Kodama. "A Nosocomial Outbreak of Food Poisoning Caused by Enterotoxigenic *Clostridium perfringens*", *Microbiol. Immunol.* 27(3):291–296 (1983).

Chapter 3

Adams, C. "Use of HACCP in Meat and Poultry Inspection *Food Tech.* (May 1990).

Adams, C. and S. Sachs "Government's Role in COmmunicating Food Safety Information to the Public", *Food Tech.* (May 1991).

Alfin-Slater, R. B., and F. X. Pi-Sunyer. "Sugar and Sugar Substitutes. Comparisions and Indications", *Postgrad. Med.* 82(23):46–50 (1987).

Anderson, J. A. "The Establishment of Common Language Concerning Adverse Reactions to Foods and Food Additives", *J. Allergy. Clin. Immunol.* 78(1):140–144 (1986).

Anderson, S.R. "Assessment of the Health Risks and Handling of PRoducts Used in Food Service Establishments", *J. Environ. Health* (Jan.–Feb. 1985).

Angus, D. S. "Comparative Mutagenicity of Two Triarylmethane Food Dyes in Salmonella, Saccharamyces, and Drosophila", *Food Chem. Toxicol.* 19(4):419–424 (1981).

Anliker, R. "Ecotoxicology of Dyestuffs—A Joint Effort by Industry", *Ecotoxicol. Environ. Safety* 3:59–74 (1979).

Archer, D. L. "The Need for Flexibility in HACCP", *Food Tech.* (May 1990).

Association of Food and Drug Officials "Retail Guidelines-Refrigerated Foods in Reduced Oxygen Packages", *Assoc. Food Drug Offic.* 54(5):80–84 (1990).

Bauman, H. "HACCP: Concept, Development, and Application", *Food Tech.* (May 1990).

Betina, V. *Citrinin and Related Substances*, Vol. 8 (Amsterdam: Elsevier/North Holland Publishing Co., 1984), p. 217–236.

Busch, L. "Biotechnilogy: Consumer Concerns about Risks and Values", *Food Tech.* (April 1991).

Campebll, W. D. "Communicating Food Safety Risks to the Public", *J. Environ. Health* (March-April 1991).

Chung, K. T., G. E. Fulk and A. W. Andrews. "Mutagenicity Testing of Some Commonly Used Dyes", *Appl. Environ. Microbiol.* 42(4):641–648 (1981).

Combes, R. D., and R. B. Haveland. "A Review of the Genotoxity of Food, Drug, and Cosmetic Colors and Other Azo, Triphenylmethane, and Xanthene Dyes", *Mutat. Res.* 98:101–248 (1982).

Cordle, M. K. "USDA Regulation of Residues in Meat and Poultry Products", *J. Anim. Sci.* 66(2):413–433 (1988).

Coye, M. J. "The Health Effects of Agricultural Production: II The Health of the Community", *Public Health Policy* 7(3):340–354 (1986).

Diamond, S., J. Prager and F. G. Freitag. "Diet and Headache. Is There a Link?", *Postgrad. Med.* 79(4):279–286 (1986).

Dixon, E. A., and G. Renyk. "Isolation, Separation, and Identification of Synthetic Food Colors", *J. Chem. Educ.* 59(1):67–69 (1982).

Ecobichon, D. J. "Pesticide Residues in Foods", *Proc. West. Pharmacol. Soc.* 29:499–502 (1986).

Felix, C. W., et al. "Utensil Sanitation", *J. Environ. Health* (Sept.-Oct. 1990).

Foulke, J. "What Happens if Packaging gets into the Food?", *FDA Consumer* (Nov. 1991).

Freydbert, N., and W. Gortner. *The Food Additives Book* (New York: Bantam Books, 1982).

Garrett, S. and M. Hudak-Roos "Use of HACCP for Seafood Surveillance and Certification", *Food Tech.* (May 1990).

Giese, J. H. "Sanitation: The Key to Food Safety and Public Health", *Food Tech.* (Dec. 1991).

Gill, C. O. and K. M. Delacey. "Growth of *Escherichia Coli* and *Salmonella typhimurium* on High-pH Beef Packaged Under Vacuum or Carbon Dioxide", *Int. J. Food Microbiol.* 13:21–30(1991).

Groth, E. "Communicating with Consumers About Food Safety and Risk Issues", *Food Tech.* (May 1991).

"Hearings Before the Subcommittee on Health and the Environment of the Committee on Interstate and Foreign Commerce House of Representatives", *Antibiotics in Animal Feed*, (Washington, DC: U.S. Government Printing Office, 1980).

Hecht, A., and J. Willis. "Sulfites: Preservatives That Can Go Wrong?", *FDA Consumer* 11:(1983).

Janssen, P. J., and C. A. van-der-Heijden. "Aspartame: Review of Recent Experimental and Observational Data", *Toxicology* 50(1):1–26 (1988).

Kalish, F. "Extending the HACCP Concept to Product Distribution", *Food Tech.* (June 1991).

Kropf, W., and M. Houben. "Harmful Food Additives", in *The Eat Safe Guide* (New York: Ashley Books, 1980).

Lecos, C. "Food Preservatives: A Fresh Report", *FDA Consumer* April (1984), p. 23–25.

Lehmann, P. "More Than You Ever Thought You Would Know About Food Additives", *FDA Consumer* Feb. (1981), p.1–4.

Leist, K. H. "Subacute Toxicity Studies of Selected Organic Colorants", *Ecotoxicol. Environ. Safety* 6:457–463 (1982).

Lener, J., and B. Bibr. "Cadmium Content in Some Foodstuffs in Respect of its Biological Effects", *Vitalstoffe Zivilisations-Krankheiten* 15(4):139–141 (1980).

Lisiecki, R., et al. "Aseptic Package Addresses a Variety of Needs", *Food Tech.* (June 1990).

London, R. "Saccharin and Aspartame. Are They Safe to Consume During Pregnancy?", *J. Reprod. Med.* 33(1):17–21 (1988).

Longer, K. "Lead Toxicity and Metabolism From Lead Sulfate Fed to Holstein Calves", *J. Dairy Sci.* 67(5):1007–1013 (1984).

Longstaff, E. "An Assessment and Categorization of the Animal Carcinogenicity Data on Selected Dyestuffs and An Extrapolation of those Data to an Evaluation of the Relative Carcinogenic Risk to Man", *Dyes Pigm.* 4:243–304 (1983).

Mahaffy, K. R. "Heavy Metal Exposure From Foods", *Environ. Health Perspec.* 12:63 (1983).

Marmion, D. M. *Handbook of U.S. Colorants for Foods, Drugs, and Cosmetics,* (New York: John Wiley & Sons, 1979).

Morbidity and Mortality Weekly Report "Evaluation of Consumer Complaints Related to Aspartame Use", 33:605–607 (1984).

National Academy of Sciences. *The Effects of Human Health of Subtherapeutic Use of Antimicrobials in Animal Feeds* (Washington, DC: U.S. Government Printing Office, 1980).

National Advisory Comittee on Microbiological Criteria for Foods. "Hazard Analysis and Critical Control System" *Int. J. Food Microbiol.* 16:1–23(1992).

Pierson, M. D. and D. A. Corlett Jr. (Eds.) HACCP Principles and Applications (New York: Van Norstrand Reinhold, 1992).

Richmond, M. H., and D. I. Smith. "The Use of Tetracycline in the Community and its Possible Relation to the Excretion of Tetracycline-Resistant Bacteria", *J. Antimicrobial Chemother.* 6:33–41 (1980).

Schaffner, D. W. "Food Service Sanitation Training Materials", *Food Tech.* (Sept. 1991).

Scherer, C. W. "Strategies for Communicating Risks to the Public", *Food Tech.* (Oct. 1991)

Simon, R. A. "Adverse Reactions to Food Additives", *N. Eng. Reg. Allergy Proc.* 7(6):533–546 (1986).

Stegink, L. D. "The Aspartame Story: A Model for the Clinical Testing of a Food Additive", *Am. J. Clin. Nutr.* 46(Suppl 1):204–215 (1987).

Stevenson, K. E. "Implementing HACCP in the Food Industry", *Food Tech.* (May 1990)

Stobbaerts, R. F., L. VanHarverbeke and M. A. Herman. "Qualitative and Quantitative Determination of Some Yellow, Orange, and Red Food Dyes by Resonance Raman Spectroscopy", *J. Food Sci.* 48:521–525 (1983).

Uno, Y. "The Toxicology of Mycotoxins", *CRC Crit. Rev. Toxicol.* 14(2):99–132 (1985).

U.S. Department of Health and Human Services, Food and Drug Administration. "Policy for Regulating Carcinogenic Chemicals in Food and Color Additives; Advance Notice of Proposed Rulemaking", *Fed. Reg.* 47:14464 (1982).

U.S. Department of Health and Human Service, Food and Drug Administration. "Sponsored Compounds in Food-Producing Animals; Proposed Rule and Notice", *Fed. Reg.* 50:45530 (1985).

U.S. Department of Health and Human Services, Food and Drug Administration, Bureau of Foods. *Toxicologic Principles for the Safety Assessment of Direct Food Additive and Color Additives Used in Food* (Washington, DC: 1982).

Wagner, G. J. "Cadmium in Wheat Grain", *J. Toxicol. Environ. Health* 13(4–6):979–989 (1984).

Walton, J. R. "Antibiotic Residues in Meat", *Br. Vet. J.* 143(6):485–486 (1987).

Watanabe, H. K., and Y. Hasimoto. "Carcinogenicities of 3-Methoxy-4-Aminoazobenzene and Related Axo Dyes in the Mouse", *Gann (Japanese Cancer Assoc.)* 73(136):136–140 (1982).

Watson, D. H. "Toxic Fungal Metabolites in Food", *CRC Crit. Rev. Food Sci. Nutr.* 22(3):177–198 (1985).

Winter, R. *A Consumers Dictionary of Food Additives* (New York: Crown Publishing, 1982).

Wolnik, K. "Cadmium in Lettuce, Peanuts, Potatoes, Sweet Corn and Wheat", *J. Agric. Food Chem.* 31(6):120–124 (1983).

Wurtman, R. "Neurochemical Changes Following High-Dose Aspartame with Dietary Carbohydrates", *N. Engl. J. Med.* 309(7):429–430 (1983).

Chapter 4

Adamis, Z., A. Antal, I. Fuzesi, J. Molnar, L. Nagy and M. Susan. "Occupational Exposure to Organophosphorus Insecticides and Synthetic Pyrethroid", *Int. Arch. Occup. Environ. Health* 56(4):299–305 (1985).

Ali, A., H. N. Nigg, J. H. Stamper, M. L. Kok-Yokomi and M. Weaver. "Diflubenzuron Application to Citrus and its Impact on Invertebrates in an Adjacent Pond", *Bull. Environ. Contam. Toxicol.* 41(5):781–790 (1988).

Babu, C. J., K. N. Panicker and P. K. Das. "Breeding of *Aedes aegypti* in Closed Septic Tanks", *Indian J. Med. Res.* 77:637 (1983).

Beausoleil, E. G. "A Review of Present Antimalaria Activities in Africa", *Bull. WHO.* 62(Suppl):13–17 (1984).

Carlson, D. B., and P. D. O'Bryan. "Mosquito Production in a Rotationally Managed Impoundment Compared to Other Management Techniques", *J. Am. Mosq. Control Assoc.* 4(2):146–151 (1988).

Cherniack, M. G. "Organophosphorus Esters and Polyneuropathy", *Ann. Intern. Med.* 104(2):264–266 (1986).

D'Ambrosio, S. M. Chlorinated Hydrocarbons: Insecticide Versus Carcinogenic Action, (Cincinnati, OH: U.S. EPA, 1983).

Flannigan, S. A., S. B. Tucker, M. M. Key, C. E. Ross, E. J. Fairchild, B. A. Grimes and R. B. Harris. "Synthetic Pyrethroid Insecticies: A Dermatologist Evaluation", *Br. J. Ind. Med.* 42(6):363–372 (1985).

Guzman, D. R., and R. C. Axtell. "Temperature and Water Quality Effects in Simulated Woodland Pools on the Infection of Culex Mosquito Larvae by *Lagenidium giganteum* (Domycetes: Lagenidiales) in North Carolina", *J. Am. Mosq. Control Assoc.*

Hogdson, E. "Development of Safer Insecticides", *Drug. Metab. Rev.* 15(5–6):881–895 (1985).

Kumari, K., R. P. Singh, and S. K. Saxena. "Movement of Carbufuran (nematicide) in Soil Columns", *Ecotoxicol. Environ. Safety* 16(1):36–44 (1988).

Liebowitz, D. P., and J. A. Kriz. "Collection and Determination of Counter (terbufos) Insecticide in Air", *Am. Ind. Hyg. Assoc. J.* 44(8):567–571 (1983).

Maloney, S. E., A. Maule and A. R. Smith. "Microbial Transformation of the Pyrethroid Insecticides: Permethrin, Deltamethrin, Fastac, Fenvalerate, and Fluvalinate", *Appl. Environ. Microbiol.* 54(11):2874–2876 (1988).

Matsumura, F. *Toxicology of Insecticides*, 2nd ed. (New York: Plenum Press, 1985).

Mulla, M. S., and H. A. Darwazeh. "Efficacy of New Insect Growth Regulators Against Mosquito Larvae in Dairy, Wastewater Lagoon", *J. Am. Mosq. Control Assoc.* 4(3):322–325 (1988).

Pandita, T. K. "Mutagenic Studies on the Insecticide Metasystox-R With Different Genetic Systems", *Mutat. Res.* 124(1):971–972 (1983).

Parry, J. E. "Control of *Aedes triseriatus* in LaCrosse, Wisconsin", *Prog. Clin. Biol. Res.* 123:355–363 (1983).

Reeves, R. R., and R. O. Pendarvis. "A New Method for the Differentiation of Naphthalene and Paradichlorobenzene Mothballs", *J. Am. Ostepath. Assoc.* 85(12):806–808 (1985).

Schaefer, C. H., and E. F. Dupres, Jr. "Recycling Suitability of Wash Waters From Mosquito Abatement Vehicles and Equipment Into Spray Diluent", *J. Am. Mosq. Control Assoc.* 4(1):1–3 (1988).

Sharma, V. P., and K. N. Mehrotra. "Malaria Resurgence in India: A Critical Study", *Soc. Sci. Med.* 22(8):835–845 (1986).

Tucker, S. B., S. A. Flannigan and C. E. Ross. "Inhibition of Cutaneous Parethesia Resulting From Synthetic Pyrethroid Exposure", *Int. J. Dermatol.* 23(10):686–689 (1984).

U. S. Department of Health and Human Services. *Insecticides: For the Control of Insects of Public Health Importance* (Washington, DC: Department of Health, 1982).

U. S. Department of Health, Education, and Welfare. *Insecticide Application Equipment* (Atlanta: Public Health Service, 1984).

Warren, M., H. C. Spencer, F. C. Churchill, V. J. Francois, R. Hippolyte and M. A. Staiger. "Assessment of Exposure to Organophosphate Insecticides During Spraying in Haiti: Monitoring of Urinary Metabolites and Blood Cholinesterases Levels", *Bull. WHO* 63(2):353–360 (1985).

Walliker, D. "The Genetic Basis of Diversity in Malaria Parasites", *Adv. Parasitol.* 22:217–259 (1983).

Chapter 5

Jones, E. C., G. H. Grow and S. C. Naiman. "Prolonged Anticoagulation in Rat Poisoning", *Pr. Med. J. Clin. Res.* 287(6385):118–119 (1983).

Kimsey, S. W., T. E. Carpenter, M. Pappaioanou and E. Lusk. "Benefit Cost Analysis of Bubonic Plague Surveillance and Control at Two Campgrounds in California, USA", *JAMA* 252(21):3005–3007 (1984).

Lipton, R. A., and E. M. Klass. "Human Infestion of a 'Superwarfarin' Rodenticide Resulting in a Prolonged Anticoagulant Effect", *J. Med. Entomol.* 22(5):499–506 (1985).

Paynter O. E., U. S. Environmental Protection Agency, Office of Pesticides Programs, Hazard Evaluation Division, Standard Evaluation Procedure. *Oncogenicity Potential: Guidance for Analysis and Evaluation of Long-Term Rodent Studies,* EPA-540/9-85-019 (Washington, DC: U.S. Environmental Protection Agency, 1985).

Schwar,T. G., D. Thompson and B. C. Nelson. "Fleas on Roof Rats in Six Areas of Los Angeles County, California: Their Potential Role in the Transmission of Plague and Murine Typhus to Human", *Am. J. Trop. Med. Hyg.* 34(2):372–379 (1985).

U. S. Department of Health and Human Services, Center for Disease Control. *Hantavirus Infection –Southwestern United States: Interim Reccommendations for Risk Reduction,* MMWR 42(RR-11, July 30, 1993).

U. S. Department of Health and Human Services, Center for Disease Control. *Laboratory Management of Agents Associated with Hantavirus Pulmonary Syndrome: Interim Biosafety Guidelines,* MMWR 43(RR-7, May 13, 1994).

Chapter 6

Adamis, Z., A. Antal, I. Fuzesi, J. Molnar, L. Nagy and M. Susan. "Occupational Exposure to Organophosphorus Insecticides and Synthetic Pyrethroid", *Int. Arch. Occup. Environ. Health* 56(4):399–305 (1985).

Balding, H. "Inert Pesticide Ingredients Toxic", *Tech. Rev.* 80:181 (1985).

Barron, M. G. "Bioconcentration", *Environ. Sci. Tech.* 24(11):1612–1618(1990).

Belluck, D. and S. Benjamin. "Pesticides and Human Health", *J. Environ. Health.* 53(7):11–12 (1990).

Brown, J., et al. "Polychlorinated Biphenyl Dechlorination in Aquatic Sediments", *Science.* 236(4802):709–712(1987).

Bull, D. *Pesticides and the Third World* (New York: Hayes & Pinkle, 1982).

Campt, D. "Pesticide Regulation is Sound, Protective, and Steady", *Chem. Eng. News* 69:44(1991).

Chaboussov, F. "How Pesticides Increase Pests", in *The Ecologist* (Wadebridge: Ecosystems Ltd., 1982).

Christensen, E. "Pesticide Regulation and International Trade", *Environment.* 32:2–3(1990).

Connolly, J. P. "Application of a Food Chain Model to Polychlorinated Biphenyl Contamination of the Lobster and Winter Flounder Food Chains in New Bedford Harbor", *Environ. Sci. Tech.* 25(4):760–770(1991).

Corning, M. C. "Pesticide Eating Bacteria", *Science News* 127:329 (1985).

Council on Scientific Affairs. "Cancer Risk of Pesticides in Agricultural Workers", *J. Am. Med. Assoc.* 260(7):959–966(1988).

Dindal, D. L., Ed. *Soil Biology as Related to Land Use Practices* (Washington, DC: U.S. EPA, WASA, DD, Office of Pesticides and Toxic Substances, 1980).

Dover, M. "Getting Off the Pesticide Treadmill", *Tech. Rev.* 88:111 (1985).

Edgerton, S. *Pesticides Analysis in Select Soil Samples* (New York: John Wiley & Sons, 1985).

Fleming, J. *Organochlorine Pesticides and PCB's: A Continuing Problem for the 1980's* (Washington, DC: U.S. Fish and Wildlife Service, 1983).

Gough, M. *Dioxin Agent Orange* (New York: Plenum Press, 1986).

Hayes, W. J. *Pesticides Studied in Man* (Baltimore: Williams & Wilkins, 1982).

Howard, P. H. *Handbook of Environmental Fate and Exposure Data for Organic Chemicals Vol. 1–3.* (Boca Raton, FL: Lewis Publishers, 1989).

Hudson, R. H. *Toxicology of Pesticides to Wildlife* (Washington: Resource Publication, 1984).

Immerman, F. W. National Urban Pesticide Usage (Cincinnati, OH: U.S. EPA, Economic Analysis Branch, Benefits and Use Division, 1984).

Jacobson, J. L. "Determinants of Polychlorinated Biphenyls, Polybrominated Biphenyls, and Dichlorodiphenyl Trichlorethane Levels in the Sera of Young Children", *Am. J. Pub. Health.* 79(10):1400–1404(1989).

Krueger, R. F. and D. J. Severn. *Regulation of Pesticide Disposal* Vol. 295 (Washington, DC: American Chemical Soc., 1984), p. 1–15.

Li, W. "The Role of Pesticides in Skin Disease", *Int. J. Dermatol.* 25(5):295–297 (1986).

Manz, A. and J. Berger. "Cancer Mortality Among Workers in Chemical Plant Contaminated with Dioxin", *The Lancet.* 338(8773):959–964(1991).

Marquis, J. K., *Contemporary Issues in Pesticides Toxicology and Pharmacology* (Basel: S. Karger, 1986).

McEwen, F. L., and G. R. Stephenson. *The Use and Significance of Pesticides in the Environment* (New York: John Wiley & Sons, 1980).

Morrill, L. G., et al. *Organic Compounds in Soil* (Ann Arbor, MI: Butterworth Ltd., 1982).

Ninety-Eighth Congress. "Government Regulation of the Pesticides Ethelene Dibromide, Joint Hearing" (Washington, DC: U.S. Government Printing Office, 1984).

Ninety-Eighth Congress. "Hearings of Pesticides Tolerance Legislation" (Washington, DC: U.S. Government Printing Office, 1984).

Ninety-Ninth Congress. "Regulation of Pesticide Residues" (Washington DC: U.S. Government Printing Office, 1986).

Norris, R. *Pills, Pesticides, and Profits* (New York: Pergamon Press, 1982).

Quensen, J. F. III. "Reductive Dechlorination of Polychlorinated Bipheyls by Anaerobic Microorganisms from Sediments", *Science.* 242(4879):752–754(1988).

Raheel, M. "Dermal Exposure to Pesticides", *J. Environ. Health.* 51(2):82–84(1988).

"Risk Assessment of Pesticides", *Chem. Eng. News* 69(1):27–55(1991).

Rosenstock, L. et al. "Chronic Central Nervous System Effects of Acute Organophosphate Pesticide Intoxication", *The Lancet.* 338(8773):223–226(1991).

Saltzman, S., Ed. *Pesticides in Soil* (New York: Van Nostrand Reinhold, 1986).

Schmitt, C. J. *National Pesticide Monitoring Program: Organochlorine Residues in Freshwater Fish* (Washington, DC: U.S. EPA, 1983).

Schoor, J. L. *Verification of a Toxic Organic Transport and Bioaccumulation Model* (Washington DC: U.S. EPA, Environmental Research, 1986).

Sericano, J. L., T. L. Wade, E. L. Atlas. "Historical Perspective on the Environmental Bioavailability of DDT and Its Derivatives to Gulf of Mexico Oysters", *Environ. Sci Tech.* 24:1541–1548(Oct. 1990).

Sims, R. *Contaminated Surface Soils* (Park Ridge, NJ: Noyes, 1986).

Smith, D. C. *Pesticides Residues in Soil* (New York: John Wiley & Sons, 1981).

Soren, J. *Aldicarb—Pesticide Contamination of Ground Water in Eastern Suffolk County, Long Island, New York* (Syosset, NY: U.S. Geological Survey, 1984).

Stehr-Green, P. et al. "A Survey of Measured Levels and Dietary Sources of Selected Organochlorine Pesticide Residues and Metabolites in Human Sera from a Rural Population", *Am. J. Pub. Health.* 78(7):828–830(1988).

U.S. Environmental Protection Agency. "Data Requirements for Pesticide Registration; Final Rule", *Fed. Reg.* 49:42856 (1984).

U.S. Environmental Protection Agency. "Inert Ingredients in Pesticide Products; Policy Statement", *Fed. Reg.* 52:13305 (1987).

U.S. Environmental Protection Agency, Office of Pesticides and Toxic Substances. *Pesticide Assessment Guidelines, Subdivision F. Hazard Evaluation: Human and Domestic Animals,* Series 83: Chronic and Long-Term Studies EPA 540/9-82-025 (Washington, DC: U.S. EPA, 1982).

U.S. Environmental Protection Agency. *Pesticides & Toxic Substances Monitoring Report* (Washington, DC: U.S. EPA, 1980).

U.S. Environmental Protection Agency. *The Federal Insecticide, Fungicide and Rodenticide Act as Amended* (Washington, DC: U.S. EPA, Office of Public Affairs, 1985).

Walsh, J. "EPA, OSHA Act to Curb Pesticide Ethylene Dibromide", *Science* 222:400 (1983).

WHO. *Field Studies in Exposure to Pesticides* (Washington DC: U.S. Printing Office, 1983).

Whorton, M. D. and D. E. Foliart. "Mutagenicity, Carcinogenicity and Reproductive Effects of Dibromochloropropane (DBCP)", *Mutat. Res.* 123(1):13–30 (1983).

Worthy, W. "Household Pesticide Hazards Examined", *Chem. Eng. News,* 69(18):50(1991).

Worthy, W. "Pesticides, Nitrates found in U.S. Wells", *Chem. Eng. News,* 69(18):46–48(1991).

Zilberman, D. "The Economics of Pestice Use and Regulation", *Science.* 253(5019):518–522(1991).

Chapter 7

Abu-Jarad, F., and J. H. Fremlin. "Effect of Internal Wall Covers on Radon Emanation Inside Houses", *Health Physics* 44(3):243–248 (1983).

Alter, H. W., and R. A. Oswald. "Results of Indoor Radon Measurements Using the Track Etch Method", *Health Physics* 45(2):425–428 (1983).

Anderson, I. "Showers Pose Risk to Health", *New Scientist* 11:23 (1986).

Angle, C. R. "Indoor Air Pollutants", *Adv. Pediatr.* 35:239–280 (1988).

Bare, J. C. "Indoor Air Pollution Source Database", *J. Air Pollut. Control Assoc.* 38(5):670–671 (1988).

Ballenger, J. J. "Some Effects of Formaldehyde on the Upper Respiratory Tract", *Laryngoscope* 94(11, pt. 1):1411–1413 (1984).

Balmot, J. L. "Monitoring of Formaldehyde in Air", *Am. Industr. Hyg. J.* 46:574–584 (1985).

Bartnett, S. G. "Wood Stove Design and Control Mode as Determinants of Efficiency, Creosote Accumulation, and Condensable Particulate Emissions", *Residential Wood and Coal Combustion* (*J. Air Pollution Control Assoc.,* 1982).

Beck, B. "Prediction of Pulmonary Toxicity of Respirable Combustion Products From Residential Wood and Coal Stoves", *Residential Wood and Coal Combustion* (*J. Air Pollut. Control Assoc.,* 1982).

Boleij, J. S. and B. Brunekreff. "Indoor Air Pollution", *Public Health Rev.* 10(2):169–198 (1982).

Breisch, S. L. "Asbestos: An Environmental Liability", *Safety and Health.* 141(5):60–62(1991).

Brown, R. "Home and Office: Shelter or Threat", *EPA J.* 13:2–5 (1987).

Brutsaert, W. F., S. A. Norton, C. T. Hess and J. S. Williams. "Geologic and Hydrologic Factors Controlling Radon-222 in Ground Water in Maine", *Ground Water* 19(4):407–417 (1981).

Burge, H. A. "Health Risks of Indoor Pollutants", *ASHRAE J.* 30:34–36(1988).

Burton, J. D. "Checklists and Questionaires Aid Investigation of Indoor Air Quality", *Occu. Health Safety.* 60(5):82–85(1991).

Burton, J. D. "Physical, Psychological Complaints can be the Result of Indoor Air Quality", *Occu. Health Safety.* 60(3):53(1991).

Canadian Center for Occupational Health and Safety. "Asbestos", *Cheminfo,* 12001-29-5(1990).

Canadian Center for Occupational Health and Safety. "Carbon Monoxide", *Cheminfo,* 630-08-0(1990).

Canadian Center for Occupational Health and Safety. "Formaldehyde", *Cheminfo,* 50-00-0(1991).

Canadian Center for Occupational Health and Safety. "Formic Acid", *Cheminfo,* 64-18-6(1991).

Canadian Center for Occupational Health and Safety. "Vinyl Chloride", *Cheminfo*, 75-01-4(1991).

Castell, J. "Mismatch between Risk and Money Spent often due to Public Opinion", *Safety and Health*. 141(2):34–36(1990).

Castell, J. "Reproductive Halth Hazards in the Workplace", *Safety and Health*. 141(6):70–72(1990).

Cohen, B. L. "Health Effects of Radon From Insulation of Buildings", *Health Physics* (39):937–941 (1980).

Cole, G. E. "Survey Aids in Assessing the Impact of IAQ on Worker Health", *Occu. Health Safety*. 60(5):38–51(1991).

Committee on Aldehydes *Formaldehyde and Other Aldehydes* (Washington, DC: National Academy Press, 1981).

Committee on Indoor Pollutants, Board of Toxicology and Environmental Health Hazards Assembly of Life Science National Research Council. *Indoor Pollutants* (Washington, DC: National Academy Press, 1981).

Committee on Science and Technology, U.S. House of Representatives 98th Congress. *Indoor Air Quality Research* (Washington, DC: U.S. Government Printing Office, 1984).

Crawford, R., D. Campbell, J. Ross. "Carbon Monoxide Poisoning in the Home: Recognition and Treatment", *Br. Med. J.* 301:977–979(1990).

"EPA Restricts Wood Preservatives", *J. Environ. Health* 47(2) (1984).

Esmen, N. A. "The Status of Indoor Air Pollution", *Environ. Health Perspect.* 62:259–265 (1985).

Evans, P. "An Integrated Approach to Indoor Air Quality", *Consult. Specify. Engin.* 10(1):48–54(1991).

Fleischer, R. L., A. Mogro-Campero and L. G. Turner. "Indoor Radon Levels in the Northeastern U.S.: Effects of Energy-Efficiency in Homes", *Health Physics* 45 (2):407–412 (1983).

Fleischer, R. L. "Theory of Passive Measurement of Radon Daughters and Working Levels by the Nuclear Track Technique", *Health Physics* 47(2):263–270 (1984).

Freeman, R., D. Perks. "Carboxyhemoglobin", *Anesthesia*. 57:45–51(1990).

Gammage, R. B., and S. V. Kaye. *Indoor Air and Human Health* (Chelsea, MI: Lewis Publishers, 1985).

George, A. C. "Passive Integrated Measurement of Indoor Radon Using Activated Carbon", *Health Phys.* 46(4):867–872 (1984).

George, A. C., E. O. Knutson and H. Franklin. "Radon and Radon Daughter Measurements in Solar Building", *Health Physics* 45(2):413–420 (1983).

Gesell, T. F. "Background Atmospheric ^{222}Rn Concentrations Outdoors and Indoors: A Review", *Health Physics* 45(2):289–302 (1983).

Gesser, H D. "Removal of Aldehydes and Acidic Pollutants from Indoor Air", *Environ. Sci. Tech.* 24(4):495–497(1990).

Godish, T. *Air Quality* (Chelsea, MI: Lewis Publishers, 1985).

Godish, T. "Mitigation of Residential Formaldehyde Contamination by Indoor Climate Control", *Am. Indust. Hyg. Assoc.* 47:792–797 (1986).

Godish, T. "Residential Formaldehyde", *J. Environ. Health*. 53(3):34(1990).

Greenfield, E. J. *House Dangerous* (New York: Vintage Books, 1987).

Grossman, D. Jr. "Ambient Levels of Formaldehyde and Formic Acid", *Environ. Sci. Tech.* 25(4):710–714(1991).

Harley, N. "Radon and Lung Cancer in Mines and Homes", *N. Engl. J. Med.* 310(23):1525–1527 (1984).

Harty, S. "Governmental Risk Management: Risk Managers need to Assess Air Qualtity Problems", *Bus. Insurance.* 25(21):12–14(1991).

Hayden, A. C., and R. W. Braaten. "Effects of Firing Rate on Domestic Wood Stove Performance", in *Residential Wood and Coal Combustion (J. Air Pollut. Control Assoc.,* 1982).

Heckerling, P., J. Leikin, A. Mturen, C. Terzian, D. Segarra, "Screening Hospital Admissions from the Emergency Department for Occult Carbon Monoxide Poisoning", *Am. J. Emergen. Med.* 8:301–304(July 1990).

Hendee, W. R., and T. C. Doege. "Origin and Health Risks of Indoor Radon", *Semin. Nucl. Med.* 18(1):3–9 (1988).

Humble, C. "Marriage to a Smoker and Lung Cancer Risk", *Am. J. Public Health* 77:598–602 (1987).

Ingersol, J. G. "A Survey of Radionuclide Contents and Radon Emanation Rates in Building Materials Used in the U.S.", *Health Physics* 45(2):363–368 (1983).

Johnson, C. F., and D. L. Coury. "Bruising and Hemophilia: Accident or Child Abuse?", *Child Abuse Negl.* 12(3):409–415 (1988).

Katzel, J. "A Common Sense Approach to Controlling Indoor Air Quality", *Plant Engin.* 45(8):32–38(1991).

Kazantzis, G. "Lead: Sources, Exposure and Possible Carcinogenicity", *IARC Sci. Publ.* (71):103–111 (1986).

Kennedy, T. "Six Steps for Improving IAQ", *J. Prop. Manage.* 56(2):44–46(1991).

Kjrgaard, S. "Human Reactions to a Mixture of Indoor Air Volatile Organic Compounds", *Atmos. Environ.* 25(8):1417–1426(1991).

Klein, G. L., and R. W. Ziering. "Environmental Control of the Home", *Clin. Rev. Allergy* 6(1):3–22 (1988).

Kriess, K. "The Sick Building Syndrome: Where is the Epidemiological Basis?", *Am. J. Pub. Health.* 80(10):1171–1173(1990).

Lais, M. "Characterization of Emissions from Building Materials, Building Occupants and Consumer Products: Laboratory Studies", *Atmosph. Environ.* 21(2):313–314 (1987).

Landman, K. A. "Diffusion of Radon Through Cracks in a Concrete Slab", *Health Physics* 43(1):65–71 (1982).

Levesque, B., E. Dewailly, and R. Lavoie. "Carbon Monoxide in Indoor Ice Skating Rinks: Evaluation of Absorption by Adult Hockey Players", *Am. J. Pub. Health Assoc.* 80:594–597(1990).

Littlejohn, C. "Carbon Monoxide Inhalation with Engine Switched Off". *Br. J. Hosp. Med.* 41:297(1989).

Malven, F. "Threat Based Definition of Health, Safety, and Welfare, in Design: An Overview", *J. Inter. Des. Edu. Res.* 16(2):5–15(1990).

Mansfield, G. "Air Quality Monitors", Safety and Health. 141(5):50–51(1991).

Marshall, E. "EPA Indicts Formaldehyde, 7 Years Later", *Science* 236:381 (1987).

McKone, T. "Human Exposure to Volatile Organic Compounds in Household Tap Water: The Indoor Inhalation Pathway", *Environ. Sci. Technol.* 21:1194–1207 (1987).

Meyer, B. "Diurnal Variations of Formaldehyde Exposure in Mobile Homes", *J. Environ. Health* 48:57–61 (1985).

Meyer, B. *Indoor Air Quality* (Reading MA: Addison-Wesley, 1983).

Morawska, L. "Influence of Sealants on 222 Radon Emanation Rate From Building Materials", *Health Physics* 44(4):416–418 (1983).

Moseley, C. "Indoor AIr Quality Problems", *J. Environ. Health.* 53(3):19–21(1990).

Munford, J. "Indoor Air Sampling and Mutagenicity Studies of Emissions from Unvented Coal Combustion", *Environ. Sci. Technol.* 21:308–311 (1987).

National Research Council. *Indoor Pollutants* (Washington, DC: National Academcy Press, 1981).

Nazaroff, W. W. "Experiments of Pollutant Transport from Soil into Residential Basements by Pressure-Driven Airflow", *Environ. Sci. Technol.* 21:459–466 (1987).

Nazazrorr, W. "Mathematical Modeling of Chemically Reactive Pollutants in Indoor Air", *Environ. Sci. Technol.* 20:924–934 (1986).

Nero, A. V. "Airborne Radionuclides and Radiation in Buildings: A Review", *Health Physics* 45(2):303–322 (1983).

Nero, A. V., Jr. "Controlling Indoor Air Pollution", *Sci. Am.* 258:42–48 (1988).

Nero, A. V. "Distribution of Airborne Radon 222 Concentration of U.S. Homes", *Science* 234:992–997 (1986).

Nero, A. V. "Indoor Radiation Exposure From 222 and Its Daughters: A View of the Issue", *Health Physics* 45(2):277–288 (1983).

Nero, A. V., et al. "Radon Concentrations and Infiltration Rates Measured in Conventional and Energy-Efficient Houses", *Health Physics* 45(2):401–405 (1983).

"New Systems Monitor AIr Pollution Exposure", *J. Environ. Health.* 52(4):338(1990).

Nishemera, H. "Reduction of NO² and NO by Rush and Other Plants", *Environ. Sci. Technol.* 20:413–416 (1986).

Nordenberg, D., R. Yip, and J. Binkin. "The Effects of Cigarette Smoking on Hemoglobin Levels and Anemia", *J. Am. Med. Assoc.* 264:1556–1559(September 1990).

Prichard, H. M., T. F. Gesell, C. T. Hess, et al. "Integrated Radon Data From Dwellings in Maine and Texas", *Health Physics* 45(2):428–432 (1983).

Rademaker, K. "Common Sense Solutions to Poor Indoor Air", *Occu. Haz.* 53(10:41–44(1991).

Raloff, J. "Toxic Showers and Bath", *Science News* 130:190 (1986).

"Reactive, Volatile Organic Compounds Major Contaminants in IAQ", *Occu. Health Safety.* 59: 106(1990).

Reasor, J. "The Composition and Dynamics of Environmental Tobacco Smoke", *J. Environ. Health* 50:109 (1987).

Repace, J. L., and A. H. Lowrey, "Indoor Air Pollution, Tobacco Smoke, and Public Health", *Science* 208:464–472 (1980).

"Residential Formaldehyde", *J. Environ. Health.* 55:34–37(1990).

Roberts, C., K. MacRae, A. Winning, L. Adams, W. Seed. "Reference Values and Prediction Equations for Normal Lung Function in a Non-Smoking White Urban Population", *Thorax.* 46:643–650(1991).

Samet, J. M., M. C. Marbury and J. D. Spengler. "Respiratory Effects of Indoor Air Pollution", *J. Allergy Clin. Immunol.* 79(5):685–700 (1987).

Samet, J. M. and J. D. Spengler. *Indoor Air Pollution: A Health Perspective.* (Baltimore, MD: John Hopkins University Press,1991).

Samson, R. A. "Occurrence of Moulds in Modern Living and Working Environments", *Eur. J. Epidemiol.* 1(1):54–61 (1985).

Schneider, C. "The Indoor Air Pollution Burden", *EPA J.* 12:14–15 (1986).

Sexton, K. "Formaldehyde Exposure Insider Mobile Homes", *Environ. Sci. Tech.* 23(8):985–988 (1989).

Shephard, R. J. *Carbon Monoxide: The Silent Killer* (Springfield, IL: Charles C Thomas, 1983).

Sheps, D., M. Herbst, A. Hinderliter, K. Adams. "Production of Arrhythmias by Elevated Carboxyhemoglobin in Patients with Coronary Artery Disease", *Ann. Internat. Med.* 113:343–351(September 1990).

Sloan, E., D. Murphy, R. Hart. "Complications and Protocol Considerations in Carbon Monoxide-Poisoned Patients who Require Hyperbaric Oxygen Therapy", *Ann. Emerg. Med.* 18:629–634(1989).

Smith, H. "Lung Cancer Risk From Indoor Exposure to Radon Daughters", *Radiology* 167:580 (1958).

Smith, S. "Controlling Sick Building Syndrome", *J. Environ. Health.* 53(3):22–23(1990).

Spengler, J. D. *Indoor Pollutants* (Washington, DC: National Academy Press, 1981).

Straden, E. "Radon in Dwellings and Lung Cancer—A Discussion", *Health Physics* 38:301–306 (1980).

Straden, E., and L. Berteig. "Radon in Dwellings and Influencing Factors", *Health Physics* 39:275–284 (1980).

Steinhausler, F., et al. "Radiation Exposure of the Respiratory Tract and Associated Carcinogenic Risk Due to Inhaled Radon Daughters", *Health Physics* 45(2):331–337 (1983).

Stowlwijk, J. *Health Effects of Indoor Air Contaminants* (Wilson, Inc., 1986).

Tartaglia, M., S. R. DiNardi and J. Ludwig. "Radon and Its Progeny in the Indoor Environment", *J. Environ. Health* 47(2):62–67(1984).

Thompson, B. "Indoor Air Quality", *J. Environ. Health.* 53(3):38–40(1990).

"Track Radon in the Workplace", *Safety and Health.* 143:42–45(1991).

Traynor, G. "Indoor Air Pollution Due to Emissions From Wood Burning Stove", *Environ. Sci. Technol.* 21:691–697 (1987).

Turiel, I. *Indoor Air Quality and Human Health* (Stanford, CA: Stanford University Press, 1985).

"Unearthing the Truth about Radon in the Workplace", *Safety and Health.* 142:80–82(1990).

U.S. Consumer Product Safety Commission and EPA. "Asbestos in the Home" (1986).

U.S. Consumer Product Safety Commission. "Labeling Asbestos-Containing Household Products; Enforcement Policy", *Fed. Reg.* 51-33910 (1986).

U.S. Consumer Product Safety Commission. no. 13, *"Carbon Monoxide"* (Sept. 1980).

U.S. Consumer Product Safety Commission. no. 97, *"Kerosene Heater"* (Dec. 1982).

U.S. Consumer Product Safety Commission. no. 34, *"Space Heaters"* (Nov. 1983).

U.S. Consumer Product Safety Commission. *Questions and Answers on Urea-Formaldehyde Foam Insulation* (July 22, 1983).

U.S. Consumer Product Safety Commission. *Unvented Gas Space Heaters* (Oct. 10, 1980).

U.S. EPA. OPA 86–005, "Radiation Reduction Methods" (Aug. 1986).

U.S. EPA. OPA-86–005, "Radon Reduction Methods" (Aug. 1986).

U.S. EPA .520/1–86–09. "Interim Indoor Radon and Radon Decay Product Measurement Protocols" (April 1986).

U.S. EPA. Residential Wood Combustion Study, 1980–1982, Executive Summary.

U.S. EPA. "Wood Stove Features and Operation Guideline for Cleaner Air", EPA 600/D-83–12.

Wadden, R. A., and P. A. Schaff. *Indoor Air Pollution* (New York: John Wiley & Sons, 1983).

Wallance, L. "Surprising Results From a New Way of Measuring Pollutants", *EPA J.* 13:15–16 (1987).

Walsh, P. J. *Indoor Air Quality* (Boca Raton, FL: CRC Press, 1984).

West, R. and S. Hack "Effects of Cigarettes on Memory Search and Subjective Ratings", *Pharmacol. Biochem. Behav.* 38:281–286(1991).

White, F. "Indoor Air Quality: What Managers can Do", *Employ. Relat. Today.* 17(2):93–101(1990).

Woolf, A., S. Fish, C. Azzara, and D. Dean, "Serious Poisonings Among Older Adults: A Study of Hospitalization and Mortality Rates in Massachusetts 1983-1985", *Am. J. Pub. Health.* 80:867–869(July 1990).

Yoshida, M., J. Adachi, T. Watabiki, Y. Tatsuno, and N. Ishida, "A Study on House Fire Victims: Age, Carboxyhemoglobin, Hydrogen Cyanide, and Hemolysis", *Foren. Sci. Internat.* 52:13–20(1991).

Zirschy, J., and L. Witherell. "Cleanup of Mercury Contamination of Thermometer Workers' Homes", *Am. Ind. Hyd. Assoc. J.* 48(1):81–84 (1987).

Chapter 8

Ackerman, J. "Monitoring Waste Nitrous Oxide. One Medical Center's Experience", *AORN J.* 41(5):895,897,898 (1985).

Altree, W. S., and J. S. Preston. "Asbestos and Other Fibre Levels in Buildings", *Ann. Occup. Hyg.* 29(3):357–363 (1985).

American Dental Association. *Guideline for Infection Control in the Dental Office and the Commercial Laboratory* (American Dental Assocciation, 1985).

Angle, C. R. "Indoor Air Pollutants", *Adv. Pediatr* 35:239–280 (1988).

Ayliffe, G. A., J. R. Babb. "Hygienic Hand Disinfection Tests in Three Laboratories", *J. Hosp. Infect.* 16:141–149(August 1990).

Bagga, B. S. R., R. A. Murphy, A. W. Anderson and I. Punwani. *Contamination of Dental Unit Cooling Water With Oral Microorganisms and Its Prevention* (American Dental Association, 1984).

Bardana, E. J., Jr., A. Montanaro and M. T. O'Hollaren. "Building-Related Illness. A Review of Available Scientific Data", *Clin. Rev. Allergy* 6(1):61–89 (1988).

Beebe, G. W. "Ionizing Radiation and Health", *Am. Sci.* 70:35–44 (1982).

Belger, A. "The Determination of Ambient Air Quality Within an Environmental Care Unit", *J. Environ. Health* 49:2880–2893 (1987).

Bennett, L. C., and S. Searl. *Communicable Disease Handbook* (New York: John Wiley & Sons, 1982).

Bennett, N. M. "Disposal of Medical Waste", *Med. J. Aust.* 149(8):400–402 (1988).

Birnbaum, D. "Analysis of Hospital Infection Surveillance Data", *Infect. Control* 5(7):332–338 (1984).

Block, S. S. *Disinfection, Sterilization, and Preservation,* 3rd ed. (Philadelphia: Lea and Febiger, 1983).

Browne, R. A. and M. A. Chernesky. "Infectious Diseases and The Anaesthetist", *Can. J. Anaesth.* 655–659(November 1988).

Burke, M. "Outpatient Bundling Regulation would be Costly for Hospitals", *Hospitals.* 65:34–37(September 1991).

Cadwallader, H. L., C. R. Bradley, and G. A. Ayliffe. "Bacterial Contamination and Frequency of Changing Ventilator CIrcuitry", *J. Hosp. Infec.* 15:65–72(January 1990).

Campos, L. L. "Measurement of the Exposure Rate Due to Low Energy X-rays Emitted From Video Display Terminals", *Int. J. Rad. Appl. Instrum. A* 39(2):173–174 (1988).

Chaddock, J. B. "Ventilation and Exhaust Air Requirements for Hospitals", *ASHRAE J.* 29:45(1987).

Cotton, D. J. "The Impact of AIDS on the Medical Care System", *JAMA* 260(4):519–523 (1988).

Cottone, J. A., E. W. Mitchell and C. H. Baker. *Proceeding of the National Symposium on Hepatitis B and the Dental Profession* (American Dental Association 1985).

Council on Dental Therapeutics. *Accepted Dental Therapeutics*, 40th ed. (Chicago: American Dental Association, 1984).

Council on Dental Therapeutics, Council on Prosthetic Services and Dental Laboratory Relations. "Guidelines for Infection Control in the Dental Office and the Commercial Dental Laboratory", *JADA* 110 (1985).

Deaderer, B. P. "Characterization of Contaminant Emissions From Indoor Sources", *Atmosph. Environ.* 21(2):279–462 (1987).

Decke, J. L. "Herpes", *Clinical Center Office of Clinical Reports and Inquiries* (Bethesda, MD: 1985).

Denis, J. L., A. Langley, and D. Lozeau. "Formal Strategy in Public Hospitals", *Long Range Planning.* 24(1):71–82(February 1991).

DiClaudio, J. "Designing a Creative Environment in a Health Care Setting", *Leadership and Org. Develop. J.* 12(143):28–31(1991).

Donham, C., S. Letsch, and B. Maple. "Health Care Indicators", *Health Care Finanacing Rev.* 12(4):141–170(1991).

Donnell, H. D., J. R. Bagby, R. G. Harmon, J. R. Brellin, H. C. Chaski, M. F. Bright, M. Van-Tuinen and R. W. Metzger. "Report of an Illness Outbreak at the Harry S. Truman State Office Building", *Am. J. Epidemiol.* 129(3):550–558 (1989).

Doxsey, K. "Toxic Substances in the Hospital Environment", *J. Nurs. Staff. Dev.* 3(1):41–42 (1987).

Ember, L. "Toxic Chemicals Levels Higher Indoors Than Out", *Chem. Engin. News* 63:22 (1985).

Eubanks, P. "Nursing Homes Change Image", *Hospitals.* 65(21):64–66(November 1991).

Federal Regulations for Long Term Care Facility Services. *Federal Register* 54(21)(February 2, 1989).

Feldman, R. G., and G. L. Ridgway. "Database Handling for Infection Control and Hospital Epidemiology", *J. Hosp. Infect.* 11(Suppl.):37–42 (1988).

Fiehn, N. E., and K. Henriksen. "Methods of Disinfection of the Water System of Dental Units by Water Chlorination", *J. Dent. Res.* 57(12):1499–1504 (1988).

Fischer, P., and Brunekreff. *Indoor NO^2 Pollution and Personal Exposure* (Dordrecht: D. Reidel Publishing Company, 1981).

Friedman, H. M., M. R. Lewis, D. M. Nemerofsky and S. A. Plotkin. "Acquisition of Cytomegalovirus Infection Among Female Employees at a Pediatric Hospital", *Pediatr. Infect. Dis.* 3(3):233–235 (1984).

Furlong, A. "Powders Added to Body Fluids Facilitate Waste Handling", *Hosp. Mater. Manage.* 16(10):2–10(October 1991).

Gammage, S. V. *Indoor Air and Human Health* (Chelsea, MI: Lewis Publishers, 1984).

Garner, J. S., and B. P. Simmons. *CDC Guideline for Isolation Precautions in Hospitals* (Atlanta, GA: Centers for Disease Control, 1983).

Glenister, H., L. Taylor, and C. Bartlett, "An Assessment of Selective Surveillance Methods of Detecting Hospital Acquired Infection", *Am. J. Med.* 121–124(September 1991).

Gofman, J. W. "Ionizing Radiation: Concepts for the Dental Assistant", *Dental Assistant* 51:17–22 (1982).

Grant, P., S. Laird, and D. Coppage, "Disposable vs. Reusable Pillows: A Case Study", *Hosp. Mater. Manage. Quat.* 11(3):32–35(February 1990).

Grosserode, M. H. and R. P. Wenzel. "The Continuing Importance of Staphylococci as Major Hospital Pathogens", *J. Hosp. Infect.* 19:3–17(September 1991).

Guidotti, T. L. "Quantitative Risk Assessment of Exposure to Airborne Asbestos in an Office Building", *Can. J. Public Health* 79(4):249–254 (1988).

Gwyther, J. "Sharps Disposal Containers and Their Use", *J. Hosp. Infec.* 15:287–294(April 1990).

Haley, R. "Infection Control Strategies Save $250,000 Annually", *Hospitals* 59(22):63–65 (1985).

Haley, R. W., D. H. Culver, J. W. White, W. M. Morgan and T. G. Emori. "The Nationwide Nosocomial Infection Rate. A New Need for Vital Statistics", *Am. J. Epidemiol.* 121(2):159–167 (1985).

Haley, R. W., W. M. Morgan, D. H. Culver, J. W. White, T. G. Emori, J. Mosser and J. M. Hughes. "Update From the SENIC Project. Hospital Infection Control: Recent Progress and Opportunities Under Prospective Payment", *Am. J. Infect. Control* 13(3):971–978 (1985).

Hard, R. "Is Regional Incineration the Answer for Waste?", *Hospitals.* 65(16):60–62(August 1991).

Harkness, G. A., D. W. Bentley, and K. J. Roghmann. "Risk Factors for Noscomial Pneumonia in the Elderly", *Am. J. Med.* 89:457–463(October 1990).

Harris, A. A., S. Levin and G. M. Trenholme. "Selected Aspects of Nosocomial Infection in the 1980's", *Am. J. Med.* 77(1B):3–10 (1984).

Hershkowitz, A. "Without a Trace: Handling Medical Waste Safely", *Tech. Rev.* 93(6):35–40(August/September 1990).

Hoffman, P. "Disinfection in Hospitals", *Nursing-Lond*, 3(3):106–108 (1986).

Holthaus, D. "Right-to-Know Rules Cover Hospital Employees", *Hospitals* 61(20):58 (1987).

Hugar, D. W. and P. S. Newman. "Incidence of Postoperative Infection in a Free-Standing Ambulatory Surgery Center", *J. Foot Surg.* 29:265–267(May 1990).

Hughes, R. T., and D. M. O'Brien. "Evaluation of Building Ventilation Systems", *Am. Ind. Hyg. Assoc. J.* 47(4):207–213 (1986).

Huth, T. S. and J. P. Burke "Infections and Antibiotic use in a Community Hospital", *Infec. Control.* 525–534(September 1991).

Jackson, M. M. "Infection Prevention Control for HIV and Other Infectious Agents", *Women's Health J.* 115–122(1990).

Jacobson, J. T., D. S. Johnson, C. A. Ross, M. T. Conti, R. S. Evans and J. P. Burke "Adapting Disease-Specific Isolation Guidelines to a Hospital Information System", *Infect. Control* 7(8):411–418 (1986).

Johnson, S. and D. N. Gerding. "Prospective, Controlled Study of Vinyl Glove Use to Interrupt Clostridium difficile Nosocomial Transmission", *Am. J. Med.* 137–140(February 1990).

Kneedler, J. A., and G. H. Dodge. *Perioperative Patient Care* (Boston: Blackwell Scientific, 1983).

Lee, J. J. and J. A. Marvin, "Infection Control in a Burn Center", *J. Burn Care Rehab.* 11(6):575–580(November 1990).

Lewey, R. "Visits to a Hospital-Based Employee Health Service", *J. Occup. Med.* 28(3):241–242 (1986).

Lightfoot, N. F., F. Ahmad, and J. Cowden. "Management of Institutional Outbreaks of Salmonella Gastroenteritis", *J. Antimicrobio.* 37–46(December 1990).

MacNeil, J. M., and G. M. Weisz. "Critical Care Nursing Stress: Another Look", *Heart-Lung* 16(3):274–277 (1987).

Mallison, G. F. "A Critical Review of Draft Manual For Infectious Waste Management", *Am. J. Infect. Control* 13(1):45–46 (1985).

Manian, F. "Comprehensive Surveillance of Surgical Wound Infections", *Infec. Cont.* 515–520(October 1990).Manzella, J. P., J. H. McConville, W. Valenti, M. A. Menegus, E. M. Swierkosz and M. Arens. "An Outbreak of Herpes Simples Virus Type I Gingivostomatitis in a Dental Hygiene Practice", *JAMA* (1984).

Marsh, L. A. "Quality Assurance Activities in a Small Community Hospital", *Quality Rev. Bull.* 9(3):77–80 (1983).

Mattia, M. A. "Hazards in the Hosptial Environment. The Sterilants: Ethylene Oxide and Formaldehyde", *Am. J. Nurs.* 83(2):240–243 (1983).

Mattila, M. K. "Job Load and Hazard Analysis: A Method for the Analysis of Workplace Conditions for Occupational Health Care", *Br. J. Ind. Med.* 41(1):656–666 (1985).

McCunney, R. J. "The Role of Building Construction and Ventilation in Indoor Air Pollution. Review of a Recurring Problem", *N.Y. State J. Med.* 87(4):203–209 (1987).

McGowan, J. E. "The Infection Control Practitioner: An Action Plan for the 1990s", *Am. J. Infect. Cont.* 35–39(February 1990).

McGowan, J. E. "New Laboratory Techniques for Hospital Infection Control", *Am. J. Med.* 91:245–251(September 1991).

Michaels, D., D. Nagin and S. Zoloth. "Occupational Hazards in the Dental Office", *N.Y. State Dent. J.* 50(5):294–296 (1984).

Morgan, M. S., and S.W. Horstman. "Systematic Investigation Needed for Tight Building Syndrome Complaints", *Occup. Health Saf.* 56(5):79–82,85–88 (1987).

Molinari, J. A., R. R. Runnells, "Role of Disinfectants in Infection Control", *Infec. Cont.* 33–37(April 1991).

Morse, D. L., M. A. Gordon, T. Matte and G. Eadie. "An Outbreak of Histoplasmosis in a Prison", *Am. J. Epidemiol.* 122(2):253–261 (1985).

Morton, B. "Contolled Hospital Humidity Curtails Airborne Disease", *Specfying Engin.* 56:58–60(1986).

Nero, A. V. "Airborne Radionuclides and Radition in Buildings: A Review", *Health Phys.* 45(2):303–322 (1983).

Moskop, J. C. "AIDS and Hospitals" The Policy Options", *Hosp. Health Serv. Admin.* ISSN 8750:159–171(Summer 1990).

Murakami, A. "Hospitals Beset by Medical Waste Disposal Problem", *Plant Engin.* 41"32(October 1987).

Neu, H. C. "Unusual Nosocomial Infection", *Dis. Mon.* 30(13):1–68 (1984).

Newman, R. P. "Air Pollution Control for Infectious Waste Incineration", *Poll. Engin.* 23:68–71(October 1991).

Occupational Safety and Health Administration. *Risk of Hepatitis B Infection for Workers in Health Care Delivery System and Suggested Methods for Risk Reduction* (Washington DC: U.S. Department of Labor, 1983).

Palmer, S. R., and B. Rowe. "Investigation of Outbreaks of Salmonella in Hospitals", *Br. Med. J. Clin. Res.* 287(6396):891–893 (1983).

Paz J. D., R. Milliken, W. T. Ingram, A. Frank and A. Atkin. "Potential Ocular Damage From Microwave Exposure During Electrosurgery: Dosimetric Survey", *J. Occup. Med.* 29(7):580–583 (1987).

Parry, M. F. "Epidemiology and Mechanisms of Antimicrobial Resistance", *AJIC.* 17–286–298(1989).

Reed, C. M., T. L. Gusafson, J. Siegel and P. Duer. "Nosocomial Transmission of Hepatitis A From a Hospital-Acquired Case", *Pediatr. Infect. Dis.* 3(4):300–303 (1984).

Ringham, S., M. Worsley and J. Archer. "Journal of Infection Control Nursing. Building Standards", *Nurs. Times* 82(38):97–98 (1986).

Robinson, G. V., B. R. Tegtmeier and J. A. Zaia. "Brief Report: Nosocomial Infection Rates in a Cancer Treatment Center", *Infect. Control* 5(6):289–294 (1984).

Runnells, R. R. *Infection Control in the West Finger Environment* (Salt Lake City: Publishers Press, 1984).

Rutala, W. A., and F. A. Sarubbi. "Management of Infectious Waste From Hospitals", *Infect. Control* 4(4):198–204 (1983).

Rulander, R., and P. Haglind. "Airborne Endotoxins and Humidifier Disease", *Clin. Allergy* 14(1):109–112 (1984).

Samet, J. M., M. C. Marbury and J. D. Spenger. "Health Effects and Sources of Indoor Air Pollution, Part II", *Am. Rev. Respir. Dis.* 137(1)221–242 (1988).

Sanford, J. P., and J. P. Luby. *Infectious Diseases* (New York: Grune & Stratton, 1982).

Sawyer, D. L., and B. L. Braddock. "Infection Control: Consultation Provides New Frontier", *Am. J. Infect. Control* 14(5)229–233 (1986).

Sebben, J. E. "Sterile Technique and the Prevention of Wound Infection in the Office", *J. Dematol. Surg. Oncol.* 54–71(December 1988).

"Sick Building Syndrome Affects People Too", *Environment.* 29:23 (1987).

Simpson, R. "Priorities For Hospital Cleaning, Disinfection, Sterilization, and Control of Infection", *Br. Med. J. Clin. Res.* 288(6434):1898–1900 (1984).

Smith, P. W., P. B. Daly, J. S. Roccaforte. "Current Status of Nosocomial Infection Control in Extendind Care Facilities", *Am. J. Med.* 281–285(September 1991).

Speller, D. C., D. C. Shanson, G. A. Ayliffe. "Acquired Immune Deficiency Syndrome: Recommendations of a Working Party of the Hospital Infection Society", *J. Hosp. Infec.* 15:17–34(January 1990).

Standfast, S. J., P. B. Michelsen, A. L. Baltch, R. P. Smith, E. K. Latham, A. B. Spellacy, R. A. Venezia and M. H. Andritz. "A Prevalence Survey of Infections in a Combined Acute and Long-Term Care Hospital", *Infect. Control* 5(4):177–184 (1984).

Stellman, J. M., S. Klitzman, G. C. Gordon and B. R. Snow. "Air Quality and Ergonomics in the Office: Survey Results and Methodologic Issues", *Am. Ind. Hyg. Assoc. J.* 45(5):286–293 (1985).

Sterling, T. D., E. Sterling and W. Dimich. "Building Illness in the White-Collar Workplace", *Int. J. Health Serv.* 13(2):277–287 (1983).

St. John, W. "Waste Management: Focus on the Food Service", *Nursing Homes and Senior Citizen Care*. 32(October 1991).

SyWassink, J. M., and L. I. Lutwick. "Risk of Hepatitis B in Dental Care Providers: A Contact Study", *JADA* 106 (1983).

Tyzack, R. "The Management of Methicillin-Resistant *Staphylococcus Aureus* in a Major Hospital", *J. Hosp. Infect.* 6(Suppl. A):195–199 (1985).

U.S. Department of Labor, OSHA Office of Occupational Medicine. OSHA Instruction CPL 2-2.36, (Washington, DC: Nov 30, 1983).

Weinstein, R. A. "Resistant Bacteria and Infection Control in the Nursing Home and Hospital", *Bull. N.Y. Acad. Med.* 63(3):337–344 (1987).

Welliver, R. C., and S. McLaughlin. "Unique Epidemiology of Nosocomial Infection in a Children's Hospital", *Am. J. Dis. Child.* 138(2):131–135 (1984).

Welty, C., S. Burstin, S. Muspratt and I. B. Tager. "Epidemiology of Tuberculous Infection in a Chronic Care Population", *Am. Rev. Resp. Dis.* 132(1):133–136 (1985).

West, K. H. "Infection Control. Disposal of Hazardous Waste: Implications for Nurses", *J. Oper. Room. Res. Inst.* 3(2):5–7 (1983).

Wilkinson, W. E. "Occupational Injury at a Midwestern Health Science Center and Teaching Hospital", *AAOHN J.* 35(8):367–376 (1987).

Wilson, S. J., and H. J. Wilson. "Mercury Vapour Levels in a Dental Hospital Environment", *Br. Dent. J.* 159(7):233–234 (1985).

Wollschlager, M. L. "Dealing with the Medical Waste Problem", *Consult. Specifying Engin.* 10(4):16–19(September 1991).

Yaffe Y., D. Jenkins, H. Mahon-Haft, W. Winkelstein, C. P. Flessel and J. J. Wesolowski. "Epidemiological Monitoring of Environmental Lead Exposures in California State Hospital", *Sci. Total. Environ.* 32(3):261–275 (1984).

Zautra, A. J., C. Eblen and K. D. Reynold. "Job Stress and Task Interest: Two Factors in Work Life Quality", *Am. J. Community Psychol.* 14(4):377–393 (1986).

Chapter 9

Ahonen, E., and U. Nousiainen. "The Sauna and Body Fluid Balance", *Ann. Clin. Res.* 20(4):257–261 (1988).

Cuddihy, R. G., W. C. Griffith and R. O. McClellan. "Potential Health and Environmental Effects of Light-Duty Vehicles", *Toxicology Rep.* LMF 89:3–25 (1981) .

Godsey, M. S., T. E. Amundson, E. C. Burgess, W. Schell, J. P. David, R. Kaslow and R. Eldeman. "Lyme Disease Ecology in Wisconsin: Distribution and Host Preference of *Ixodes dammini*, and Prevalence of Antibody to *Borrelia burgdorferi* in Small Mammals", *Am. J. Trop. Med. Hyg.* 37(1):180–187 (1987).

Gong, H. J., F. Bedi and S. M. Horvath. "Inhaled Albuterol Does Not Protect Against Ozone Toxicity in Nonasthmatic Athletes", *Arch. Environ. Health* 43(1):46–53 (1988).

Gregory, D. W., and W. Schaffner. "Pseudomonas Infections Associated with Hot Tubs and Other Environments", *Infect. Dis. Clin. North Am.* 1(3):635–648 (1987).

Grobler, S. R., L. S. Maresky and R. J. Rossouw. "Blood Lead Levels of South African Long-Distance Road-Runners", *Arch. Environ. Health* 41(3):155–158 (1986).

Holtan, N. R. "Giardiasis. A Crimp in the Life-Style of Campers, Travelers, and Others", *Postgrad. Med.* 83(5):54–56, 59–61 (1988).

Kauppihen, K., and I. Vuori. "Man in the Sauna", *Ann. Clin. Res.* 18(4):173–185 (1986).

Linn, W. S., E. L. Avol, D. A. Shamoo, R. C. Peng, C. E. Spier, M. N. Smith and J. D. Hackney. "Effect of Metaproterenol Sulfate on Mild Asthmatics' Response to Sulfur Dioxide Exposure and Exercise", *Arch. Environ. Health* 43(6):399–406 (1988).

Marwick, C. "Olympic Athletes May Face Extra Challenge — Pollution [news]", *JAMA* 251(19):2495–2497 (1984).

Morrison, A. B. "Recreational Water Quality and Human Health", *Can. J. Public Health* 75(1):13–14 (1984).

Shephard, R. J. "Athletic Performance and Urban Air Pollution", *Can. Med. Assoc. J.* 131(2):105–109 (1984).

Solomon, S. L. "Host Factors in Whirlpool-Associated *Pseudomonas aeruginosa* Skin Disease", *Infect. Control* 6(10):402–406 (1985).

Tunnicliff, B., and S. K. Brickler. "Recreational Water Quality Analyses of the Colorado River Corridor in Grand Canyon", *Appl. Environ. Microbiol.* 48(5):909–917 (1984).

Chapter 10

Abrams, H. K. "Credibility in the TLV Process [letter]", *Am. J. Ind. Med.* 13(5):609–610 (1988).

"Acute Occupational Exposure to Sulfur Dioxide—Missouri", *MMWR* 32(41):541–542 (1983).

Alessio, L. "Relationship Between 'Chelatable Lead' and the Indicator of Exposure and Effect in Current and Past Occupational Exposure", *Sci. Total Environ.* 71(3):292–299 (1988).

Akesson, B., M. Bengtsson and I. Floren. "Visual Disturbances After Industrial Triethyl-amine Exposure", *Int. Arch. Occup. Environ. Health* 57(4):297–302 (1986).

Allison, W. W. "Industrial Hygiene [letter]", *Am. Ind. Hyg. Assoc. J.* 48(4):A244, A246 (1987).

Angerer, J. "Occupational Chronic Exposure to Organic Solvents", *Int. Arch. Occup. Environ. Health* 56(4):307–321 (1985).

Ashford, N. A. "Policy Considerations for Human Monitoring in the Workplace", *J. Occup. Med.* 28(8):563–568 (1986).

Atuhaire, L. K., J. J. Campbell and M. Jones. "Specific Causes of Death in Miners and Ex-miners of the Rhondda Fach 1950–80", *Br. J. Ind. Med.* 43:497–499 (1986).

Austin, W. B., and C. F. Phillips. "Development and Implementation of a Health Surveil-lance System", *Am. Ind. Hyg. Assoc. J.* 44(9):6538–6542 (1983).

Baird, D. D. "Occupational Exposure to Heat or Noise and Reduced Fertility [letter]", *JAMA* 253(18):2643–2644 (1985).

Baker, D. G. "Radiology, Is There an Occupational Hazard?", *Am. Ind. Hyg. Assoc. J.* 49(1):17–20 (1988).

Baloch, U. K. "Problems Associated With the Use of Chemicals by Agricultural Workers", *Basic Life Sci.* 34:63–78 (1985).

Baram, M. S. "The Right to Know and the Duty to Disclose Hazard Information", *Am. J. Public Health* 74(4):385–390 (1984).

Becklake, M. R. "Use of Information: Environmental Standards, Assessments of Risk, Prevention and Clinical Implication", *Med. J. Aust.* 143(2):63–64 (1985).

Blackwell, R. P. "The Effects of Low-Level Radiofrequency and Microwave Radiation on Brain Tissue and Animal Behavior", *Int. J. Radiat. Biol. Relat. Stud. Phys., Chem. Med.* 50(5):761–787 (1986).

Blot, W. J., and J. F. Fraumeni. "Cancer Among Shipyard Workers", *Branbury Rep.* 9:37–49 (1981).

Borak, T. B. "Radomization of Grab-Sampling Strategies for Estimating the Annual Exposure of U Miners to Rn Daughters", *Health Phys.* 50(4):465–472 (1986).

Bransford, J. S. "Proliferation of New Chemicals Increasing Need for Databases", *Occup. Health Saf.* 53(4):32–35 (1984).

Brewster, M. A. "Biomarkers of Xenobiotic Exposures", *Ann. Clin. Lab. Sci.* 18(4):306–317 (1988).

Brix, K. "Environmental and Occupational Hazards to the Fetus", *Medicine* 27(9):577–583 (1982).

Bunn, W. B. "Right-to-Know Laws and Evaluation of Toxicologic Data", *Ann. Intern. Med.* 103(6 pt 1):947–949 (1985).

Burgess, J. E. "Occupational Health in Agriculture", *Curr. Approaches Occup. Health* 2:62–85 (1982).

Burgess, W. A. *Recognition of Health Hazards in Industry* (New York: Wiley-Interscience Publication, 1981).

Burton, J. *Industrial Ventilation* (Ive Inc. 1986).

Cabrera, I. T. *Encyclopedia of Occupational Health and Safety* (Washington, DC: Government Printing Office, 1983), p. 1712–1715.

Castleman, B. I., and V. Navarro. "International Mobility of Hazardous Products, Industries, and Wastes", *Annu. Rev. Public Health* 8:1–19 (1987).

Castren, J. "On the Microwave Exposure", *Acta Ophthalmol.* 60(4):647–654 (1982).

Checkoway, H. "Methods of Treatment of Exposure Data in Occupational Epidemiology", *Med. Lav.* 77(1):48–73 (1986).

Cheremisinoff, P. N., and F. Ellerbusch. *Guide for Industrial Noise Control* (Ann Arbor, MI: Ann Arbor Science Publishers, 1982).

Christenson, W. N., C. F. Brennan and J. DeVito. "Even 'Healthy' Environments Can Harbor Many Hidden Hazards", *Occup. Health Saf.* 54(3):40–42, 46 (1985).

Corn, M. "Assessment and Control of Environmental Exposure", *J. Allergy Clin. Immunol.* 72(3):231–241 (1983).

Costa, M. "Principles of Industrial Toxicology", *Clin. Podiatr. Med. Surg.* 4(3):559–570 (1987).

Cox, C., W. E. Murray and E. P. Foley. "Occupational Exposures to Radiofrequency Radiation", *Am. Industr. Hyg. Assoc. J.* 43:149–153 (1982).

Cullen, A. P. "Industrial Nonionizing Radiation and Contact Lenses", *Can. J. Pub. Health* 73:251–254 (1982).

Davies, J. E. "A Global Need: Farm Worker Safety", *Am. J. Ind. Med.* 13(6):725–729 (1988).

Delaine, J. "Lead Oxide Emission—A Case History", *Ann. Occup. Hyg.* 30(2):257–261 (1986).

DeSantis, V., and I. Londo. "Health Effects of Airborne Effluents Released From the Nuclear Cycle", *Environ. Res.* 37(2):300–312 (1985).

Division of Labor Statistics & Research. Occupational Injuries and Illnesses Survey (San Francisco, CA: 1984).

Dooms, A. E. Goossens, K. M. Debusschere, D. M. Gevers, K. M. Dupre, H. J. Degreef, J. P. Loncke and J. E. Snauwaert. "Contact Dermatitis Caused by Airborne Agents. A Review and Case Reports", *J. Am. Acad. Dermatol.* 15(1):1–10 (1986).

Dreyer, N., and E. Friedlander. "Identifying the Health Risks From Very Low-Dose Sparsely Ionizing Radiation", *Am. J. Pub. Health* 72:585–587 (1982).

Drury, C. G. "Injury Potential of Industrial Jobs", *Semin. Occup. Med.* 2:41–49 (1987).

Dye, and E. White. "Environmental Hazards in the Work Setting: Their Effect on Women of Child-Bearing Age", *AAOHN J.* 34(2):76–78 (1986).

Dukes, F. N., Dobos and A. Henschel. *Recommended Heat Stress Standards* (Washington DC: Government Printing Office, 1980).

Ecker, M. D., and N. J. Bramesco. *Radiation* (New York: Vintage Books, 1981).

Edgerton, T. R. *Method Development for the Assessment of Possible Human Exposure to Pesticides and Industrial Chemicals* (Research Triangle Park, NC: U.S. EPA, 1981).

Eisenbud, M. *Environmental Radioactivity from Natural Industrial and Military Sources*, 3rd ed. (Troy, MO: Academic Press, 1987).

Fischbein, A., and Y. Lerman. "Environmental and Occupational Factors in General Medical Practices", *Mt. Sinai J. Med. N.Y.* 50(6):468–475 (1983).

Fletcher, R. A. "A Review of Persoanl Portable Monitors and Samplers for Airborne Particles", *J. Air Pollut. Control Assoc.* 34(10):1014–1016 (1984).

Flodin, U., M. Fredribsson, B. Persson, L. Hardell and O. Axelson. "Background Radiation, Electrical Work, and Some Other Exposures Associated with Acute Myeloid Leukemia in a Case", *Arch. Environ. Health* 41:77–84 (1986).

Fomenko, B. S. "Effect of Microwave Radiation (340 and 900 MHz) on Different Structural Levels of Erthrocyte Membranes", *Bioelectromagnetics* 6(3):305–312 (1985).

Frazier, T. "NIOSH Occupational Health and Hazard Surveillance Systems", *J. Toxicol. Clin. Toxicol.* 21(2):201–209 (1983–1984).

Gardner, W. A. *Current Approaches to Occupational Health 2* (London: Weight, PSG, 1982).

Goldman, M. "Ionizing Radiation and its Risks", *West. J. Med.* 137:540–547 (1982).

Goldman, R. H. "General Occupational Health History and Examinations", *J. Occup. Med.* 28(10):967–974 (1986).

Gramicidin, A. "Effects of Ionizing Radiation on Artificial Lipid Membranes", *Int. J. Radiat. Biol. Rel. Studies in Phy. Chem. Med.* 51:265–286 (1987).

Grandjean, P. "Indirect Exposures: The Significance of Bystanders at Work and at Home", *Ame. Ind. Hyg. Assoc.* 47:819–824 (1986).

Grazer, R. E., and H. W. Meislin. "A Nine-Year Evaluation of Emergency Department Personnel Exposure to Ionizing Radiation", *Ann. Emerg. Med.* 16:340–342 (1987).

Green, J. "Detecting the Hypersusceptible Workers; Genetics and Politics in Industrial Medicine", *Int. J. Health Serv.* 13(2):247–264 (1983).

Gun, R. A. "National Institute of Occupational Health and Safety", *Community Health Stud.* 8(3 Suppl):3–7 (1984).

Halder, C. A., G. S. VanGorp, N.S . Hatoum and T. M. Warne. "Gasoline Vapor Exposures, Part I, Characteriziaton of Workplace Exposures", *Am. Ind. Hyg. Assoc. J.* 47(3):164–172 (1986).

Halperin, W. E., and T. M. Frazier. "Surveillance for the Effects of Workplace Exposure", *Annu. Rev. Public Health* 6:419–432 (1985).

Hamburger, S. "Occupational Exposure to Nonionizing Radiation and an Association with Heart Disease", *J. Chron. Dis.* 36:791–802 (1983).

Harm, W. *Biological Effects of Ultraviolet Radiation* (New Rochelle, NY: Cambridge University Press, 1980).

Hart, D. L., and M. Jaraidei. "Effect of Lumbar Posture on Lifting", *Spine* 12:138–145 (1987).

Hema, P., K. K. Pushpavathi and P. P. Reddy. "Cytogenetic Damage in Lymphocytes of Rubber Industry Workers", *Env. Res.* 40(1):199–201 (1986).

Henderson, J., H. W. Baker and P. J. Hanna. "Occupation-Related Male Infertility: A Review", *Clin. Reprod. Fertil.* 4(2):87–106 (1986).

Hughes, D. "Ionizing Radiation: A Guide to Regulations", *Occup. Health* 38:362–366 1986.

Hutton, W. C. "Diurnal Variations in Stresses on the Lumbar Spine", *Spine* 12:130–137 (1987).

Inoue, O., K. Seiji, M. Kashara, H. Nakatsuka, T. Watanabe, S. G. Yin, G. L. Li, C. Jin, S. X. Cai, X. Z. Wang, et al. "Quantitative Relation of Urinary Phenol Levels to Breathzone Benzene Concentratons: A Factory Survey", *Br. J. Ind. Med.* 42(1):692–697 (1986).

International Labour Organization. *Occupational Exposure Limits For Airborne Toxic Substances*, 2nd ed. (Geneva: International Labour Office, 1980).

Irwin, J. D., and E. R. Graff. *Industrial Noise and Vibration Control* (Englewood Cliffs, NJ: Prentice-Hall, 1982).

Jarost, D. "Health Complaints", *Can. J. Pub. Health* 77:132–135 (1986).

Johnson, M. L. "Human Effects Following Exposure to Ionizing Radiation", *Arch. Dermatol.* 122:1380–1382 (1986).

Leads from the MMWR "Cytotoxicity of Volcanic Ash: Assessing the Risk for Pneumonconiosis", *JAMA* 255(20):2727, 2731 (1986).

Kanna Pilly, G. M., R. K. Woolf and P. B. DeNee. "Generation Characteriziations and Inhalation of Ultrafine Aerosols", *Ann. Occup. Hyg.* 26:77–81 (1982).

Karrh, B. W. "Reproductive Hazards in the Workplace", *Del. Med. J.* 58(3):203–204 (1986).

Kauppinen, T. P., T. J. Partanen, M. M. Nurminen, J. I. Nickels, S. G. Hernberg, T. R. Hakulinen, E. I. Pukkala and E. T. Savonen. "Respiratory Cancers and Chemical Exposures in the Wood Indsutry: A Nested Case-Control Study", *Br. J. Ind. Med.* 43(2):84–90 (1986).

Koplin, A. N. "Right-to-Know: Implications of New Jersey's Law", *J. Public Health Policy* 5(4):538–549 (1984).

Korn, R. J., D. W. Dockery, F. E. Speizer, J. H. Ware and B. G. Ferris. "Occupational Exposures and Chronic Respiratory Symptoms. A Population-Based Study", *Am. Rev. Respir. Dis.* 136(2):298–304 (1987).

Lammintausta, K., and H. I. Maibach. "Dermatologic Considerations in Worker Fitness Evaluation", *State Art Rev. Occup. Med.* 3(2):341–350 (1988).

Laufer, A. "Assessment of Safety Performance Measures at Construction Sites", *J. Constr. Eng. Manage.* 112:530,542 (1986).

Lauwerys, R. *Industrial Chemical Exposure: Guidelines for Biological Monitoring* (Cal. Biomedical, 1983).

Laverghetta, T. *Practical Microwaves* (Indiana: Howard W. Sams and Company, Inc., 1984).

Lecea, J. "Injuries in Small Press Work", *Semin. Occup. Med.* 2:69–70 (1987).

Logan, D. C. "Reproduction and the Workplace: An Industry Perspective", *State of Art Rev.: Occup. Med.* 1(3):473–481 (1986).

Logue, N. J., and S. Hamburger. "Congenitial Anomalies and Paternal Occupational Exposure to Shortwave, Microwave, Infrared, and Acoustic Radition", *J. Occup. Med.* 27:(1985)451–452.

Lohman, W. *Low Back Pain in Nurses* (Washington, DC: Government Printing Office, 1987).

Loomis, T. A. *Essentials of Toxicology* (Philadelphia: Lea & Febriger, 1978).

Lowry, L. K. "Biological Exposure Index as a Complement to the TLV", *J. Occup. Med.* 28(8):578–582 (1986).

Malek, R. F., J. M. Daise, and B. S. Cohen. "The Effect of Aerosol on Estimates of Inhalation Exposure to Airborne Styrene", *Am. Ind. Hyg. Assoc. J.* 47(9):524–529 (1986).

McAtt, F. L. "Reducing Repetitive Motions", *Semin. Occup. Med.* 2:73–74 (1987).

McDonald, J. C. *Recent Advances in Occupational Health* (London: Churchill Livingston, 1981).

McMichael, A. J. "Carcinogenicity of Benzene, Toluene and Xylene: Epidemiological and Experimental Evidence", *IARC Sci. Publ.* 85:3–18 (1988).

Merz, B. "Multiple Efforts Directed at Defining, Eliminating Excess Radiation", *Med. News Perspect.* 258:577 (1987).

Moson, R. R. *Occupational Epidemiology* (Boca Raton, FL: CRC Press, 1980).

Morgan, W., and P. Ravens. "Prediction of Distress for Individuals Wearing Industrial Respirators", *Am. Ind. Hyg. Assoc. J.* 46:363–368 (1985).

Morrow, P. E. "Toxicological Data on Nox: An Overview", *J. Toxicol. Environ. Health* 13(2–3):205–257 (1984).

Murphy, M. J. "Environmental Risk Assessment of Industrial Facilities: Techniques, Regulatory Initiatives and Insurance", *Sci. Total Environ.* 51:185–196 (1986).

NIOSH Criteria Document. Occupational Exposure to Hot Environments, NIOSH Publ. No. 72-10269 (1972).

National Safety Council. *Fundamentals of Industrial Hygiene*, (6th Printing 1983).

National Toxicology Program. *Technical Report on the Toxicology and Carcinogenesis of Benzene* (Bethesda, MD: National Institutes of Health Publications, 1986).

National Toxicology Program. *Technical Report on the Toxicology and Carcinogenesis of Tetrachloroethylene* (Bethesda, MD: National Institutes of Health Publication, 1986).

National Toxicology Program. *Technical Report on the Toxicology and Carcinogenesis of Trichloroethylene* (Bethesda, MD: National Institutes of Health Publication, 1986).

National Toxicology Program. *Technical Report on the Toxicology and Carcinogenesis of 1,1,1-Trichloroethane* (Bethesda, MD: National Institutes of Health Publication, 1986).

National Toxicology Program. *Technical Report on the Toxicology and Carcinogenesis of Vinyl Chloride* (Betheseda, MD: National Institutes of Health Publication, 1986).

Nickson, K. "Occupation: Nurse Occupational Hazard: Radiation", *Can. Nurse* 80:30–31 (1984).

Nussey, C., and B. Worthington. "The Use of Systematic Reliability Assessment Techniques for the Evaluation and Design of Mining Electronic Systems", *J. Occup. Accidents* 7:1–17 (1985).

Occupational Exposure to Ionizing Radiation in the United States (Washington, DC: Office of Radiation Programs, U.S. Environmental Protection Agency, 1984).

"Occupational Fatality Following Exposure to Hydrogen Sulfide—Nebraska", *MMWR* 35(33):533–535 (1986).

Olishifski, J. B., and E. R. Hartford. *Industrial Noise and Hearing Conservation* (Chicago: National Safety Council, 1975).

Oser, L. J. "Extent of Industrial Exposure to Epichlorohydrin, Vinyl Fluoride, Vinyl Bromide and Ethylene Dibromide", *Am. Indust. Hygiene Assoc. J.* 41:461–468 (1980).

Ott, M. G., H. G. Scharnweber and R. R. Langner. "Mortality Experience of 161 Employees Exposed to Ethylene Dibromide in Two Production Units", *Br. J. Indust. Med.* 37:163–168 (1979).

Parkinson, D. K., and M. J. Greenan. "Establishment for Medical Surveillance in Industry: Problems and Procedures", *J. Occup. Med.* 28(8):772–777 (1986).

Paull, J. M. "The Origin and Basis of Threshold Limit Values", *Am. J. Ind. Med.* 5(3):227–238 (1984).

Pellizzari, E. D., R. A. Zweidinger and L. S. Sheldon. "Breath Sampling", *IARC Sci. Publ.* (68):399–411 (1985).

Peters, J. M., D. H. Garabrant, W. E. Wright, L. Bernstein and T. M. Mack. "Uses of a Cancer Registry in the Assessment of Occupational Cancer Risks", *Natl. Cancer Inst. Monogr.* 69:157–161 (1985).

Petreas, M. "A Laboratory Evaluation of Two Methods of Measuring Low Levels of Formaldehyde in Indoor Air", *Am. Indust. Hygi. Assoc. J.* 47:276–280 (1986).

Phillip, R. "Cadmium Risk Assessment of an Exposed Occupational Population", *J. Soc. Med.* 78(4):328–333 (1985).

Poch, D. I. *Radiation Alert* (New York: Doubleday & Company, 1985).

Pollack, H. "Medical Aspects of Exposure to Radiofrequency Radiation Including Microwaves", *South. Med. J.* 76:(6)759 (1983).

"Polychlorinated Biphenyl Transformer Incident—New Mexico", *MMWR* 34(36):557–559 (1985).

"Radiation", *Occup. Health Nurs.* 32:263–264 (1984).

Ratoff, J. "Microwaves: Hints of Low-Dose Hazards", *Sci. News* 126:(7)103 (1984).

Rice, C., R. L. Harris, J. C. Lumsden and M. J. Symons. "Reconstruction of Silica Exposure in the North Carolina Dusty Trades", *Am. Ind. Hyg. Assoc. J.* 45(10):689–696 (1984).

Riggin, R. M. "Determination of Volatile Organic Compounds in Ambient Air Using Tenax Adsorption and Gas Chromatography/Mass Spectrometry", *IARC Sci. Publ.* (68):269–289 (1985).

Robertson, J. M., J. F. Diaz, I. M. Fyfe and T. H. Ingalls. "A Cross-sectional Study of Pulmonary Function in Carbon Black Workers in the United States", *Am. Ind. Hyg. Assoc. J.* 49(4):161–166 (1988).

Ross, D. "Case History: An Acid Fumes Hazard", *Occup. Health Lond.* 37(4):187–188 (1985).

Rycroft, R. J. "Environmental Aspects of Occupational Dermatology", *Derm. Beruf. Umwelt.* 34(6):157–159 (1986).

Sagripanti, J. L. "DNA Structural Changes Caused by Microwave Radiation", *Int. J. Rad. Biol. Related Stud. Phys. Chem. Med.* 50(1)47–50 (1986).

Saltzman, B. E. "Variability and Bias in the Analyses of Industrial Hygiene Samples", *Am. Ind. Hyg. Assoc. J.* 46(3):134–141 (1985).

Samuels, S. W. "Medical Surveillance: Biological, Social, and Ethical Parameters", *J. Occup. Med.* 28(8):572–577 (1986).

Sass, R. "What's in a Name? The Occupational Hygienists Problem With Threshold Limit Values", *Am. J. Ind. Med.* 14(3):355–363 (1988).

Schilling, R. S. "The Role of Medical Examination in Protecting Worker Health", *J. Occup. Med.* 28(8):553–537 (1986).

Schwartz, D. A., and J. P. LoGerfo. "Congenital Limb Reduction Defects in the Agricultural Setting", *Am. J. Public Health* 78(6):645–648 (1988).

Sever, L. E. "Epidemiologic Approaches to Reproductive Hazards of the Workplace", *Birth Defects: Original Article Series* 18(3A):33–38 (1982).

Singh, N. P., M. E. Bennett, D. B. Wrenn, and G. Saccomanno. "Concentrations of ^{210}Pb and Its States of Equilibrium With ^{238}U, ^{234}U, and ^{230}Th in U Miners' Lungs", *Health Physics*, 51:501–507 (1986).

Sinks, T. and S. Newman. "Surveillance of Work-Related Cold Injuries Using Workers", *J. Occup. Med.* 29:504–509 (1987).

Sittig, M. *Hazardous and Toxic Effects of Industrial Chemicals* (New Jersey: Noyes Data, 1979).

Smith, J. M., and R. D. Costlow. "Recognition, Evaluation, and Control of Chemical Embryotoxins in the Workplace", *Fundament. Appl. Toxicol.* 5(4):626–633 (1985).

Smith, R. G. "Cummings Award Address. Occupational Health Standard Setting in the United States", *Am. Ind. Hyg. Assoc. J.* 46(1):541–546 (1985).

Solomon, C. J. "Occupational Health Problems in High Tech Industries", *Occup. Health Nurs.* 32(5):262–265 (1984).

Spear, R. C., S. Selvin and M. Francis. "The Influence of Averaging Time on the Distribution of Exposures", *Am. Ind. Hyg. Assoc. J.* 47(6):365–368 (1986).

Stallard, C. W. "Use of SOHIO's Health Information System", *J. Occup. Med.* 28(8):684–686 (1986).

Stinnett, S. *Mutagenistic Testing of Industrial Wastes From Representative Organic Chemical, Industries* (Ada, OK: U.S. EPA, 1981).

Thorpe, J. J. "Occupational Medicine in the Petroleum Industry: An Historical Perspective", *State Art Rev. Occup. Med.* 3(3):371–390 (1988).

Thrane, K. E., and H. Stray. "Organic Air Pollutants in an Aluminum Reduction Plant", *Sci. Total Environ.* 53(1–2):111–131 (1986).

Ungers, L. J., and J. H. Jones. "Industrial Hygiene and Control Technology Assessment of Ion Implantation Operations", *Am. Ind. Hyg. Assoc. J.* 47(10):607–614 (1986).

U.S. Congress, Office of Technology Assessment. *Preventing Illness and Injury in the Workplace* (Washington, DC: U.S. Government Printing Office, 1985).

U.S. Congress, Office of Technology Assessment. *Reproductive Health Hazards in the Workplace* (Washington, DC: U.S. Government Printing Office, 1985).

U.S. Department of Health and Human Services. *Handbook for Industrial Noise Control* (Cincinnati, OH: 1981).

U.S. Department of Health and Human Services. *Hot Environment* (Washington, DC: U.S. Government Printing Office, 1986).

U.S. Department of Health and Human Services, National Institute of Occupational Safety and Health, Division of Standards Development and Technology Transfer. "Evaluation of Epidemiologic Studies Examining the Lung Cancer Mortality of Underground Uranium Miners", report prepared for U.S. Mine Safety and Health Administration (May 9, 1985).

U.S. Department of Health and Human Services. *Occupational Exposure to Hot Environments* (Washington, DC: U.S. Government Printing Office, 1986).

U.S. Department of Health and Human Services. *Recognition of Occupational Health Hazards* (Washington, DC: Government Pritning Office, 1982).

U.S. Department of Health and Human Services. *Survey of Hearing Conservation Programs in Industry* (Washington, DC: U.S. Government Printing Office, 1982).

U.S. Department of Labor, Mine Safety and Health Administration. "Ionizing Radiation Standards for Underground Metal and Nonmetal Mines; Proposed Rule", *Fed. Reg.* 51:45678 (1986).

U.S. Department of Labor, Occupational Safety and Health Administration. "Identification, Classification and Regulation of Potential Occupational Carcinogens", *Fed. Reg.* 45:5002, (1980.)

U.S. Department of Labor. *OSHA Programs* (Washington, DC: U.S. Government Printing Office, 1985).

U.S. Department of Labor. *Job Hazard Analysis* (Washington, DC: U.S. Government Printing Office, 1986).

U.S. Environmental Protection Agency. "Asbestos; Proposed Mining and Import Restrictions and Proposed Manufacturing Importation and Processing Prohibitions", *Fed. Reg.* 51:3738, (1986).

U.S. Environmental Protection Agency. *Mortality and Tumor Incidence Among Ferochromium Workers* (1983).

Upton, A. C. "The Biological Effects of Low-Level Ionizing Radiation", *Sci. Am.* 236:41–49 (1982).

Upton, A. C. "Environmental Standards for Ionizing Radiation: Theoretical Basis for Dose-Respoxe Curves", *Environmental Health Perspectives* 52:31–39 (1983).

Utidjian, W. O., and H. M. Karten. "Retrospective Evaluation of Reproductive Performance of Workers Exposed to Ethylene Dibromide", *J. Occup. Med.* 21: 98–102 (1979).

Voelz, G. L. "Ionizing Radiation", *Occup. Health Saf.* 51:34–37 (1982).

Weil, C. S. "1984 Stokinger Lecture. Some Questions and Opinions on Issues in Toxicology and Risk Assessment", *Am. Ind. Hyg. Assoc. J.* 45(1):663–670 (1984).

Weisburger, J. H., and G. M. Williams. "Chemical Carcinogens", in *Toxicology: The Basic Science of Poisons,* 2nd ed. J. Doull, C. D. Klassen, and M. O. Amdur, Eds. (New York: Macmillan, 1980).

Wick, J. L. "Workplace Design Changes", *Sem. Occup. Med.* 2:75–77 (1987).

Williams, P. L., and J. Burson. *Industrial Toxicology: Safety and Health Applications in the Workplace* (New York: Van Nostrand Reinhold, 1985).

Witt, M. *White Book, Noise Control, A Guide For Workers and Employers* (New York: Holt, Reinehart & Winston, 1985).

Wood, J. L. "Right-to-Know Laws: Maintaining Compliance", *Occup. Health Saf.* 56(3):20–21,24,27–29 (1987).

WHO Offset Publ. "Evaluation of Exposure to Airborne Particles in the Work Environment", (80):1–75 (1984).

Yodaiken, R.E. "Medical Screening and Biological Monitoring for the Effects of Exposure in the Workplace. Surveillance, Monitoring, and Regulatory Concerns", *J. Occu. Med.* 28(8):569–571 (1986).

Chapter 11

Abromowitz, M. *Microscope Basics and Beyond,* Vol. 1, (Lake Success, NY: Olympus Corporation, 1985).

Abromowitz, M. *Contrast Methods in Microscopy: Transmitted Light,* Vol. 2, (Lake Success, NY: Olympus Corporation, 1987).

American Conference of Governmental Industrial Hygienists *Industrial Ventilation: A Manual of Recommended Practice,* 21st ed. (Cincinnati, OH: ACGIH,1992)

American Industrial Hygiene Association, *Fundamentals of Analytical Procedures in Industrial Hygiene,* (Akron, OH: AIHA, 1987).

Berger, E. H., W. D. Ward, J. C. Morrill, and L. H. Royster (Eds.) *Noise and Hearing Conservation Manual,* (Akron, OH: AIHA, 1986).

Bisesi, M. S. and J. P. Kohn. *Industry Hygiene Evaluation Methods.* (Boca Raton, FL: Lewis Publishers, 1995).

Gollnick, D. A. *Basic Radiation Protection Technology,* 3rd ed. (Altadena, CA: Pacific Radiation Corporation, 1994).

Hering, S. *Air Sampling Instruments for Evaluation of Atmospheric Contaminants,* 7th ed. (Cincinnati, OH: American Conference of Govenmental Industrial Hygienists, 1989).

Kenkel, J. *Analytical Chemistry Refresher Manual* (Boca Raton, FL: Lewis Publishers, 1992).

Lodge, J. P., Jr. (Ed.) *Methods of Air Sampling and Analysis,* 3rd ed., Intersociety Committee: APCA/ACS/AIChE/APWA/ASME/AOAC/HPS/ISA. (Boca Raton, FL: Lewis Publishers, 1989).

Malansky, C. J. and S. P. Malansky. *Air Monitoring Instrumentation: A Manual for Emergency, Investigatory, and Remedial Responders,* (New York: Van Nostrand Reinhold, 1993).

Occupational Safety and Health Administration. *OSHA Industrial Hygiene Technical Manual,* (Chicago, IL: Commerce Clearing House, 1993).

Plog, B. A. (Ed.) *Fundamentals of Industrial Hygiene,* 3rd ed. (Chicago, IL: National Safety Council, 1988).

Talty, J. T. *Industrial Hygiene Engineering: Recognition, Measurement, Evaluation, and Control,* (Park Ridge, NJ: Noyes Data Corporation, 1988).

Wight, G. D. *Fundamentals of Air Sampling,* (Boca Raton, FL: Lewis Publishers, 1994).

Index

VOLUME I

Index

VOLUME II